# STUDENT'S
# SOLUTIONS MANUAL

### BEVERLY FUSFIELD

# TRIGONOMETRY
## NINTH EDITION

# Margaret L. Lial
*American River College*

# John Hornsby
*University of New Orleans*

# David I. Schneider
*University of Maryland*

PEARSON

Addison
Wesley

Boston  San Francisco  New York
London  Toronto  Sydney  Tokyo  Singapore  Madrid
Mexico City  Munich  Paris  Cape Town  Hong Kong  Montreal

Reproduced by Pearson Addison-Wesley from electronic files supplied by the author.

Copyright © 2009 Pearson Education, Inc.
Publishing as Pearson Addison-Wesley, 75 Arlington Street, Boston, MA 02116.

All rights reserved. No part of this publication may be reproduced, stored in a retrieval system, or transmitted, in any form or by any means, electronic, mechanical, photocopying, recording, or otherwise, without the prior written permission of the publisher. Printed in the United States of America.

ISBN-13: 978-0-321-53040-0
ISBN-10: 0-321-53040-3

2 3 4 5 6 BB 10 09 08

# CONTENTS

# Appendices

# Chapter 1

## Trigonometric Functions

### Section 1.1: Angles

1. $30°$

    (a) $90° - 30° = 60°$

    (b) $180° - 30° = 150°$

3. $45°$

    (a) $90° - 45° = 45°$

    (b) $180° - 45° = 135°$

5. $54°$

    (a) $90° - 54° = 36°$

    (b) $180° - 54° = 126°$

7. $1°$

    (a) $90° - 1° = 89°$

    (b) $180° - 1° = 179°$

9. $14°20'$

    (a) $90° - 14°20' = 89°60' - 14°20' = 75°40'$

    (b) $180° - 14°20' = 179°60' - 14°20'$
    $= 165°40'$

11. $20°10'30''$

    (a) $90° - 20°10'30'' = 89°59'60'' - 20°10'30''$
    $= 69°49'30''$

    (b) $180° - 20°10'30''$
    $= 179°59'60'' - 20°10'30''$
    $= 159°49'30''$

13. The two angles form a straight angle.
    $7x + 11x = 180 \Rightarrow 18x = 180 \Rightarrow x = 10$
    The measures of the two angles are
    $(7x)° = \left[7(10)\right]° = 70°$ and
    $(11x)° = \left[11(10)\right]° = 110°$.

15. The two angles form a right angle.
    $4y + 2y = 90 \Rightarrow 6y = 90 \Rightarrow y = 15$
    The two angles have measures of
    $(4y)° = \left[4(15)\right]° = 60°$ and
    $(2y)° = \left[2(15)\right]° = 30°$.

17. The two angles form a straight angle.
    $(-4x) + (-14x) = 180 \Rightarrow -18x = 180 \Rightarrow$
    $x = -10$
    The measures of the two angles are
    $(-4x)° = \left[-4(-10)\right]° = 40°$ and
    $(-14x)° = \left[-14(-10)\right]° = 140°$.

19. The sum of the measures of two
    supplementary angles is $180°$.
    $(10x + 7) + (7x + 3) = 180$
    $17x + 10 = 180$
    $17x = 170 \Rightarrow x = 10$
    The measures of the two angles are
    $(10x + 7)° = \left[10(10) + 7\right]° = (100 + 7)° = 107°$
    and $(7x + 3)° = \left[7(10) + 3\right]°$
    $= (70 + 3)° = 73°$.

21. The sum of the measures of two
    complementary angles is $90°$.
    $(9x + 6) + 3x = 90 \Rightarrow 12x + 6 = 90 \Rightarrow$
    $12x = 84 \Rightarrow x = 7$
    The measures of the two angles are
    $(9x + 6)° = \left[9(7) + 6\right]° = (63 + 6)° = 69°$ and
    $(3x)° = \left[3(7)\right]° = 21°$.

23. Let $x$ = the measure of the angle.
    If the angle is its own complement, then we
    have $x + x = 90 \Rightarrow 2x = 90 \Rightarrow x = 45$
    Thus, a $45°$ angle is its own complement.

25. $\dfrac{25\,\text{minutes}}{60\,\text{minutes}} = \dfrac{x}{360°}$

    $x = \dfrac{25}{60}(360) = 25(6) = 150°$

27. At 15 minutes after the hour, the minute hand
    is $\frac{1}{4}$ the way around, so the hour hand is $\frac{1}{4}$ of
    the way between the 3 and 4. Thus, the hour
    hand is located 16.25 minutes past 12. The
    minute hand is 15 minutes after the 12. The
    smaller angle formed by the hands of the clock
    can be found by solving the proportion
    $\dfrac{(16.25 - 15)\,\text{minutes}}{60\,\text{minutes}} = \dfrac{x}{360°}$.

*(continued on next page)*

*(continued from page 1)*

$$\frac{(16.25-15)\,\text{minutes}}{60\,\text{minutes}} = \frac{x}{360°} \Rightarrow \frac{1.25}{60} = \frac{x}{360} \Rightarrow$$

$$x = \frac{1.25}{60}(360) = 1.25(6) = 7.5° = 7°30'$$

**29.** If an angle measures $x$ degrees and two angles are complementary if their sum is $90°$, then the complement of an angle of $x°$ is $(90-x)°$.

**31.** The first negative angle coterminal with $x$ between $0°$ and $60°$ is $(x-360)°$.

**33.** 
$$\begin{array}{r} 62°\ 18' \\ +21°\ 41' \\ \hline 83°\ 59' \end{array}$$

**35.** $71°18' - 47°\ 29' = 70°\ 78' - 47°\ 29'$

$$\begin{array}{r} 70°\ 78' \\ -47°\ 29' \\ \hline 23°\ 49' \end{array}$$

**37.** $90° - 51°\ 28' = 89°\ 60' - 51°\ 28'$

$$\begin{array}{r} 89°\ 60' \\ -51°\ 28' \\ \hline 38°\ 32' \end{array}$$

**39.** $180° - 119°\ 26' = 179°\ 60' - 119°\ 26'$

$$\begin{array}{r} 179°\ 60' \\ -119°\ 26' \\ \hline 60°\ 34' \end{array}$$

**41.** $26°20' + 18°17' - 14°10' = 44°37' - 14°10'$
$$= 30°27'$$

**43.** $90° - 72°\ 58'\ 11'' = 89°\ 59'\ 60'' - 72°\ 58'\ 11''$

$$\begin{array}{r} 89°\ 59'\ 60'' \\ -72°\ 58'\ 11'' \\ \hline 17°\ \ 1'\ 49'' \end{array}$$

**45.** $35°30' = 35° + \frac{30}{60}° = 35° + .5° = 35.5°$

**47.** $112°15' = 112° + \frac{15}{60}° = 112° + .25° = 112.25°$

**49.** $-60°12' = -\left(60° + \frac{12}{60}°\right) = -\left(60° + .2°\right) = -60.2°$

**51.** $20°54'00'' = 20° + \frac{54}{60}° = 20° + .900° = 20.9°$

**53.** $91°35'54'' = 91° + \frac{35}{60}° + \frac{54}{3600}°$
$$\approx 91° + .5833° + .0150° \approx 91.598°$$

**55.** $274°18'\ 59'' = 274° + \frac{18}{60}° + \frac{59}{3600}°$
$$\approx 274° + .3000° + .0164°$$
$$\approx 274.316°$$

**57.** $39.25° = 39° + .25° = 39° + .25(60')$
$$= 39° + 15' + 0'' = 39°15'00''$$

**59.** $126.76° = 126° + .76° = 126° + .76(60')$
$$= 126° + 45.6' = 126° + 45' + .6'$$
$$= 126° + 45' + .6(60'')$$
$$= 126° + 45' + 36'' = 126°45'36''$$

**61.** $-18.515° = -(18° + .515°)$
$$= -(18° + .515(60'))$$
$$= -(18° + 30.9') = -(18° + 30' + .9')$$
$$= -(18° + 30' + .9(60''))$$
$$= -(18° + 30' + 54'') = -18°30'54''$$

**63.** $31.4296° = 31° + .4296° = 31° + .4296(60')$
$$= 31° + 25.776' = 31° + 25' + .776'$$
$$= 31° + 25' + .776(60'')$$
$$= 31°25'46.56'' \approx 31°\ 25'\ 47''$$

**65.** $89.9004° = 89° + .9004° = 89° + .9004(60')$
$$= 89° + 54.024' = 89° + 54' + .024'$$
$$= 89° + 54' + .024(60'')$$
$$= 89°54'1.44'' \approx 89°54'1''$$

**67.** $178.5994° = 178° + .5994°$
$$= 178° + .5994(60')$$
$$= 178° + 35.964'$$
$$= 178° + 35' + .964'$$
$$= 178° + 35' + .964(60'')$$
$$= 178°35'57.84'' \approx 178°35'58''$$

**69.** $32°$ is coterminal with $360° + 32° = 392°$.

**71.** $26°30'$ is coterminal with
$$360° + 26°30' = 386°30'.$$

**73.** $-40°$ is coterminal with $360° + (-40°) = 320°$.

**75.** $-125°$ is coterminal with
$$360° + (-125°) = 235°.$$

**77.** $361°$ is coterminal with $361° - 360° = 1°$.

**79.** $-361°$ is coterminal with
$$-361° + 2(360°) = 359°.$$

**81.** $539°$ is coterminal with $539° - 360° = 179°$.

**83.** $850°$ is coterminal with
$$850° - 2(360°) = 850° - 720° = 130°.$$

**85.** 5280° is coterminal with
$5280° - 14 \cdot 360° = 5280° - 5040° = 240°.$

**87.** −5280° is coterminal with
$-5280° + 15 \cdot 360° = -5280° + 5400° = 120°.$

In exercises 89–91, answers may vary.

**89.** 90° is coterminal with
$$90° + 360° = 450°$$
$$90° + 2(360°) = 810°$$
$$90° - 360° = -270°$$
$$90° - 2(360°) = -630°$$

**91.** 0° is coterminal with
$$0° + 360° = 360°$$
$$0° + 2(360°) = 720°$$
$$0° - 360° = -360°$$
$$0° - 2(360°) = -720°$$

**93.** 30°
A coterminal angle can be obtained by adding an integer multiple of 360°.
$30° + n \cdot 360°$

**95.** 135°
A coterminal angle can be obtained by adding an integer multiple of 360°.
$135° + n \cdot 360°$

**97.** −90°
A coterminal angle can be obtained by adding an integer multiple of 360°.
$-90° + n \cdot 360°$

**99.** 0°
A coterminal angle can be obtained by adding integer multiple of 360°.
$0° + n \cdot 360° = n \cdot 360°$

**101.** The answers to Exercises 99 and 100 give the same set of angles since 0° is coterminal with $0° + 360° = 360°.$

For Exercises 103–113, angles other than those given are possible.

**103.**

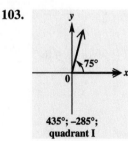

435°; −285°;
**quadrant I**

75° is coterminal with 75° + 360° = 435° and 75° − 360° = −285°. These angles are in quadrant I.

**105.**

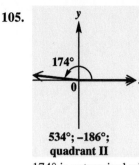

534°; −186°;
**quadrant II**

174° is coterminal with 174° + 360° = 534° and 174° − 360° = −186°. These angles are in quadrant II.

**107.**

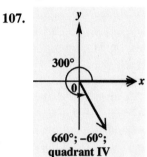

660°; −60°;
**quadrant IV**

300° is coterminal with 300° + 360° = 660° and 300° − 360° = −60°. These angles are in quadrant IV.

**109.**

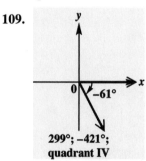

299°; −421°;
**quadrant IV**

−61° is coterminal with −61° + 360° = 299° and −61° − 360° = −421°. These angles are in quadrant IV.

**111.**

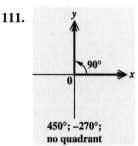

450°; −270°;
**no quadrant**

90° is coterminal with 90° + 360° = 450° and 90° − 360° = −270°. These angles are not in a quadrant.

**113.**

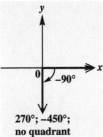

**270°; −450°;
no quadrant**

−90° is coterminal with −90° + 360° = 270°
and −90° − 360° = −450°. These angles are
not in a quadrant.

**115.**

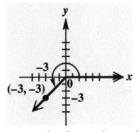

Points: $(0,0)$ and $(-3,-3)$

$$r = \sqrt{(-3-0)^2 + (-3-0)^2}$$
$$= \sqrt{(-3)^2 + (-3)^2}$$
$$= \sqrt{9+9} = \sqrt{18} = \sqrt{9 \cdot 2} = 3\sqrt{2}$$

**117.**

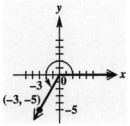

Points: $(0,0)$ and $(-3,-5)$

$$r = \sqrt{(-3-0)^2 + (-5-0)^2}$$
$$= \sqrt{(-3)^2 + (-5)^2} = \sqrt{9+25} = \sqrt{34}$$

**119.**

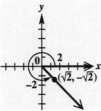

Points: $(0,0)$ and $\left(\sqrt{2}, -\sqrt{2}\right)$

$$r = \sqrt{\left(\sqrt{2}-0\right)^2 + \left(-\sqrt{2}-0\right)^2}$$
$$= \sqrt{\left(\sqrt{2}\right)^2 + \left(-\sqrt{2}\right)^2} = \sqrt{2+2} = \sqrt{4} = 2$$

**121.**

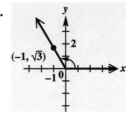

Points: $(0,0)$ and $\left(-1, \sqrt{3}\right)$

$$r = \sqrt{(-1-0)^2 + \left(\sqrt{3}-0\right)^2}$$
$$= \sqrt{(-1)^2 + \left(\sqrt{3}\right)^2} = \sqrt{1+3} = \sqrt{4} = 2$$

**123.**

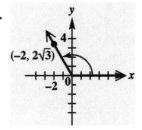

Points: $(0,0)$ and $\left(-2, 2\sqrt{3}\right)$

$$r = \sqrt{(-2-0)^2 + \left(2\sqrt{3}-0\right)^2}$$
$$= \sqrt{(-2)^2 + \left(2\sqrt{3}\right)^2} = \sqrt{4+12} = \sqrt{16} = 4$$

**125.**

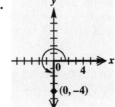

Points: $(0,0)$ and $(0,-4)$

$$r = \sqrt{(0-0)^2 + (-4-0)^2}$$
$$= \sqrt{(0)^2 + (-4)^2} = \sqrt{0+16} = \sqrt{16} = 4$$

**127.** 45 revolutions per min $= \frac{45}{60}$ revolution per sec

$= \frac{3}{4}$ revolution per sec

A turntable will make $\frac{3}{4}$ revolution in 1 sec.

**129.** 600 rotations per min $= \frac{600}{60}$ rotations per sec

$\qquad\qquad\qquad\quad = 10$ rotations per sec

$\qquad\qquad\qquad\quad = 5$ rotations per $\frac{1}{2}$ sec

$\qquad\qquad\qquad\quad = 5(360°)$ per $\frac{1}{2}$ sec

$\qquad\qquad\qquad\quad = 1800°$ per $\frac{1}{2}$ sec

A point on the edge of the tire will move $1800°$ in $\frac{1}{2}$ sec.

**131.** $75°$ per min $= 75°(60)$ per hr $= 4500°$ per hr

$\qquad\qquad\quad = \frac{4500°}{360°}$ rotations per hr

$\qquad\qquad\quad = 12.5$ rotations per hr

The pulley makes $12.5$ rotations in $1$ hr.

**133.** The earth rotates $360°$ in $24$ hr. $360°$ is equal to $360(60') = 21,600'$.

$$\frac{24\,\text{hr}}{21,600'} = \frac{x}{1'} \Rightarrow$$

$$x = \frac{24}{21,600}\,\text{hr} = \frac{24}{21,600}(60\,\text{min})$$

$$= \frac{1}{15}\,\text{min} = \frac{1}{15}(60\,\text{sec}) = 4\,\text{sec}$$

It should take the motor $4$ sec to rotate the telescope through an angle of $1$ min.

## Section 1.2: Angle Relationships and Similar Triangles

**1.** $m\angle 1 = 55°$

$m\angle 1 + m\angle 2 = 120° \Rightarrow m\angle 2 = 120° - 55° = 65°$

$m\angle 1 + m\angle 2 + m\angle 3 = 180° \Rightarrow$

$55° + 65° + m\angle 3 = 180° \Rightarrow m\angle 3 = 60°$

$m\angle 4 = m\angle 2 = 60°$

$m\angle 5 = m\angle 3 = 60°$

$m\angle 6 = 120°$

$m\angle 7 + 120° = 180° \Rightarrow m\angle 7 = 60°$

$m\angle 8 = m\angle 7 = 60°$

$m\angle 9 = 55°$

$m\angle 10 = m\angle 9 = 55°$

**3.** The two indicated angles are vertical angles, so their measures are equal.

$5x - 129 = 2x - 21 \Rightarrow 3x = 108 \Rightarrow x = 36$

$5(36) - 129 = 51$ and $2(36) - 21 = 51$, so both angles measure $51°$.

**5.** The three angles are the interior angles of a triangle, so the sum of their measures is $180°$.

$x + (x + 20) + (210 - 3x) = 180$

$\qquad\qquad 230 - x = 180 \Rightarrow x = 50$

$50 + 20 = 70$ and $210 - 3(50) = 60$, so the three angles measure $50°$, $60°$, and $70°$.

**7.** The three angles are the interior angles of a triangle, so the sum of their measures is $180°$.

$(2x - 120) + \left(\frac{1}{2}x + 15\right) + (x - 30) = 180$

$\qquad\qquad\qquad\qquad \frac{7}{2}x - 135 = 180$

$\qquad\qquad\qquad\qquad\qquad \frac{7}{2}x = 315$

$\qquad\qquad\qquad\qquad\qquad\quad x = 90$

$2(90) - 120 = 60, \frac{1}{2}(90) + 15 = 60$, and $90 - 30 = 60$, so the three angles each measure $60°$.

**9.** In a triangle, the measure of an exterior angle equals the sum of the measures of the non-adjacent interior angles. Thus,

$(6x + 3) + (4x - 3) = 9x + 12$

$\qquad\qquad\quad 10x = 9x + 12$

$\qquad\qquad\qquad x = 12$

$6(12) + 3 = 75, 4(12) - 3 = 45$, and

$9(12) + 12 = 120$, so the three angles measure $45°$, $75°$, and $120°$.

**11.** Since the two angles are alternate interior angles, their measures are equal.

$2x - 5 = x + 22 \Rightarrow x = 27$

$2(27) - 5 = 49$ and $27 + 22 = 49$, so both angles measure $49°$.

**13.** Since the two angles are interior angles on the same side of the transversal, the sum of their measures is $180°$.

$(x + 1) + (4x - 56) = 180$

$\qquad\quad 5x - 55 = 180 \Rightarrow 5x = 235 \Rightarrow x = 47$

$47 + 1 = 48$ and $4(47) - 56 = 132$, so the angles measure $48°$ and $132°$.

**15.** Let $x$ = the measure of the third angle. Then

$37 + 52 + x = 180 \Rightarrow x = 91$

The third angle of the triangle measures $91°$.

**17.** Let $x$ = the measure of the third angle. Then

$147°12' + 30°19' + x = 180°$

$\qquad\qquad 177°31' + x = 180°$

$\qquad\qquad 177°31' + x = 179°60' \Rightarrow x = 2°29'$

The third angle of the triangle measures $2°29'$.

**19.** Let $x$ = the measure of the third angle. Then

$74.2° + 80.4° + x = 180° \Rightarrow x = 25.4°$

The third angle of the triangle measures $25.4°$.

**21.** Let $x$ = the measure of the third angle. Then
$$51°20'14'' + 106°10'12'' + x = 180°$$
$$157°30'26'' + x = 180°$$
$$157°30'26'' + x = 179°59'60''$$
$$x = 22°29'34''$$
The third angle of the triangle measures $22°29'34''$.

**23.** A triangle cannot have angles of measures 85° and 100°. The sum of the measures of these two angles is 85° + 100°=185°, which exceeds 180°.

**25.** The triangle has a right angle, but each side has a different measure. The triangle is a right triangle and a scalene triangle.

**27.** The triangle has three acute angles and three equal sides, so it is acute and equilateral.

**29.** The triangle has a right angle and three unequal sides, so it is right and scalene.

**31.** The triangle has a right angle and two equal sides, so it is right and isosceles.

**33.** The triangle has one obtuse angle and three unequal sides, so it is obtuse and scalene.

**35.** The triangle has three acute angles and two equal sides, so it is acute and isosceles.

**37.–39.** Answers will vary.

**41.** Corresponding angles are $A$ and $P$, $B$ and $Q$, $C$ and $R$. Corresponding sides are $AC$ and $PR$, $BC$ and $QR$, $AB$ and $PQ$.

**43.** Corresponding angles are $A$ and $C$, $E$ and $D$, $ABE$ and $CBD$. Corresponding sides are $EB$ and $DB$, $AB$ and $CB$, $AE$ and $CD$

**45.** Since angle $Q$ corresponds to angle $A$, the measure of angle $Q$ is 42°. Since angles $A$, $B$, and $C$ are interior angles of a triangle, the sum of their measures is 180°.
$$m\angle A + m\angle B + m\angle C = 180°$$
$$42° + m\angle B + 90° = 180°$$
$$132° + m\angle B = 180°$$
$$m\angle B = 48°$$
Since angle $R$ corresponds to angle $B$, the measure of angle R is 48°.

**47.** Since angle $B$ corresponds to angle $K$, the measure of angle $B$ is 106°. Angles $A$, $B$, and $C$ are interior angles of a triangle, so the sum of their measures is 180°.
$$m\angle A + m\angle B + m\angle C = 180°$$
$$m\angle A + 106° + 30° = 180°$$
$$m\angle A + 136° = 180°$$
$$m\angle A = 44°$$
Since angle $M$ corresponds to angle $A$, the measure of angle $M$ is 44°.

**49.** Angles $X$, $Y$, and $Z$ are interior angles of a triangle, so the sum of their measures is 180°.
$$m\angle X + m\angle Y + m\angle Z = 180°$$
$$m\angle X + 90° + 38° = 180°$$
$$m\angle X + 128° = 180°$$
$$m\angle X = 52°$$
Since angle $M$ corresponds to angle $X$, the measure of angle $M$ is 52°.

In Exercises 51–55, corresponding sides of similar triangles are proportional. Other proportions are possible in solving these exercises.

**51.** $\dfrac{25}{10} = \dfrac{a}{8} \Rightarrow 8(25) = 10a \Rightarrow 200 = 10a \Rightarrow 20 = a$

$\dfrac{25}{10} = \dfrac{b}{6} \Rightarrow 6(25) = 10b \Rightarrow 150 = 10b \Rightarrow 15 = b$

**53.** $\dfrac{6}{12} = \dfrac{a}{12} \Rightarrow a = 6$

$\dfrac{6}{12} = \dfrac{b}{15} \Rightarrow \dfrac{1}{2} = \dfrac{b}{15} \Rightarrow b = \dfrac{15}{2} = 7\dfrac{1}{2}$

**55.** $\dfrac{6}{9} = \dfrac{4}{x} \Rightarrow \dfrac{2}{3} = \dfrac{4}{x} \Rightarrow 2x = 12 \Rightarrow x = 6$

**57.** Let $x$ = the height of the tree.
The triangle formed by the tree and its shadow is similar to the triangle formed by the stick and its shadow.

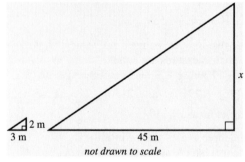
not drawn to scale

$\dfrac{x}{2} = \dfrac{45}{3} \Rightarrow \dfrac{x}{2} = \dfrac{15}{1} \Rightarrow x = 30$
The tree is 30 m high.

**59.** Let $x$ = the middle side of the actual triangle (in meters); $y$ = the longest side of the actual triangle (in meters).

The triangles in the photograph and the piece of land are similar. The shortest side on the land corresponds to the shortest side on the photograph.

$$\frac{400 \text{ m}}{4 \text{ cm}} = \frac{x}{5 \text{ cm}} \Rightarrow \frac{100}{1} = \frac{x}{5} \Rightarrow x = 500 \text{ and}$$

$$\frac{400 \text{ m}}{4 \text{ cm}} = \frac{y}{7 \text{ cm}} \Rightarrow \frac{100}{1} = \frac{y}{7} \Rightarrow y = 700$$

The other two sides are 500 m and 700 m long.

**61.** Let $x$ = the height of the building.
The triangle formed by the house and its shadow is similar to the triangle formed by the building and its shadow.

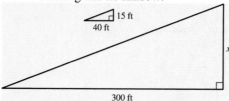

$$\frac{15}{40} = \frac{x}{300} \Rightarrow 40x = 4500 \Rightarrow x = 112.5$$

The building is 112.5 ft tall.

**63.** $\frac{x}{50} = \frac{100+120}{100} \Rightarrow \frac{x}{50} = \frac{220}{100} \Rightarrow \frac{x}{50} = \frac{11}{5} \Rightarrow$
$5x = 550 \Rightarrow x = 110$

**65.** $\frac{c}{100} = \frac{10+90}{90} \Rightarrow \frac{c}{100} = \frac{100}{90} \Rightarrow \frac{c}{100} = \frac{10}{9} \Rightarrow$
$9c = 1000 \Rightarrow c = \dfrac{1000}{9} \approx 111.1$

**67.** In these two similar quadrilaterals, the largest of the three shortest sides of the first quadrilateral (32 cm) corresponds to the smaller of the two longest sides of the second quadrilateral (48 cm). The following diagram is not drawn to scale.

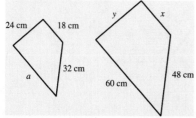

Let $a$ = the length of the longest side of the first quadrilateral; $x$ = the length of the shortest side of the second quadrilateral; $y$ = the other unknown length.

Corresponding sides are in proportion.

$$\frac{a}{60} = \frac{32}{48} \Rightarrow \frac{a}{60} = \frac{2}{3} \Rightarrow 3a = 120 \Rightarrow a = 40 \text{ cm}$$

$$\frac{24}{y} = \frac{32}{48} \Rightarrow \frac{24}{y} = \frac{2}{3} \Rightarrow 2y = 72 \Rightarrow y = 36 \text{ cm}$$

$$\frac{18}{x} = \frac{32}{48} \Rightarrow \frac{18}{x} = \frac{2}{3} \Rightarrow 2x = 54 \Rightarrow y = 27 \text{ cm}$$

**69. (a)** Let $D_s$ = the distance from the Earth to the sun; $d_s$ = the diameter of the sun, $D_m$ = the distance from the Earth to the moon; and $d_m$ = the diameter of the moon.

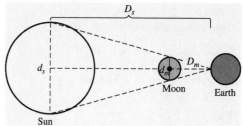

$$\frac{D_s}{D_m} = \frac{d_s}{d_m}$$

$$\frac{94,500,000}{D_m} = \frac{865,000}{2159}$$

$$865,000 D_m = 94,500,000 \cdot 2159$$

$$D_m = \frac{94,500,000 \cdot 2159}{865,000}$$

$$\approx 236,000 \text{ mi}$$

**(b)** No, a total solar eclipse cannot occur every time. The moon must be less than 236,000 miles away from Earth for an eclipse to occur, and sometimes it is farther than this.

**71. (a)** Let $D_s$ = the distance from Mars to the sun; $d_s$ = the diameter of the sun, $D_m$ = the distance from Mars to Phobos; and $d_m$ = the diameter of Phobos.

$$\frac{D_s}{D_m} = \frac{d_s}{d_m}$$

$$\frac{142,000,000}{D_m} = \frac{865,000}{17.4}$$

$$865,000 D_m = 142,000,000 \cdot 17.4$$

$$D_m = \frac{142,000,000 \cdot 17.4}{865,000}$$

$$\approx 2900 \text{ mi}$$

**(b)** No, Phobos does not come close enough to the surface of Mars.

**73. (a)** The thumb covers about 2 arc degrees or about 120 arc minutes. This is $\frac{31}{120}$ or approximately $\frac{1}{4}$ of the thumb would cover the moon.

**(b)** $20° + 10° = 30°$
The stars are 30 arc degrees apart.

**75.** Since the two triangles are similar, the corresponding angles have the same measure. Note that the angle with measure $(10y+8)°$ and the angle with measure $58°$ are vertical angles, and thus, are equal.
$10y + 8 = 58 \Rightarrow 10y = 50 \Rightarrow y = 5$.
The two triangles are similar, so the corresponding sides are in proportion.
$$\frac{x-y}{x+y} = \frac{6}{18} \Rightarrow \frac{x-y}{x+y} = \frac{1}{3} \Rightarrow$$
$$3(x-y) = x+y \Rightarrow 3x - 3y = x+y \Rightarrow$$
$$2x = 4y \Rightarrow x = 2y$$
Substituting 5 for $y$, we have
$x = 2y \Rightarrow 2(5) = 10$. Therefore, $x = 10$ and $y = 5$.

## Chapter 1 Quiz
### (Sections 1.1–1.2)

**1.** $19°$
**(a)** $90° - 19° = 71°$

**(b)** $180° - 19° = 161°$

**3.** The two angles form a right angle.
$$(5x-1) + 2x = 90 \Rightarrow 7x - 1 = 90 \Rightarrow$$
$$7x = 91 \Rightarrow x = 13$$
The measures of the two angles are
$(5x-1)° = [5(13)-1] = 64°$ and
$2x = 2(13) = 26°$.

**5.** The two marked angles are supplements, so their sum is $180°$.
$$(-14x+18) + (-6x+2) = 180$$
$$-20x + 20 = 180$$
$$-20x = 160 \Rightarrow x = -8$$
The measures of the angles are
$-14(-8) + 18 = 130°$ and $-6(-8) + 2 = 50°$.

**7. (a)** $410°$ is coterminal with
$410° - 360° = 50°$.

**(b)** $-60°$ is coterminal with
$-60° + 360° = 300°$.

**(c)** $890°$ is coterminal with
$890° - 2(360°) = 890° - 720° = 170°$.

**(d)** $57°$ is coterminal with $57° + 360° = 417°$.

**9.**
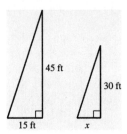

Using similar triangles, we have
$$\frac{45}{15} = \frac{30}{x} \Rightarrow \frac{3}{1} = \frac{30}{x} \Rightarrow 3x = 30 \Rightarrow x = 10.$$
The pole's shadow is 10 ft.

## Section 1.3: Trigonometric Functions

**1.**
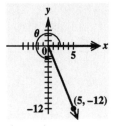

$(5, -12)$
$x = 5$, $y = -12$, and
$$r = \sqrt{x^2 + y^2} = \sqrt{5^2 + (-12)^2}$$
$$= \sqrt{25 + 144} = \sqrt{169} = 13$$
$$\sin\theta = \frac{y}{r} = \frac{-12}{13} = -\frac{12}{13}; \quad \cos\theta = \frac{x}{r} = \frac{5}{13}$$
$$\tan\theta = \frac{y}{x} = \frac{-12}{5} = -\frac{12}{5}$$
$$\cot\theta = \frac{x}{y} = \frac{5}{-12} = -\frac{5}{12}$$
$$\sec\theta = \frac{r}{x} = \frac{13}{5}$$
$$\csc\theta = \frac{r}{y} = \frac{13}{-12} = -\frac{13}{12}$$

**3.**
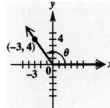

$(-3, 4)$
$x = -3$, $y = 4$ and
$$r = \sqrt{x^2 + y^2} = \sqrt{(-3)^2 + 4^2}$$
$$= \sqrt{9 + 16} = \sqrt{25} = 5$$

$\sin\theta = \frac{y}{r} = \frac{4}{5}$; $\cos\theta = \frac{x}{r} = \frac{-3}{5} = -\frac{3}{5}$

$\tan\theta = \frac{y}{x} = \frac{4}{-3} = -\frac{4}{3}$

$\cot\theta = \frac{x}{y} = \frac{-3}{4} = -\frac{3}{4}$

$\sec\theta = \frac{r}{x} = \frac{5}{-3} = -\frac{5}{3}$; $\csc\theta = \frac{r}{y} = \frac{5}{4}$

5. 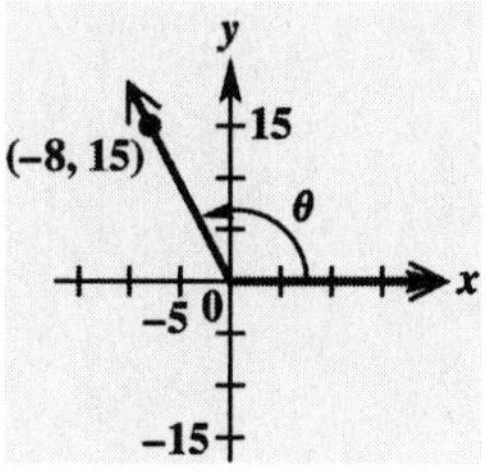


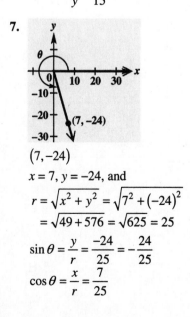

$(-8, 15)$

$x = -8$, $y = 15$, and

$r = \sqrt{x^2 + y^2} = \sqrt{(-8)^2 + 15^2}$
$= \sqrt{64 + 225} = \sqrt{289} = 17$

$\sin\theta = \frac{y}{r} = \frac{15}{17}$

$\cos\theta = \frac{x}{r} = \frac{-8}{17} = -\frac{8}{17}$

$\tan\theta = \frac{y}{x} = \frac{15}{-8} = -\frac{15}{8}$

$\cot\theta = \frac{x}{y} = \frac{-8}{15} = -\frac{8}{15}$

$\sec\theta = \frac{r}{x} = \frac{17}{-8} = -\frac{17}{8}$

$\csc\theta = \frac{r}{y} = \frac{17}{15}$

7. 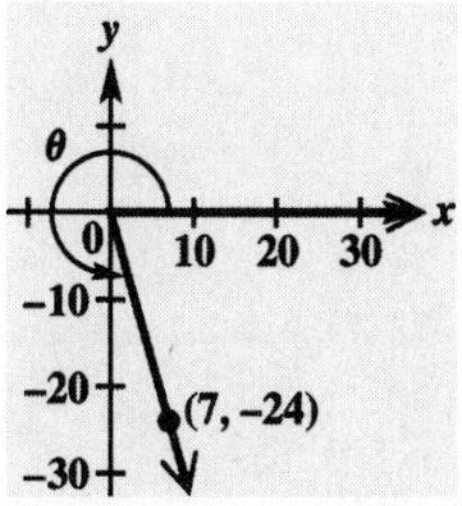


$(7, -24)$

$x = 7$, $y = -24$, and

$r = \sqrt{x^2 + y^2} = \sqrt{7^2 + (-24)^2}$
$= \sqrt{49 + 576} = \sqrt{625} = 25$

$\sin\theta = \frac{y}{r} = \frac{-24}{25} = -\frac{24}{25}$

$\cos\theta = \frac{x}{r} = \frac{7}{25}$

$\tan\theta = \frac{y}{x} = \frac{-24}{7} = -\frac{24}{7}$

$\cot\theta = \frac{x}{y} = \frac{7}{-24} = -\frac{7}{24}$

$\sec\theta = \frac{r}{x} = \frac{25}{7}$

$\csc\theta = \frac{r}{y} = \frac{25}{-24} = -\frac{25}{24}$

9. 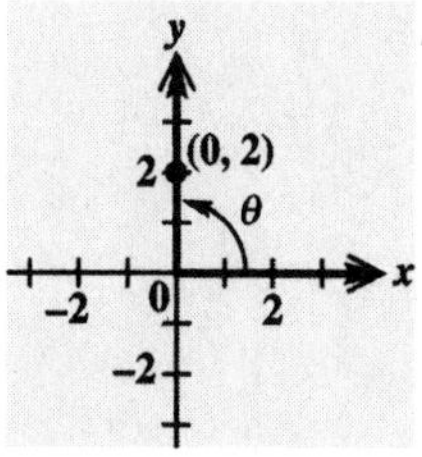


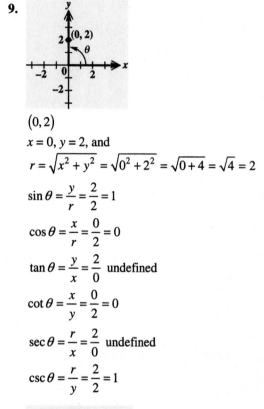

$(0, 2)$

$x = 0$, $y = 2$, and

$r = \sqrt{x^2 + y^2} = \sqrt{0^2 + 2^2} = \sqrt{0 + 4} = \sqrt{4} = 2$

$\sin\theta = \frac{y}{r} = \frac{2}{2} = 1$

$\cos\theta = \frac{x}{r} = \frac{0}{2} = 0$

$\tan\theta = \frac{y}{x} = \frac{2}{0}$ undefined

$\cot\theta = \frac{x}{y} = \frac{0}{2} = 0$

$\sec\theta = \frac{r}{x} = \frac{2}{0}$ undefined

$\csc\theta = \frac{r}{y} = \frac{2}{2} = 1$

11. 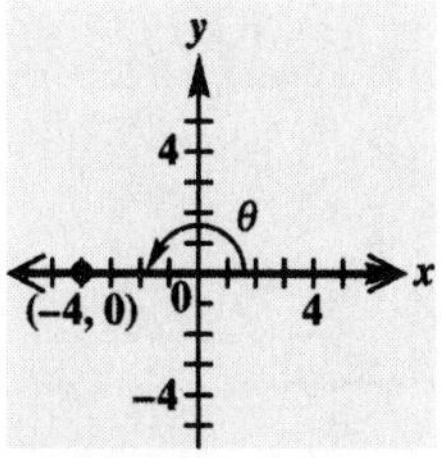


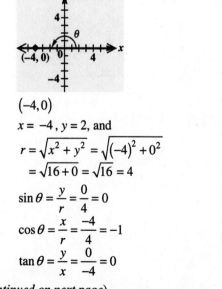

$(-4, 0)$

$x = -4$, $y = 2$, and

$r = \sqrt{x^2 + y^2} = \sqrt{(-4)^2 + 0^2}$
$= \sqrt{16 + 0} = \sqrt{16} = 4$

$\sin\theta = \frac{y}{r} = \frac{0}{4} = 0$

$\cos\theta = \frac{x}{r} = \frac{-4}{4} = -1$

$\tan\theta = \frac{y}{x} = \frac{0}{-4} = 0$

*(continued on next page)*

(*continued from page 9*)

$$\cot\theta = \frac{x}{y} = \frac{-4}{0} \text{ undefined}$$

$$\sec\theta = \frac{r}{x} = \frac{4}{-4} = -1$$

$$\csc\theta = \frac{r}{y} = \frac{4}{0} \quad \text{undefined}$$

$$\tan\theta = \frac{y}{x} = \frac{\sqrt{3}}{1} = \sqrt{3}$$

$$\cot\theta = \frac{x}{y} = \frac{1}{\sqrt{3}} = \frac{1}{\sqrt{3}}\cdot\frac{\sqrt{3}}{\sqrt{3}} = \frac{\sqrt{3}}{3}$$

$$\sec\theta = \frac{r}{x} = \frac{2}{1} = 2$$

$$\csc\theta = \frac{r}{y} = \frac{2}{\sqrt{3}} = \frac{2}{\sqrt{3}}\cdot\frac{\sqrt{3}}{\sqrt{3}} = \frac{2\sqrt{3}}{3}$$

**13.**

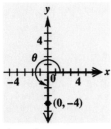

$$(0,-4)$$

$x = 0$, $y = -4$, and

$$r = \sqrt{x^2 + y^2} = \sqrt{0^2 + (-4)^2}$$
$$= \sqrt{0+16} = \sqrt{16} = 4$$

$$\sin\theta = \frac{y}{r} = \frac{-4}{4} = -1$$

$$\cos\theta = \frac{x}{r} = \frac{0}{4} = 0$$

$$\tan\theta = \frac{y}{x} = \frac{-4}{0} \text{ undefined}$$

$$\cot\theta = \frac{x}{y} = \frac{0}{-4} = 0$$

$$\sec\theta = \frac{r}{x} = \frac{4}{0} \text{ undefined}$$

$$\csc\theta = \frac{r}{y} = \frac{4}{-4} = -1$$

**17.**

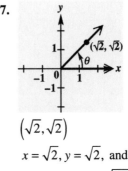

$$\left(\sqrt{2}, \sqrt{2}\right)$$

$x = \sqrt{2}$, $y = \sqrt{2}$, and

$$r = \sqrt{x^2 + y^2} = \sqrt{\left(\sqrt{2}\right)^2 + \left(\sqrt{2}\right)^2}$$
$$= \sqrt{2+2} = \sqrt{4} = 2$$

$$\sin\theta = \frac{y}{r} = \frac{\sqrt{2}}{2} \; ; \; \cos\theta = \frac{x}{r} = \frac{\sqrt{2}}{2}$$

$$\tan\theta = \frac{y}{x} = \frac{\sqrt{2}}{\sqrt{2}} = 1 \; ; \; \cot\theta = \frac{x}{y} = \frac{\sqrt{2}}{\sqrt{2}} = 1$$

$$\sec\theta = \frac{r}{x} = \frac{2}{\sqrt{2}} = \frac{2}{\sqrt{2}}\cdot\frac{\sqrt{2}}{\sqrt{2}} = \frac{2\sqrt{2}}{2} = \sqrt{2}$$

$$\csc\theta = \frac{r}{y} = \frac{2}{\sqrt{2}} = \frac{2}{\sqrt{2}}\cdot\frac{\sqrt{2}}{\sqrt{2}} = \frac{2\sqrt{2}}{2} = \sqrt{2}$$

**15.**

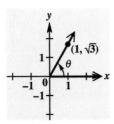

$$\left(1, \sqrt{3}\right)$$

$x = 1$, $y = \sqrt{3}$, and

$$r = \sqrt{x^2 + y^2} = \sqrt{1^2 + \left(\sqrt{3}\right)^2}$$
$$= \sqrt{1+3} = \sqrt{4} = 2$$

$$\sin\theta = \frac{y}{r} = \frac{\sqrt{3}}{2} \; ; \; \cos\theta = \frac{x}{r} = \frac{1}{2}$$

**19.**

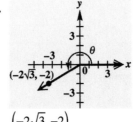

$$\left(-2\sqrt{3}, -2\right)$$

$x = -2\sqrt{3}$, $y = -2$, and

$$r = \sqrt{x^2 + y^2} = \sqrt{\left(-2\sqrt{3}\right)^2 + (-2)^2}$$
$$= \sqrt{12+4} = \sqrt{16} = 4$$

$$\sin\theta = \frac{y}{r} = \frac{-2}{4} = -\frac{1}{2}$$

$$\cos\theta = \frac{x}{r} = \frac{-2\sqrt{3}}{4} = -\frac{\sqrt{3}}{2}$$

$$\tan\theta = \frac{y}{x} = \frac{-2}{-2\sqrt{3}} = \frac{1}{\sqrt{3}} = \frac{\sqrt{3}}{3}$$

$$\cot\theta = \frac{x}{y} = \frac{-2\sqrt{3}}{-2} = \sqrt{3}$$

$$\sec\theta = \frac{r}{x} = \frac{4}{-2\sqrt{3}} = -\frac{2}{\sqrt{3}} = -\frac{2\sqrt{3}}{3}$$

$$\csc\theta = \frac{r}{y} = \frac{4}{-2} = -2$$

**21.** Answers will vary. For any nonquadrantal angle $\theta$, a point on the terminal side of $\theta$ will be of the form $(x, y)$ where $x, y \neq 0$.

Now $\sin\theta = \frac{y}{r}$ and $\csc\theta = \frac{r}{y}$ both exist and are simply reciprocals of each other, and hence will have the same sign.

**23.** Since $\cot\theta$ is undefined, and $\cot\theta = \frac{x}{y}$, where $(x, y)$ is a point on the terminal side of $\theta$, $y = 0$ and $x$ can be any nonzero number. Therefore $\tan\theta = 0$.

In Exercises 25–43, $r = \sqrt{x^2 + y^2}$, which is positive.

**25.** In quadrant II, $x$ is negative, so $\frac{x}{r}$ is negative.

**27.** In quadrant IV, $x$ is positive and $y$ is negative, so $\frac{y}{x}$ is negative.

**29.** In quadrant II, $y$ is positive, so $\frac{y}{r}$ is positive.

**31.** In quadrant IV, $x$ is positive, so $\frac{x}{r}$ is positive.

**33.** In quadrant II, $x$ is negative and $y$ is positive, so $\frac{x}{y}$ is negative.

**35.** In quadrant III, $x$ is negative and $y$ is negative, so $\frac{y}{x}$ is positive.

**37.** In quadrant IV, $x$ is positive and $y$ is negative, so $\frac{x}{y}$ is negative.

**39.** In quadrant I, $x$ is positive and $y$ is positive, so $\frac{x}{y}$ is positive.

**41.** In quadrant I, $y$ is positive, so $\frac{y}{r}$ is positive.

**43.** In quadrant I, $x$ is positive, so $\frac{r}{x}$ is positive.

**45.** Since $x \geq 0$, the graph of the line $2x + y = 0$ is shown to the right of the $y$-axis. A point on this line is $(1, -2)$ since $2(1) + (-2) = 0$. The corresponding value of $r$ is
$$r = \sqrt{1^2 + (-2)^2} = \sqrt{1 + 4} = \sqrt{5}.$$

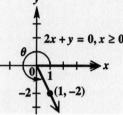

$$\sin\theta = \frac{y}{r} = \frac{-2}{\sqrt{5}} = -\frac{2}{\sqrt{5}} \cdot \frac{\sqrt{5}}{\sqrt{5}} = -\frac{2\sqrt{5}}{5}$$

$$\cos\theta = \frac{x}{r} = \frac{1}{\sqrt{5}} = \frac{1}{\sqrt{5}} \cdot \frac{\sqrt{5}}{\sqrt{5}} = \frac{\sqrt{5}}{5}$$

$$\tan\theta = \frac{y}{x} = \frac{-2}{1} = -2$$

$$\cot\theta = \frac{x}{y} = \frac{1}{-2} = -\frac{1}{2}$$

$$\sec\theta = \frac{r}{x} = \frac{\sqrt{5}}{1} = \sqrt{5}$$

$$\csc\theta = \frac{r}{y} = \frac{\sqrt{5}}{-2} = -\frac{\sqrt{5}}{2}$$

**47.** Since $x \leq 0$, the graph of the line $-6x - y = 0$ is shown to the left of the $y$-axis. A point on this graph is $(-1, 6)$ since $-6(-1) - 6 = 0$. The corresponding value of $r$ is
$$r = \sqrt{(-1)^2 + 6^2} = \sqrt{1 + 36} = \sqrt{37}.$$

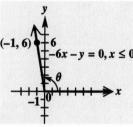

(*continued on next page*)

(*continued from page 11*)

$$\sin\theta = \frac{y}{r} = \frac{6}{\sqrt{37}} = \frac{6}{\sqrt{37}}\cdot\frac{\sqrt{37}}{\sqrt{37}} = \frac{6\sqrt{37}}{37}$$

$$\cos\theta = \frac{x}{r} = \frac{-1}{\sqrt{37}} = -\frac{1}{\sqrt{37}}\cdot\frac{\sqrt{37}}{\sqrt{37}} = -\frac{\sqrt{37}}{37}$$

$$\tan\theta = \frac{y}{x} = \frac{6}{-1} = -6$$

$$\cot\theta = \frac{x}{y} = \frac{-1}{6} = -\frac{1}{6}$$

$$\sec\theta = \frac{r}{x} = \frac{\sqrt{37}}{-1} = -\sqrt{37}$$

$$\csc\theta = \frac{r}{y} = \frac{\sqrt{37}}{6}$$

**49.** Since $x \le 0$, the graph of the line $-4x + 7y = 0$ is shown to the left of the $y$-axis. A point on this line is $(-7, -4)$ since $-4(-7) + 7(-4) = 0$. The corresponding value of $r$ is $r = \sqrt{(-7)^2 + (-4)^2} = \sqrt{49 + 16} = \sqrt{65}$.

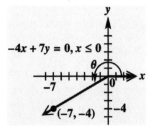

$$\sin\theta = \frac{y}{r} = \frac{-4}{\sqrt{65}} = -\frac{4}{\sqrt{65}}\cdot\frac{\sqrt{65}}{\sqrt{65}} = -\frac{4\sqrt{65}}{65}$$

$$\cos\theta = \frac{x}{r} = \frac{-7}{\sqrt{65}} = -\frac{7}{\sqrt{65}}\cdot\frac{\sqrt{65}}{\sqrt{65}} = -\frac{7\sqrt{65}}{65}$$

$$\tan\theta = \frac{y}{x} = \frac{-4}{-7} = \frac{4}{7}$$

$$\cot\theta = \frac{x}{y} = \frac{-7}{-4} = \frac{7}{4}$$

$$\sec\theta = \frac{r}{x} = \frac{\sqrt{65}}{-7} = -\frac{\sqrt{65}}{7}$$

$$\csc\theta = \frac{r}{y} = \frac{\sqrt{65}}{-4} = -\frac{\sqrt{65}}{4}$$

**51.** Since $x \ge 0$, the graph of the line $x + y = 0$ is shown to the right of the $y$-axis. A point on this line is $(2, -2)$ since $2 + (-2) = 0$. The corresponding value of $r$ is
$$r = \sqrt{2^2 + (-2)^2} = \sqrt{4 + 4} = \sqrt{8} = 2\sqrt{2}.$$

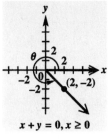

$$\sin\theta = \frac{y}{r} = \frac{-2}{2\sqrt{2}} = -\frac{1}{\sqrt{2}}\cdot\frac{\sqrt{2}}{\sqrt{2}} = -\frac{\sqrt{2}}{2}$$

$$\cos\theta = \frac{x}{r} = \frac{2}{2\sqrt{2}} = \frac{1}{\sqrt{2}}\cdot\frac{\sqrt{2}}{\sqrt{2}} = \frac{\sqrt{2}}{2}$$

$$\tan\theta = \frac{y}{x} = \frac{-2}{2} = -1$$

$$\cot\theta = \frac{x}{y} = \frac{2}{-2} = -1$$

$$\sec\theta = \frac{r}{x} = \frac{2\sqrt{2}}{2} = \sqrt{2}$$

$$\csc\theta = \frac{r}{y} = \frac{2\sqrt{2}}{-2} = -\sqrt{2}$$

**53.** Since $x \le 0$, the graph of the line $-\sqrt{3}x + y = 0$ is shown to the left of the $y$-axis. A point on this line is $\left(-1, -\sqrt{3}\right)$ since $-\sqrt{3}(-1) - \sqrt{3} = \sqrt{3} - \sqrt{3} = 0$ The corresponding value of $r$ is
$$r = \sqrt{(-1)^2 + \left(-\sqrt{3}\right)^2} = \sqrt{1 + 3} = \sqrt{4} = 2.$$

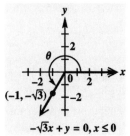

$$\sin\theta = \frac{y}{r} = \frac{-\sqrt{3}}{2} = -\frac{\sqrt{3}}{2}$$

$$\cos\theta = \frac{x}{r} = \frac{-1}{2} = -\frac{1}{2}$$

$$\tan\theta = \frac{y}{x} = \frac{-\sqrt{3}}{-1} = \sqrt{3}$$

$$\cot\theta = \frac{x}{y} = \frac{-1}{-\sqrt{3}} = \frac{1}{\sqrt{3}}\cdot\frac{\sqrt{3}}{\sqrt{3}} = \frac{\sqrt{3}}{3}$$

$\sec\theta = \dfrac{r}{x} = \dfrac{2}{-1} = -2$

$\csc\theta = \dfrac{r}{y} = \dfrac{2}{-\sqrt{3}} = -\dfrac{2}{\sqrt{3}} \cdot \dfrac{\sqrt{3}}{\sqrt{3}} = -\dfrac{2\sqrt{3}}{3}$

Use the figure below to help solve exercises 55–83.

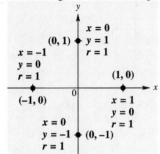

**55.**  $\cos 90°$

$\cos 90° = \dfrac{x}{r} = \dfrac{0}{1} = 0$

**57.**  $\tan 180°$

$\tan 180° = \dfrac{y}{x} = \dfrac{0}{-1} = 0$

**59.**  $\sec 180°$

$\sec 180° = \dfrac{r}{x} = \dfrac{1}{-1} = -1$

**61.**  $\sin\left(-270°\right)$

The quadrantal angle $\theta = -270°$ is coterminal with $-270° + 360° = 90°$.

$\sin\left(-270°\right) = \sin 90 = \dfrac{y}{r} = \dfrac{1}{1} = 1$

**63.**  $\cot 540°$

The quadrantal angle $\theta = 540°$ is coterminal with $540° - 360° = 180°$.

$\cot 540° = \cot 180° = \dfrac{x}{y} = \dfrac{-1}{0}$ undefined

**65.**  $\csc\left(-450°\right)$

The quadrantal angle $\theta = -450°$ is coterminal with $720° - 450° = 270°$.

$\csc\left(-450°\right) = \csc 270° = \dfrac{r}{y} = \dfrac{1}{-1} = -1$

**67.**  $\sin 1800°$

The quadrantal angle $\theta = 1800°$ is coterminal with $1800° - 5\left(360°\right) = 1800° - 1800° = 0°$

$\sin 1800° = \sin 0° = \dfrac{y}{r} = \dfrac{0}{1} = 0$

**69.**  $\csc 1800°$

The quadrantal angle $\theta = 1800°$ is coterminal with $1800° - 5\left(360°\right) = 1800° - 1800° = 0°$

$\csc 1800° = \csc 0° = \dfrac{r}{y} = \dfrac{1}{0}$ undefined

**71.**  $\sec 1800°$

The quadrantal angle $\theta = 1800°$ is coterminal with $1800° - 5\left(360°\right) = 1800° - 1800° = 0°$

$\sec 1800° = \sec 0° = \dfrac{r}{x} = \dfrac{1}{1} = 1$

**73.**  $\cos 90° + 3\sin 270°$

$\cos 90° = \dfrac{x}{r} = \dfrac{0}{1} = 0$

$\sin 270° = \dfrac{y}{r} = \dfrac{-1}{1} = -1$

$\cos 90° + 3\sin 270° = 0 + 3\left(-1\right) = -3$

**75.**  $3\sec\ 180° - 5\tan\ 360°$

$\sec 180° = \dfrac{r}{x} = \dfrac{1}{-1} = -1$ and

$\tan 360° = \tan 0° = \dfrac{y}{x} = \dfrac{0}{1} = 0$

$3\sec\ 180° - 5\tan\ 360° = 3\left(-1\right) - 5\left(0\right)$
$= -3 - 0 = -3$

**77.**  $\tan 360° + 4\sin 180° + 5\cos^2 180°$

$\tan 360° = \tan 0° = \dfrac{y}{x} = \dfrac{0}{1} = 0,$

$\sin 180° = \dfrac{y}{r} = \dfrac{0}{1} = 0,$ and

$\cos 180° = \dfrac{x}{r} = \dfrac{-1}{1} = -1$

$\tan 360° + 4\sin 180° + 5\cos^2 180°$
$= 0 + 4\left(0\right) + 5\left(-1\right)^2 = 0 + 0 + 5\left(1\right) = 5$

**79.**  $\sin^2 180° + \cos^2 180°$

$\sin 180° = \dfrac{y}{r} = \dfrac{0}{1} = 0$ and

$\cos 180° = \dfrac{x}{r} = \dfrac{-1}{1} = -1$

$\sin^2 180° + \cos^2 180° = 0^2 + \left(-1\right)^2 = 0 + 1 = 1$

**81.** $\sec^2 180° - 3\sin^2 360° + \cos 180°$

$\sec 180° = \dfrac{r}{x} = \dfrac{1}{-1} = -1,$

$\sin 360° = \sin 0° = \dfrac{y}{r} = \dfrac{0}{1} = 0$ and

$\cos 180° = \dfrac{x}{r} = \dfrac{-1}{1} = -1$

$\sec^2 180° - 3\sin^2 360° + \cos 180°$

$\quad = (-1)^2 - 3(0)^2 + (-1) = 1 - 1 = 0$

**83.** $-2\sin^4 0° + 3\tan^2 0°$

$\sin 0° = \dfrac{y}{r} = \dfrac{0}{1} = 0; \tan 0° = \dfrac{y}{x} = \dfrac{0}{1} = 0$

$-2\sin^4 0° + 3\tan^2 0° = -2(0)^4 + 3(0)^2 = 0$

**85.** $\cos\left[(2n+1)\cdot 90°\right]$

This angle is a quadrantal angle whose terminal side lies on either the positive part of the $y$-axis or the negative part of the $y$-axis. Any point on these terminal sides would have the form $(0, k),$ where $k$ is any real number, $k \neq 0.$

$\cos\left[(2n+1)\cdot 90°\right] = \dfrac{x}{r} = \dfrac{0}{\sqrt{0^2 + k^2}}$

$\quad = \dfrac{0}{\sqrt{k^2}} = \dfrac{0}{|k|} = 0$

**87.** $\tan\left[n\cdot 180°\right]$

The angle is a quadrantal angle whose terminal side lies on either the positive part of the $x$-axis or the negative part of the $x$-axis. Any point on these terminal sides would have the form $(k, 0),$ where $k$ is any real number, $k \neq 0.$

$\tan\left[n\cdot 180°\right] = \dfrac{y}{x} = \dfrac{0}{k} = 0$

**89.** $\sin\left[270° + n\cdot 360°\right]$

This angle is a quadrantal angle that is coterminal with $\theta = 270°.$ $\sin 270° = -1,$ so $\sin\left[270° + n\cdot 360°\right] = -1.$

**91.** $\cot\left[(2n+1)\cdot 90°\right]$

This angle is a quadrantal angle whose terminal side lies on either the positive part of the $y$-axis or the negative part of the $y$-axis. Any point on these terminal sides would have the form $(0, k),$ where $k$ is any real number, $k \neq 0.$

$\cot\left[(2n+1)\cdot 90°\right] = \dfrac{x}{y} = \dfrac{0}{k} = 0$

**93.** $\sec\left[(2n+1)\cdot 90°\right]$

This angle is a quadrantal angle whose terminal side lies on either the positive part of the $y$-axis or the negative part of the $y$-axis. Any point on these terminal sides would have the form $(0, k),$ where $k$ is any real number, $k \neq 0.$

$\sec\left[(2n+1)\cdot 90°\right] = \dfrac{r}{x} = \dfrac{\sqrt{0^2 + k^2}}{0}$ undefined

**95.** Using a calculator, $\sin 15° = .258819045$ and $\cos 75° = .258819045.$ We can conjecture that the sines and cosines of complementary angles are equal. Trying another pair of complementary angles we obtain $\sin 30° = \cos 60° = .5.$ Therefore, our conjecture appears to be true.

**97.** Using a calculator, $\sin 10° = .173648178$ and $\sin(-10°) = -.173648178.$ We can conjecture that the sines of an angle and its negative are opposites of each other. Using a circle, an angle $\theta$ having the point $(x, y)$ on its terminal side has a corresponding angle $-\theta$ with a point $(x, -y)$ on its terminal side. From the definition of sine, $\sin(-\theta) = \dfrac{-y}{r} = -\dfrac{y}{r}$ and $\sin\theta = \dfrac{y}{r}.$ The sines are negatives of each other.

In Exercises 99–103, make sure your calculator is in the modes indicated in the instructions.

**99.** Use the TRACE feature to move around the circle in quadrant I to find $T = 20$. We see that cos 20° is about .940, and sin 20° is about .342.

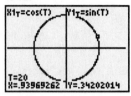

**101.** Use the TRACE feature to move around the circle in quadrant I. We see that sin 35° ≈ .574, so $T = 35°$.

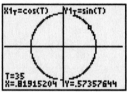

**103.** As $T$ increases from 0° to 90°, the cosine decreases and the sine increases.

## Section 1.4: Using the Definitions of the Trigonometric Functions

**1.** $\sec \theta = \dfrac{1}{\cos \theta} = \dfrac{1}{\frac{2}{3}} = \dfrac{3}{2}$

**3.** $\csc \theta = \dfrac{1}{\sin \theta} = \dfrac{1}{-\frac{3}{7}} = -\dfrac{7}{3}$

**5.** $\cot \theta = \dfrac{1}{\tan \theta} = \dfrac{1}{5}$

**7.** $\cos \theta = \dfrac{1}{\sec \theta} = \dfrac{1}{-\frac{5}{2}} = -\dfrac{2}{5}$

**9.** $\sin \theta = \dfrac{1}{\csc \theta} = \dfrac{1}{\frac{\sqrt{8}}{2}} = \dfrac{2}{\sqrt{8}} = \dfrac{2}{2\sqrt{2}} \cdot \dfrac{\sqrt{2}}{\sqrt{2}} = \dfrac{\sqrt{2}}{2}$

**11.** $\tan \theta = \dfrac{1}{\cot \theta} = \dfrac{1}{-2.5} = -.4$

**13.** $\sin \theta = \dfrac{1}{\csc \theta} = \dfrac{1}{1.42716327} \approx .70069071$

**15.** No; Since $\sin \theta = \dfrac{1}{\csc \theta}$, if $\sin \theta > 0$, then $\csc \theta > 0$.

**17.** Since tan 90° is undefined, it does not have a reciprocal.

**19.** A 74° angle in standard position lies in quadrant I, so all its trigonometric functions are positive.

**21.** A 218° angle in standard position lies in quadrant III, so its sine, cosine, secant, and cosecant are negative, while its tangent and cotangent are positive.

**23.** A 178° angle in standard position lies in quadrant II, so its sine and cosecant are positive, while its cosine, secant, tangent and cotangent are negative.

**25.** A −80° angle in standard position lies in quadrant IV, so its cosine and secant are positive, while its sine, cosecant, tangent and cotangent are negative.

**27.** An 845° angle in standard position is coterminal with a 125° angle, and thus, lies in quadrant II. So its sine and cosecant are positive, while its cosine, secant, tangent and cotangent are negative.

**29.** A −345° angle in standard position lies in quadrant I, so all its trigonometric functions are positive.

**31.** Since $\sin \theta > 0$, $\csc \theta$ is also greater than 0. The functions are greater than 0 (positive) in quadrants I and II.

**33.** $\cos \theta > 0$ in quadrants I and IV, while $\sin \theta > 0$ in quadrants I and II. Both conditions are met only in quadrant I.

**35.** $\tan \theta < 0$ in quadrants II and IV, while $\cos \theta < 0$ in quadrants II and III. Both conditions are met only in quadrant II.

**37.** $\sec \theta > 0$ in quadrants I and IV, while $\csc \theta > 0$ in quadrants I and II. Both conditions are met only in quadrant I.

**39.** $\sec \theta < 0$ in quadrants II and III, while $\csc \theta < 0$ in quadrants III and IV. Both conditions are met only in quadrant III.

**41.** Since $\sin \theta < 0$, $\csc \theta$ is also less than 0. The functions are less than 0 (negative) in quadrants III and IV.

**43.** The answers to exercises 33 and 37 are the same because functions in exercise 37 are the reciprocals of the functions in exercise 33.

**45.** Impossible because the range of $\sin\theta$ is $[-1, 1]$.

**47.** Possible because the range of $\cos\theta$ is $[-1, 1]$.

**49.** Possible because the range of $\tan\theta$ is $(-\infty, \infty)$.

**51.** Impossible because the range of $\sec\theta$ is $(-\infty, -1] \cup [1, \infty)$.

**53.** Possible because the range of $\csc\theta$ is $(-\infty, -1] \cup [1, \infty)$.

**55.** Possible because the range of $\cot\theta$ is $(-\infty, \infty)$.

**57.** $\sin\theta = \dfrac{1}{2}$ is possible because the range of $\sin\theta$ is $[-1, 1]$. Furthermore, when $\sin\theta = \dfrac{1}{2}$, $\csc\theta = \dfrac{1}{\frac{1}{2}} = 2$. Thus, $\sin\theta = \dfrac{1}{2}$ and $\csc\theta = 2$ is possible.

**59.** $\cos\theta = -2$ is impossible because the range of $\cos\theta$ is $[-1, 1]$.

**61.** If $\sin\theta = \dfrac{3}{5}$, then $y = 3$ and $r = 5$. So
$$r^2 = x^2 + y^2 \Rightarrow 5^2 = x^2 + 3^2 \Rightarrow 25 = x^2 + 9 \Rightarrow$$
$$16 = x^2 \Rightarrow \pm 4 = x. \ \theta \text{ is in quadrant II, so}$$
$x = -4$. Therefore, $\cos\theta = -\dfrac{4}{5}$. Alternatively, use the identity $\sin^2\theta + \cos^2\theta = 1$:
$$\left(\dfrac{3}{5}\right)^2 + \cos^2\theta = 1 \Rightarrow$$
$$\dfrac{9}{25} + \cos^2\theta = 1 \Rightarrow \cos^2\theta = \dfrac{16}{25} \Rightarrow \cos\theta = \pm\dfrac{4}{5}$$
$\theta$ is in quadrant II, so $\cos\theta$ is negative. Thus, $\cos\theta = -\dfrac{4}{5}$.

**63.** If $\cot\theta = -\dfrac{1}{2}$ and $\theta$ is in quadrant IV, then $x = 1$ and $y = -2$. So
$$r^2 = x^2 + y^2 \Rightarrow r^2 = 1^2 + (-2)^2 \Rightarrow$$
$$r^2 = 1 + 4 \Rightarrow r^2 = 5 \Rightarrow r = \sqrt{5}$$
Therefore, $\csc\theta = \dfrac{r}{y} = -\dfrac{\sqrt{5}}{2}$. Alternatively, use the identity $1 + \cot^2\theta = \csc^2\theta$:
$$1 + \left(-\dfrac{1}{2}\right)^2 = \csc^2\theta \Rightarrow 1 + \dfrac{1}{4} = \csc^2\theta \Rightarrow$$
$$\dfrac{5}{4} = \csc^2\theta \Rightarrow \pm\dfrac{\sqrt{5}}{2} = \csc\theta. \ \theta \text{ is in quadrant}$$
IV, so $\csc\theta$ is negative. Thus, $\csc\theta = -\dfrac{\sqrt{5}}{2}$

**65.** If $\sin\theta = \dfrac{1}{2}$, then $y = 1$ and $r = 2$. So
$$r^2 = x^2 + y^2 \Rightarrow 2^2 = x^2 + 1^2 \Rightarrow 4 = x^2 + 1 \Rightarrow$$
$$3 = x^2 \Rightarrow \pm\sqrt{3} = x$$
$\theta$ is in quadrant II, so $x = -\sqrt{3}$. Therefore,
$$\tan\theta = -\dfrac{1}{\sqrt{3}} = -\dfrac{1}{\sqrt{3}}\cdot\dfrac{\sqrt{3}}{\sqrt{3}} = -\dfrac{\sqrt{3}}{3}.$$
Alternatively, using the identity $\sin^2\theta + \cos^2\theta = 1$ gives
$$\left(\dfrac{1}{2}\right)^2 + \cos^2\theta = 1 \Rightarrow \cos^2\theta = \dfrac{3}{4} \Rightarrow$$
$$\cos\theta = \pm\dfrac{\sqrt{3}}{2}. \text{ Since } \theta \text{ is in quadrant II,}$$
$\cos\theta = -\dfrac{\sqrt{3}}{2}$. Then, using the identity
$$\tan\theta = \dfrac{\sin\theta}{\cos\theta} = \dfrac{\frac{1}{2}}{-\frac{\sqrt{3}}{2}} = -\dfrac{1}{\sqrt{3}}$$
$$= -\dfrac{1}{\sqrt{3}}\cdot\dfrac{\sqrt{3}}{\sqrt{3}} = -\dfrac{\sqrt{3}}{3}$$

**67.** Using the identity $1 + \cot^2\theta = \csc^2\theta$ gives
$$1 + \cot^2\theta = (-3.5891420)^2 \Rightarrow$$
$$\cot^2\theta = 11.8819403 \Rightarrow \cot\theta \approx \pm 3.44701905$$
Since $\theta$ is in quadrant III,
$\cot\theta \approx 3.44701905$.

For Exercises 69–73, remember that $r$ is always positive.

**69.** $\tan \theta = -\dfrac{15}{8} = \dfrac{15}{-8}$, with $\theta$ in quadrant II

$\tan \theta = \dfrac{y}{x}$ and $\theta$ is in quadrant II, so let $y = 15$ and $x = -8$.

$x^2 + y^2 = r^2 \Rightarrow (-8)^2 + 15^2 = r^2 \Rightarrow$
$64 + 225 = r^2 \Rightarrow 289 = r^2 \Rightarrow r = 17$

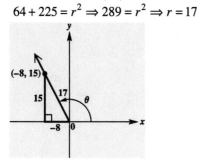

$\sin \theta = \dfrac{y}{r} = \dfrac{15}{17}$; $\cos \theta = \dfrac{x}{r} = \dfrac{-8}{17} = -\dfrac{8}{17}$

$\tan \theta = \dfrac{y}{x} = \dfrac{15}{-8} = -\dfrac{15}{8}$

$\cot \theta = \dfrac{x}{y} = \dfrac{-8}{15} = -\dfrac{8}{15}$

$\sec \theta = \dfrac{r}{x} = \dfrac{17}{-8} = -\dfrac{17}{8}$; $\csc \theta = \dfrac{r}{y} = \dfrac{17}{15}$

**71.** $\sin \theta = \dfrac{\sqrt{5}}{7}$, with $\theta$ in quadrant I

$\sin \theta = \dfrac{y}{r}$ and $\theta$ in quadrant I, so let $y = \sqrt{5}, r = 7$.

$x^2 + y^2 = r^2 \Rightarrow x^2 + (\sqrt{5})^2 = 7^2 \Rightarrow$
$x^2 + 5 = 49 \Rightarrow x^2 = 44 \Rightarrow$
$x = \pm\sqrt{44} \Rightarrow x = \pm 2\sqrt{11}$

$\theta$ is in quadrant I, so $x = 2\sqrt{11}$.

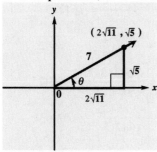

*Drawing not to scale*

$\sin \theta = \dfrac{y}{r} = \dfrac{\sqrt{5}}{7}$

$\cos \theta = \dfrac{x}{r} = \dfrac{2\sqrt{11}}{7}$

$\tan \theta = \dfrac{y}{x} = \dfrac{\sqrt{5}}{2\sqrt{11}} = \dfrac{\sqrt{5}}{2\sqrt{11}} \cdot \dfrac{\sqrt{11}}{\sqrt{11}} = \dfrac{\sqrt{55}}{22}$

$\cot \theta = \dfrac{x}{y} = \dfrac{2\sqrt{11}}{\sqrt{5}} = \dfrac{2\sqrt{11}}{\sqrt{5}} \cdot \dfrac{\sqrt{5}}{\sqrt{5}} = \dfrac{2\sqrt{55}}{5}$

$\sec \theta = \dfrac{r}{x} = \dfrac{7}{2\sqrt{11}} = \dfrac{7}{2\sqrt{11}} \cdot \dfrac{\sqrt{11}}{\sqrt{11}} = \dfrac{7\sqrt{11}}{22}$

$\csc \theta = \dfrac{r}{y} = \dfrac{7}{\sqrt{5}} = \dfrac{7}{\sqrt{5}} \cdot \dfrac{\sqrt{5}}{\sqrt{5}} = \dfrac{7\sqrt{5}}{5}$

**73.** $\cot \theta = \dfrac{\sqrt{3}}{8}$, with $\theta$ in quadrant I

$\cot \theta = \dfrac{x}{y}$ and $\theta$ is in quadrant I, so let $x = \sqrt{3}$ and $y = 8$.

$x^2 + y^2 = r^2 \Rightarrow (\sqrt{3})^2 + (8)^2 = r^2 \Rightarrow$
$3 + 64 = r^2 \Rightarrow 67 = r^2 \Rightarrow \sqrt{67} = r$

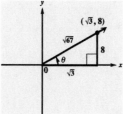

$\sin \theta = \dfrac{y}{r} = \dfrac{8}{\sqrt{67}} = \dfrac{8}{\sqrt{67}} \cdot \dfrac{\sqrt{67}}{\sqrt{67}} = \dfrac{8\sqrt{67}}{67}$

$\cos \theta = \dfrac{x}{r} = \dfrac{\sqrt{3}}{\sqrt{67}} = \dfrac{\sqrt{3}}{\sqrt{67}} \cdot \dfrac{\sqrt{67}}{\sqrt{67}}$
$= \dfrac{\sqrt{3}\sqrt{67}}{67} = \dfrac{\sqrt{201}}{67}$

$\tan \theta = \dfrac{y}{x} = \dfrac{8}{\sqrt{3}} = \dfrac{8}{\sqrt{3}} \cdot \dfrac{\sqrt{3}}{\sqrt{3}} = \dfrac{8\sqrt{3}}{3}$

$\cot \theta = \dfrac{x}{y} = \dfrac{\sqrt{3}}{8}$

$\sec \theta = \dfrac{r}{x} = \dfrac{\sqrt{67}}{\sqrt{3}} = \dfrac{\sqrt{67}}{\sqrt{3}} \cdot \dfrac{\sqrt{3}}{\sqrt{3}}$
$= \dfrac{\sqrt{67}\sqrt{3}}{3} = \dfrac{\sqrt{201}}{3}$

$\csc \theta = \dfrac{r}{y} = \dfrac{\sqrt{67}}{8}$

**75.** $\sin\theta = \dfrac{\sqrt{2}}{6}$, given that $\cos\theta < 0$

$\sin\theta$ is positive and $\cos\theta$ is negative when $\theta$ is in quadrant II.

$\sin\theta = \dfrac{y}{r}$ and $\theta$ in quadrant II, so let

$y = \sqrt{2}, r = 6.$

$x^2 + y^2 = r^2 \Rightarrow x^2 + \left(\sqrt{2}\right)^2 = 6^2 \Rightarrow$

$x^2 + 2 = 36 \Rightarrow x^2 = 34 \Rightarrow$

$x = \pm\sqrt{34}$

$\theta$ is in quadrant II, so $x = -\sqrt{34}.$

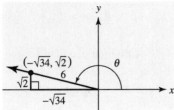

$\sin\theta = \dfrac{y}{r} = \dfrac{\sqrt{2}}{6}$ ; $\cos\theta = \dfrac{x}{r} = -\dfrac{\sqrt{34}}{6}$

$\tan\theta = \dfrac{y}{x} = -\dfrac{\sqrt{2}}{\sqrt{34}} = -\dfrac{\sqrt{2}}{\sqrt{34}} \cdot \dfrac{\sqrt{34}}{\sqrt{34}} = -\dfrac{\sqrt{68}}{34}$

$= -\dfrac{2\sqrt{17}}{34} = -\dfrac{\sqrt{17}}{17}$

$\cot\theta = \dfrac{x}{y} = -\dfrac{\sqrt{34}}{\sqrt{2}} = -\dfrac{\sqrt{34}}{\sqrt{2}} \cdot \dfrac{\sqrt{2}}{\sqrt{2}} = -\dfrac{\sqrt{68}}{2}$

$= -\dfrac{2\sqrt{17}}{2} = -\sqrt{17}$

$\sec\theta = \dfrac{r}{x} = -\dfrac{6}{\sqrt{34}} = -\dfrac{6}{\sqrt{34}} \cdot \dfrac{\sqrt{34}}{\sqrt{34}} = -\dfrac{6\sqrt{34}}{34}$

$= -\dfrac{3\sqrt{34}}{17}$

$\csc\theta = \dfrac{r}{y} = \dfrac{6}{\sqrt{2}} = \dfrac{6}{\sqrt{2}} \cdot \dfrac{\sqrt{2}}{\sqrt{2}} = \dfrac{6\sqrt{2}}{2} = 3\sqrt{2}$

**77.** $\sec\theta = -4$, given that $\sin\theta > 0$

$\sec\theta$ is negative and $\sin\theta$ is positive when $\theta$ is in quadrant II.

$\sec\theta = \dfrac{r}{x}$ and $\theta$ in quadrant II, so let

$x = -1, r = 4.$

$x^2 + y^2 = r^2 \Rightarrow (-1)^2 + y^2 = 4^2 \Rightarrow$

$1 + y^2 = 16 \Rightarrow y^2 = 15 \Rightarrow y = \pm\sqrt{15}$

$\theta$ is in quadrant II, so $y = \sqrt{15}.$

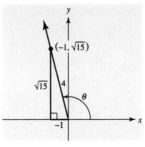

$\sin\theta = \dfrac{y}{r} = \dfrac{\sqrt{15}}{4}$ ; $\cos\theta = \dfrac{x}{r} = -\dfrac{1}{4}$

$\tan\theta = \dfrac{y}{x} = -\dfrac{\sqrt{15}}{1} = -\sqrt{15}$

$\cot\theta = \dfrac{x}{y} = -\dfrac{1}{\sqrt{15}} = -\dfrac{1}{\sqrt{15}} \cdot \dfrac{\sqrt{15}}{\sqrt{15}} = -\dfrac{\sqrt{15}}{15}$

$\sec\theta = \dfrac{r}{x} = -4$

$\csc\theta = \dfrac{r}{y} = \dfrac{4}{\sqrt{15}} = \dfrac{4}{\sqrt{15}} \cdot \dfrac{\sqrt{15}}{\sqrt{15}} = \dfrac{4\sqrt{15}}{15}$

**79.** $\sin\theta = .164215 = \dfrac{.164215}{1}$, with $\theta$ in quadrant II

$\sin\theta = \dfrac{y}{r}$ and $\theta$ is in quadrant II, let

$y = .164215$ and $r = 1.$

$x^2 + y^2 = r^2 \Rightarrow x^2 + (.164215)^2 = 1^2 \Rightarrow$

$x^2 + .026966 = 1 \Rightarrow x^2 = .973034 \Rightarrow$

$x \approx \pm\sqrt{.973034} \Rightarrow x \approx \pm.986425$

$\theta$ is in quadrant II, so, $x = -.986425.$

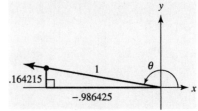

$\sin\theta = \dfrac{y}{r} = \dfrac{.164215}{1} = .164215$

$\cos\theta = \dfrac{x}{r} \approx \dfrac{-.986425}{1} = -.986425$

$\tan\theta = \dfrac{y}{x} \approx \dfrac{.164215}{-.986425} \approx -.166475$

$\cot\theta = \dfrac{x}{y} \approx \dfrac{-.986425}{.164215} \approx -6.00691$

$\sec\theta = \dfrac{r}{x} \approx \dfrac{1}{-.986425} \approx -1.01376$

$\csc\theta = \dfrac{r}{y} = \dfrac{1}{.164215} \approx 6.08958$

**81.**  $x^2 + y^2 = r^2 \Rightarrow \dfrac{x^2 + y^2}{y^2} = \dfrac{r^2}{y^2} \Rightarrow$

$\dfrac{x^2}{y^2} + \dfrac{y^2}{y^2} = \dfrac{r^2}{y^2} \Rightarrow \left(\dfrac{x}{y}\right)^2 + 1 = \left(\dfrac{r}{y}\right)^2 \Rightarrow$

$1 + \left(\dfrac{x}{y}\right)^2 = \left(\dfrac{r}{y}\right)^2$

Since $\cot\theta = \dfrac{x}{y}$ and $\csc\theta = \dfrac{r}{y}$, we have

$1 + \left(\cot\theta\right)^2 = \left(\csc\theta\right)^2$ or $1 + \cot^2\theta = \csc^2\theta$.

**83.**  The statement is false. For example,
$\sin 180° + \cos 180° = 0 + (-1) = -1 \neq 1$.

**85.**  $90° < \theta < 180° \Rightarrow 180° < 2\theta < 360°$, so $2\theta$ lies in either quadrant III or IV. Thus, $\sin 2\theta$ is negative.

**87.**  $90° < \theta < 180° \Rightarrow 270° < \theta + 180° < 360°$, so $\theta + 180°$ lies in quadrant IV. Thus, $\cot(\theta + 180°)$ is negative.

**89.**  $-90° < \theta < 90° \Rightarrow -45° < \dfrac{\theta}{2} < 45°$, so $\dfrac{\theta}{2}$ lies in either quadrant I or quadrant IV. Thus $\cos\dfrac{\theta}{2}$ is positive.

**91.**  $-90° < \theta < 90° \Rightarrow 90° > -\theta > -90° \Rightarrow$ $-90° < -\theta < 90°$, so $-\theta$ lies in either quadrant I or quadrant IV. Thus $\sec(-\theta)$ is positive.

**93.**  $\tan(3\theta - 4°) = \dfrac{1}{\cot(5\theta - 8°)} \Rightarrow$
$\tan(3\theta - 4°) = \tan(5\theta - 8°)$
The second equation is true if $3\theta - 4° = 5\theta - 8°$, so solving this equation will give a value (but not the only value) for which the given equation is true.
$3\theta - 4° = 5\theta - 8° \Rightarrow 4° = 2\theta \Rightarrow \theta = 2°$

**95.**  $\sin(4\theta + 2°)\csc(3\theta + 5°) = 1 \Rightarrow$
$\sin(4\theta + 2°) = \dfrac{1}{\csc(3\theta + 5°)} \Rightarrow$
$\sin(4\theta + 2°) = \sin(3\theta + 5°)$
The third equation is true if $4\theta + 2° = 3\theta + 5°$, so solving this equation will give a value (but not the only value) for which the given equation is true.
$4\theta + 2° = 3\theta + 5° \Rightarrow \theta = 3°$

**97.**  In quadrant II, the cosine is negative and the sine is positive.

## Chapter 1 Review Exercises

**1.**  The complement of $35°$ is $90° - 35° = 55°$.
The supplement of $35°$ is $180° - 35° = 145°$.

**3.**  $-174°$. is coterminal with
$-174° + 360° = 186°$

**5.**  The sum of the measures of the interior angles of a triangle is $180°$.
$60° + y + 90° = 180°$
$\qquad 150° + y = 180° \Rightarrow y = 30°$
Substitute $30°$ for $y$, and solve for $x$:
$\qquad 30° + (x + y) + 90° = 180°$
$\quad 30° + (x + 30°) + 90° = 180°$
$\qquad\qquad 150° + x = 180° \Rightarrow x = 30°$
Thus, $x = y = 30°$.

**7.**  650 rotations per min $= \dfrac{650}{60}$ rotations per sec
$= \dfrac{65}{6}$ rotations per sec
$= 26$ rotations per $2.4$ sec
$= 26(360°)$ per $2.4$ sec $= 9360°$ per $2.4$ sec
A point of the edge of the propeller will move $9360°$ in $2.4$ sec.

**9.**  $119°8'3'' = 119° + \dfrac{8}{60}° + \dfrac{3}{3600}°$
$\approx 119° + .1333° + .0008°$
$\approx 119.134°$

**11.**  $275.1005 = 275° + .1005(60') = 275° + 6.03'$
$= 275°\ 6' + .03' = 275°6' + .03(60'')$
$= 275°6'1.8'' \approx 275°6'2''$

**13.**  The three angles are the interior angles of a triangle. Hence, the sum of their measures is $180°$.
$4x + (5x + 5) + (4x - 20) = 180$
$\qquad\qquad 13x - 15 = 180$
$\qquad\qquad\qquad 13x = 195 \Rightarrow x = 15$
$4(15) = 60, 5(15) + 5 = 80$, and
$4(15) - 20 = 40$, so the measures of the angles are $40°, 60°$, and $80°$.

**15.** Assuming $PQ$ and $BA$ are parallel, $\triangle PCQ$ is similar to $\triangle ACB$ since the measure of $\angle PCQ$ is equal to the measure of $\angle ACB$ (they are vertical angles). The corresponding sides of similar triangles are proportional, so $\dfrac{PQ}{BA} = \dfrac{PC}{AC}$. Thus, we have $\dfrac{PQ}{BA} = \dfrac{PC}{AC} \Rightarrow$

$$\frac{1.25 \text{ mm}}{BA} = \frac{150 \text{ mm}}{30 \text{ km}} \Rightarrow BA = \frac{1.25 \cdot 30}{150} \Rightarrow$$

$BA = .25$ km

**17.** Since angle $R$ corresponds to angle $P$, the measure of angle $R$ is 82°. Since angle $M$ corresponds to angle $S$, the measure of angle $M$ is 86°. Since angle $N$ corresponds to angle $Q$, the measure of angle $N$ is 12°. Note: 12°+82°+86° =180°.

**19.** The large triangle is equilateral, so the smaller triangle is also equilateral, and $p = q = 7$.

**21.** $\dfrac{k}{6} = \dfrac{12+9}{9} \Rightarrow k = \dfrac{6 \cdot 21}{9} = 14$

**23.** Let $x =$ the shadow of the 30-ft tree.

$$\frac{20}{8} = \frac{30}{x} \Rightarrow 20x = 240 \Rightarrow x = 12 \text{ ft}$$

**25.** $x = 1,\ y = -\sqrt{3}$ and

$$r = \sqrt{x^2 + y^2} = \sqrt{1^2 + \left(-\sqrt{3}\right)^2}$$
$$= \sqrt{1+3} = \sqrt{4} = 2$$

$$\sin \theta = \frac{y}{r} = \frac{-\sqrt{3}}{2} = -\frac{\sqrt{3}}{2}$$

$$\cos \theta = \frac{x}{r} = \frac{1}{2}$$

$$\tan \theta = \frac{y}{x} = \frac{-\sqrt{3}}{1} = -\sqrt{3}$$

$$\cot \theta = \frac{x}{y} = \frac{1}{-\sqrt{3}} = -\frac{1}{\sqrt{3}} \cdot \frac{\sqrt{3}}{\sqrt{3}} = -\frac{\sqrt{3}}{3}$$

$$\sec \theta = \frac{r}{x} = \frac{2}{1} = 2$$

$$\csc \theta = \frac{r}{y} = \frac{2}{-\sqrt{3}} = -\frac{2}{\sqrt{3}} \cdot \frac{\sqrt{3}}{\sqrt{3}} = -\frac{2\sqrt{3}}{3}$$

**27.** $(3, -4)$

$x = 3,\ y = -4$ and

$$r = \sqrt{3^2 + (-4)^2} = \sqrt{9+16} = \sqrt{25} = 5$$

$$\sin \theta = \frac{y}{r} = \frac{-4}{5} = -\frac{4}{5}$$

$$\cos \theta = \frac{x}{r} = \frac{3}{5}$$

$$\tan \theta = \frac{y}{x} = \frac{-4}{3} = -\frac{4}{3}$$

$$\cot \theta = \frac{x}{y} = \frac{3}{-4} = -\frac{3}{4}$$

$$\sec \theta = \frac{r}{x} = \frac{5}{3}$$

$$\csc \theta = \frac{r}{y} = \frac{5}{-4} = -\frac{5}{4}$$

**29.** $(-8, 15)$

$x = -8,\ y = 15$, and

$$r = \sqrt{(-8)^2 + 15^2} = \sqrt{64 + 225} = \sqrt{289} = 17$$

$$\sin \theta = \frac{y}{r} = \frac{15}{17}; \quad \cos \theta = \frac{x}{r} = \frac{-8}{17} = -\frac{8}{17}$$

$$\tan \theta = \frac{y}{x} = \frac{15}{-8} = -\frac{15}{8}$$

$$\cot \theta = \frac{x}{y} = \frac{-8}{15} = -\frac{8}{15}$$

$$\sec \theta = \frac{r}{x} = \frac{17}{-8} = -\frac{17}{8}; \quad \csc \theta = \frac{r}{y} = \frac{17}{15}$$

**31.** $\left(6\sqrt{3}, -6\right)$

$x = 6\sqrt{3}$, $y = -6$, and

$$r = \sqrt{\left(6\sqrt{3}\right)^2 + (-6)^2} = \sqrt{108 + 36} = \sqrt{144} = 12$$

$$\sin \theta = \frac{y}{r} = \frac{-6}{12} = -\frac{1}{2}$$

$$\cos \theta = \frac{x}{r} = \frac{6\sqrt{3}}{12} = \frac{\sqrt{3}}{2}$$

$$\tan \theta = \frac{y}{x} = \frac{-6}{6\sqrt{3}} = -\frac{1}{\sqrt{3}} \cdot \frac{\sqrt{3}}{\sqrt{3}} = -\frac{\sqrt{3}}{3}$$

$$\cot \theta = \frac{x}{y} = \frac{6\sqrt{3}}{-6} = -\sqrt{3}$$

$$\sec \theta = \frac{r}{x} = \frac{12}{6\sqrt{3}} = \frac{2}{\sqrt{3}} \cdot \frac{\sqrt{3}}{\sqrt{3}} = \frac{2\sqrt{3}}{3}$$

$$\csc \theta = \frac{r}{y} = \frac{12}{-6} = -2$$

**33.** If the terminal side of a quadrantal angle lies along the $y$-axis, a point on the terminal side would be of the form $(0, k)$, where $k$ is a real number, $k \neq 0$ .

$$\sin\theta = \frac{y}{r} = \frac{k}{r}$$

$$\cos\theta = \frac{x}{r} = \frac{0}{r} = 0$$

$$\tan\theta = \frac{y}{x} = \frac{k}{0} \quad \text{undefined}$$

$$\cot\theta = \frac{x}{y} = \frac{0}{k} = 0$$

$$\sec\theta = \frac{r}{x} = \frac{r}{0} \quad \text{undefined}$$

$$\csc\theta = \frac{r}{y} = \frac{r}{k}$$

The tangent and secant are undefined.

**35.**

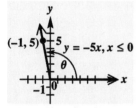

$(-1, 5)$, $y = -5x, x \le 0$

**37.**

| $\theta$ | 180° |
| --- | --- |
| $\sin\theta$ | 0 |
| $\cos\theta$ | −1 |
| $\tan\theta$ | 0 |
| $\cot\theta$ | undefined |
| $\sec\theta$ | −1 |
| $\csc\theta$ | undefined |

**39.** (a) Impossible because the range of $\sec\theta$ is $(-\infty, -1] \cup [1, \infty)$ .

   (b) Possible because the range of $\tan\theta$ is $(-\infty, \infty)$ .

   (c) Possible because the range of $\csc\theta$ is $(-\infty, -1] \cup [1, \infty)$ .

**41.** $\cos\theta = -\dfrac{5}{8}$ given that $\theta$ is in quadrant III

$\cos\theta = \dfrac{x}{r} = \dfrac{-5}{8}$ and $\theta$ in quadrant III, so let $x = -5, r = 8$.

$x^2 + y^2 = r^2 \Rightarrow (-5)^2 + y^2 = 8^2 \Rightarrow$
$25 + y^2 = 64 \Rightarrow y^2 = 39 \Rightarrow y = \pm\sqrt{39}$

$\theta$ is in quadrant III, so $y = -\sqrt{39}$.

$$\sin\theta = \frac{y}{r} = \frac{-\sqrt{39}}{8} = -\frac{\sqrt{39}}{8}$$

$$\cos\theta = \frac{x}{r} = \frac{-5}{8} = -\frac{5}{8}$$

$$\tan\theta = \frac{y}{x} = \frac{-\sqrt{39}}{-5} = \frac{\sqrt{39}}{5}$$

$$\cot\theta = \frac{x}{y} = \frac{-5}{-\sqrt{39}} = \frac{5}{\sqrt{39}} \cdot \frac{\sqrt{39}}{\sqrt{39}} = \frac{5\sqrt{39}}{39}$$

$$\sec\theta = \frac{r}{x} = \frac{8}{-5} = -\frac{8}{5}$$

$$\csc\theta = \frac{r}{y} = \frac{8}{-\sqrt{39}} = -\frac{8}{\sqrt{39}}$$
$$= -\frac{8}{\sqrt{39}} \cdot \frac{\sqrt{39}}{\sqrt{39}} = -\frac{8\sqrt{39}}{39}$$

**43.** $\sec\theta = -\sqrt{5}$ given that $\theta$ is in quadrant II
$\theta$ in quadrant II $\Rightarrow x < 0, y > 0$ , so

$$\sec\theta = -\sqrt{5} = \frac{r}{x} = \frac{\sqrt{5}}{-1} \Rightarrow x = -1, r = \sqrt{5}$$

$x^2 + y^2 = r^2 \Rightarrow (-1)^2 + y^2 = (\sqrt{5})^2 \Rightarrow$
$1 + y^2 = 5 \Rightarrow y^2 = 4 \Rightarrow y = 2$

$$\sin\theta = \frac{y}{r} = \frac{2}{\sqrt{5}} = \frac{2}{\sqrt{5}} \cdot \frac{\sqrt{5}}{\sqrt{5}} = \frac{2\sqrt{5}}{5}$$

$$\cos\theta = \frac{x}{r} = \frac{-1}{\sqrt{5}} = -\frac{1}{\sqrt{5}} \cdot \frac{\sqrt{5}}{\sqrt{5}} = -\frac{\sqrt{5}}{5}$$

$$\tan\theta = \frac{y}{x} = \frac{2}{-1} = -2$$

$$\cot\theta = \frac{x}{y} = \frac{-1}{2} = -\frac{1}{2}$$

$$\sec\theta = \frac{r}{x} = \frac{\sqrt{5}}{-1} = -\sqrt{5}$$

$$\csc\theta = \frac{r}{y} = \frac{\sqrt{5}}{2}$$

**45.** $\sec\theta = \dfrac{5}{4}$ given that $\theta$ is in quadrant IV

$\theta$ in quadrant IV $\Rightarrow x>0, y<0$, so

$\sec\theta = \dfrac{r}{x} = \dfrac{5}{4} \Rightarrow x = 4, r = 5$

$x^2 + y^2 = r^2 \Rightarrow (4)^2 + y^2 = 5^2 \Rightarrow$
$16 + y^2 = 25 \Rightarrow y^2 = 9 \Rightarrow y = -3$

$\sin\theta = \dfrac{y}{r} = \dfrac{-3}{5} = -\dfrac{3}{5}$; $\cos\theta = \dfrac{x}{r} = \dfrac{4}{5}$

$\tan\theta = \dfrac{y}{x} = \dfrac{-3}{4} = -\dfrac{3}{4}$

$\cot\theta = \dfrac{x}{y} = \dfrac{4}{-3} = -\dfrac{4}{3}$

$\sec\theta = \dfrac{r}{x} = \dfrac{5}{4}$

$\csc\theta = \dfrac{r}{y} = \dfrac{5}{-3} = -\dfrac{5}{3}$

**47.** The triangles are similar, so
$$\dfrac{20}{30} = \dfrac{x}{100-x} \Rightarrow 20(100-x) = 30x \Rightarrow$$
$2000 - 20x = 30x \Rightarrow 2000 = 50x \Rightarrow 40 = x$
The lifeguard will enter the water 40 yards east of his original position.

**49.** Let $x =$ the depth of the crater Autolycus. Then
$$\dfrac{x}{11,000} = \dfrac{1.3}{1.5} \Rightarrow 1.5x = 1.3(11,000) \Rightarrow$$
$1.5x = 14,300 \Rightarrow x \approx 9500$
Autoclycus is about 9500 feet deep.

## Chapter 1 Test

**1.** $67°$

(a) complement: $90° - 67° = 23°$

(b) supplement: $180° - 67° = 113°$

**2.** The two angles are supplements, so their sum is $180°$.
$(7x+19) + (2x-1) = 180 \Rightarrow 9x + 18 = 180 \Rightarrow$
$\qquad\qquad\qquad\qquad 9x = 162 \Rightarrow x = 18$
$7(18) + 19 = 145; 2(18) - 1 = 35$
The measures of the angles are $145°$ and $35°$.

**3.** The two angles are complements, so their sum is $90°$ $(-8x+30) + (-3x+5) = 90$
$\qquad\qquad\qquad -11x + 35 = 90$
$\qquad\qquad\qquad -11x = 55 \Rightarrow x = -5$
$-8(-5) + 30 = 70; -3(-5) + 5 = 20$
The angles measure $20°$ and $70°$.

**4.** The angles are vertical angles, so their measures are equal.
$4x - 30 = 5x - 70 \Rightarrow 40 = x$
$4(40) - 30 = 130; 5(40) - 70 = 130$
The angles each measure $130°$.

**5.** The angles are alternate interior angles, so their measures are equal.
$8x + 14 = 10x - 10 \Rightarrow 24 = 2x \Rightarrow 12 = x$
$8(12) + 14 = 110; 10(12) - 10 = 110$
The angles each measure $110°$.

**6.** The angles are interior angles of a triangle, so the sum of the three angles is $180°$.
$(2x+18) + (20x+10) + (32-2x) = 180$
$\qquad\qquad\qquad\qquad 20x + 60 = 180$
$\qquad\qquad\qquad\qquad 20x = 120$
$\qquad\qquad\qquad\qquad\qquad x = 6$
$2(6) + 18 = 30; 20(6) + 10 = 130; 32 - 2(6) = 20$
The three angles measure $30°$, $130°$, and $20°$.

**7.** (a) $74°18'36'' = 74° + \dfrac{18}{60}° + \dfrac{36}{3600}°$
$\qquad\qquad\qquad \approx 74° + .3° + .01° = 74.31°$

(b) $45.2025° = 45° + .2025°$
$\qquad\qquad = 45° + .2025(60')$
$\qquad\qquad = 45° + 12.15'$
$\qquad\qquad = 45° + 12' + .15'$
$\qquad\qquad = 45° + 12' + .15(60'')$
$\qquad\qquad = 45° + 12' + 09''$
$\qquad\qquad = 45°12'9''$

**8.** (a) $390°$ is coterminal with $390° - 360° = 30°$.

(b) $-80°$ is coterminal with $-80° + 360° = 280°$.

(c) $810°$ is coterminal with $810° - 2(360°) = 810° - 720° = 90°$.

**9.** $\dfrac{450(360°)}{1 \text{ min}} = \dfrac{450(360°)}{60 \text{ sec}} = \dfrac{450(6°)}{\text{sec}}$
$\qquad = 2700°/\text{sec}$

A point on the tire rotates $2700°$ in one second.

**10.** $\dfrac{8}{30} = \dfrac{x}{40} \Rightarrow \dfrac{8 \cdot 40}{30} = x \Rightarrow x = 10\frac{2}{3}$

The shadow of the 40-ft pole is $10\frac{2}{3}$ ft, or 10 ft, 8 in.

**11.** $\dfrac{10}{25} = \dfrac{x}{20} \Rightarrow x = \dfrac{10 \cdot 20}{25} = 8$
$\dfrac{10}{25} = \dfrac{y}{15} \Rightarrow y = \dfrac{10 \cdot 15}{25} = 6$

**12.**

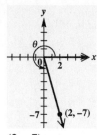

$(2, -7)$

$x = 2, y = -7$

$r = \sqrt{x^2 + y^2} = \sqrt{2^2 + (-7)^2} = \sqrt{4 + 49} = \sqrt{53}$

$\sin\theta = \dfrac{y}{r} = \dfrac{-7}{\sqrt{53}} = -\dfrac{7}{\sqrt{53}} \cdot \dfrac{\sqrt{53}}{\sqrt{53}} = -\dfrac{7\sqrt{53}}{53}$

$\cos\theta = \dfrac{x}{r} = \dfrac{2}{\sqrt{53}} = \dfrac{2}{\sqrt{53}} \cdot \dfrac{\sqrt{53}}{\sqrt{53}} = \dfrac{2\sqrt{53}}{53}$

$\tan\theta = \dfrac{y}{x} = \dfrac{-7}{2} = -\dfrac{7}{2}$; $\cot\theta = \dfrac{x}{y} = \dfrac{2}{-7} = -\dfrac{2}{7}$

$\sec\theta = \dfrac{r}{x} = \dfrac{\sqrt{53}}{2}$

$\csc\theta = \dfrac{r}{y} = \dfrac{\sqrt{53}}{-7} = -\dfrac{\sqrt{53}}{7}$

**13.**

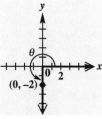

$(0, -2)$

$x = 0, y = -2$

$r = \sqrt{x^2 + y^2} = \sqrt{0^2 + (-2)^2} = \sqrt{0 + 4} = \sqrt{4} = 2$

$\sin\theta = \dfrac{y}{r} = \dfrac{-2}{2} = -1$

$\cos\theta = \dfrac{x}{r} = \dfrac{0}{2} = 0$

$\tan\theta = \dfrac{y}{x} = \dfrac{-2}{0}$ undefined

$\cot\theta = \dfrac{x}{y} = \dfrac{0}{-2} = 0$

$\sec\theta = \dfrac{r}{x} = \dfrac{2}{0}$ undefined

$\csc\theta = \dfrac{r}{y} = \dfrac{2}{-2} = -1$

**14.** Since $x \le 0$, the graph of the line $3x - 4y = 0$ is shown to the left of the $y$-axis. A point on this graph is $(-4, -3)$ since

$3(-4) - 4(-3) = 0$. The corresponding value of $r$ is

$r = \sqrt{(-4)^2 + (-3)^2} = \sqrt{16 + 9} = \sqrt{25} = 5.$

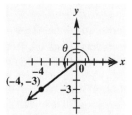

$3x - 4y = 0, x \;\; 0$

$(-4, -3)$

$x = -4, y = -3$

$\sin\theta = \dfrac{y}{r} = -\dfrac{3}{5}$; $\cos\theta = \dfrac{x}{r} = -\dfrac{4}{5}$

$\tan\theta = \dfrac{y}{x} = \dfrac{3}{4}$; $\cot\theta = \dfrac{x}{y} = \dfrac{4}{3}$

$\sec\theta = \dfrac{r}{x} = -\dfrac{5}{4}$; $\csc\theta = \dfrac{r}{y} = -\dfrac{5}{3}$

**15.**

| $\theta$ | 90° | −360° | 630° |
|---|---|---|---|
| $\sin\theta$ | 1 | 0 | 1 |
| $\cos\theta$ | 0 | 1 | 0 |
| $\tan\theta$ | undefined | 0 | Undefined |
| $\cot\theta$ | 0 | undefined | 0 |
| $\sec\theta$ | undefined | 1 | Undefined |
| $\csc\theta$ | 1 | undefined | 1 |

**16.** If the terminal side of a quadrantal angle lies on the negative part of the $x$-axis, any point on the terminal side would have the form $(k, 0)$, where $k$ is any real number $< 0$.

$\sin\theta = \dfrac{y}{r} = \dfrac{0}{r} = 0$; $\cos\theta = \dfrac{x}{r} = \dfrac{k}{r}$

$\tan\theta = \dfrac{y}{x} = \dfrac{0}{k} = 0$; $\cot\theta = \dfrac{x}{y} = \dfrac{k}{0}$ undefined

$\sec\theta = \dfrac{r}{x} = \dfrac{r}{k}$; $\csc\theta = \dfrac{r}{y} = \dfrac{r}{0}$ undefined

Thus, the cotangent and the cosecant would be undefined.

**17. (a)** $\cos\theta > 0$ in quadrants I and IV, while $\tan\theta > 0$ in quadrants I and III. So, both conditions are met only in quadrant I.

**(b)** $\sin\theta < 0$ in quadrants III and IV. Since $\csc\theta$ is the reciprocal of $\sin\theta < 0$, $\csc\theta < 0$ also in quadrants III and IV. Thus, both conditions are met in quadrants III and IV.

**(c)** $\cot\theta > 0$ in quadrants I and III, while $\cos < 0$ in quadrants II and III. Both conditions are met only in quadrant III.

**18. (a)** Impossible because the range of $\sin\theta$ is $[-1, 1]$.

**(b)** Possible because the range of $\sec\theta$ is $(-\infty, -1] \cup [1, \infty)$.

**(c)** Possible because the range of $\tan\theta$ is $(-\infty, \infty)$.

**19.** $\cos\theta = -\dfrac{7}{12} \Rightarrow \sec\theta = -\dfrac{12}{7}$

**20.** $\sin\theta = \dfrac{3}{7}$ with $\theta$ in quadrant II

$\theta$ in quadrant II $\Rightarrow x < 0, y > 0$

$\sin\theta = \dfrac{y}{r} = \dfrac{3}{7} \Rightarrow y = 3, r = 7$

$r^2 = x^2 + y^2 \Rightarrow 7^2 = x^2 + 3^2 \Rightarrow 49 = x^2 + 9 \Rightarrow$
$x^2 = 40 \Rightarrow x = -\sqrt{40} = -2\sqrt{10}$

$\cos\theta = \dfrac{x}{r} = \dfrac{-2\sqrt{10}}{7} = -\dfrac{2\sqrt{10}}{7}$

$\tan\theta = \dfrac{y}{x} = \dfrac{3}{-2\sqrt{10}} = -\dfrac{3}{2\sqrt{10}} \cdot \dfrac{\sqrt{10}}{\sqrt{10}} = -\dfrac{3\sqrt{10}}{20}$

$\cot\theta = \dfrac{x}{y} = \dfrac{-2\sqrt{10}}{7}$

$\sec\theta = \dfrac{r}{x} = \dfrac{7}{-2\sqrt{10}} = -\dfrac{7}{2\sqrt{10}} \cdot \dfrac{\sqrt{10}}{\sqrt{10}} = -\dfrac{7\sqrt{10}}{20}$

$\csc\theta = \dfrac{r}{y} = \dfrac{7}{3}$

# Chapter 2

## Acute Angles and Right Triangles

### Section 2.1: Trigonometric Functions of Acute Angles

**1.** $\sin A = \dfrac{\text{side opposite}}{\text{hypotenuse}} = \dfrac{21}{29}$

$\cos A = \dfrac{\text{side adjacent}}{\text{hypotenuse}} = \dfrac{20}{29}$

$\tan A = \dfrac{\text{side opposite}}{\text{side adjacent}} = \dfrac{21}{20}$

**3.** $\sin A = \dfrac{\text{side opposite}}{\text{hypotenuse}} = \dfrac{n}{p}$

$\cos A = \dfrac{\text{side adjacent}}{\text{hypotenuse}} = \dfrac{m}{p}$

$\tan A = \dfrac{\text{side opposite}}{\text{side adjacent}} = \dfrac{n}{m}$

For Exercises 5–9, refer to the Function Values of Special Angles chart on page 54 of the text.

**5.** C; $\sin 30° = \dfrac{1}{2}$

**7.** B; $\tan 45° = 1$

**9.** E; $\csc 60° = \dfrac{1}{\sin 60°} = \dfrac{1}{\frac{\sqrt{3}}{2}} = \dfrac{2}{\sqrt{3}}$

$= \dfrac{2}{\sqrt{3}} \cdot \dfrac{\sqrt{3}}{\sqrt{3}} = \dfrac{2\sqrt{3}}{3}$

**11.** $a = 5, b = 12$

$c^2 = a^2 + b^2 \Rightarrow c^2 = 5^2 + 12^2 \Rightarrow c^2 = 169 \Rightarrow$
$c = 13$

$\sin B = \dfrac{\text{side opposite}}{\text{hypotenuse}} = \dfrac{b}{c} = \dfrac{12}{13}$

$\cos B = \dfrac{\text{side adjacent}}{\text{hypotenuse}} = \dfrac{a}{c} = \dfrac{5}{13}$

$\tan B = \dfrac{\text{side opposite}}{\text{side adjacent}} = \dfrac{b}{a} = \dfrac{12}{5}$

$\cot B = \dfrac{\text{side adjacent}}{\text{side opposite}} = \dfrac{a}{b} = \dfrac{5}{12}$

$\sec B = \dfrac{\text{hypotenuse}}{\text{side adjacent}} = \dfrac{c}{a} = \dfrac{13}{5}$

$\csc B = \dfrac{\text{hypotenuse}}{\text{side opposite}} = \dfrac{c}{b} = \dfrac{13}{12}$

**13.** $a = 6, c = 7$

$c^2 = a^2 + b^2 \Rightarrow 7^2 = 6^2 + b^2 \Rightarrow$
$49 = 36 + b^2 \Rightarrow 13 = b^2 \Rightarrow \sqrt{13} = b$

$\sin B = \dfrac{\text{side opposite}}{\text{hypotenuse}} = \dfrac{b}{c} = \dfrac{\sqrt{13}}{7}$

$\cos B = \dfrac{\text{side adjacent}}{\text{hypotenuse}} = \dfrac{a}{c} = \dfrac{6}{7}$

$\tan B = \dfrac{\text{side opposite}}{\text{side adjacent}} = \dfrac{b}{a} = \dfrac{\sqrt{13}}{6}$

$\cot B = \dfrac{\text{side adjacent}}{\text{side opposite}} = \dfrac{a}{b} = \dfrac{6}{\sqrt{13}}$

$= \dfrac{6}{\sqrt{13}} \cdot \dfrac{\sqrt{13}}{\sqrt{13}} = \dfrac{6\sqrt{13}}{13}$

$\sec B = \dfrac{\text{hypotenuse}}{\text{side adjacent}} = \dfrac{c}{a} = \dfrac{7}{6}$

$\csc B = \dfrac{\text{hypotenuse}}{\text{side opposite}} = \dfrac{c}{b} = \dfrac{7}{\sqrt{13}}$

$= \dfrac{7}{\sqrt{13}} \cdot \dfrac{\sqrt{13}}{\sqrt{13}} = \dfrac{7\sqrt{13}}{13}$

**15.** $a = 3, c = 5$

$c^2 = a^2 + b^2 \Rightarrow 5^2 = 3^2 + b^2 \Rightarrow$
$25 = 9 + b^2 \Rightarrow 16 = b^2 \Rightarrow 4 = b$

$\sin B = \dfrac{\text{side opposite}}{\text{hypotenuse}} = \dfrac{b}{c} = \dfrac{4}{5}$

$\cos B = \dfrac{\text{side adjacent}}{\text{hypotenuse}} = \dfrac{a}{c} = \dfrac{3}{5}$

$\tan B = \dfrac{\text{side opposite}}{\text{side adjacent}} = \dfrac{b}{a} = \dfrac{4}{3}$

$\cot B = \dfrac{\text{side adjacent}}{\text{side opposite}} = \dfrac{a}{b} = \dfrac{3}{4}$

$\sec B = \dfrac{\text{hypotenuse}}{\text{side adjacent}} = \dfrac{c}{a} = \dfrac{5}{3}$

$\csc B = \dfrac{\text{hypotenuse}}{\text{side opposite}} = \dfrac{c}{b} = \dfrac{5}{4}$

**17.** $\sin\theta = \cos(90° - \theta); \cos\theta = \sin(90° - \theta);$
$\tan\theta = \cot(90° - \theta); \cot\theta = \tan(90° - \theta);$
$\sec\theta = \csc(90° - \theta); \csc\theta = \sec(90° - \theta)$

**19.** $\sec 39° = \csc(90° - 39°) = \csc 51°$

**21.** $\sec(\theta + 15°) = \csc\left[90° - (\theta + 15°)\right]$
$= \csc(75° - \theta)$

**23.** $\cot(\theta - 10°) = \tan\left[90° - (\theta - 10°)\right]$
$= \tan(100° - \theta)$

**25.** $\sin 38.7° = \cos(90° - 38.7°) = \cos 51.3°$

For exercises 27–35, if the functions in the equations are cofunctions, then the equations are true if the sum of the angles is 90°.

**27.** $\tan \alpha = \cot(\alpha + 10°) \Rightarrow$
$\alpha + (\alpha + 10°) = 90° \Rightarrow 2\alpha + 10° = 90° \Rightarrow$
$2\alpha = 80° \Rightarrow \alpha = 40°$

**29.** $\sin(2\theta + 10°) = \cos(3\theta - 20°)$
$(2\theta + 10°) + (3\theta - 20°) = 90°$
$5\theta - 10° = 90°$
$5\theta = 100° \Rightarrow \theta = 20°$

**31.** $\tan(3\beta + 4°) = \cot(5\beta - 10°)$
$(3\beta + 4°) + (5\beta - 10°) = 90°$
$8\beta - 6° = 90°$
$8\beta = 96° \Rightarrow \beta = 12°$

**33.** $\sin(\theta - 20°) = \cos(2\theta + 5°)$
$(\theta - 20°) + (2\theta + 5°) = 90°$
$3\theta - 15° = 90°$
$3\theta = 105° \Rightarrow \theta = 35°$

**35.** $\sec(3\beta + 10°) = \csc(\beta + 8°)$
$(3\beta + 10°) + (\beta + 8°) = 90°$
$4\beta + 18° = 90°$
$4\beta = 72° \Rightarrow \beta = 18°$

**37.** $\sin 50° > \sin 40°$
In the interval from 0° to 90°, as the angle increases, so does the sine of the angle, so $\sin 50° > \sin 40°$ is true.

**39.** $\sin 46° < \cos 46°$
Using the cofunction identity,
$\cos 46° = \sin(90° - 46°) = \sin 44°$. In the interval from 0° to 90°, as the angle increases, so does the sine of the angle, so
$\sin 46° < \sin 44° \Rightarrow \sin 46° < \cos 44°$ is false.

**41.** $\tan 41° < \cot 41°$
Using the cofunction identity,
$\cot 41° = \tan(90° - 41°) = \tan 49°$. In the interval from 0° to 90°, as the angle increases, the tangent of the angle increases, so
$\tan 41° < \tan 49° \Rightarrow \tan 41° < \cot 41°$ is true.

**43.** $\sec 60° > \sec 30°$
In the interval from 0° to 90°, as the angle increases, the cosine of the angle decreases, so the secant of the angle increases. Thus, $\sec 60° > \sec 30°$ is true.

Use the following figures for exercises 45–59.

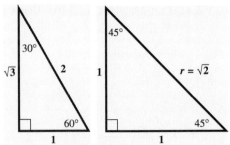

**45.** $\tan 30° = \dfrac{\text{side opposite}}{\text{side adjacent}} = \dfrac{1}{\sqrt{3}}$
$= \dfrac{1}{\sqrt{3}} \cdot \dfrac{\sqrt{3}}{\sqrt{3}} = \dfrac{\sqrt{3}}{3}$

**47.** $\sin 30° = \dfrac{\text{side opposite}}{\text{hypotenuse}} = \dfrac{1}{2}$

**49.** $\sec 30° = \dfrac{\text{hypotenuse}}{\text{side adjacent}} = \dfrac{2}{\sqrt{3}}$
$= \dfrac{2}{\sqrt{3}} \cdot \dfrac{\sqrt{3}}{\sqrt{3}} = \dfrac{2\sqrt{3}}{3}$

**51.** $\csc 45° = \dfrac{\text{hypotenuse}}{\text{side opposite}} = \dfrac{\sqrt{2}}{1} = \sqrt{2}$

**53.** $\cos 45° = \dfrac{\text{side adjacent}}{\text{hypotenuse}} = \dfrac{1}{\sqrt{2}}$
$= \dfrac{1}{\sqrt{2}} \cdot \dfrac{\sqrt{2}}{\sqrt{2}} = \dfrac{\sqrt{2}}{2}$

**55.** $\tan 45° = \dfrac{\text{side opposite}}{\text{side adjacent}} = \dfrac{1}{1} = 1$

**57.** $\sin 60° = \dfrac{\text{side opposite}}{\text{hypotenuse}} = \dfrac{\sqrt{3}}{2}$

**59.** $\tan 60° = \dfrac{\text{side opposite}}{\text{side adjacent}} = \dfrac{\sqrt{3}}{1} = \sqrt{3}$

**61.**

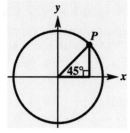

**63.** The legs of the right triangle provide the coordinates of $P$, $\left(2\sqrt{2}, 2\sqrt{2}\right)$.

**65.** $Y_1$ is $\sin x$ and $Y_2$ is $\tan x$.

| | |
|---|---|
| $\sin 0° = 0$ | $\tan 0° = 0$ |
| $\sin 30° = .5$ | $\tan 30° \approx .57735$ |
| $\sin 45° \approx .70711$ | $\tan 45° = 1$ |
| $\sin 60° \approx .86603$ | $\tan 60° = 1.7321$ |
| $\sin 90° = 1$ | $\tan 90°$ : undefined |

**67.** Since $\sin 60° = \dfrac{\sqrt{3}}{2}$ and $60°$ is between $0°$ and $90°$, $A = 60°$.

**69.** The point of intersection is (.70710678, .70710678). This corresponds to the point $\left(\dfrac{\sqrt{2}}{2}, \dfrac{\sqrt{2}}{2}\right)$.

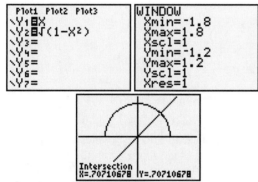

These coordinates are the sine and cosine of $45°$.

**71.**

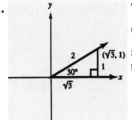

The line passes through $(0, 0)$ and $\left(\sqrt{3}, 1\right)$. The slope is change in $y$ over the change in $x$. Thus,

$$m = \frac{1}{\sqrt{3}} = \frac{1}{\sqrt{3}} \cdot \frac{\sqrt{3}}{\sqrt{3}} = \frac{\sqrt{3}}{3}$$

and the equation of the line is $y = \dfrac{\sqrt{3}}{3}x$.

**73.** One point on the line $y = \sqrt{3}x$ is the origin $(0,0)$. Let $(x, y)$ be any other point on this line. Then, by the definition of slope,
$$m = \frac{y-0}{x-0} = \frac{y}{x} = \sqrt{3},$$ but also, by the definition of tangent, $\tan \theta = \sqrt{3}$. Because $\tan 60° = \sqrt{3}$, the line $y = \sqrt{3}x$ makes a $60°$ angle with the positive $x$-axis (See exercise 70).

**75. (a)** Each of the angles of the equilateral triangle has measure $\frac{1}{3}(180°) = 60°$.

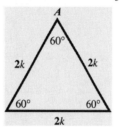

**(b)** The perpendicular bisects the opposite side so the length of each side opposite each $30°$ angle is $k$.

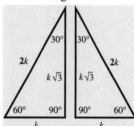

**(c)** Let $x$ = the length of the perpendicular. Then apply the Pythagorean theorem.
$$x^2 + k^2 = (2k)^2 \Rightarrow x^2 + k^2 = 4k^2 \Rightarrow$$
$$x^2 = 3k^2 \Rightarrow x = \sqrt{3}k$$
The length of the perpendicular is $\sqrt{3}k$.

**(d)** In a $30°$-$60°$ right triangle, the hypotenuse is always $\underline{2}$ times as long as the shorter leg, and the longer leg has a length that is $\sqrt{3}$ times as long as that of the shorter leg. Also, the shorter leg is opposite the $\underline{30°}$ angle, and the longer leg is opposite the $\underline{60°}$ angle.

**77.** Apply the relationships between the lengths of the sides of a $30° - 60°$ right triangle first to the triangle on the left to find the values of $y$ and $x$, and then to the triangle on the right to find the values of $z$ and $w$. In the $30° - 60°$ right triangle, the side opposite the $30°$ angle is $\frac{1}{2}$ the length of the hypotenuse. The longer leg is $\sqrt{3}$ times the shorter leg.

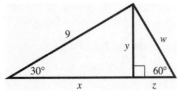

Thus, we have

$$y = \frac{1}{2}(9) = \frac{9}{2} \text{ and } x = y\sqrt{3} = \frac{9\sqrt{3}}{2}$$

$$y = z\sqrt{3}, \text{ so } z = \frac{y}{\sqrt{3}} = \frac{\frac{9}{2}}{\sqrt{3}} = \frac{9\sqrt{3}}{6} = \frac{3\sqrt{3}}{2},$$

and $w = 2z$, so $w = 2\left(\frac{3\sqrt{3}}{2}\right) = 3\sqrt{3}$

**79.** Apply the relationships between the lengths of the sides of a $45° - 45°$ right triangle to the triangle on the left to find the values of $p$ and $r$. In the $45° - 45°$ right triangle, the sides opposite the $45°$ angles measure the same. The hypotenuse is $\sqrt{2}$ times the measure of a leg.

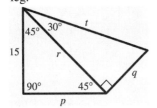

Thus, we have $p = 15$ and $r = p\sqrt{2} = 15\sqrt{2}$

Apply the relationships between the lengths of the sides of a $30° - 60°$ right triangle next to the triangle on the right to find the values of $q$ and $t$. In the $30° - 60°$ right triangle, the side opposite the $60°$ angle is $\sqrt{3}$ times as long as the side opposite to the $30°$ angle. The length of the hypotenuse is 2 times as long as the shorter leg (opposite the $30°$ angle). Thus, we have $r = q\sqrt{3} \Rightarrow$

$$q = \frac{r}{\sqrt{3}} = \frac{15\sqrt{2}}{\sqrt{3}} = \frac{15\sqrt{2}}{\sqrt{3}} \cdot \frac{\sqrt{3}}{\sqrt{3}} = 5\sqrt{6} \text{ and}$$

$$t = 2q = 2\left(5\sqrt{6}\right) = 10\sqrt{6}$$

**81.** Since $A = \frac{1}{2}bh$, we have

$$A = \frac{1}{2} \cdot s \cdot s = \frac{1}{2}s^2 \text{ or } A = \frac{s^2}{2}.$$

**83.** Answers will vary.

## Section 2.2: Trigonometric Functions of Non-Acute Angles

**1.** C; $180° - 98° = 82°$
($98°$ is in quadrant II)

**3.** A; $-135° + 360° = 225°$ and
$225° - 180° = 45°$
($225°$ is in quadrant III)

**5.** D; $750° - 2 \cdot 360° = 30°$
($30°$ is in quadrant I)

**7.** 2 is a good choice for $r$ because in a $30° - 60°$ right triangle, the hypotenuse is twice the length of the shorter side (the side opposite to the $30°$ angle). By choosing 2, one avoids introducing a fraction (or decimal) when determining the length of the shorter side. Choosing any even positive integer for $r$ would have this result; however, 2 is the most convenient value.

**9.** Answers will vary.

| | $\theta$ | $\sin \theta$ | $\cos \theta$ | $\tan \theta$ | $\cot \theta$ | $\sec \theta$ | $\csc \theta$ |
|---|---|---|---|---|---|---|---|
| **11.** | $45°$ | $\dfrac{\sqrt{2}}{2}$ | $\dfrac{\sqrt{2}}{2}$ | $1$ | $1$ | $\sqrt{2}$ | $\sqrt{2}$ |
| **13.** | $120°$ | $\dfrac{\sqrt{3}}{2}$ | $\begin{aligned}\cos 120° \\ = -\cos 60° \\ = -\dfrac{1}{2}\end{aligned}$ | $-\sqrt{3}$ | $\begin{aligned}\cot 120° \\ = -\cot 60° \\ = -\dfrac{\sqrt{3}}{3}\end{aligned}$ | $\begin{aligned}\sec 120° \\ = -\sec 60° \\ = -2\end{aligned}$ | $\dfrac{2\sqrt{3}}{3}$ |

| $\theta$ | $\sin\theta$ | $\cos\theta$ | $\tan\theta$ | $\cot\theta$ | $\sec\theta$ | $\csc\theta$ |
|---|---|---|---|---|---|---|
| **15.** $150°$ | $\sin 150°$ $=\sin 30°$ $=\dfrac{1}{2}$ | $-\dfrac{\sqrt{3}}{2}$ | $-\dfrac{\sqrt{3}}{3}$ | $\cot 150°$ $=-\cot 30°$ $=-\sqrt{3}$ | $\sec 150°$ $=-\sec 30°$ $=-\dfrac{2\sqrt{3}}{3}$ | $2$ |
| **17.** $240°$ | $-\dfrac{\sqrt{3}}{2}$ | $-\dfrac{1}{2}$ | $\tan 240°$ $=\tan 60°$ $=\sqrt{3}$ | $\cot 240°$ $=\cot 60°$ $=\dfrac{\sqrt{3}}{3}$ | $-2$ | $-\dfrac{2\sqrt{3}}{3}$ |

**19.** To find the reference angle for $315°$, sketch this angle in standard position.

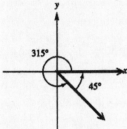

The reference angle is $360° - 315° = 45°$.
Since $315°$ lies in quadrant IV, the sine, tangent, cotangent, and cosecant are negative.

$$\sin 315° = -\sin 45° = -\frac{\sqrt{2}}{2}$$
$$\cos 315° = \cos 45° = \frac{\sqrt{2}}{2}$$
$$\tan 315° = -\tan 45° = -1$$
$$\cot 315° = -\cot 45° = -1$$
$$\sec 315° = \sec 45° = \sqrt{2}$$
$$\csc 315° = -\csc 45° = -\sqrt{2}$$

**21.** To find the reference angle for $-300°$, sketch this angle in standard position.

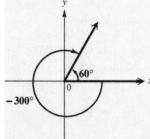

The reference angle for $-300°$ is $-300° + 360° = 60°$. Because $-300°$ lies in quadrant I, the values of all of its trigonometric functions will be positive, so these values will be identical to the trigonometric function values for $60°$. See the Function Values of Special Angles table on page 54.)

$$\sin\left(-300°\right) = \sin 60° = \frac{\sqrt{3}}{2}$$
$$\cos\left(-300°\right) = \cos 60° = \frac{1}{2}$$
$$\tan\left(-300°\right) = \tan 60° = \sqrt{3}$$
$$\cot\left(-300°\right) = \cot 60° = \frac{\sqrt{3}}{3}$$
$$\sec\left(-300°\right) = \sec 60° = 2$$
$$\csc\left(-300°\right) = \csc 60° = \frac{2\sqrt{3}}{3}$$

**23.** To find the reference angle for $480°$, sketch this angle in standard position.

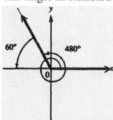

$480°$ is coterminal with $480° - 360° = 120°$. The reference angle is $180° - 120° = 60°$. Because $480°$ lies in quadrant II, the cosine, tangent, cotangent, and secant are negative.

$$\sin\left(480°\right) = \sin 60° = \frac{\sqrt{3}}{2}$$
$$\cos\left(480°\right) = -\cos 60° = -\frac{1}{2}$$
$$\tan\left(480°\right) = -\tan 60° = -\sqrt{3}$$
$$\cot\left(480°\right) = -\cot 60° = -\frac{\sqrt{3}}{3}$$
$$\sec\left(80°\right) = -\sec 60° = -2$$
$$\csc\left(480°\right) = \csc 60° = \frac{2\sqrt{3}}{3}$$

**25.** To find the reference angle for 570° sketch this angle in standard position.

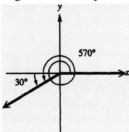

570° is coterminal with 570° − 360° = 210°. The reference angle is 210° − 180° = 30°. Since 570° lies in quadrant III, the sine, cosine, secant, and cosecant are negative.

$$\sin 570° = -\sin 30° = -\frac{1}{2}$$

$$\cos 570° = -\cos 30° = -\frac{\sqrt{3}}{2}$$

$$\tan 570° = \tan 30° = \frac{\sqrt{3}}{3}$$

$$\cot 570° = \cot 30° = \sqrt{3}$$

$$\sec 570° = -\sec 30° = -\frac{2\sqrt{3}}{3}$$

$$\csc 570° = -\csc 30° = -2$$

**27.** 1305° is coterminal with 1305° − 3 · 360° = 1305° = 1080° = 225°. The reference angle is 225° − 180° = 45°. Since 1305° lies in quadrant III, the sine, cosine, and secant and cosecant are negative.

$$\sin 1305° = -\sin 45° = -\frac{\sqrt{2}}{2}$$

$$\cos 1305° = -\cos 45° = -\frac{\sqrt{2}}{2}$$

$$\tan 1305° = \tan 45° = 1$$

$$\cot 1305° = \cot 45° = 1$$

$$\sec 1305° = -\sec 45° = -\sqrt{2}$$

$$\csc 1305° = -\csc 45° = -\sqrt{2}$$

**29.** 2670° is coterminal with 2670° − 7 · 360° = 2670° + 2520° = 150°. The reference angle is 180° − 150° = 30°. Since 2670° lies in quadrant II, the cosine, tangent, cotangent, and secant are negative.

$$\sin 2670° = \sin 30° = \frac{1}{2}$$

$$\cos 2670° = -\cos 30° = -\frac{\sqrt{3}}{2}$$

$$\tan 2670° = -\tan 30° = -\frac{\sqrt{3}}{3}$$

$$\cot 2670° = -\cot 30° = -\sqrt{3}$$

$$\sec 2670° = -\sec 30° = -\frac{2\sqrt{3}}{3}$$

$$\csc 2670° = \csc 30° = 2$$

**31.** −510° is coterminal with −510° + 2 · 360° = −510° + 720° = 210°. The reference angle is 210° − 180° = 30°. Since −510° lies in quadrant III, the sine, cosine, and secant and cosecant are negative.

$$\sin(-510°) = -\sin 30° = -\frac{1}{2}$$

$$\cos(-510°) = -\cos 30° = -\frac{\sqrt{3}}{2}$$

$$\tan(-510°) = \tan 30° = \frac{\sqrt{3}}{3}$$

$$\cot(-510°) = \cot 30° = \sqrt{3}$$

$$\sec(-510°) = -\sec 30° = -\frac{2\sqrt{3}}{3}$$

$$\csc(-510°) = -\csc 30° = -2$$

**33.** −1290° is coterminal with −1290° + 4 · 360° = −1290° + 1440° = 150°. The reference angle is 180° − 150° = 30°. Since −1290° lies in quadrant II, the cosine, tangent, cotangent, and secant are negative.

$$\sin 2670° = \sin 30° = \frac{1}{2}$$

$$\cos 2670° = -\cos 30° = -\frac{\sqrt{3}}{2}$$

$$\tan 2670° = -\tan 30° = -\frac{\sqrt{3}}{3}$$

$$\cot 2670° = -\cot 30° = -\sqrt{3}$$

$$\sec 2670° = -\sec 30° = -\frac{2\sqrt{3}}{3}$$

$$\csc 2670° = \csc 30° = 2$$

**35.** −1860° is coterminal with −1860° + 6 · 360° = −1860° + 2160° = 300°. The reference angle is 360° − 300° = 60°. Since −1860° lies in quadrant IV, the sine, tangent, cotangent, and cosecant are negative.

$$\sin(-1860°) = -\sin 60° = -\frac{\sqrt{3}}{2}$$

$$\cos(-1860°) = \cos 60° = \frac{1}{2}$$

$$\tan(-1860°) = -\tan 60° = -\sqrt{3}$$

$$\cot(-1860°) = -\cot 60° = -\frac{\sqrt{3}}{3}$$

$$\sec(-1860°) = \sec 60° = 2$$

$$\csc(-1860°) = -\csc 60° = -\frac{2\sqrt{3}}{3}$$

**37.** Since $-510°$ is coterminal with an angle of
$-510° + 2 \cdot 360° = -510° + 720° = 210°,$ it lies
in quadrant III. Its reference angle is
$210° - 180° = 30°.$ Since the cosine is
negative in quadrant III, we have

$$\cos(-510°) = -\cos 30° = -\frac{\sqrt{3}}{2}.$$

**39.** Since $1500°$ is coterminal with an angle of
$1500° - 4 \cdot 360° = 1500° - 1440° = 60°,$ it lies
in quadrant I. Because $1500°$ lies in quadrant
I, the values of all of its trigonometric
functions will be positive, so

$$\sin 1500° = \sin 60° = \frac{\sqrt{3}}{2}.$$

**41.** Since $-855°$ is coterminal with
$-855° + 3 \cdot 360° = -855° + 1080° = 225°,$ it
lies in quadrant III. Its reference angle is
$225° - 180° = 45°.$ Since the cosecant is
negative in quadrant III, we have.

$$\csc(-855°) = -\csc 45° = -\sqrt{2}$$

**43.** Since $3015°$ is coterminal with
$3015° - 8 \cdot 360° = 3015° - 2880° = 135°,$ it
lies in quadrant II. Its reference angle is
$180° - 135° = 45°.$ Since the tangent is
negative in quadrant II, we have
$\tan 3015° = -\tan 45° = -1.$

**45.** $\sin(30° + 60°) \overset{?}{=} \sin 30° \cdot \cos 60° + \sin 60° \cdot \cos 30°$
Evaluate each side to determine whether this
equation is true or false.
$\sin(30° + 60°) = \sin 90° = 1$ and
$\sin 30° \cdot \cos 60° + \sin 60° \cdot \cos 30°$
$$= \frac{1}{2} \cdot \frac{1}{2} + \frac{\sqrt{3}}{2} \cdot \frac{\sqrt{3}}{2} = \frac{1}{4} + \frac{3}{4} = 1$$
Since, $1 = 1,$ the statement is true.

**47.** $\cos 60° \overset{?}{=} 2\cos 30°$
Evaluate each side to determine whether this
statement is true or false.
$$\cos 60° = \frac{1}{2} \text{ and } 2\cos 30° = 2\left(\frac{\sqrt{3}}{2}\right) = \sqrt{3}$$
Since $\frac{1}{2} \neq \sqrt{3},$ the statement is false.

**49.** $\sin 120° \overset{?}{=} \sin 180° \cdot \cos 60° - \sin 60° \cdot \cos 180°$
Evaluate each side to determine whether this
statement is true or false.

$$\sin 120° = \frac{\sqrt{3}}{2} \text{ and}$$
$$\sin 180° \cdot \cos 60° - \sin 60° \cdot \cos 180°$$
$$= 0\left(\frac{1}{2}\right) - \left(\frac{\sqrt{3}}{2}\right)(-1) = 0 - \left(-\frac{\sqrt{3}}{2}\right) = \frac{\sqrt{3}}{2}$$
Since $\frac{\sqrt{3}}{2} = \frac{\sqrt{3}}{2},$ the statement is true.

**51.** $\sin^2 45° + \cos^2 45° \overset{?}{=} 1$
$$\left(\frac{\sqrt{2}}{2}\right)^2 + \left(\frac{\sqrt{2}}{2}\right)^2 = \frac{2}{4} + \frac{2}{4} = 1$$
Since $1 = 1,$ the statement is true.

**53.** $\cos(30° + 60°) \overset{?}{=} \cos 30° + \cos 60°$
Evaluate each side to determine whether this
statement is true or false.
$\cos(30° + 60°) = \cos 90° = 0$ and
$$\cos 30° + \cos 60° = \frac{\sqrt{3}}{2} + \frac{1}{2} = \frac{\sqrt{3}+1}{2}$$
Since $0 \neq \frac{\sqrt{3}+1}{2},$ the statement is false.

**55.** $150°$ is in quadrant II, so the reference angle is
$180° - 150° = 30°.$
$$\cos 30° = \frac{x}{r} \Rightarrow x = r\cos 30° = 6 \cdot \frac{\sqrt{3}}{2} = 3\sqrt{3}$$
and $\sin 30° = \frac{y}{r} \Rightarrow y = r\sin 30° = 6 \cdot \frac{1}{2} = 3$
Since $150°$ is in quadrant II, the $x$- coordinate
will be negative. The coordinates of $P$ are
$\left(-3\sqrt{3}, 3\right).$

**57.** For every angle $\theta$, $\sin^2 \theta + \cos^2 \theta = 1.$ Since
$\left(\frac{3}{4}\right)^2 + \left(\frac{2}{3}\right)^2 = \frac{9}{16} + \frac{4}{9} = \frac{145}{144} \neq 1,$ there is no
angle $\theta$ for which $\cos \theta = \frac{2}{3}$ and $\sin \theta = \frac{3}{4}.$

**59.** If $\theta$ is in the interval $(90°, 180°),$ then
$90° < \theta < 180° \Rightarrow 45° < \frac{\theta}{2} < 90°.$ Thus $\frac{\theta}{2}$
lies in quadrant I, and $\cos \frac{\theta}{2}$ is positive.

**61.** If $\theta$ is in the interval $(90°, 180°),$ then
$90° < \theta < 180° \Rightarrow 270° < \theta + 180° < 360°.$
Thus $\theta + 180°$ lies in quadrant IV, and
$\sec(\theta + 180°)$ is positive.

**63.** If $\theta$ is in the interval $(90°, 180°)$, then
$90° < \theta < 180° \Rightarrow -90° > -\theta > -180° \Rightarrow$
$-180° < -\theta < -90°$
Since 180° is coterminal with
$-180° + 360° = 180°$ and $-90°$ is coterminal
with $-90° + 360° = 270°$, $-\theta$ lies in quadrant
III, and $\sin(-\theta)$ is negative.

**65.** $\theta$ and $\theta + n \cdot 360°$ are coterminal angles, so
the cosine of each of these will result in the
same value.

**67.** The reference angle for 115° is
$180° - 115° = 65°$. Since 115° is in quadrant II
the cosine is negative. Sin $\theta$ decreases on the
interval (90°, 180°) from 1 to 0. Therefore,
sin 115° is closest to .9.

**69.** When $\theta = 45°$, $\sin\theta = \cos\theta = \dfrac{\sqrt{2}}{2}$. Sine and
cosine are both positive in quadrant I and both
negative in quadrant III. Since
$\theta + 180° = 45° + 180° = 225°$, 45° is the
quadrant I angle, and 225° is the quadrant III
angle.

**71.** $\sin\theta = \dfrac{1}{2}$
Since $\sin\theta$ is positive, $\theta$ must lie in
quadrants I or II. Since one angle, namely
30°, lies in quadrant I, that angle is also the
reference angle, $\theta'$. The angle in quadrant II
will be $180° - \theta' = 180° - 30° = 150°$.

**73.** $\tan\theta = -\sqrt{3}$
Since $\tan\theta$ is negative, $\theta$ must lie in
quadrants II or IV. Since the absolute value of
$\tan\theta$ is $\sqrt{3}$, the reference angle, $\theta'$ must be
60°. The quadrant II angle $\theta$ equals
$180° - \theta' = 180° - 60° = 120°$, and the
quadrant IV angle $\theta$ equals
$360° - \theta' = 360° - 60° = 300°$.

**75.** $\cos\theta = \dfrac{\sqrt{2}}{2}$
Since $\cos\theta$ is positive, $\theta$ must lie in
quadrants I or IV. Since one angle, namely
45°, lies in quadrant I, that angle is also the
reference angle, $\theta'$. The angle in quadrant IV
will be $360° - \theta' = 360° - 45° = 315°$.

**77.** $\csc\theta = -2$
Since $\csc\theta$ is negative, $\theta$ must lie in
quadrants III or IV. The absolute value of
$\csc\theta$ is 2, so the reference angle, $\theta'$, is 30°.
The angle in quadrant III will be
$180° + \theta' = 180° + 30° = 210°$, and the quadrant
IV angle is $360° - \theta' = 360° - 30° = 330°$.

**79.** $\tan\theta = \dfrac{\sqrt{3}}{3}$
Since $\tan\theta$ is positive, $\theta$ must lie in
quadrants I or III. Since one angle, namely
30°, lies in quadrant I, that angle is also the
reference angle, $\theta'$. The angle in quadrant III
will be $180° + \theta' = 180° + 30° = 210°$.

**81.** $\csc\theta = -\sqrt{2}$
Since $\csc\theta$ is negative, $\theta$ must lie in
quadrants III or IV. Since the absolute value
of $\csc\theta$ is $\sqrt{2}$, the reference angle, $\theta'$ must
be 45°. The quadrant III angle $\theta$ equals
$180° + \theta' = 180° + 45° = 225°$. and the
quadrant IV angle $\theta$ equals
$360° - \theta' = 360° - 45° = 315°$.

## Section 2.3: Finding Trigonometric Function Values Using a Calculator

**1.** The CAUTION at the beginning of this section
suggests verifying that a calculator is in
degree mode by finding <u>sin</u> 90°. If the
calculator is in degree mode, the display
should be <u>1</u>.

**3.** To find values of the cotangent, secant, and
cosecant functions with a calculator, it is
necessary to find the <u>reciprocal</u> of the
<u>reciprocal</u> function value.

For Exercises 5–21, be sure your calculator is in
degree mode. If your calculator accepts angles in
degrees, minutes, and seconds, it is not necessary to
change angles to decimal degrees.
Keystroke sequences may vary on the type and/or
model of calculator being used. Screens shown will
be from a TI-83 Plus calculator. To obtain the degree
(°) and (′) symbols, go to the ANGLE menu (2nd
APPS).

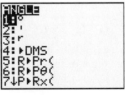

For Exercises 5–15, the calculation for decimal degrees is indicated for calculators that do not accept degree, minutes, and seconds.

**5.** $\sin 38°42' \approx .6252427$

$38°42' = \left(38 + \frac{42}{60}\right)° = 38.7°$

**7.** $\sec 13°15' \approx 1.0273488$

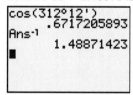

$13°15' = \left(13 + \frac{15}{60}\right)° = 13.25°$

**9.** $\cot 183°48' \approx 15.055723$

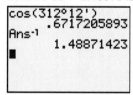

$183°48' = \left(183 + \frac{48}{60}\right)° = 183.8°$

**11.** $\sec 312°12' \approx 1.4887142$

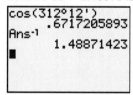

$312°12' = \left(312 + \frac{12}{60}\right)° = 312.2°$

**13.** $\sin\left(-317°36'\right) \approx .6743024$

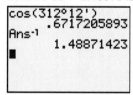

$-317°36' = -\left(317 + \frac{36}{60}\right)° = -317.6°$

**15.** $\cos\left(-15'\right) \approx .9999905$

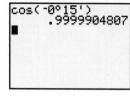

$-15' = -\left(\frac{15}{60}\right)° = -.25°$

**17.** $\dfrac{1}{\cot 23.4°} = \tan 23.4° \approx .4327386$

**19.** $\dfrac{\cos 77°}{\sin 77°} = \cot 77° \approx .2308682$

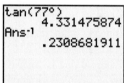

**21.** $\cot\left(90° - 4.72°\right) = \tan 4.72° \approx .0825664$

**23.** $\tan\theta = 1.4739716 \Rightarrow \theta \approx 55.845496°$

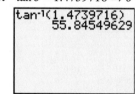

**25.** $\sin\theta = .27843196 \Rightarrow \theta \approx 16.166641°$

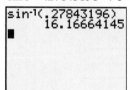

**27.** $\cot \theta = 1.2575516 \Rightarrow \theta \approx 38.491580°$

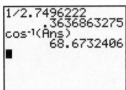

**29.** $\sec \theta = 2.7496222 \Rightarrow \theta \approx 68.673241°$

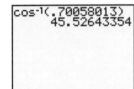

**31.** $\cos \theta = .70058013 \Rightarrow \theta \approx 45.526434°$

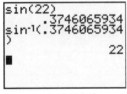

**33.** If the calculator allowed an angle $\theta$ where $0° \le \theta < 360°$, then we would need to find an angle within this interval that is coterminal with 2000° by subtracting a multiple of 360°: $2000° - 5 \cdot 360° = 2000° - 1800° = 200°$. If the calculator was more restrictive on evaluating angles (such as $0° \le \theta < 90°$), then reference angles would need to be used.

**35.** Find the sine of 22°.

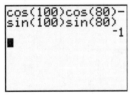

$A = .3746065934°$

**37.** $\cos 100° \cos 80° - \sin 100° \sin 80° = -1$

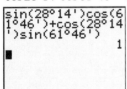

**39.** $\cos 28°14' \cos 61°46' - \sin 28°14' \sin 61°46' = 1$

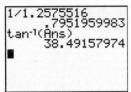

**41.** $2 \sin 25°13' \cos 25°13' - \sin 50°26' = 0$

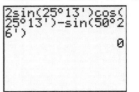

**43.** The figure for this exercise indicates a right triangle. Because we are not considering the time involved in detecting the speed of the car, we will consider the speeds as sides of the right triangle. Given angle $\theta$, $\cos \theta = \dfrac{r}{a}$. Thus, the speed that the radar detects is $r = a \cos \theta$.

**45.** $\sin 10° + \sin 10° \overset{?}{=} \sin 20°$
Using a calculator gives $\sin 10° \approx .34729636$ and $\sin 20° \approx .34202014$. Thus, the statement is false.

**47.** $\sin 50° \overset{?}{=} 2 \sin 25° \cos 25°$
Using a calculator gives $\sin 50° \approx .76604444$ and $2 \sin 25° \cos 25° \approx .76604444$. Thus, the statement is true.

**49.** $\cos 40° \overset{?}{=} 1 - 2 \sin^2 80°$
Using a calculator gives $\cos 40° \approx .76604444$ and $1 - 2 \sin^2 80° \approx -.93969262$. Thus, the statement is false.

**51.** $\sin 39°48' + \cos 39°48' \overset{?}{=} 1$
Using a calculator gives $\sin 39°48' + \cos 39°48' \approx 1.40839322 \ne 1$. Thus, the statement is false.

**53.** $\cos(30° + 20°) \overset{?}{=} \cos 30° \cos 20° - \sin 30° \sin 20°$
Using a calculator gives $\cos(30° + 20°) \approx .64278761$ and $\cos 30° \cos 20° - \sin 30° \sin 20° \approx .64278761$. Thus, the statement is true.

**55.** $1 + \cot^2 42.5° \overset{?}{=} \csc^2 42.5°$
Using a calculator gives $1 + \cot^2 42.5° \approx 2.1909542$ and $\csc^2 42.5° \approx 2.1909542$ Thus, the statement is true.

**57.** $F = W \sin \theta$
$F = 2100 \sin 1.8° \approx 65.96$ lb

**59.** $F = W \sin \theta$

$$-130 = 2600 \sin \theta \Rightarrow \frac{-130}{2600} = \sin \theta \Rightarrow$$

$$-.05 = \sin \theta \Rightarrow \theta = \sin^{-1}(-.05) \approx -2.87°$$

**61.** $F = W \sin \theta$

$$150 = 3000 \sin \theta \Rightarrow \frac{150}{3000} = \sin \theta \Rightarrow$$

$$.05 = \sin \theta \Rightarrow \theta = \sin^{-1}(.05) \approx 2.87°$$

**63.** $R = \dfrac{V^2}{g(f + \tan \theta)}$

**(a)** Since 45 mph = 66 ft/sec,
$V = 66$, $\theta = 3°$, $g = 32.2$, and $f = .14$,
we have

$$R = \frac{V^2}{g(f + \tan \theta)} = \frac{66^2}{32.2(.14 + \tan 3°)}$$
$$\approx 703 \text{ ft}$$

**(b)** Since there are 5280 ft in one mile and
3600 sec in one min, we have

$$70 \text{ mph} = 70 \text{ mph} \cdot \frac{1 \text{ hr}}{3600 \text{ sec}} \cdot \frac{5280 \text{ ft}}{1 \text{ mi}}$$

$$= 102\frac{2}{3} \text{ ft per sec}$$
$$\approx 102.67 \text{ ft per sec}$$

Since $V = 102.67$, $\theta = 3°$, $g = 32.2$, and
$f = .14$, we have

$$R = \frac{V^2}{g(f + \tan \theta)} \approx \frac{102.67^2}{32.2(.14 + \tan 3°)}$$
$$\approx 1701 \text{ ft}$$

**(c)** Intuitively, increasing $\theta$ would make it
easier to negotiate the curve at a higher
speed much like is done at a race track.
Mathematically, a larger value of $\theta$
(acute) will lead to a larger value for
$\tan \theta$. If $\tan \theta$ increases, then the ratio
determining $R$ will *decrease*. Thus, the
radius can be smaller and the curve
sharper if $\theta$ is increased.

$$R = \frac{V^2}{g(f + \tan \theta)} = \frac{66^2}{32.2(.14 + \tan 4°)}$$
$$\approx 644 \text{ ft}$$

$$R = \frac{V^2}{g(f + \tan \theta)} \approx \frac{102.67^2}{32.2(.14 + \tan 4°)}$$
$$\approx 1559 \text{ ft}$$

As predicted, both values are less.

**65.** **(a)** $\theta_1 = 46°$, $\theta_2 = 31°$, $c_1 = 3 \times 10^8$ m per sec

$$\frac{c_1}{c_2} = \frac{\sin \theta_1}{\sin \theta_2} \Rightarrow c_2 = \frac{c_1 \sin \theta_2}{\sin \theta_1} \Rightarrow$$

$$c_2 = \frac{(3 \times 10^8)(\sin 31°)}{\sin 46°} \approx 2 \times 10^8$$

Since $c_1$ is only given to one significant
digit, $c_2$ can only be given to one
significant digit. The speed of light in the
second medium is about $2 \times 10^8$ m per
sec.

**(b)** $\theta_1 = 39°$, $\theta_2 = 28°$, $c_1 = 3 \times 10^8$ m per sec

$$\frac{c_1}{c_2} = \frac{\sin \theta_1}{\sin \theta_2} \Rightarrow c_2 = \frac{c_1 \sin \theta_2}{\sin \theta_1} \Rightarrow$$

$$c_2 = \frac{(3 \times 10^8)(\sin 28°)}{\sin 39°} \approx 2 \times 10^8$$

Since $c_1$ is only given to one significant
digit, $c_2$ can only be given to one
significant digit. The speed of light in the
second medium is about $2 \times 10^8$ m per
sec.

**67.** $\theta_1 = 90°$, $c_1 = 3 \times 10^8$ m per sec, and
$c_2 = 2.254 \times 10^8$

$$\frac{c_1}{c_2} = \frac{\sin \theta_1}{\sin \theta_2} \Rightarrow \sin \theta_2 = \frac{c_2 \sin \theta_1}{c_1}$$

$$\sin \theta_2 = \frac{(2.254 \times 10^8)(\sin 90°)}{3 \times 10^8}$$

$$= \frac{2.254 \times 10^8 (1)}{3 \times 10^8} = \frac{2.254}{3} \Rightarrow$$

$$\theta_2 = \sin^{-1}\left(\frac{2.254}{3}\right) \approx 48.7°$$

**69.** **(a)** Let

$$V_1 = 55 \text{ mph} = 55 \text{ mph} \cdot \frac{1 \text{ hr}}{3600 \text{ sec}} \cdot \frac{5280 \text{ ft}}{1 \text{ mi}}$$

$$= 80\frac{2}{3} \text{ ft per sec} \approx 80.67 \text{ ft per sec},$$

and

$$V_2 = 30 \text{ mph} = 30 \text{ mph} \cdot \frac{1 \text{ hr}}{3600 \text{ sec}} \cdot \frac{5280 \text{ ft}}{1 \text{ mi}}$$

$$= 44 \text{ ft per sec}$$

(*continued on next page*)

*(continued from page 35)*

Also, let $\theta = 3.5°$, $K_1 = .4$, and $K_2 = .02$.

$$D = \frac{1.05\left(V_1^2 - V_2^2\right)}{64.4\left(K_1 + K_2 + \sin\theta\right)}$$

$$= \frac{1.05\left(80.67^2 - 44^2\right)}{64.4\left(.4 + .02 + \sin 3.5°\right)} \approx 155 \text{ ft}$$

**(b)** Let $V_1 \approx 80.67$ ft per sec,
$V_2 = 44$ ft per sec, $\theta = -2°$,
$K_1 = .4$, and $K_2 = .02$.

$$D = \frac{1.05\left(V_1^2 - V_2^2\right)}{64.4\left(K_1 + K_2 + \sin\theta\right)}$$

$$= \frac{1.05\left(80.67^2 - 44^2\right)}{64.4\left[.4 + .02 + \sin\left(-2°\right)\right]} \approx 194 \text{ ft}$$

## Chapter 2 Quiz
**(Sections 2.1–2.3)**

**1.** $\sin A = \dfrac{\text{side opposite}}{\text{hypotenuse}} = \dfrac{24}{40} = \dfrac{3}{5}$

$\cos A = \dfrac{\text{side adjacent}}{\text{hypotenuse}} = \dfrac{32}{40} = \dfrac{4}{5}$

$\tan A = \dfrac{\text{side opposite}}{\text{side adjacent}} = \dfrac{24}{32} = \dfrac{3}{4}$

**3.** $\sin 30° = \dfrac{w}{36} \Rightarrow w = 36\sin 30° = 36 \cdot \dfrac{1}{2} = 18$

$\cos 30° = \dfrac{x}{36} \Rightarrow x = 36\cos 30° = 36 \cdot \dfrac{\sqrt{3}}{2} = 18\sqrt{3}$

$\tan 45° = \dfrac{w}{y} \Rightarrow 1 = \dfrac{18}{y} \Rightarrow y = 18$

$\sin 45° = \dfrac{w}{z} \Rightarrow \dfrac{\sqrt{2}}{2} = \dfrac{18}{z} \Rightarrow z = \dfrac{36}{\sqrt{2}} = 18\sqrt{2}$

**5.** $180° - 135° = 45°$, so the reference angle is $45°$. The original angle ($135°$) lies in quadrant II, so the sine and cosecant are positive, while the remaining trigonometric functions are negative.

$\sin 135° = \dfrac{\sqrt{2}}{2}$ ; $\cos 135° = -\dfrac{\sqrt{2}}{2}$

$\tan 135° = -1$ ; $\cot 135° = -1$

$\sec 135° = -\sqrt{2}$ ; $\csc 135° = \sqrt{2}$

**7.** $1020°$ is coterminal with $1020° - 720° = 300°$. Since this lies in quadrant IV, the reference angle is $360° - 300° = 60°$. In quadrant IV, the cosine and secant are positive, while the remaining trigonometric functions are negative.

$\sin 1020° = -\sin 60° = -\dfrac{\sqrt{3}}{2}$

$\cos 1020° = \cos 60° = \dfrac{1}{2}$

$\tan 1020° = -\tan 60° = -\sqrt{3}$

$\cot 1020° = -\cot 60° = -\dfrac{\sqrt{3}}{3}$

$\sec 1020° = \sec 60° = 2$

$\csc 1020° = -\csc 60° = -\dfrac{2\sqrt{3}}{3}$

**9.** $\sec\theta = -\sqrt{2}$
Since $\sec\theta$ is negative, $\theta$ must lie in quadrants II or III. Since the absolute value of $\sec\theta$ is $\sqrt{2}$, the reference angle, $\theta'$ must be $45°$. The quadrant II angle $\theta$ equals $180° - \theta' = 180° - 45° = 135°$, and the quadrant III angle $\theta$ equals $180° + \theta' = 180° + 45° = 225°$.

**11.** $\sec\left(-212°12'\right) \approx -1.18176327$

```
cos(-212°12')
       -.8461931661
Ans⁻¹
       -1.181763266
■
```

**13.** $\csc\theta = 2.3861147 \Rightarrow \theta \approx 24.777233°$

```
1/2.3861147
       .4190913371
sin⁻¹(Ans)
       24.77723309
```

**15.** The statement is true. Using the cofunction identity, $\tan\left(90° - 35°\right) = \cot 35°$.

## Section 2.4: Solving Right Triangles

1.  22,894.5 to 22,895.5

3.  8958.5 to 8959.5

5.  Answers will vary.
    It would be cumbersome to write 2 as 2.00 or
    2.000, for example, if the measurements had 3
    or 4 significant digits (depending on the
    problem). In the formula, it is understood that
    2 is an exact value. Since the radius
    measurement, 54.98 cm, has four significant
    digits, an appropriate answer would be 345.4
    cm.

7.  If $h$ is the actual height of a building and the
    height is measured as 58.6 ft, then
    $\left|h - 58.6\right| \le .05$.

9.  $A = 36°20'$, $c = 964$ m

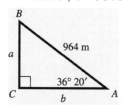

$A + B = 90° \Rightarrow B = 90° - A \Rightarrow$
$\quad B = 90° - 36°20'$
$\quad\quad = 89°60' - 36°20' = 53°40'$

$\sin A = \dfrac{a}{c} \Rightarrow \sin 36°20' = \dfrac{a}{964} \Rightarrow$
$\quad a = 964 \sin 36°20' \approx 571 \text{ m}$ (rounded to
three significant digits)

$\cos A = \dfrac{b}{c} \Rightarrow \cos 36°20' = \dfrac{b}{964} \Rightarrow$
$\quad b = 964 \cos 36°20' \approx 777 \text{ m}$ (rounded to
three significant digits)

11. $N = 51.2°$, $m = 124$ m

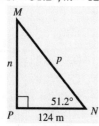

$M + N = 90° \Rightarrow M = 90° - N \Rightarrow$
$\quad M = 90° - 51.2° = 38.8°$

$\tan N = \dfrac{n}{m} \Rightarrow \tan 51.2° = \dfrac{n}{124} \Rightarrow$
$\quad n = 124 \tan 51.2° \approx 154 \text{ m}$ (rounded to
three significant digits)

$\cos N = \dfrac{m}{p} \Rightarrow \cos 51.2° = \dfrac{124}{p} \Rightarrow$

$p = \dfrac{124}{\cos 51.2°} \approx 198 \text{ m}$ (rounded to three
significant digits)

12. $X = 47.8°$, $z = 89.6$ cm

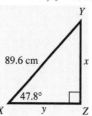

$Y + X = 90° \Rightarrow Y = 90° - X \Rightarrow$
$\quad Y = 90° - 47.8° = 42.2°$

$\sin X = \dfrac{x}{z} \Rightarrow \sin 47.8° = \dfrac{x}{89.6} \Rightarrow$
$\quad x = 89.6 \sin 47.8° \approx 66.4 \text{ cm}$ (rounded to
three significant digits)

$\cos X = \dfrac{y}{z} \Rightarrow \cos 47.8° = \dfrac{y}{89.6} \Rightarrow$
$\quad y = 89.6 \cos 47.8° \approx 60.2 \text{ cm}$ (rounded to
three significant digits)

13. $B = 42.0892°$, $b = 56.851$

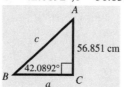

$A + B = 90° \Rightarrow A = 90° - B \Rightarrow$
$\quad A = 90° - 42.0892° = 47.9108°$

$\sin B = \dfrac{b}{c} \Rightarrow \sin 42.0892° = \dfrac{56.851}{c} \Rightarrow$

$c = \dfrac{56.851}{\sin 42.0892°} \approx 84.816 \text{ cm}$
(rounded to five significant digits)

$\tan B = \dfrac{b}{a} \Rightarrow \tan 42.0892° = \dfrac{56.851}{a} \Rightarrow$

$a = \dfrac{56.851}{\tan 42.0892°} \approx 62.942 \text{ cm}$
(rounded to five significant digits)

15. $a = 12.5$, $b = 16.2$

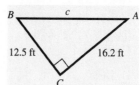

Using the Pythagorean theorem, we have
$$a^2 + b^2 = c^2 \Rightarrow 12.5^2 + 16.2^2 = c^2 \Rightarrow$$
$418.69 = c^2 \Rightarrow c \approx 20.5$ ft (rounded to three significant digits)

$$\tan A = \frac{a}{b} \Rightarrow \tan A = \frac{12.5}{16.2} \Rightarrow$$
$$A = \tan^{-1} \frac{12.5}{16.2} \approx 37.6540°$$
$$\approx 37° + (.6540 \cdot 60)' \approx 37°39' \approx 37°40'$$
(rounded to three significant digits)

$$\tan B = \frac{b}{a} \Rightarrow \tan B = \frac{16.2}{12.5} \Rightarrow$$
$$B = \tan^{-1} \frac{16.2}{12.5} \approx 52.3460°$$
$$\approx 52° + (.3460 \cdot 60)' \approx 52°21'$$
$$\approx 52°20' \text{ (rounded to three significant}$$
$$\text{digits)}$$

**17.** No; You need to have at least one side to solve the triangle.

**19.** Answers will vary. If you know one acute angle, the other acute angle may be found by subtracting the given acute angle from 90°. If you know one of the sides, then choose two of the trigonometric ratios involving sine, cosine or tangent that involve the known side in order to find the two unknown sides.

**21.** $A = 28.0°$, $c = 17.4$ ft

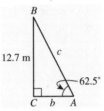

$$A + B = 90°$$
$$B = 90° - A$$
$$B = 90° - 28.0° = 62.0°$$

$$\sin A = \frac{a}{c} \Rightarrow \sin 28.0° = \frac{a}{17.4} \Rightarrow$$
$$a = 17.4 \sin 28.0° \approx 8.17 \text{ ft} \quad \text{(rounded to}$$
three significant digits)

$$\cos A = \frac{b}{c} \Rightarrow \cos 28.00° = \frac{b}{17.4} \Rightarrow$$
$$b = 17.4 \cos 28.00° \approx 15.4 \text{ ft} \quad \text{(rounded to}$$
three significant digits)

**23.** Solve the right triangle with $B = 73.0°$, $b = 128$ in. and $C = 90°$

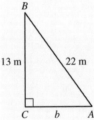

$$A + B = 90° \Rightarrow A = 90° - B \Rightarrow$$
$$A = 90° - 73.0° = 17.0°$$

$$\tan B° = \frac{b}{a} \Rightarrow \tan 73.0° = \frac{128}{a} \Rightarrow$$
$$a = \frac{128}{\tan 73.0°} \Rightarrow a = 39.1 \text{ in (rounded to}$$
three significant digits)

$$\sin B° = \frac{b}{c} \Rightarrow \sin 73.0° = \frac{128}{c} \Rightarrow$$
$$c = \frac{128}{\sin 73.0°} \Rightarrow c = 134 \text{ in (rounded to}$$
three significant digits)

**25.** $A = 62.5°$, $a = 12.7$ m

$$A + B = 90° \Rightarrow B = 90° - A \Rightarrow$$
$$B = 90° - 62.5° = 27.5°$$

$$\tan A = \frac{a}{b} \Rightarrow \tan 62.5° = \frac{12.7}{b} \Rightarrow$$
$$b = \frac{12.7}{\tan 62.5°} \approx 6.61 \text{ m (rounded to}$$
three significant digits)

$$\sin A = \frac{a}{c} \Rightarrow \sin 62.5° = \frac{12.7}{c} \Rightarrow$$
$$c = \frac{12.7}{\sin 62.5°} \approx 14.3 \text{ m (rounded to}$$
three significant digits)

**27.** $a = 13$ m, $c = 22$m

$$c^2 = a^2 + b^2 \Rightarrow 22^2 = 13^2 + b^2 \Rightarrow$$
$$484 = 169 + b^2 \Rightarrow 315 = b^2 \Rightarrow b \approx 18 \text{ m}$$
(rounded to two significant digits)

We will determine the measurements of both $A$ and $B$ by using the sides of the right triangle. In practice, once you find one of the measurements, subtract it from 90° to find the other.

$$\sin A = \frac{a}{c} \Rightarrow \sin A = \frac{13}{22} \Rightarrow$$

$$A \approx \sin^{-1}\left(\frac{13}{22}\right) \approx 36.2215^\circ \approx 36^\circ \text{ (rounded}$$

to two significant digits)

$$\cos B = \frac{b}{c} \Rightarrow \cos B = \frac{13}{22} \Rightarrow$$

$$B \approx \cos^{-1}\left(\frac{13}{22}\right) \approx 53.7784^\circ \approx 54^\circ$$

(rounded to two significant digits)

**29.** $a = 76.4$ yd, $b = 39.3$ yd

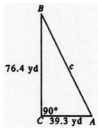

$$c^2 = a^2 + b^2 \Rightarrow c = \sqrt{a^2 + b^2}$$

$$= \sqrt{(76.4)^2 + (39.3)^2} = \sqrt{5836.96 + 1544.49}$$

$$= \sqrt{7381.45} \approx 85.9 \text{ yd (rounded to three}$$

significant digits)

We will determine the measurements of both $A$ and $B$ by using the sides of the right triangle. In practice, once you find one of the measurements, subtract it from $90^\circ$ to find the other.

$$\tan A = \frac{a}{b} \Rightarrow \tan A = \frac{76.4}{39.3} \Rightarrow$$

$$A \approx \tan^{-1}\left(\frac{76.4}{39.3}\right) \approx 62.7788^\circ$$

$$\approx 62^\circ + (.7788 \cdot 60)' \approx 62^\circ 47' \approx 62^\circ 50'$$

(rounded to three significant digits)

$$\tan B = \frac{b}{a} \Rightarrow \tan B = \frac{39.3}{76.4} \Rightarrow$$

$$B \approx \tan^{-1}\left(\frac{39.3}{76.4}\right) \approx 27.2212^\circ$$

$$\approx 27^\circ + (.2212 \cdot 60)' \approx 27^\circ 13' \approx 27^\circ 10'$$

(rounded to three significant digits)

**31.** $a = 18.9$ cm, $c = 46.3$ cm

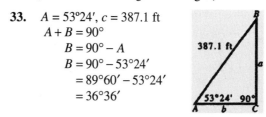

$$c^2 = a^2 + b^2 \Rightarrow 46.3^2 = 18.9^2 + b^2 \Rightarrow$$

$$2143.69 = 357.21 + b^2 \Rightarrow 1786.48 = b^2 \Rightarrow$$

$$b \approx 42.3 \text{ cm (rounded to three}$$

significant digits)

$$\sin A = \frac{a}{c} \Rightarrow \sin A = \frac{18.9}{46.3} \Rightarrow$$

$$A \approx \sin^{-1}\left(\frac{18.9}{46.3}\right) \approx 24.09227^\circ$$

$$\approx 24^\circ + (.09227 \cdot 60)' \approx 24^\circ 06' \approx 24^\circ 10'$$

(rounded to three significant digits)

$$\cos B = \frac{a}{c} \Rightarrow \cos B = \frac{18.9}{46.3} \Rightarrow$$

$$B \approx \cos^{-1}\left(\frac{18.9}{46.3}\right) \approx 65.9077^\circ$$

$$\approx 65^\circ + (.9077 \cdot 60)' \approx 65^\circ 54' \approx 65^\circ 50'$$

(rounded to three significant digits)

**33.** $A = 53^\circ 24'$, $c = 387.1$ ft

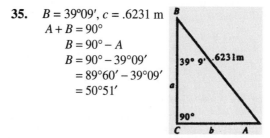

$$A + B = 90^\circ$$

$$B = 90^\circ - A$$

$$B = 90^\circ - 53^\circ 24'$$

$$= 89^\circ 60' - 53^\circ 24'$$

$$= 36^\circ 36'$$

$$\sin A = \frac{a}{c} \Rightarrow \sin 53^\circ 24' = \frac{a}{387.1} \Rightarrow$$

$$a = 387.1 \sin 53^\circ 24' \approx 310.8 \text{ ft (rounded}$$

to four significant digits)

$$\cos A = \frac{b}{c} \Rightarrow \cos 53^\circ 24' = \frac{b}{387.1} \Rightarrow$$

$$b = 387.1 \cos 53^\circ 24' \approx 230.8 \text{ ft (rounded}$$

to four significant digits)

**35.** $B = 39^\circ 09'$, $c = .6231$ m

$$A + B = 90^\circ$$

$$B = 90^\circ - A$$

$$B = 90^\circ - 39^\circ 09'$$

$$= 89^\circ 60' - 39^\circ 09'$$

$$= 50^\circ 51'$$

$$\sin B = \frac{b}{c} \Rightarrow \sin 39^\circ 09' = \frac{b}{.6231} \Rightarrow$$

$$b = .6231 \sin 39^\circ 09' \approx .3934 \text{ m (rounded}$$

to four significant digits)

$$\cos B = \frac{a}{c} \Rightarrow \cos 39^\circ 09' = \frac{a}{.6231} \Rightarrow$$

$$a = .6231 \cos 39^\circ 09' \approx .4832 \text{ m (rounded}$$

to four significant digits)

**37.** The angle of elevation from $X$ to $Y$ (with $Y$ above $X$) is the acute angle formed by ray $XY$ and a horizontal ray with endpoint at $X$.

**39.** Answers will vary. The angle of elevation and the angle of depression are measured between the line of sight and a horizontal line. So, in the diagram, lines $AD$ and $CB$ are both horizontal. Hence, they are parallel. The line formed by $AB$ is a transversal and angles $DAB$ and $ABC$ are alternate interior angle and thus have the same measure.

**41.** $\sin 43°50' = \dfrac{d}{13.5}$
$d = 13.5 \sin 43°50' \approx 9.3496000$
The ladder goes up the wall 9.35 m. (rounded to three significant digits)

**43.** Let $x$ represent the horizontal distance between the two buildings and $y$ represent the height of the portion of the building across the street that is higher than the window.

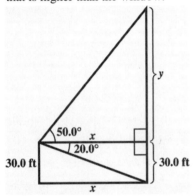

$\tan 20.0° = \dfrac{30.0}{x} \Rightarrow x = \dfrac{30.3}{\tan 20.0°} \approx 82.4$

$\tan 50.0° = \dfrac{y}{x} \Rightarrow$

$y = x \tan 50.0° = \left( \dfrac{30.0}{\tan 20.0°} \right) \tan 50.0°$

height $= y + 30.0$

$= \left( \dfrac{30.0}{\tan 20.0°} \right) \tan 50.0° + 30.0$

$\approx 128.2295$

The height of the building across the street is about 128 ft. (rounded to three significant digits)

**45.** The altitude of an isosceles triangle bisects the base as well as the angle opposite the base. The two right triangles formed have interior angles which have the same measure. The lengths of the corresponding sides also have the same measure. Since the altitude bisects the base, each leg (base) of the right triangles is $\dfrac{42.36}{2} = 21.18$ in.

Let $x =$ the length of each of the two equal sides of the isosceles triangle.

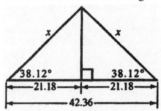

$\cos 38.12° = \dfrac{21.18}{x} \Rightarrow x \cos 38.12° = 21.18 \Rightarrow$

$x = \dfrac{21.18}{\cos 38.12°} \approx 26.921918$

The length of each of the two equal sides of the triangle is 26.92 in. (rounded to four significant digits)

**47.** In order to find the angle of elevation, $\theta$, we need to first find the length of the diagonal of the square base. The diagonal forms two isosceles right triangles. Each angle formed by a side of the square and the diagonal measures $45°$.

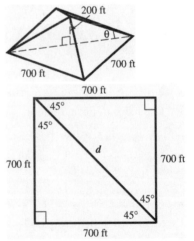

By the Pythagorean theorem,
$700^2 + 700^2 = d^2 \Rightarrow 2 \cdot 700^2 = d^2 \Rightarrow$
$d = \sqrt{2 \cdot 700^2} \Rightarrow d = 700\sqrt{2}$

Thus, length of the diagonal is $700\sqrt{2}$ ft. To to find the angle, $\theta$, we consider the following isosceles triangle.

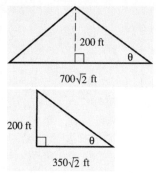

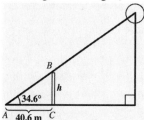

The height of the pyramid bisects the base of this triangle and forms two right triangles. We can use one of these triangles to find the angle of elevation, $\theta$.

$$\tan\theta = \frac{200}{350\sqrt{2}} \approx .4040610178$$

$$\theta \approx \tan^{-1}(.4040610178) \approx 22.0017$$

Rounding this figure to two significant digits, we have $\theta \approx 22°$.

**49.** Let $h$ represent the height of the tower. In triangle $ABC$ we have

$$\tan 34.6° = \frac{h}{40.6}$$

$$h = 40.6\tan 34.6° \approx 28.0081$$

The height of the tower is 28.0 m. (rounded to three significant digits)

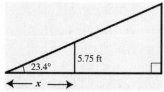

**51.** Let $x =$ the length of the shadow cast by Diane Carr.

$$\tan 23.4° = \frac{5.75}{x}$$

$$x\tan 23.4° = 5.75$$

$$x = \frac{5.75}{\tan 23.4°} \approx 13.2875$$

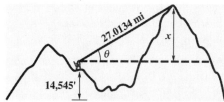

The length of the shadow cast by Diane Carr is 13.3 ft. (rounded to three significant digits)

**53.** Let $x =$ the height of the taller building; $h =$ the difference in height between the shorter and taller buildings; $d =$ the distance between the buildings along the ground.

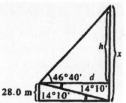

$$\frac{28.0}{d} = \tan 14°10' \Rightarrow 28.0 = d\tan 14°10' \Rightarrow$$

$$d = \frac{28.0}{\tan 14°10'} \approx 110.9262493 \text{ m}$$

(We hold on to these digits for the intermediate steps.) To find $h$, solve

$$\frac{h}{d} = \tan 46°40'$$

$$h = d\tan 46°40' \approx (110.9262493)\tan 46°40'$$

$$\approx 117.5749$$

Thus, the value of $h$ rounded to three significant digits is 118 m. Since

$$x = h + 28.0 = 118 + 28.0 \approx 146\,\text{m},$$ the height

of the taller building is 146 m.

**55. (a)** Let $x =$ the height of the peak above 14,545 ft.

Since the diagonal of the right triangle formed is in miles, we must first convert this measurement to feet. Since there are 5280 ft in one mile, we have the length of the diagonal is $27.0134(5280) =$ 142,630.752. Find the value of $x$ by

solving $\sin 5.82° = \dfrac{x}{142,630.752}$

$$x = 142,630.752\sin 5.82°$$

$$\approx 14,463.2674$$

Thus, the value of $x$ rounded to five significant digits is 14,463 ft. Thus, the total height is about

$14,545 + 14,463 = 29,008 \approx 29,000$ ft.

**(b)** The curvature of the earth would make the peak appear shorter than it actually is. Initially the surveyors did not think Mt. Everest was the tallest peak in the Himalayas. It did not look like the tallest peak because it was farther away than the other large peaks.

## Section 2.5: Further Applications of Right Triangles

**1.** It should be shown as an angle measured clockwise from due north.

**3.** A sketch is important to show the relationships among the given data and the unknowns.

**5.** (−4, 0)

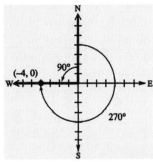

The bearing of the airplane measured in a clockwise direction from due north is 270°. The bearing can also be expressed as N 90° W, or S 90° W.

**7.** (−5, 5)

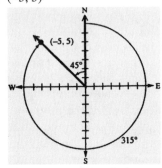

The bearing of the airplane measured in a clockwise direction from due north is 315°. The bearing can also be expressed as N 45° W.

**9.** (0, 4)

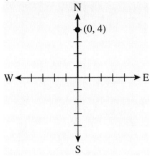

The bearing of the airplane measured in a clockwise direction from due north is 0°. The bearing can also be expressed as N 0° E or N 0° W.

**11.** (2, −2)

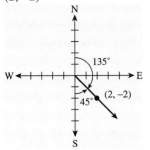

The bearing of the airplane measured in a clockwise direction from due north is 135°. The bearing can also be expressed as S 45° E.

**13.** All points whose bearing from the origin is 240° lie in quadrant III.

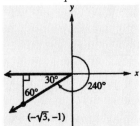

The reference angle, $\theta'$, is 30°. For any point, $(x, y)$ on the ray $\dfrac{x}{r} = -\cos\theta'$ and

$\dfrac{y}{r} = -\sin\theta'$, where $r$ is the distance from the point to the origin. Let $r = 2$, so

$\dfrac{x}{r} = -\cos\theta'$

$x = -r\cos\theta' = -2\cos 30° = -2 \cdot \dfrac{\sqrt{3}}{2} = -\sqrt{3}$

$\dfrac{y}{r} = -\sin\theta'$

$y = -r\sin\theta' = -2\sin 30° = -2 \cdot \dfrac{1}{2} = -1$

Thus, a point on the ray is $\left(-\sqrt{3},-1\right)$. Since the ray contains the origin, the equation is of the form $y = mx$. Substituting the point $\left(-\sqrt{3},-1\right)$, we have $-1 = m\left(-\sqrt{3}\right) \Rightarrow$

$m = \frac{-1}{-\sqrt{3}} = \frac{1}{\sqrt{3}} \cdot \frac{\sqrt{3}}{\sqrt{3}} = \frac{\sqrt{3}}{3}$. Thus, the equation of

the ray is $y = \frac{\sqrt{3}}{3}x, x \le 0$ (since the ray lies in quadrant III).

**15.** Let $x =$ the distance the plane is from its starting point. In the figure, the measure of angle $ACB$ is

$40° + \left(180° - 130°\right) = 40° + 50° = 90°$.

Therefore, triangle $ACB$ is a right triangle.

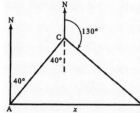

Since $d = rt$, the distance traveled in 1.5 hr is (1.5 hr)(110 mph) = 165 mi. The distance traveled in 1.3 hr is (1.3 hr)(110 mph) = 143 mi.

Using the Pythagorean theorem, we have $x^2 = 165^2 + 143^2 \Rightarrow x^2 = 27,225 + 20,449 \Rightarrow$ $x^2 = 47,674 \Rightarrow x \approx 218.3438$

The plane is 220 mi from its starting point. (rounded to two significant figures)

**17.** Let $x =$ distance the ships are apart. In the figure, the measure of angle $CAB$ is $130° - 40° = 90°$. Therefore, triangle $CAB$ is a right triangle.

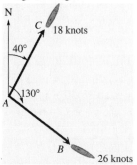

Since $d = rt$, the distance traveled by the first ship in 1.5 hr is (1.5 hr)(18 knots) = 27 nautical mi and the second ship is (1.5hr)(26 knots) = 39 nautical mi.

Applying the Pythagorean theorem, we have $x^2 = 27^2 + 39^2 \Rightarrow x^2 = 729 + 1521 \Rightarrow$ $x^2 = 2250 \Rightarrow x = \sqrt{2250} \approx 47.4342$

The ships are 47 nautical mi apart. (rounded to 2 significant digits)

**19.** Draw triangle $WDG$ with $W$ representing Winston-Salem, $D$ representing Danville, and $G$ representing Goldsboro. Name any point $X$ on the line due south from $D$.

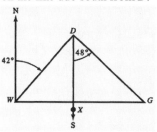

Since the bearing from $W$ to $D$ is 42° (equivalent to N 42° E), angle $WDX$ measures 42°. Since angle $XDG$ measures 48°, the measure of angle $D$ is 42° + 48° = 90°. Thus, triangle $WDG$ is a right triangle.

Using $d = rt$ and the Pythagorean theorem, we

have $WG = \sqrt{\left(WD\right)^2 + \left(DG\right)^2}$

$= \sqrt{\left[65\left(1.1\right)\right]^2 + \left[65\left(1.8\right)\right]^2}$

$WG = \sqrt{71.5^2 + 117^2} = \sqrt{5112.25 + 13,689}$

$= \sqrt{18,8001.25} \approx 137$

The distance from Winston-Salem to Goldsboro is approximately 140 mi. (rounded to two significant digits)

**21.** Let $x =$ distance between the two ships.

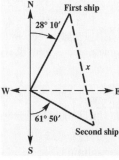

The angle between the bearings of the ships is $180° - \left(28°10' + 61°50'\right) = 90°$. The triangle formed is a right triangle. The distance traveled at 24.0 mph is (4 hr)(24.0 mph) = 96 mi. The distance traveled at 28.0 mph is (4 hr)(28.0 mph) = 112 mi.

*(continued on next page)*

(*continued from page 43*)

Applying the Pythagorean theorem we have
$$x^2 = 96^2 + 112^2 \Rightarrow x^2 = 9216 + 12,544 \Rightarrow$$
$$x^2 = 21,760 \Rightarrow x = \sqrt{21,760} \approx 147.5127$$
The ships are 148 mi apart. (rounded to three significant digits)

23. $ax = b + cx \Rightarrow ax - cx = b \Rightarrow x(a-c) = b \Rightarrow$
$$x = \frac{b}{a-c}$$

25. From exercise 24, we have
$$y = \tan\theta(x-a) \Rightarrow y = \tan 35°(x-25)$$

27. Algebraic solution:
Let $x$ = the side adjacent to 49.2° in the smaller triangle.

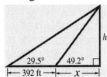

In the larger right triangle, we have
$$\tan 29.5° = \frac{h}{392 + x} \Rightarrow h = (392 + x)\tan 29.5°.$$
In the smaller right triangle, we have
$$\tan 49.2° = \frac{h}{x} \Rightarrow h = x\tan 49.2°.$$
Substituting, we have
$$x\tan 49.2° = (392 + x)\tan 29.5°$$
$$x\tan 49.2° = 392\tan 29.5°$$
$$+ x\tan 29.5°$$
$$x\tan 49.2° - x\tan 29.5° = 392\tan 29.5°$$
$$x(\tan 49.2° - \tan 29.5°) = 392\tan 29.5°$$
$$x = \frac{392\tan 29.5°}{\tan 49.2° - \tan 29.5°}$$
Now substitute this expression for $x$ in the equation for the smaller triangle to obtain
$$h = x\tan 49.2°$$
$$h = \frac{392\tan 29.5°}{\tan 49.2° - \tan 29.5°} \cdot \tan 49.2°$$
$$\approx 433.4762 \approx 433 \text{ ft (rounded to three}$$
significant digits.
Graphing calculator solution:
The first line considered is $y = (\tan 29.5°)x$
and the second is $y = (\tan 29.5°)(x - 392)$.

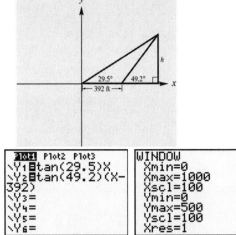

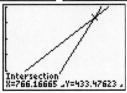

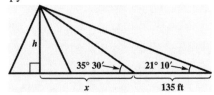

The height of the triangle is 433 ft (rounded to three significant digits.

29. Let $x$ = the distance from the closer point on the ground to the base of height $h$ of the pyramid.

In the larger right triangle, we have
$$\tan 21°10' = \frac{h}{135 + x} \Rightarrow h = (135 + x)\tan 21°10'$$
In the smaller right triangle, we have
$$\tan 35°30' = \frac{h}{x} \Rightarrow h = x\tan 35°30'.$$
Substitute for $h$ in this equation, and solve for $x$ to obtain the following.
$$(135 + x)\tan 21°10' = x\tan 35°30'$$
$$135\tan 21°10' + x\tan 21°10' = x\tan 35°30'$$
$$135\tan 21°10' = x\tan 35°30' - x\tan 21°10'$$
$$135\tan 21°10' = x(\tan 35°30' - \tan 21°10')$$
$$\frac{135\tan 21°10'}{\tan 35°30' - \tan 21°10'} = x$$
Substitute for $x$ in the equation for the smaller triangle.
$$h = \frac{135\tan 21°10'}{\tan 35°30' - \tan 21°10'}\tan 35°30'$$
$$\approx 114.3427$$
The height of the pyramid is 114 ft. (rounded to three significant digits)

**31.** Let $x$ = the height of the antenna; $h$ = the height of the house.

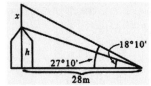

In the smaller right triangle, we have

$$\tan 18°10' = \frac{h}{28} \Rightarrow h = 28 \tan 18°10'.$$

In the larger right triangle, we have

$$\tan 27°10' = \frac{x+h}{28} \Rightarrow x+h = 28 \tan 27°10' \Rightarrow$$
$$x = 28 \tan 27°10' - h$$
$$x = 28 \tan 27°10' - 28 \tan 18°10'$$
$$\approx 5.1816$$

The height of the antenna is 5.18 m. (rounded to three significant digits)

**33. (a)** From the figure in the text,

$$d = \frac{b}{2}\cot\frac{\alpha}{2} + \frac{b}{2}\cot\frac{\beta}{2}$$
$$= \frac{b}{2}\left(\cot\frac{\alpha}{2} + \cot\frac{\beta}{2}\right)$$

**(b)** Using the result of part (a), let
$\alpha = 37'48''$, $\beta = 42'3''$, and $b = 2.000$

$$d = \frac{b}{2}\left(\cot\frac{\alpha}{2} + \cot\frac{\beta}{2}\right) \Rightarrow$$
$$d = \frac{2.000}{2}\left(\cot\frac{37'48''}{2} + \cot\frac{42'3''}{2}\right)$$
$$= \cot .315° + \cot .3504166667$$
$$\approx 345.3951$$

The distance between the two point $P$ and $Q$ is about 345.4 cm.

**35.** Let $x$ = the minimum distance that a plant needing full sun can be placed from the fence.

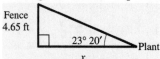

$$\tan 23°20' = \frac{4.65}{x} \Rightarrow x \tan 23°20' = 4.65 \Rightarrow$$
$$x = \frac{4.65}{\tan 23°20'} \approx 10.7799$$

The minimum distance is 10.8 ft. (rounded to three significant digits)

**37. (a)** If $\theta = 37°$, then $\dfrac{\theta}{2} = \dfrac{37°}{2} = 18.5°$.

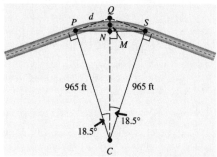

To find the distance between $P$ and $Q$, $d$, we first note that angle $QPC$ is a right angle. Hence, triangle $QPC$ is a right triangle and we can solve

$$\tan 18.5° = \frac{d}{965}$$
$$d = 965 \tan 18.5° \approx 322.8845$$

The distance between $P$ and $Q$, is 320 ft. (rounded to two significant digits)

**(b)** Since we are dealing with a circle, the distance between $M$ and $C$ is $R$. If we let $x$ be the distance from $N$ to $M$, then the distance from $C$ to $N$ will be $R - x$.

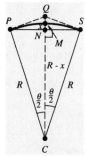

Since triangle $CNP$ is a right triangle, we can set up the following equation.

$$\cos\frac{\theta}{2} = \frac{R-x}{R} \Rightarrow R\cos\frac{\theta}{2} = R - x \Rightarrow$$
$$x = R - R\cos\frac{\theta}{2} \Rightarrow x = R\left(1 - \cos\frac{\theta}{2}\right)$$

**39. (a)** $\theta \approx \dfrac{57.3S}{R} = \dfrac{57.3(336)}{600} = 32.088°$

$$d = R\left(1 - \cos\frac{\theta}{2}\right)$$
$$= 600\left(1 - \cos 16.044°\right) \approx 23.3702 \text{ ft}$$

The distance is 23 ft. (rounded to two significant digits)

**(b)** $\theta \approx \dfrac{57.3S}{R} = \dfrac{57.3(485)}{600} = 46.3175°$

$d = R\left(1 - \cos\dfrac{\theta}{2}\right)$

$= 600(1 - \cos 23.15875°)$

$\approx 48.3488$

The distance is 48 ft. (rounded to two significant digits)

**(c)** The faster the speed, the more land needs to be cleared on the inside of the curve.

## Chapter 2: Review Exercises

**1.** $\sin A = \dfrac{\text{side opposite}}{\text{hypotenuse}} = \dfrac{60}{61}$

$\cos A = \dfrac{\text{side adjacent}}{\text{hypotenuse}} = \dfrac{11}{61}$

$\tan A = \dfrac{\text{side opposite}}{\text{side adjacent}} = \dfrac{60}{11}$

$\cot A = \dfrac{\text{side adjacent}}{\text{side opposite}} = \dfrac{11}{60}$

$\sec A = \dfrac{\text{hypotenuse}}{\text{side adjacent}} = \dfrac{61}{11}$

$\csc A = \dfrac{\text{hypotenuse}}{\text{side opposite}} = \dfrac{61}{60}$

**3.** $\sin 4\beta = \cos 5\beta$

Sine and cosine are cofunctions, so the sum of the angles is 90°.

$4\beta + 5\beta = 90° \Rightarrow 9\beta = 90° \Rightarrow \beta = 10°$

**5.** $\tan(5x + 11°) = \cot(6x + 2°)$

Tangent and cotangent are cofunctions, so the sum of the angles is 90°.

$(5x + 11°) + (6x + 2°) = 90°$

$11x + 13° = 90°$

$11x = 77° \Rightarrow x = 7°$

**7.** $\sin 46° < \sin 58°$

In the interval from 0° to 90°, as the angle increases, so does the sine of the angle, so $\sin 46° < \sin 58°$ is true.

**9.** $\sec 48° \geq \cos 42°$

Using the reciprocal identity,

$\sec 48° = \dfrac{1}{\cos 48°}$. Since 48° and 42° are in quadrant I, sec 48° and cos 42° are both positive. Thus

$0 < \cos 48° < 1 \Rightarrow \dfrac{1}{\cos 48°} = \sec 48° > 1$. Since

$0 < \cos 42° < 1$, $\sec 48° \geq \cos 42°$ is true.

**11.** The sum of the measures of angles $A$ and $B$ is 90°, and, thus, they are complementary angles. Since sine and cosine are cofunctions, we have $\sin B = \cos(90° - B) = \cos A$.

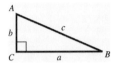

**13.** 1020° is coterminal with $1020° - 2 \cdot 360° = 300°$. The reference angle is $360° - 300° = 60°$. Because 1020° lies in quadrant IV, the sine, tangent, cotangent, and cosecant are negative.

$\sin 1020° = -\sin 60° = -\dfrac{\sqrt{3}}{2}$

$\cos 1020° = \cos 60° = \dfrac{1}{2}$

$\tan 1020° = -\tan 60° = -\sqrt{3}$

$\cot 1020° = -\cot 60° = -\dfrac{\sqrt{3}}{3}$

$\sec 1020° = \sec 60° = 2$

$\csc 1020° = -\csc 60° = -\dfrac{2\sqrt{3}}{3}$

**15.** –1470° is coterminal with $-1470° + 5 \cdot 360° = 330°$. This angle lies in quadrant IV. The reference angle is $360° - 330° = 30°$. Since –1470° is in quadrant IV, the sine, tangent, cotangent, and cosecant are negative.

$\sin(-1470°) = -\sin 30° = -\dfrac{1}{2}$

$\cos(-1470°) = \cos 30° = \dfrac{\sqrt{3}}{2}$

$\tan(-1470°) = -\tan 30° = -\dfrac{\sqrt{3}}{3}$

$\cot(-1470°) = -\cot 30° = -\sqrt{3}$

$\sec(-1470°) = \sec 30° = \dfrac{2\sqrt{3}}{3}$

$\csc(-1470°) = -\csc 30° = -2$

**17.** $\cos\theta = -\dfrac{1}{2}$

Since $\cos\theta$ is negative, $\theta$ must lie in quadrants II or III. Since the absolute value of $\cos\theta$ is $\dfrac{1}{2}$, the reference angle, $\theta'$ must be 60°. The quadrant II angle $\theta$ equals $180° - \theta' = 180° - 60° = 120°$, and the quadrant III angle $\theta$ equals $180° + \theta' = 180° + 60° = 240°$.

**19.** $\sec\theta = -\dfrac{2\sqrt{3}}{3}$

Since $\sec\theta$ is negative, $\theta$ must lie in quadrants II or III. Since the absolute value of $\sec\theta$ is $\dfrac{2\sqrt{3}}{3}$, the reference angle, $\theta'$ must be 30°. The quadrant II angle $\theta$ equals $180° - \theta' = 180° - 30° = 150°,$ and the quadrant III angle $\theta$ equals $180° + \theta' = 180° + 30° = 210°.$

**21.** $\tan^2 120° - 2\cot 240° = \left(-\sqrt{3}\right)^2 - 2\left(\dfrac{\sqrt{3}}{3}\right)$

$$= 3 - \dfrac{2\sqrt{3}}{3}$$

**23.** **(a)** $(-3, -3)$

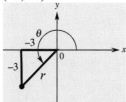

The distance from the origin is $r$:

$r = \sqrt{x^2 + y^2} \Rightarrow r = \sqrt{(-3)^2 + (-3)^2} \Rightarrow$
$r = \sqrt{9+9} \Rightarrow r = \sqrt{18} \Rightarrow r = 3\sqrt{2}$

$\sin\theta = \dfrac{y}{r} = -\dfrac{3}{3\sqrt{2}} = -\dfrac{1}{\sqrt{2}} \cdot \dfrac{\sqrt{2}}{\sqrt{2}} = -\dfrac{\sqrt{2}}{2}$

$\cos\theta = \dfrac{x}{r} = -\dfrac{3}{3\sqrt{2}} = -\dfrac{1}{\sqrt{2}} \cdot \dfrac{\sqrt{2}}{\sqrt{2}} = -\dfrac{\sqrt{2}}{2}$

$\tan\theta = \dfrac{y}{x} = \dfrac{-3}{-3} = 1$

**(b)** $\left(1, -\sqrt{3}\right)$

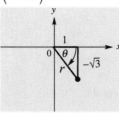

The distance from the origin is $r$:

$r = \sqrt{x^2 + y^2} = \sqrt{(1)^2 + \left(-\sqrt{3}\right)^2}$

$\quad = \sqrt{1+3} = \sqrt{4} = 2$

$\sin\theta = \dfrac{y}{r} = \dfrac{-\sqrt{3}}{2} = -\dfrac{\sqrt{3}}{2}$ ; $\cos\theta = \dfrac{x}{r} = \dfrac{1}{2}$

$\tan\theta = \dfrac{y}{x} = \dfrac{-\sqrt{3}}{1} = -\sqrt{3}$

For the exercises in this section, be sure your calculator is in degree mode. If your calculator accepts angles in degrees, minutes, and seconds, it is not necessary to change angles to decimal degrees. Keystroke sequences may vary on the type and/or model of calculator being used. Screens shown will be from a TI-83 Plus calculator. To obtain the degree (°) and (′) symbols, go to the ANGLE menu (2nd APPS).

For Exercises 25–27, the calculation for decimal degrees is indicated for calculators that do not accept degree, minutes, and seconds.

**25.** $\sec 222°30' \approx -1.3563417$

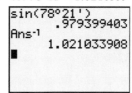

$222°30' = \left(222 + \frac{30}{60}\right)° = 222.5°$

**27.** $\csc 78°21' \approx 1.0210339$

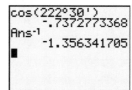

$78°21' = \left(78 + \frac{21}{60}\right)° = 78.35°$

**29.** $\tan 11.7689° \approx .20834446$

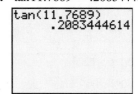

**31.** $\sin\theta = .8254121$

$\theta \approx 55.673870°$

**33.**   $\cos \theta = .97540415$

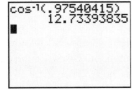

$\theta \approx 12.733938°$

**35.**   $\tan \theta = 1.9633124$

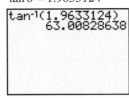

$\theta \approx 63.008286°$

**37.**   Since the value of $\sin \theta$ is positive in quadrants I and II, the two angles in $[0°, 360°)$ are approximately $47.1°$ and $180° - 47.1° = 132.9°$.

**39.**   $\sin 50° + \sin 40° \overset{?}{=} \sin 90°$

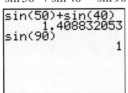

Using a calculator gives $\sin 50° + \sin 40° = 1.408832053$ while $\sin 90° = 1$. Thus, the statement is false.

**41.**   $\sin 240° \overset{?}{=} 2 \sin 120° \cos 120°$

$\sin 240° = -\sin 60° = -\dfrac{\sqrt{3}}{2}$ and

$2 \sin 120° \cos 120° = 2 \sin 60° (-\cos 60°)$

$\qquad = 2 \left( \dfrac{\sqrt{3}}{2} \right) \left( \dfrac{1}{2} \right) = -\dfrac{\sqrt{3}}{2}$

Thus, the statement is true.

**43.**   No, $\cot 25° = \dfrac{1}{\tan 25°} \neq \tan^{-1} 25°$.

**45.**   Since $\cos 1997°$ and $\sin 1997°$ are both negative, $1997°$ must lie in quadrant III.

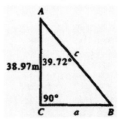

**47.**   $A = 58° \, 30', \;\; c = 748$

$A + B = 90° \Rightarrow B = 90° - A \Rightarrow$
$\quad B = 90° - 58°30' = 89°60' - 58°30'$
$\qquad = 31°30'$

$\sin A = \dfrac{a}{c} \Rightarrow \sin 58°30' = \dfrac{a}{748} \Rightarrow$
$\quad a = 748 \sin 58°30' \approx 638$ (rounded to three significant digits)

$\cos A = \dfrac{b}{c} \Rightarrow \cos 58°30' = \dfrac{b}{748} \Rightarrow$
$\quad b = 748 \cos 58°30' \approx 391$ (rounded to three significant digits)

**49.**   $A = 39.72°, \;\; b = 38.97$ m

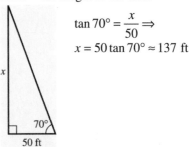

$A + B = 90° \Rightarrow B = 90° - A \Rightarrow$
$\quad B = 90° - 39.72° = 50.28°$

$\tan A = \dfrac{a}{b} \Rightarrow \tan 39.72° = \dfrac{a}{38.97} \Rightarrow$
$\quad a = 38.97 \tan 39.72° \approx 32.38$ m (rounded to four significant digits)

$\cos A = \dfrac{b}{c} \Rightarrow \cos 39.72° = \dfrac{38.97}{c} \Rightarrow$
$c \cos 39.72° = 38.97 \Rightarrow$

$\qquad c = \dfrac{38.97}{\cos 39.72°} \approx 50.66$ m

(rounded to five significant digits)

**51.**   Let $x$ = the height of the tree.

$\tan 70° = \dfrac{x}{50} \Rightarrow$
$x = 50 \tan 70° \approx 137$ ft

**53.** Let $x$ = height of the tower.

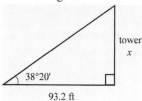

$$\tan 38°20' = \frac{x}{93.2}$$
$$x = 93.2 \tan 38°20'$$
$$x \approx 73.6930$$

The height of the tower is 73.7 ft. (rounded to three significant digits)

**55.** Let $x$ = length of the diagonal

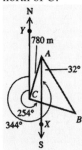

$$\cos 35.65° = \frac{15.24}{x}$$
$$x = \frac{15.24}{\cos 35.65°} \approx 18.7548$$

The length of the diagonal is 18.75 cm (rounded to three significant digits).

**57.** Draw triangle $ABC$ and extend the north-south lines to a point $X$ south of $A$ and $S$ to a point $Y$, north of $C$.

Angle $ACB = 344° - 254° = 90°$, so $ABC$ is a right triangle.
Angle $BAX = 32°$ since it is an alternate interior angle to $32°$.
Angle $YCA = 360° - 344° = 16°$
Angle $XAC = 16°$ since it is an alternate interior angle to angle $YCA$.
Angle $BAC = 32° + 16° = 48°$.
In triangle $ABC$,
$$\cos A = \frac{AC}{AB} \Rightarrow \cos 48° = \frac{780}{AB} \Rightarrow$$
$$AB \cos 48° = 780 \Rightarrow AB = \frac{780}{\cos 48°} \approx 1165.6917$$

The distance from $A$ to $B$ is 1200 m. (rounded to two significant digits)

**59.** Suppose $A$ is the car heading south at 55 mph, $B$ is the car heading west, and point $C$ is the intersection from which they start. After two hours, using $d = rt$, $AC = 55(2) = 110$. Angle $ACB$ is a right angle, so triangle $ACB$ is a right triangle. The bearing of $A$ from $B$ is 324°, so angle $CAB = 360° - 324° = 36°$.

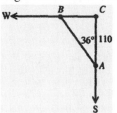

$$\cos \angle CAB = \frac{AC}{AB} \Rightarrow \cos 36° = \frac{110}{AB} \Rightarrow$$
$$AB = \frac{110}{\cos 36°} \approx 135.9675$$

The distance from $A$ to $B$ is about 140 mi (rounded to two significant digits).

**61.** Answers will vary.

**63.** $h = R\left( \dfrac{1}{\cos\left(\frac{180T}{P}\right)} - 1 \right)$

**(a)** Let $R = 3955$ mi, $T = 25$ min, $P = 140$ min.

$$h = R\left( \frac{1}{\cos\left(\frac{180T}{P}\right)} - 1 \right)$$
$$h = 3955\left( \frac{1}{\cos\left(\frac{180 \cdot 25}{140}\right)} - 1 \right) \approx 715.9424$$

The height of the satellite is approximately 716 mi.

**(b)** Let $R = 3955$ mi, $T = 30$ min, $P = 140$ min.

$$h = R\left( \frac{1}{\cos\left(\frac{180T}{P}\right)} - 1 \right)$$
$$h = 3955\left( \frac{1}{\cos\left(\frac{180 \cdot 30}{140}\right)} - 1 \right) \approx 1103.6349$$

The height of the satellite is approximately 1104 mi.

### Chapter 2: Chapter Test

1.  $\sin A = \dfrac{\text{side opposite}}{\text{hypotenuse}} = \dfrac{12}{13}$

    $\cos A = \dfrac{\text{side adjacent}}{\text{hypotenuse}} = \dfrac{5}{13}$

    $\tan A = \dfrac{\text{side opposite}}{\text{side adjacent}} = \dfrac{12}{5}$

    $\cot A = \dfrac{\text{side adjacent}}{\text{side opposite}} = \dfrac{5}{12}$

    $\sec A = \dfrac{\text{hypotenuse}}{\text{side adjacent}} = \dfrac{13}{5}$

    $\csc A = \dfrac{\text{hypotenuse}}{\text{side opposite}} = \dfrac{13}{12}$

2.  Apply the relationships between the lengths of the sides of a $30° - 60°$ right triangle first to the triangle on the right to find the values of $y$ and $w$. In the $30° - 60°$ right triangle, the side opposite the $60°$ angle is $\sqrt{3}$ times as long as the side opposite to the $30°$ angle. The length of the hypotenuse is 2 times as long as the shorter leg (opposite the $30°$ angle).

    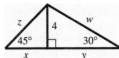

    Thus, we have $y = 4\sqrt{3}$ and $w = 2(4) = 8$.

    Apply the relationships between the lengths of the sides of a $45° - 45°$ right triangle next to the triangle on the left to find the values of $x$ and $z$. In the $45° - 45°$ right triangle, the sides opposite the $45°$ angles measure the same. The hypotenuse is $\sqrt{2}$ times the measure of a leg. Thus, we have $x = 4$ and $z = 4\sqrt{2}$

3.  $\sin(B + 15°) = \cos(2B + 30°)$

    Since sine and cosine are cofunctions, the sum of the angles is 90°. So,
    $$(B + 15°) + (2B + 30°) = 90°$$
    $$3B + 45° = 90°$$
    $$3B = 45° \Rightarrow B = 15°$$

4.  (a)  $\sin 24° < \sin 48°$
        In the interval from 0° to 90°, as the angle increases, so does the sine of the angle, so $\sin 24° < \sin 48°$ is true.

    (b)  $\cos 24° < \cos 48°$
        In the interval from 0° to 90°, as the angle increases, so the cosine of the angle decreases, so $\cos 24° < \cos 48°$ is false.

    (c)  $\cos(60° + 30°)$
        $$= \cos 60° \cos 30° - \sin 60° \sin 30°$$
        $$\cos(60° + 30°) = \cos 90° = 0$$
        $$\cos 60° \cos 30° - \sin 60° \sin 30°$$
        $$= \frac{1}{2}\left(\frac{\sqrt{3}}{2}\right) - \frac{\sqrt{3}}{2}\left(\frac{1}{2}\right) = 0$$

        Thus, the statement is true.

5.  A 240° angle lies in quadrant III, so the reference angle is $240° - 180° = 60°$. Since 240° is in quadrant III, the sine, cosine, secant, and cosecant are negative.

    $$\sin 240° = -\sin 60° = -\frac{\sqrt{3}}{2}$$
    $$\cos 240° = -\cos 60° = -\frac{1}{2}$$
    $$\tan 240° = \tan 60° = \sqrt{3}$$
    $$\cot 240° = \cot 60° = \frac{1}{\sqrt{3}} = \frac{\sqrt{3}}{3}$$
    $$\sec 240° = -\sec 60° = -2$$
    $$\csc 240° = -\csc 60° = -\frac{2}{\sqrt{3}} = -\frac{2\sqrt{3}}{3}$$

6.  $-135°$ is coterminal with $-135° + 360° = 225°$. This angle lies in quadrant III. The reference angle is $225° - 180° = 45°$. Since $-135°$ is in quadrant III, the sine, cosine, secant, and cosecant are negative.

    $$\sin(-135°) = -\sin 45° = -\frac{\sqrt{2}}{2}$$
    $$\cos(-35°) = -\cos 45° = -\frac{\sqrt{2}}{2}$$
    $$\tan(-135°) = \tan 45° = 1$$
    $$\cot(-135°) = \cot 45° = 1$$
    $$\sec(-135°) = -\sec 45° = -\sqrt{2}$$
    $$\csc(-135°) = -\csc 45° = -\sqrt{2}$$

7.  990° is coterminal with $990° - 2 \cdot 360° = 270°$, which is the reference angle.
    $$\sin 990° = \sin 270° = -1$$
    $$\cos 990° = \cos 270° = 0$$
    $$\tan 990° = \tan 270° \text{ undefined}$$
    $$\cot 990° = \cot 270° = 0$$
    $$\sec 990° = \sec 270° \text{ undefined}$$
    $$\csc 990° = \csc 270° = -1$$

8.  $\cos\theta = -\dfrac{\sqrt{2}}{2}$

Since $\cos\theta$ is negative, $\theta$ must lie in quadrant II or quadrant III. The absolute value of $\cos\theta$ is $\dfrac{\sqrt{2}}{2}$, so $\theta' = 45°$. The quadrant II angle $\theta$ equals $180° - \theta' = 180° - 45° = 135°$, and the quadrant III angle $\theta$ equals $180° + \theta' = 180° + 45° = 225°$.

9.  $\csc\theta = -\dfrac{2\sqrt{3}}{3}$

Since $\csc\theta$ is negative, $\theta$ must lie in quadrant III or quadrant IV. The absolute value of $\csc\theta$ is $\dfrac{2\sqrt{3}}{3}$, so $\theta' = 60°$. The quadrant III angle $\theta$ equals $180° + \theta' = 180° + 60° = 240°$, and the quadrant IV angle $\theta$ equals $360° - \theta' = 360° - 60° = 300°$.

10.  $\tan\theta = 1 \Rightarrow \theta = 45°$ or $\theta = 225°$

11.  $\tan\theta = 1.6778490$

Since $\cot\theta = \dfrac{1}{\tan\theta} = (\tan\theta)^{-1}$, we can use division or the inverse key (multiplicative inverse).

12.  (a)  $\sin 78°21' \approx .97939940$

$78°21' = \left(78 + \frac{21}{60}\right)° = 78.35°$

(b)  $\tan 117.689° \approx -1.9056082$

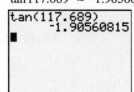

(c)  $\sec 58.9041° \approx 1.9362432$

13.  $\sin\theta = .27843196$

$\theta \approx 16.16664145°$

14.  $A = 58°30', a = 748$

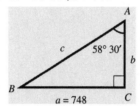

$A + B = 90° \Rightarrow B = 90° - A \Rightarrow$
$B = 90° - 58°30' = 31°30'$

$\tan A = \dfrac{a}{b} \Rightarrow \tan 58°30' = \dfrac{748}{b} \Rightarrow$

$b = \dfrac{748}{\tan 58°30'} \approx 458$ (rounded to three significant digits

$\sin A = \dfrac{a}{c} \Rightarrow \sin 58°30' = \dfrac{748}{c} \Rightarrow$

$c = \dfrac{748}{\sin 58°30'} \approx 877$ (rounded to three significant digits

15.  Let $\theta$ = the measure of the angle that the guy wire makes with the ground.

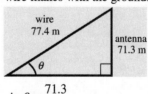

$\sin\theta = \dfrac{71.3}{77.4}$

$\theta = \sin^{-1}\left(\dfrac{71.3}{77.4}\right) \approx 67.1° \approx 67°10'$

**16.** Let $x$ = the height of the flagpole.

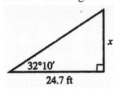

$$\tan 32°10' = \frac{x}{24.7}$$
$$x = 24.7 \tan 32°10' \approx 15.5344$$

The flagpole is approximately 15.5 ft high. (rounded to three significant digits)

**17.** Let $h$ = the height of the top of mountain above the cabin. Then $2000 + h$ = the height of the mountain.

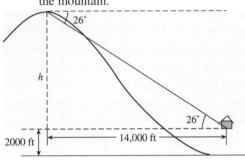

$$\tan 26° = \frac{h}{14,000} \Rightarrow h \approx 6800 \text{ (rounded to two}$$

significant digits). Thus, the height of the mountain is about $6800 + 2000 = 8800$ ft.

**18.** Let $x$ = distance the ships are apart.
In the figure, the measure of angle $CAB$ is $122° - 32° = 90°$. Therefore, triangle $CAB$ is a right triangle.

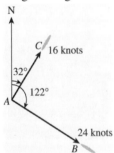

Since $d = rt$, the distance traveled by the first ship in 2.5 hr is
(2.5 hr)(16 knots) = 40 nautical mi and the second ship is
(2.5hr)(24 knots) = 60 nautical mi.
Applying the Pythagorean theorem, we have
$$x^2 = 40^2 + 60^2 \Rightarrow x^2 = 1600 + 3600 \Rightarrow$$
$$x^2 = 5200 \Rightarrow x = \sqrt{5200} \approx 72.111$$
The ships are 72 nautical mi apart. (rounded to 2 significant digits)

**19.** Draw triangle $ACB$ and extend north-south lines from points $A$ and $C$. Angle $ACD$ is $62°$ (alternate interior angles of parallel lines cut by a transversal have the same measure), so Angle $ACB$ is $62° + 28° = 90°$.

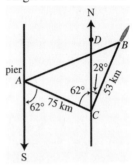

Since angle $ACB$ is a right angle, use the Pythagorean theorem to find the distance from $A$ to $B$.
$$(AB)^2 = 75^2 + 53^2 \Rightarrow (AB)^2 = 5625 + 2809 \Rightarrow$$
$$(AB)^2 = 8434 \Rightarrow AB = \sqrt{8434} \approx 91.8368$$

It is 92 km from the pier to the boat, rounded to two significant digits.

**20.** Let $x$ = the side adjacent to $52.5°$ in the smaller triangle.

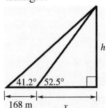

In the larger triangle, we have
$$\tan 41.2° = \frac{h}{168 + x} \Rightarrow h = (168 + x)\tan 41.2°.$$
In the smaller triangle, we have
$$\tan 52.5° = \frac{h}{x} \Rightarrow h = x \tan 52.5°.$$
Substitute for $h$ in this equation to solve for $x$.
$$(168 + x)\tan 41.2° = x \tan 52.5°$$
$$168 \tan 41.2° + x \tan 41.2° = x \tan 52.5°$$
$$168 \tan 41.2° = x \tan 52.5° - x \tan 41.2°$$
$$168 \tan 41.2° = x(\tan 52.5° - \tan 41.2°)$$
$$\frac{168 \tan 41.2°}{\tan 52.5° - \tan 41.2°} = x$$
Substituting for $x$ in the equation for the smaller triangle gives
$$h = x \tan 52.5°$$
$$h = \frac{168 \tan 41.2° \tan 52.5°}{\tan 52.5° - \tan 41.2°} \approx 448.0432$$
The height of the triangle is approximately 448 m. (rounded to three significant digits)

# Chapter 3

## Radian Measure and Circular Functions

### Section 3.1: Radian Measure

1. Since $\theta$ is in quadrant I, $0 < \theta < \dfrac{\pi}{2}$. Since $\dfrac{\pi}{2} \approx 1.57$, 1 is the only integer value in the interval. Thus, the radian measure of $\theta$ is 1 radian.

3. Since $\theta$ is in quadrant II, $\dfrac{\pi}{2} < \theta < \pi$. Since $\dfrac{\pi}{2} \approx 1.57$ and $\pi \approx 3.14$, 2 and 3 are the only integers in the interval. Since $\theta$ is closer to $\pi$, the radian measure of $\theta$ is 3 radians.

5. Since $\theta$ is an angle in quadrant III drawn in a clockwise direction, $-\pi < \theta < -\dfrac{\pi}{2}$. Also $-\pi \approx -3.14$ and $-\dfrac{\pi}{2} \approx -1.57$, so $-2$ and $-3$ are the only integers in the interval. $\theta$ is closer to $-\pi$, so the radian measure of $\theta$ is $-3$ radians.

7. $60° = 60\left(\dfrac{\pi}{180} \text{ radian}\right) = \dfrac{\pi}{3}$ radians

9. $90° = 90\left(\dfrac{\pi}{180} \text{ radian}\right) = \dfrac{\pi}{2}$ radians

11. $150° = 150\left(\dfrac{\pi}{180} \text{ radian}\right) = \dfrac{5\pi}{6}$ radians

13. $-300° = -300\left(\dfrac{\pi}{180} \text{ radian}\right) = -\dfrac{5\pi}{3}$ radians

15. $450° = 450\left(\dfrac{\pi}{180} \text{ radian}\right) = \dfrac{5\pi}{2}$ radians

17. $1800° = 1800°\left(\dfrac{\pi}{180°} \text{ radian}\right) = 10\pi$ radians

19. Multiply the degree measure by $\dfrac{\pi}{180}$ radian and reduce. Your answer will be in radians. Leave the answer as a multiple of $\pi$, unless otherwise directed.

21.–23. Answers will vary.

25. $\dfrac{\pi}{3} = \dfrac{\pi}{3}\left(\dfrac{180°}{\pi}\right) = 60°$

27. $\dfrac{7\pi}{4} = \dfrac{7\pi}{4}\left(\dfrac{180°}{\pi}\right) = 315°$

29. $\dfrac{11\pi}{6} = \dfrac{11\pi}{6}\left(\dfrac{180°}{\pi}\right) = 330°$

31. $-\dfrac{\pi}{6} = -\dfrac{\pi}{6}\left(\dfrac{180°}{\pi}\right) = -30°$

33. $\dfrac{7\pi}{10} = \dfrac{7\pi}{10}\left(\dfrac{180°}{\pi}\right) = 126°$

35. $-\dfrac{4\pi}{15} = -\dfrac{4\pi}{15}\left(\dfrac{180°}{\pi}\right) = -48°$

37. $\dfrac{17\pi}{20} = \dfrac{17\pi}{20}\left(\dfrac{180°}{\pi}\right) = 153°$

39. $-5\pi = -5\pi\left(\dfrac{180°}{\pi}\right) = -900°$

41. $39° = 39\left(\dfrac{\pi}{180} \text{ radian}\right) \approx .68$ radian

43. $42.5° = 42.5\left(\dfrac{\pi}{180} \text{ radian}\right) \approx .742$ radians

45. $139°10' = \left(139 + \tfrac{10}{60}\right)°$
$\approx 139.1666667\left(\dfrac{\pi}{180} \text{ radian}\right)$
$\approx 2.43$ radians

47. $64.29° = 64.29\left(\dfrac{\pi}{180} \text{ radian}\right) \approx 1.122$ radians

**49.** $56°25' = \left(56 + \frac{25}{60}\right)°$

$\approx 56.41666667\left(\frac{\pi}{180} \text{ radian}\right)$

$\approx .9847 \text{ radians}$

**51.** $47.6925° = 47.6925\left(\frac{\pi}{180} \text{ radian}\right)$

$\approx .832391 \text{ radian}$

**53.** $2 \text{ radians} = 2\left(\frac{180°}{\pi}\right) \approx 114.591559°$

$= 114° + (.591559 \cdot 60)'$

$\approx 114°35'$

**55.** $1.74 \text{ radians} = 1.74\left(\frac{180°}{\pi}\right) \approx 99.69465635°$

$= 99° + (.69465635 \cdot 60)'$

$\approx 99°42'$

**57.** $.3417 \text{ radian} = .3417\left(\frac{180°}{\pi}\right) \approx 19.57796786°$

$= 19° + (.57796786 \cdot 60)'$

$= 19°35'$

**59.** $-5.01095 \text{ radian} = -5.01095\left(\frac{180°}{\pi}\right)$

$\approx -287.1062864°$

$= -\left[287° + (.1062864 \cdot 60)'\right]$

$\approx -287°06'$

**61.** Without the degree symbol on the 30, it is assumed that 30 is measured in radians. Thus, the approximate value of sin 30 is $-.98803$, not $\frac{1}{2}$.

**63.** $\sin\frac{\pi}{3} = \sin\left(\frac{\pi}{3} \cdot \frac{180°}{\pi}\right) = \sin 60° = \frac{\sqrt{3}}{2}$

**65.** $\tan\frac{\pi}{4} = \tan\left(\frac{\pi}{4} \cdot \frac{180°}{\pi}\right) = \tan 45° = 1$

**67.** $\sec\frac{\pi}{6} = \sec\left(\frac{\pi}{6} \cdot \frac{180°}{\pi}\right) = \sec 30° = \frac{2\sqrt{3}}{3}$

**69.** $\sin\frac{\pi}{2} = \sin\left(\frac{\pi}{2} \cdot \frac{180°}{\pi}\right) = \sin 90° = 1$

**71.** $\tan\frac{5\pi}{3} = \tan\left(\frac{5\pi}{3} \cdot \frac{180°}{\pi}\right) = \tan 300°$

$= -\tan 60° = -\sqrt{3}$

**73.** $\sin\left(\frac{5\pi}{6}\right) = \sin\left(\frac{5\pi}{6} \cdot \frac{180°}{\pi}\right) = \sin 150°$

$= \sin 30° = \frac{1}{2}$

**75.** $\cos 3\pi = \cos\left(3\pi \cdot \frac{180°}{\pi}\right) = \cos 540°$

$= \cos(540° - 360°) = \cos 180° = -1$

**77.** $\sin\left(-\frac{8\pi}{3}\right) = \sin\left(-\frac{8\pi}{3} \cdot \frac{180°}{\pi}\right)$

$= \sin(-480°)$

$= \sin(-480° + 2 \cdot 360°)$

$= \sin 240° = -\sin 60° = -\frac{\sqrt{3}}{2}$

**79.** $\sin\left(-\frac{7\pi}{6}\right) = \sin\left(-\frac{7\pi}{6} \cdot \frac{180°}{\pi}\right)$

$= \sin(-210°)$

$= \sin(-210° + 360°)$

$= \sin 150° = \sin 30° = \frac{1}{2}$

**81.** $\tan\left(-\frac{14\pi}{3}\right) = \tan\left(-\frac{14\pi}{3} \cdot \frac{180°}{\pi}\right)$

$= \tan(-840°)$

$= \tan(-840° + 3 \cdot 360) = \tan 240°$

$= \tan 60° = \sqrt{3}$

**83.** Begin the calculation with the blank next to 30°, and then proceed counterclockwise from there.

$30° = 30\left(\frac{\pi}{180} \text{ radian}\right) = \frac{\pi}{6} \text{ radians}$

$\frac{\pi}{4} \text{ radians} = \frac{\pi}{4}\left(\frac{180°}{\pi}\right) = 45°$

$60° = 60\left(\frac{\pi}{180} \text{ radian}\right) = \frac{\pi}{3} \text{ radians}$

$\frac{2\pi}{3} \text{ radians} = \frac{2\pi}{3}\left(\frac{180°}{\pi}\right) = 120°$

$\frac{3\pi}{4} \text{ radians} = \frac{3\pi}{4}\left(\frac{180°}{\pi}\right) = 135°$

$150° = 150\left(\frac{\pi}{180} \text{ radian}\right) = \frac{5\pi}{6} \text{ radians}$

$180° = 180\left(\frac{\pi}{180} \text{ radian}\right) = \pi \text{ radians}$

$210° = 210\left(\frac{\pi}{180} \text{ radian}\right) = \frac{7\pi}{6} \text{ radians}$

$225° = 225\left(\dfrac{\pi}{180}\text{ radian}\right) = \dfrac{5\pi}{4}$ radians

$\dfrac{4\pi}{3}$ radians $= \dfrac{4\pi}{3}\left(\dfrac{180°}{\pi}\right) = 240°$

$\dfrac{5\pi}{3}$ radians $= \dfrac{5\pi}{3}\left(\dfrac{180°}{\pi}\right) = 300°$

$315° = 315\left(\dfrac{\pi}{180}\text{ radian}\right) = \dfrac{7\pi}{4}$ radians

$330° = 330\left(\dfrac{\pi}{180}\text{ radian}\right) = \dfrac{11\pi}{6}$ radians

**85. (a)** In 24 hours, the hour hand will rotate twice around the clock. One complete rotation is $2\pi$ radians, the two rotations will measure $2 \cdot 2\pi = 4\pi$ radians.

**(b)** In 4 hours, the hour heand will rotate $\dfrac{4}{12} = \dfrac{1}{3}$ of the way around the clock, which is $\dfrac{1}{3} \cdot 2\pi = \dfrac{2\pi}{3}$ radians.

**87.** In each rotation around Earth, the space vehicle would rotate $2\pi$ radians.

**(a)** In 2.5 orbits, the space vehicle travels $2.5 \cdot 2\pi = 5\pi$ radians.

**(b)** In $\dfrac{4}{3}$ of an orbit, the space vehicle travels $\dfrac{4}{3} \cdot 2\pi = \dfrac{8\pi}{3}$ radians.

## Section 3.2: Applications of Radian Measure

**Connections** (page 112)

Answers will vary. The longitude at Greenwich is 0°.

**Exercises**

**1.** $r = 4,\ \theta = \dfrac{\pi}{2}$

$s = r\theta = 4\left(\dfrac{\pi}{2}\right) = 2\pi$

**3.** $r = 16,\ \theta = \dfrac{5\pi}{4}$

$s = r\theta = 16\left(\dfrac{5\pi}{4}\right) = 20\pi$

**5.** $s = 3\pi,\ \theta = \dfrac{\pi}{2}$

$s = r\theta \Rightarrow r = \dfrac{s}{\theta} = \dfrac{3\pi}{\frac{\pi}{2}} = 3\pi \cdot \dfrac{2}{\pi} = 6$

**7.** $r = 3,\ s = 3$

$s = r\theta \Rightarrow \theta = \dfrac{s}{r} = \dfrac{3}{3} = 1$

**9.** $s = 20,\ r = 10$

$s = r\theta \Rightarrow \theta = \dfrac{s}{r} = \dfrac{20}{10} = 2$

**11.** $r = 12.3$ cm, $\theta = \dfrac{2\pi}{3}$ radians

$s = r\theta = 12.3\left(\dfrac{2\pi}{3}\right) = 8.2\pi \approx 25.8$ cm

**13.** $r = 1.38$ ft, $\theta = \dfrac{5\pi}{6}$ radians

$s = r\theta = 1.38\left(\dfrac{5\pi}{6}\right)$

$= 1.15\pi$ ft $\approx 3.61$ ft  (rounded to three significant digits)

**15.** $r = 4.82$ m, $\theta = 60°$

Converting $\theta$ to radians, we have

$\theta = 60° = 60\left(\dfrac{\pi}{180}\text{ radian}\right) = \dfrac{\pi}{3}$ radians.

Thus, the arc is

$s = r\theta = 4.82\left(\dfrac{\pi}{3}\right) = \dfrac{4.82\pi}{3} \approx 5.05$ m.

(rounded to three significant digits)

**17.** $r = 15.1$ in., $\theta = 210°$

Converting $\theta$ to radians, we have

$\theta = 210° = 210\left(\dfrac{\pi}{180}\text{ radian}\right) = \dfrac{7\pi}{6}$ radians.

Thus, the arc is

$s = r\theta = 15.1\left(\dfrac{7\pi}{6}\right) = \dfrac{105.7\pi}{6} \approx 55.3$ in.

(rounded to three significant digits)

**19.** The formula for arc length is $s = r\theta$. Substituting $2r$ for $r$ we obtain $s = (2r)\theta = 2(r\theta)$. The length of the arc is doubled.

For Exercises 21–26, note that since 6400 has two significant digits and the angles are given to the nearest degree, we can have only two significant digits in the answers.

**21.** 9° N, 40° N

$$\theta = 40° - 9° = 31° = 31\left(\frac{\pi}{180} \text{ radian}\right)$$

$$= \frac{31\pi}{180} \text{ radian}$$

$$s = r\theta = 6400\left(\frac{31\pi}{180}\right) \approx 3500 \text{ km}$$

**23.** 41° N, 12° S
12° S = −12° N

$$\theta = 41° - (-12°) = 53° = 53\left(\frac{\pi}{180} \text{ radian}\right)$$

$$= \frac{53\pi}{180} \text{ radian}$$

$$s = r\theta = 6400\left(\frac{53\pi}{180}\right) \approx 5900 \text{ km}$$

**25.** $r = 6400$ km, $s = 1200$ km

$$s = r\theta \Rightarrow 1200 = 6400\theta \Rightarrow \theta = \frac{1200}{6400} = \frac{3}{16}$$

Converting $\frac{3}{16}$ radian to degrees, we have

$$\theta = \frac{3}{16}\left(\frac{180°}{\pi}\right) \approx 11°. \text{ The north-south}$$

distance between the two cities is 11°.
Let $x$ = the latitude of Madison.
$x - 33° = 11° \Rightarrow x = 44°$ N
The latitude of Madison is 44° N.

**27.** The arc length on the smaller gear is

$$s = r\theta = 3.7\left(300 \cdot \frac{\pi}{180}\right) = 3.7\left(\frac{5\pi}{3}\right)$$

$$= \frac{18.5\pi}{3} \text{ cm}$$

An arc with this length on the larger gear corresponds to an angle measure $\theta$ where

$$s = r\theta \Rightarrow \frac{18.5\pi}{3} = 7.1\theta \Rightarrow \frac{18.5\pi}{21.3} = \theta \Rightarrow$$

$$\theta = \frac{18.5\pi}{21.3} \cdot \frac{180}{\pi} \approx 156°$$

The larger gear will rotate through approximately 156°.

**29.** A rotation of

$$\theta = 60.0\left(\frac{\pi}{180} \text{ radian}\right) = \frac{\pi}{3} \text{ radians on the}$$

smaller wheel moves through an arc length of

$$s = r\theta = 5.23\left(\frac{\pi}{3}\right) = \frac{5.23\pi}{3} \text{ cm. (holding on}$$

to more digits for the intermediate steps)
Since both wheels move together, the larger

wheel moves $\frac{5.23\pi}{3}$ 5.48 cm, which rotates it

through an angle $\theta$, where

$$\frac{5.23\pi}{3} = 8.16\theta$$

$$\theta = \frac{5.23\pi}{24.48} \text{ radian}$$

$$= \frac{5.23\pi}{24.48}\left(\frac{180°}{\pi}\right) \approx 38.5°$$

The larger wheel rotates through 38.5°.

**31.** The arc length $s$ represents the distance traveled by a point on the rim of a wheel. Since the two wheels rotate together, $s$ will be the same for both wheels.
For the smaller wheel,

$$\theta = 80° = 80\left(\frac{\pi}{180}\right) = \frac{4\pi}{9} \text{ radians and}$$

$$s = r\theta = 11.7\left(\frac{4\pi}{9}\right) = 5.2\pi \text{ cm.}$$

For the larger wheel,

$$\theta = 50° = 50\left(\frac{\pi}{180} \text{ radian}\right) = \frac{5\pi}{18} \text{ radian.}$$

Thus, we can solve

$$s = r\theta \Rightarrow 5.2\pi = r\left(\frac{5\pi}{18}\right) \Rightarrow$$

$$r = 5.2\pi \cdot \frac{18}{5\pi} = 18.72$$

The radius of the larger wheel is 18.7 cm. (rounded to 3 significant digits)

**33. (a)** The number of inches lifted is the arc length in a circle with $r = 9.27$ in. and $\theta = 71°50'$.

$$71°50' = \left(71 + \frac{50}{60}\right)\left(\frac{\pi}{180°}\right)$$

$$s = r\theta \Rightarrow$$

$$s = 9.27\left(71 + \frac{50}{60}\right)\left(\frac{\pi}{180°}\right) \approx 11.6221$$

The weight will rise 11.6 in. (rounded to three significant digits)

**(b)** When the weight is raised 6 in., we have
$$s = r\theta \Rightarrow 6 = 9.27\theta \Rightarrow$$
$$\theta = \frac{6}{9.27} \text{ radian} = \frac{6}{9.27}\left(\frac{180°}{\pi}\right)$$
$$\approx 37.0846° = 37° + .0846\,(60') \approx 37°5'$$

The pulley must be rotated through $37°5'$.

**35.** A rotation of
$$\theta = 180\left(\frac{\pi}{180} \text{ radian}\right) = \pi \text{ radians. The chain}$$

moves a distance equal to half the arc length of the larger gear. So, for the large gear and pedal, $s = r\theta \Rightarrow 4.72\pi$. Thus, the chain moves $4.72\pi$ in. The small gear rotates through an angle as follows.
$$\theta = \frac{s}{r} \Rightarrow \theta = \frac{4.72\pi}{1.38} \approx 3.42\pi$$

$\theta$ for the wheel and $\theta$ for the small gear are the same, or $3.42\pi$. So, for the wheel, we have
$$s = r\theta \Rightarrow r = 13.6\,(3.42\pi) \approx 146.12$$

The bicycle will move 146 in. (rounded to three significant digits)

**37.** Let $t$ = the length of the train.
$t$ is approximately the arc length subtended by $3°\,20'$. First convert $\theta = 3°20'$ to radians.
$$\theta = 3°20' = \left(3 + \tfrac{20}{60}\right)° = 3\tfrac{1}{3}°$$
$$= \left(3\tfrac{1}{3}\right)\left(\frac{\pi}{180} \text{ radian}\right) = \left(\frac{10}{3}\right)\left(\frac{\pi}{180} \text{ radian}\right)$$
$$= \frac{\pi}{54} \text{ radian}$$

The length of the train is
$$t = r\theta \Rightarrow t = 3.5\left(\frac{\pi}{54}\right) \approx .20 \text{ km long.}$$

(rounded to two significant digits)

**39.** $r = 6,\ s = 2\pi$
$$s = r\theta \Rightarrow 2\pi = 6\theta \Rightarrow \theta = \frac{2\pi}{6} = \frac{\pi}{3}$$
$$A = \frac{1}{2}r^2\theta \Rightarrow$$
$$A = \frac{1}{2}(6)^2\left(\frac{\pi}{3}\right) = \frac{1}{2}(36)\left(\frac{\pi}{3}\right) = 6\pi$$

**41.** $r = 12,\ s = 12\pi$
$$s = r\theta \Rightarrow 12\pi = 12\theta \Rightarrow \theta = \frac{12\pi}{12} = \pi$$
$$A = \frac{1}{2}r^2\theta \Rightarrow$$
$$A = \frac{1}{2}(12)^2\,\pi = 72\pi$$

**43.** $A = 6\pi$ sq units, $r = 6$
$$A = \frac{1}{2}r^2\theta \Rightarrow 6\pi = \frac{1}{2}(6)^2\,\theta \Rightarrow$$
$$6\pi = \frac{1}{2}(36)\theta \Rightarrow 6\pi = 18\theta \Rightarrow$$
$$\theta = \frac{6\pi}{18} = \frac{\pi}{3} \text{ radian}$$
$$\frac{\pi}{3} \text{ radian} = \frac{\pi}{3} \cdot \frac{180}{\pi} = 60°$$

The measure of the central angle is $60°$.

**45.** $A = 3$ sq units, $r = 2$
$$A = \frac{1}{2}r^2\theta \Rightarrow 3 = \frac{1}{2}(2)^2\,\theta \Rightarrow 3 = \frac{1}{2}(4)\theta \Rightarrow$$
$$3 = 2\theta \Rightarrow \theta = \frac{3}{2} = 1.5 \text{ radians}$$

In Exercises 47–53, we will be rounding to the nearest tenth.

**47.** $r = 29.2$ m, $\theta = \dfrac{5\pi}{6}$ radians
$$A = \frac{1}{2}r^2\theta \Rightarrow$$
$$A = \frac{1}{2}(29.2)^2\left(\frac{5\pi}{6}\right) = \frac{1}{2}(852.64)\left(\frac{5\pi}{6}\right)$$
$$\approx 1116.1032$$

The area of the sector is $1116.1 \text{ m}^2$. ($1120 \text{ m}^2$ rounded to three significant digits)

**49.** $r = 30.0$ ft, $\theta = \dfrac{\pi}{2}$ radians
$$A = \frac{1}{2}r^2\theta \Rightarrow$$
$$A = \frac{1}{2}(30.0)^2\left(\frac{\pi}{2}\right) = \frac{1}{2}(900)\left(\frac{\pi}{2}\right) = 225\pi$$
$$\approx 706.8583$$

The area of the sector is $706.9 \text{ ft}^2$. ($707 \text{ ft}^2$ rounded to three significant digits)

**51.** $r = 12.7$ cm, $\theta = 81°$

The formula $A = \dfrac{1}{2}r^2\theta$ requires that $\theta$ be measured in radians. Converting $81°$ to radians, we have
$$\theta = 81\left(\frac{\pi}{180} \text{ radian}\right) = \frac{9\pi}{20} \text{ radians. Since}$$
$$A = \frac{1}{2}(12.7)^2\left(\frac{9\pi}{20}\right) = \frac{1}{2}(161.29)\left(\frac{9\pi}{20}\right) \approx 114.0092,$$

the area of the sector is $114.0 \text{ cm}^2$. ($114 \text{ cm}^2$ rounded to three significant digits)

**53.** $r = 40.0$ mi, $\theta = 135°$

The formula $A = \dfrac{1}{2}r^2\theta$ requires that $\theta$ be measured in radians. Converting $135°$ to radians, we have

$$\theta = 135\left(\dfrac{\pi}{180} \text{ radian}\right) = \dfrac{3\pi}{4} \text{ radians.}$$

$$A = \dfrac{1}{2}(40.0)^2\left(\dfrac{3\pi}{4}\right) = \dfrac{1}{2}(1600)\left(\dfrac{3\pi}{4}\right)$$
$$= 600\pi \approx 1884.9556$$

The area of the sector is $1885.0$ mi$^2$.
($1880$ mi$^2$ rounded to three significant digits)

**55.** $A = 16$ in.$^2$, $r = 3.0$ in.

$$A = \dfrac{1}{2}r^2\theta \Rightarrow 16 = \dfrac{1}{2}(3)^2\,\theta \Rightarrow 16 = \dfrac{9}{2}\theta \Rightarrow$$
$$\theta = 16\cdot\dfrac{2}{9} = \dfrac{32}{9} \approx 3.6 \text{ radians}$$

(rounded to two significant digits)

**57. (a)** The central angle in degrees measures $\dfrac{360°}{27} = 13\frac{1}{3}°$. Converting to radians, we have the following.

$$13\frac{1}{3}° = \left(13\frac{1}{3}\right)\left(\dfrac{\pi}{180} \text{ radian}\right)$$
$$= \left(\dfrac{40}{3}\right)\left(\dfrac{\pi}{180} \text{ radian}\right) = \dfrac{2\pi}{27} \text{ radian}$$

**(b)** Since $C = 2\pi r$, and $r = 76$ ft, we have $C = 2\pi(76) = 152\pi \approx 477.5221$. The circumference is $478$ ft.

**(c)** Since $r = 76$ ft and $\theta = \dfrac{2\pi}{27}$, we have

$$s = r\theta = 76\left(\dfrac{2\pi}{27}\right) = \dfrac{152\pi}{27} \approx 17.6860.$$

Thus, the length of the arc is $17.7$ ft. Note: If this measurement is approximated to be $\dfrac{160}{9}$, then the approximated value would be $17.8$ ft, rounded to three significant digits.

**(d)** The area of a sector with $r = 76$ ft and $\theta = \dfrac{2\pi}{27}$ is $A = \dfrac{1}{2}r^2\theta \Rightarrow$

$$A = \dfrac{1}{2}(76^2)\dfrac{2\pi}{27} = \dfrac{1}{2}(5776)\dfrac{2\pi}{27} = \dfrac{5776\pi}{27}$$
$$\approx 672.0681 \approx 672 \text{ ft}^2$$

**59. (a)**

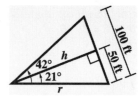

The triangle formed by the central angle and the chord is isosceles. Therefore, the bisector of the central angle is also the perpendicular bisector of the chord.

$$\sin 21° = \dfrac{50}{r} \Rightarrow r = \dfrac{50}{\sin 21°} \approx 140 \text{ ft}$$

**(b)** $r = \dfrac{50}{\sin 21°}$; $\theta = 42°$

Converting $\theta$ to radians, we have

$$42\left(\dfrac{\pi}{180} \text{ radians}\right) = \dfrac{7\pi}{30} \text{ radians. Solving}$$

for the arc length, we have
$$s = r\theta \Rightarrow$$
$$s = \dfrac{50}{\sin 21°}\cdot\dfrac{7\pi}{30} = \dfrac{35\pi}{3\sin 21°} \approx 102 \text{ ft}$$

**(c)**

The area of the portion of the circle can be found by subtracting the area of the triangle from the area of the sector. From the figure in part (a), we have

$$\tan 21° = \dfrac{50}{h} \text{ so } h = \dfrac{50}{\tan 21°}.$$

$$A_{\text{sector}} = \dfrac{1}{2}r^2\theta \Rightarrow$$
$$A_{\text{sector}} = \dfrac{1}{2}\left(\dfrac{50}{\sin 21°}\right)^2\left(\dfrac{7\pi}{30}\right) \approx 7135 \text{ ft}^2$$

and

$$A_{\text{triangle}} = \dfrac{1}{2}bh \Rightarrow$$
$$A_{\text{triangle}} = \dfrac{1}{2}(100)\left(\dfrac{50}{\tan 21°}\right) \approx 6513 \text{ ft}^2$$

The area bounded by the arc and the chord is $7135 - 6513 = 622$ ft$^2$.

**61.** Use the Pythagorean theorem to find the hypotenuse of the triangle, which is also the radius of the sector of the circle.

$$r^2 = 30^2 + 40^2 \Rightarrow r^2 = 900 + 1600 \Rightarrow$$
$$r^2 = 2500 \Rightarrow r = 50$$

The total area of the lot is the sum of the areas of the triangle and the sector.

Converting $\theta = 60°$ to radians, we have

$$60\left(\frac{\pi}{180} \text{ radian}\right) = \frac{\pi}{3} \text{ radians.}$$

$$A_{\text{triangle}} = \frac{1}{2}bh = \frac{1}{2}(30)(40) = 600 \text{ yd}^2$$

$$A_{\text{sector}} = \frac{1}{2}r^2\theta = \frac{1}{2}(50)^2\left(\frac{\pi}{3}\right)$$

$$= \frac{1}{2}(2500)\left(\frac{\pi}{3}\right) = \frac{1250\pi}{3} \text{ yd}^2$$

Total area

$$A_{\text{triangle}} + A_{\text{sector}} = 600 + \frac{1250\pi}{3} \approx 1908.9969$$

or $1900 \text{ yd}^2$, rounded to two significant digits.

**63.** Converting $\theta = 7°12' = \left(7 + \frac{12}{60}\right)° = 7.2°$ to radians, we have

$$7.2\left(\frac{\pi}{180} \text{ radian}\right) = \frac{7.2\pi}{180} = \frac{\pi}{25} \text{ radian.}$$

Solving for the radius with the arc length formula, we have

$$s = r\theta \Rightarrow 496 = r \cdot \frac{\pi}{25} \Rightarrow$$

$$r = 496 \cdot \frac{25}{\pi} = \frac{12,400}{\pi} \approx 3947.0426$$

Thus, the radius is approximately 3950 mi. (rounded to three significant digits) Using this approximate radius, we can find the circumference of the Earth. Since $C = 2\pi r$, we have $C \approx 2\pi(3950) \approx 24,800$. Thus, the approximate circumference is 24,800 mi. (rounded to three significant digits)

**65.** The base area is $A_{\text{sector}} = \frac{1}{2}r^2\theta$. Thus, the

volume is $V = \frac{1}{2}r^2\theta h$ or $V = \frac{r^2\theta h}{2}$, where $\theta$ is in radians.

**67.** $L$ is the arc length, so $L = r\theta$. Thus, $r = \frac{L}{\theta}$.

**69.** $d = r - h \Rightarrow d = r - r\cos\frac{\theta}{2} \Rightarrow$

$$d = r\left(1 - \cos\frac{\theta}{2}\right)$$

**71.** If we let $r' = 2r$, then

$$A_{\text{sector}} = \frac{1}{2}(r')^2\theta = \frac{1}{2}(2r)^2\theta \quad \text{Thus, the area,}$$

$$= \frac{1}{2}(4r^2)\theta = 4\left(\frac{1}{2}r^2\theta\right)$$

$\frac{1}{2}r^2\theta$, is quadrupled.

## Section 3.3: The Unit Circle and Circular Functions

**Connections (page 124)**

Answers will vary.

**Exercises**

**1.** An angle of $\theta = \frac{\pi}{2}$ radians intersects the unit circle at the point $(0,1)$.

   **(a)** $\sin\theta = y = 1$

   **(b)** $\cos\theta = x = 0$

   **(c)** $\tan\theta = \frac{y}{x} = \frac{1}{0}$; undefined

**3.** An angle of $\theta = 2\pi$ radians intersects the unit circle at the point $(1,0)$.

   **(a)** $\sin\theta = y = 0$

   **(b)** $\cos\theta = x = 1$

   **(c)** $\tan\theta = \frac{y}{x} = \frac{0}{1} = 0$

**5.** An angle of $\theta = -\pi$ radians intersects the unit circle at the point $(-1,0)$.

   **(a)** $\sin\theta = y = 0$

   **(b)** $\cos\theta = x = -1$

   **(c)** $\tan\theta = \frac{y}{x} = \frac{0}{-1} = 0$

For Exercises 7–21, use the following copy of Figure 12 on page 119 of the text.

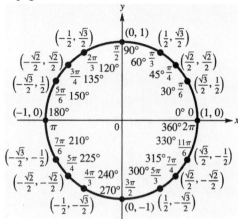

**7.** Since $\dfrac{7\pi}{6}$ is in quadrant III, the reference

angle is $\dfrac{7\pi}{6} - \pi = \dfrac{7\pi}{6} - \dfrac{6\pi}{6} = \dfrac{\pi}{6}$. In quadrant III, the sine is negative. Thus,

$\sin\dfrac{7\pi}{6} = -\sin\dfrac{\pi}{6} = -\dfrac{1}{2}$. Converting $\dfrac{7\pi}{6}$ to

degrees, we have $\dfrac{7\pi}{6} = \dfrac{7}{6}(180°) = 210°$. The

reference angle is $210° - 180° = 30°$. Thus,

$\sin\dfrac{7\pi}{6} = \sin 210° = -\sin 30° = -\dfrac{1}{2}$.

**9.** Since $\dfrac{3\pi}{4}$ is in quadrant II, the reference angle

is $\pi - \dfrac{3\pi}{4} = \dfrac{4\pi}{4} - \dfrac{3\pi}{4} = \dfrac{\pi}{4}$. In quadrant II, the

tangent is negative. Thus,

$\tan\dfrac{3\pi}{4} = -\tan\dfrac{\pi}{4} = -1.$

Converting $\dfrac{3\pi}{4}$ to degrees, we have

$\dfrac{3\pi}{4} = \dfrac{3}{4}(180°) = 135°$. The reference angle is

$180° - 135° = 45°$. Thus,

$\tan\dfrac{3\pi}{4} = \tan 135° = -\tan 45° = -1.$

**11.** Since $\dfrac{11\pi}{6}$ is in quadrant IV, the reference

angle is $2\pi - \dfrac{11\pi}{6} = \dfrac{12\pi}{6} - \dfrac{11\pi}{6} = \dfrac{\pi}{6}$. In

quadrant IV, the cosecant is negative. Thus,

$\csc\dfrac{11\pi}{6} = -\csc\dfrac{\pi}{6} = -2.$

Converting $\dfrac{11\pi}{6}$ to degrees, we have

$\dfrac{11\pi}{6} = \dfrac{11}{6}(180°) = 330°$. The reference angle

is $360° - 330° = 30°$. Thus,

$\csc\dfrac{11\pi}{6} = \csc 330° = -\csc 30° = -2.$

**13.** $-\dfrac{4\pi}{3}$ is coterminal with

$-\dfrac{4\pi}{3} + 2\pi = -\dfrac{4\pi}{3} + \dfrac{6\pi}{3} = \dfrac{2\pi}{3}$. Since $\dfrac{2\pi}{3}$ is

in quadrant II, the reference angle is

$\pi - \dfrac{2\pi}{3} = \dfrac{3\pi}{3} - \dfrac{2\pi}{3} = \dfrac{\pi}{3}$. In quadrant II, the

cosine is negative. Thus,

$\cos\left(-\dfrac{4\pi}{3}\right) = \cos\dfrac{2\pi}{3} = -\cos\dfrac{\pi}{3} = -\dfrac{1}{2}.$

Converting $\dfrac{2\pi}{3}$ to degrees, we have

$\dfrac{2\pi}{3} = \dfrac{2}{3}(180°) = 120°$. The reference angle is

$180° - 120° = 60°$. Thus,

$\cos\left(-\dfrac{4\pi}{3}\right) = \cos\dfrac{2\pi}{3} = \cos 120°$

$= -\cos 60° = -\dfrac{1}{2}$

**15.** Since $\dfrac{7\pi}{4}$ is in quadrant IV, the reference

angle is $2\pi - \dfrac{7\pi}{4} = \dfrac{8\pi}{4} - \dfrac{7\pi}{4} = \dfrac{\pi}{4}$. In

quadrant IV, the cosine is positive. Thus,

$\cos\dfrac{7\pi}{4} = \cos\dfrac{\pi}{4} = \dfrac{\sqrt{2}}{2}$. Converting $\dfrac{7\pi}{4}$ to

degrees, we have $\dfrac{7\pi}{4} = \dfrac{7}{4}(180°) = 315°$. The

reference angle is $360° - 315° = 45°$. Thus,

$\cos\dfrac{7\pi}{4} = \cos 315° = \cos 45° = \dfrac{\sqrt{2}}{2}.$

**17.** $-\dfrac{4\pi}{3}$ is coterminal with

$-\dfrac{4\pi}{3} + 2\pi = -\dfrac{4\pi}{3} + \dfrac{6\pi}{3} = \dfrac{2\pi}{3}$. Since $\dfrac{2\pi}{3}$ is

in quadrant II, the reference angle is

$\pi - \dfrac{2\pi}{3} = \dfrac{3\pi}{3} - \dfrac{2\pi}{3} = \dfrac{\pi}{3}$. In quadrant II, the

sine is positive. Thus,

$\sin\left(-\dfrac{4\pi}{3}\right) = \sin\dfrac{2\pi}{3} = \sin\dfrac{\pi}{3} = \dfrac{\sqrt{3}}{2}$.

Converting $\dfrac{2\pi}{3}$ to degrees, we have

$\dfrac{2\pi}{3} = \dfrac{2}{3}(180°) = 120°$. The reference angle is

$180° - 120° = 60°$. Thus,

$\sin\left(-\dfrac{4\pi}{3}\right) = \sin\dfrac{2\pi}{3} = \sin 120° = \sin 60° = \dfrac{\sqrt{3}}{2}$

**19.** $\dfrac{23\pi}{6}$ is coterminal with

$\dfrac{23\pi}{6} - 2\pi = \dfrac{23\pi}{6} - \dfrac{12\pi}{6} = \dfrac{11\pi}{6}$. Since $\dfrac{11\pi}{6}$ is

in quadrant IV, the reference angle is

$2\pi - \dfrac{11\pi}{6} = \dfrac{12\pi}{6} - \dfrac{11\pi}{6} = \dfrac{\pi}{6}$. In quadrant IV,

the secant is positive. Thus,

$\sec\dfrac{23\pi}{6} = \sec\dfrac{11\pi}{6} = \sec\dfrac{\pi}{6} = \dfrac{2\sqrt{3}}{3}$. Converting

$\dfrac{11\pi}{6}$ to degrees, we have

$\dfrac{11\pi}{6} = \dfrac{11}{6}(180°) = 330°$. The reference angle is

$360° - 330° = 30°$. Thus,

$\sec\dfrac{23\pi}{6} = \sec\dfrac{11\pi}{6} = \sec 330° = \sec 30° = \dfrac{2\sqrt{3}}{3}$.

**21.** Since $\dfrac{5\pi}{6}$ is in quadrant II, the reference angle

is $\pi - \dfrac{5\pi}{6} = \dfrac{6\pi}{6} - \dfrac{5\pi}{6} = \dfrac{\pi}{6}$. In quadrant II, the

tangent is negative. Thus,

$\tan\dfrac{5\pi}{6} = -\tan\dfrac{\pi}{6} = -\dfrac{\sqrt{3}}{3}$. Converting $\dfrac{5\pi}{6}$ to

degrees, we have $\dfrac{5\pi}{6} = \dfrac{5}{6}(180°) = 150°$.

The reference angle is $180° - 150° = 30°$.

Thus, $\tan\dfrac{5\pi}{6} = \tan 150° = -\tan 30° = -\dfrac{\sqrt{3}}{3}$.

For Exercises 23–34, 49–54, and 61–70, your calculator must be set in radian mode. Keystroke sequences may vary based on the type and/or model of calculator being used. As in Example 3, we will set the calculator to show four decimal digits.

**23.** $\sin .6109 \approx .5736$

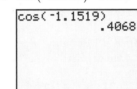

**25.** $\cos(-1.1519) \approx .4068$

**27.** $\tan 4.0203 \approx 1.2065$

**29.** $\csc(-9.4946) \approx 14.3338$

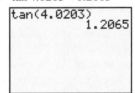

**31.** $\sec 2.8440 \approx -1.0460$

**33.** $\cot 6.0301 \approx -3.8665$

```
tan(6.0301)
           -.2586
Ans⁻¹
           -3.8665
```

**35.** $\cos .8 \approx .7$          **37.** $\sin 2 \approx .9$

**39.** $\sin 3.8 \approx -.6$

**41.** $\cos \theta = -.65 \Rightarrow x = -.65 \Rightarrow \theta \approx 2.3$ radians  or
$\theta \approx 4.0$ radians

**43.** $\sin \theta = -.7 \Rightarrow y = .7 \Rightarrow \theta \approx .8$ radians  or
$\theta \approx 2.4$ radians

**45.** $\cos 2$

$\dfrac{\pi}{2} \approx 1.57$  and  $\pi \approx 3.14$,  so  $\dfrac{\pi}{2} < 2 < \pi$. Thus,
an angle of 2 radians is in quadrant II. (The
figure for Exercises 35–38 also shows that 2
radians is in quadrant II.) Because values of
the cosine function are negative in quadrant II,
$\cos 2$ is negative.

**47.** $\sin 5$

$\dfrac{3\pi}{2} \approx 4.71$  and  $2\pi \approx 6.28$,  so  $\dfrac{3\pi}{2} < 5 < 2\pi$.
Thus, an angle of 5 radians is in quadrant IV.
(The figure for Exercises 35 – 38 also shows
that 5 radians is in quadrant IV.) Because
values of the sine function are negative in
quadrant IV, $\sin 5$ is negative.

**49.** $\tan 6.29$
$2\pi \approx 6.28$ and

$2\pi + \dfrac{\pi}{2} = \dfrac{4\pi}{2} + \dfrac{\pi}{2} = \dfrac{5\pi}{2} \approx 7.85$,  so

$2\pi < 6.29 < \dfrac{5\pi}{2}$. Notice that $2\pi$ is

coterminal with 0 and $\dfrac{5\pi}{2}$ is coterminal with

$\dfrac{\pi}{2}$. Thus, an angle of 6.29 radians is in

quadrant I. Because values of the tangent
function are positive in quadrant I, $\tan 6.29$ is
positive.

**51.** $\sin \theta = y = \dfrac{\sqrt{2}}{2}$;  $\cos \theta = x = \dfrac{\sqrt{2}}{2}$

$\tan \theta = \dfrac{y}{x} = \dfrac{\frac{\sqrt{2}}{2}}{\frac{\sqrt{2}}{2}} = 1$;  $\cot \theta = \dfrac{x}{y} = \dfrac{\frac{\sqrt{2}}{2}}{\frac{\sqrt{2}}{2}} = 1$

$\sec \theta = \dfrac{1}{x} = \dfrac{1}{\frac{\sqrt{2}}{2}} = \dfrac{2}{\sqrt{2}} = \dfrac{2}{\sqrt{2}} \cdot \dfrac{\sqrt{2}}{\sqrt{2}} = \sqrt{2}$

$\csc \theta = \dfrac{1}{y} = \dfrac{1}{\frac{\sqrt{2}}{2}} = \dfrac{2}{\sqrt{2}} = \dfrac{2}{\sqrt{2}} \cdot \dfrac{\sqrt{2}}{\sqrt{2}} = \sqrt{2}$

**53.** $\sin \theta = y = -\dfrac{12}{13}$;  $\cos \theta = x = \dfrac{5}{13}$

$\tan \theta = \dfrac{y}{x} = \dfrac{-\frac{12}{13}}{\frac{5}{13}} = -\dfrac{12}{13}\left(\dfrac{13}{5}\right) = -\dfrac{12}{5}$

$\cot \theta = \dfrac{x}{y} = \dfrac{\frac{5}{13}}{-\frac{12}{13}} = \dfrac{5}{13}\left(-\dfrac{13}{12}\right) = -\dfrac{5}{12}$

$\sec \theta = \dfrac{1}{x} = \dfrac{1}{\frac{5}{13}} = \dfrac{13}{5}$;  $\csc \theta = \dfrac{1}{y} = \dfrac{1}{-\frac{12}{13}} = -\dfrac{13}{12}$

**55.** $\tan s = .2126 \Rightarrow s \approx .2095$

```
tan⁻¹(.2126)
           .2095
```

**57.** $\sin s = .9918 \Rightarrow s \approx 1.4426$

```
sin⁻¹(.9918)
          1.4426
```

**59.** $\sec s = 1.0806 \Rightarrow s \approx .3887$

```
1/1.0806
          .9254
cos⁻¹(Ans)
          .3887
```

**61.** $\left[\dfrac{\pi}{2}, \pi\right]$; $\sin s = \dfrac{1}{2}$

Recall that $\sin \dfrac{\pi}{6} = \dfrac{1}{2}$ and in quadrant II, $\sin s$
is positive. Therefore,

$\sin\left(\pi - \dfrac{\pi}{6}\right) = \sin \dfrac{5\pi}{6} = \dfrac{1}{2}$,  so  $s = \dfrac{5\pi}{6}$.

**63.** $\left[\pi, \dfrac{3\pi}{2}\right]$; $\tan s = \sqrt{3}$

Recall that $\tan \dfrac{\pi}{3} = \sqrt{3}$ and in quadrant III,

$\tan s$ is positive. Therefore,

$\tan\left(\pi + \dfrac{\pi}{3}\right) = \tan\dfrac{4\pi}{3} = \sqrt{3}$, so $s = \dfrac{4\pi}{3}$.

**65.** $\left[\dfrac{3\pi}{2}, 2\pi\right]$; $\tan s = -1$

Recall that $\tan\dfrac{\pi}{4} = 1$ and in quadrant IV, $\tan s$

is negative. Therefore,

$\tan\left(2\pi - \dfrac{\pi}{4}\right) = \tan\dfrac{7\pi}{4} = -1$, so $s = \dfrac{7\pi}{4}$.

**67.** $s$ = the length of an arc on the unit circle = 2.5

$x = \cos s \Rightarrow x = \cos 2.5$

$y = \sin s \Rightarrow y = \sin 2.5$

```
cos(2.5)
            -.8011
sin(2.5)
             .5985
```

$(x, y) = (-.8011, .5985)$

**69.** $s = -7.4$

$x = \cos s \Rightarrow x = \cos(-7.4)$

$y = \sin s \Rightarrow y = \sin(-7.4)$

```
cos(-7.4)
             .4385
sin(-7.4)
            -.8987
```

$(x, y) = (.4385, -.8987)$

**71.** $s = 51$

```
cos(51)
             .7422
sin(51)
             .6702
```

Since cosine and sine are both positive, an angle of 51 radians lies in quadrant I.

**73.** $s = 65$

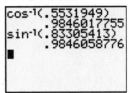

Since cosine is negative and sine is positive, an angle of 65 radians lies in quadrant II.

**75.** $(x, y) = (.55319149, .83305413) \Rightarrow$

$\cos s = .55319149$ and $\sin s = .83305413 \Rightarrow$

$s \approx .9846$

```
cos-1(.5531949)
        .9846017755
sin-1(.83305413)
        .9846058776
```

**77.** **(a)** New Orleans has latitude $L = 30°$ or .5236 radians. Since the day and time have not changed, $D = -.1425$ and $\omega = .7854$

$\sin\theta = \cos D \cos L \cos\omega + \sin D \sin L$

$\sin\theta$

$= \cos(-.1425)\cos(.5236)\cos(.7854)$

$\qquad + \sin(-.1425)\sin(.5236)$

$\theta = \sin^{-1}\big[\cos(-.1425)\cos(.5236)\cos(.7854)$

$\qquad\qquad + \sin(-.1425)\sin(.5236)\big]$

$\approx .5647$ radians or $32.4°$

**(b)** Answers will vary.

**79.** $t = 60 - 30\cos\dfrac{x\pi}{6}$

**(a)** January: $x = 0$;

$t = 60 - 30\cos\dfrac{0 \cdot \pi}{6} = 60 - 30\cos 0$

$= 60 - 30(1) = 60 - 30 = 30°$

**(b)** April: $x = 3$;

$t = 60 - 30\cos\dfrac{3 \cdot \pi}{6} = 60 - 30\cos\dfrac{\pi}{2}$

$= 60 - 30(0) = 60°$

**(c)** May: $x = 4$;

$t = 60 - 30\cos\dfrac{4 \cdot \pi}{6} = 60 - 30\cos\dfrac{2\pi}{3}$

$= 60 - 30\left(-\dfrac{1}{2}\right) = 75°$

**(d)** June $x = 5$;

$$t = 60 - 30\cos\frac{5 \cdot \pi}{6} = 60 - 30\cos\frac{5\pi}{6}$$
$$= 60 - 30\left(-\frac{\sqrt{3}}{2}\right) = 60 + 15\sqrt{3} \approx 86°$$

**(e)** August $x = 7$;

$$t = 60 - 30\cos\frac{7 \cdot \pi}{6} = 60 - 30\cos\frac{7\pi}{6}$$
$$= 60 - 30\left(-\frac{\sqrt{3}}{2}\right) = 60 + 15\sqrt{3} \approx 86°$$

**(f)** October $x = 9$;

$$t = 60 - 30\cos\frac{9 \cdot \pi}{6} = 60 - 30\cos\frac{3\pi}{2}$$
$$= 60 - 30(0) = 60°$$

**(d)** June 15 is day 166 $(31 + 28 + 31 + 30 + 31 + 15 = 166)$

$$T(166) = 37\sin\left[\frac{2\pi}{365}(166 - 101)\right] + 25$$
$$= 37\sin\left[\frac{130\pi}{365}\right] + 25 \approx 58°F$$

**(e)** September 1 is day 244 $(31 + 28 + 31 + 30 + 31 + 30 + 31 + 31 + 1 = 244)$

$$T(244) = 37\sin\left[\frac{2\pi}{365}(244 - 101)\right] + 25$$
$$= 37\sin\left[\frac{286\pi}{365}\right] + 25 \approx 48°F$$

**(f)** October 31 is day 304 $(31 + 28 + 31 + 30 + 31 + 30 + 31 + 31 + 30 + 31 = 304)$

$$T(304) = 37\sin\left[\frac{2\pi}{365}(304 - 101)\right] + 25$$
$$= 37\sin\left[\frac{406\pi}{365}\right] + 25 \approx 12°F$$

## Chapter 3: Quiz
**(Sections 3.1–3.3)**

**1.** $225° = 225\left(\frac{\pi}{180} \text{ radian}\right) = \frac{5\pi}{4}$ radians

**3.** $\frac{5\pi}{3} = \frac{5\pi}{3}\left(\frac{180°}{\pi}\right) = 300°$

**5.** $r = 300$, $s = 450$

$$s = r\theta \Rightarrow 450 = 300\theta \Rightarrow \theta = \frac{450}{300} = 1.5$$

**7.** Since $\frac{7\pi}{4}$ is in quadrant IV, the reference angle is $2\pi - \frac{7\pi}{4} = \frac{8\pi}{4} - \frac{7\pi}{4} = \frac{\pi}{4}$. In quadrant IV, the cosine is positive. Thus,

$$\cos\frac{7\pi}{4} = \cos\frac{\pi}{4} = \frac{\sqrt{2}}{2}.$$

Converting $\frac{7\pi}{4}$ to degrees, we have

$\frac{7\pi}{4} = \frac{7}{4}(180°) = 315°$. The reference angle is $180° - 315° = 45°$. Thus,

$$\cos\frac{7\pi}{4} = \cos 315° = \cos 45° = \frac{\sqrt{2}}{2}.$$

**9.** $3\pi$ is coterminal with $3\pi - 2\pi = \pi$. So $\tan 3\pi = \tan \pi = 0$
Converting $3\pi$ to degrees, we have

$3\pi\left(\frac{180}{\pi}\right) = 540°$. The reference angle is

$540° - 360° = 180°$. Thus
$\tan 3\pi = \tan 540° = \tan 180° = 0$.

## Section 3.4: Linear and Angular Speed

**1.** The circumference of the unit circle is $2\pi$.
$\omega = 1$ radian per sec, $\theta = 2\pi$ radians

$$\omega = \frac{\theta}{t} \Rightarrow 1 = \frac{2\pi}{t} \Rightarrow t = 2\pi \text{ sec}$$

**3.** $r = 20$ cm, $\omega = \frac{\pi}{12}$ radian per sec, $t = 6$ sec

**(a)** $\omega = \frac{\theta}{t} \Rightarrow \frac{\pi}{12} = \frac{\theta}{6} \Rightarrow \theta = \frac{\pi}{2}$ radians

**(b)** $s = r\theta \Rightarrow s = 20 \cdot \frac{\pi}{2} = 10\pi$ cm

**(c)** $v = \frac{r\theta}{t} \Rightarrow v = \frac{20 \cdot \frac{\pi}{2}}{6} = \frac{10\pi}{6} = \frac{5\pi}{3}$ cm per sec

**5.** $\omega = \frac{2\pi}{3}$ radians per sec, $t = 3$ sec

$$\omega = \frac{\theta}{t} \Rightarrow \frac{2\pi}{3} = \frac{\theta}{3} \Rightarrow \theta = 2\pi \text{ radians}$$

**7.** $\theta = \dfrac{3\pi}{4}$ radians, $t = 8$ sec

$\omega = \dfrac{\theta}{t} \Rightarrow \theta = \dfrac{\frac{3\pi}{4}}{8} = \dfrac{3\pi}{4} \cdot \dfrac{1}{8} = \dfrac{3\pi}{32}$ radian per sec

**9.** $\theta = \dfrac{2\pi}{9}$ radian, $\omega = \dfrac{5\pi}{27}$ radian per min

$\omega = \dfrac{\theta}{t} \Rightarrow \dfrac{5\pi}{27} = \dfrac{\frac{2\pi}{9}}{t} \Rightarrow \dfrac{5\pi}{27} = \dfrac{2\pi}{9t} \Rightarrow$

$45\pi t = 54\pi \Rightarrow t = \dfrac{54\pi}{45\pi} = \dfrac{6}{5}$ min

**11.** $\theta = 3.871142$ radians, $t = 21.4693$ sec

$\omega = \dfrac{\theta}{t}$

$\omega = \dfrac{3.871142}{21.4693} \approx .180311$ radian per sec

**13.** $r = 12$ m, $\omega = \dfrac{2\pi}{3}$ radians per sec

$v = r\omega \Rightarrow v = 12\left(\dfrac{2\pi}{3}\right) = 8\pi$ m per sec

**15.** $v = 9$ m per sec, $r = 5$ m

$v = r\omega \Rightarrow 9 = 5\omega \Rightarrow \omega = \dfrac{9}{5}$ radians per sec

**17.** $v = 107.692$ m per sec, $r = 58.7413$ m

$v = r\omega \Rightarrow 107.692 = 58.7413\omega \Rightarrow$

$\omega = \dfrac{107.692}{58.7413} \approx 1.83333$ radians per sec

**19.** $r = 6$ cm, $\omega = \dfrac{\pi}{3}$ radians per sec, $t = 9$ sec

$s = r\omega t \Rightarrow s = 6\left(\dfrac{\pi}{3}\right)(9) = 18\pi$ cm

**21.** $s = 6\pi$ cm, $r = 2$ cm, $\omega = \dfrac{\pi}{4}$ radians per sec

$s = r\omega t \Rightarrow 6\pi = 2\left(\dfrac{\pi}{4}\right)t \Rightarrow 6\pi = \left(\dfrac{\pi}{2}\right)t \Rightarrow$

$t = 6\pi\left(\dfrac{2}{\pi}\right) = 12$ sec

**23.** $s = \dfrac{3\pi}{4}$ km, $r = 2$ km, $t = 4$ sec

$s = r\omega t \Rightarrow \dfrac{3\pi}{4} = 2\omega \cdot 4 \Rightarrow$

$\dfrac{3\pi}{4} = 8\omega \Rightarrow \omega = \dfrac{3\pi}{4} \cdot \dfrac{1}{8} = \dfrac{3\pi}{32}$ radian per sec

**25.** The hour hand of a clock moves through an angle of $2\pi$ radians (one complete revolution) in 12 hours, so

$\omega = \dfrac{\theta}{t} = \dfrac{2\pi}{12} = \dfrac{\pi}{6}$ radian per hr.

**27.** The minute hand makes one revolution per hour. Each revolution is $2\pi$ radians, so we have $\omega = 2\pi(1) = 2\pi$ radians per hr . There are 60 minutes in 1 hour, so $\omega = \dfrac{2\pi}{60} = \dfrac{\pi}{30}$ radians per min.

**29.** The minute hand of a clock moves through an angle of $2\pi$ radians in 60 min, and at the tip of the minute hand, $r = 7$ cm, so we have

$v = \dfrac{r\theta}{t} \Rightarrow v = \dfrac{7(2\pi)}{60} = \dfrac{7\pi}{30}$ cm per min

**31.** The flywheel making 42 rotations per min turns through an angle $42(2\pi) = 84\pi$ radians in 1 minute with $r = 2$ m. So,

$v = \dfrac{r\theta}{t} \Rightarrow v = \dfrac{2(84\pi)}{1} = 168\pi$ m per min

**33.** At 500 rotations per min, the propeller turns through an angle of $\theta = 500(2\pi) = 1000\pi$ radians in 1 min with $r = \dfrac{3}{2} = 1.5$ m, we have

$v = \dfrac{r\theta}{t} \Rightarrow v = \dfrac{1.5(1000\pi)}{1} = 1500\pi$ m per min.

**35.** At 215 revolutions per minute, the bicycle tire is moving $215(2\pi) = 430\pi$ radians per min. This is the angular velocity $\omega$. The linear velocity of the bicycle is

$v = r\omega = 13(430\pi) = 5590\pi$ in. per min.

Convert this to miles per hour:

$v = \dfrac{5590\pi \text{ in.}}{\text{min}} \cdot \dfrac{60 \text{ min}}{\text{hr}} \cdot \dfrac{1 \text{ ft}}{12 \text{ in.}} \cdot \dfrac{1 \text{ mi}}{5280 \text{ ft}}$

$\approx 16.6$ mph

**37. (a)** $\theta = \dfrac{1}{365}(2\pi) = \dfrac{2\pi}{365}$ radian

**(b)** $\omega = \dfrac{2\pi}{365}$ radian per day

$= \dfrac{2\pi}{365} \cdot \dfrac{1}{24}$ radian per hr

$= \dfrac{\pi}{4380}$ radian per hr

(c)  $v = r\omega$

$$v = (93{,}000{,}000)\left(\frac{\pi}{4380}\right) \approx 67{,}000 \text{ mph}$$

**39.** (a) Since $s = 56$ cm of belt go around in $t = 18$ sec, the linear velocity is

$$v = \frac{s}{t} \Rightarrow v = \frac{56}{18} = \frac{28}{9} \approx 3.1 \text{ cm per sec}$$

(b) Since the 56 cm belt goes around in 18 sec, we have

$$v = r\omega \Rightarrow \frac{56}{18} = (12.96)\omega \Rightarrow$$

$$\frac{28}{9} = (12.96)\omega \Rightarrow$$

$$\omega = \frac{\frac{28}{9}}{12.96} \approx .24 \text{ radian per sec}$$

**41.**  $\omega = (152)(2\pi) = 304\pi$ radians per min

$$= \frac{304\pi}{60} \text{ radians per sec}$$

$$= \frac{76\pi}{15} \text{ radians per sec}$$

$$v = r\omega \Rightarrow 59.4 = r\left(\frac{76\pi}{15}\right) \Rightarrow$$

$$r = 59.4\left(\frac{15}{76\pi}\right) \approx 3.73 \text{ cm}$$

Expressing the velocity $v = 30$ mph in ft per sec, we have

$$v = 30 \text{ mph} = \frac{30}{3600} \text{ mi per sec}$$

$$= \frac{(30)5280}{3600} \text{ ft per sec} = 44 \text{ ft per sec}$$

$$v = \frac{s}{t} \Rightarrow 44 = \frac{400\pi}{t} \Rightarrow 44t = 400\pi \Rightarrow$$

$$t = \frac{400\pi}{44} = \frac{100\pi}{11} \approx 29 \text{ sec}$$

**43.** In one minute, the propeller makes 5000 revolutions. Each revolution is $2\pi$ radians, so we have $5000(2\pi) = 10{,}000\pi$ radians per min. There are 60 sec in a minute, so

$$\omega = \frac{10{,}000\pi}{60} = \frac{500\pi}{3} \approx 523.6 \text{ radians per sec}$$

## Chapter 3: Review Exercises

**1.** An angle of 1° is $\frac{1}{360}$ of the way around a circle. 1 radian is the measurement of an angle with its vertex at the center of a circle that intercepts an arc on the circle equal in length to the radius of the circle. Thus, 1 radian is $\frac{1}{2\pi}$ of the way around the circle. Since $\frac{1}{360} < \frac{1}{2\pi}$, the angular measurement of 1 radian is larger than 1°.

**3.** To find a coterminal angle, add or subtract multiples of $2\pi$. Three of the many possible answers are $1 + 2\pi$, $1 + 4\pi$, and $1 + 6\pi$.

**5.**  $45° = 45\left(\frac{\pi}{180}\text{ radian}\right) = \frac{\pi}{4}\text{ radians}$

**7.**  $175° = 175\left(\frac{\pi}{180}\text{ radian}\right) = \frac{35\pi}{36}\text{ radians}$

**9.**  $800° = 800\left(\frac{\pi}{180}\text{ radian}\right) = \frac{40\pi}{9}\text{ radians}$

**11.**  $\frac{5\pi}{4} = \frac{5\pi}{4}\left(\frac{180°}{\pi}\right) = 225°$

**13.**  $\frac{8\pi}{3} = \frac{8\pi}{3}\left(\frac{180°}{\pi}\right) = 480°$

**15.**  $-\frac{11\pi}{18} = -\frac{11\pi}{18}\left(\frac{180°}{\pi}\right) = -110°$

**17.** Since $\frac{15}{60} = \frac{1}{4}$ rotation, we have

$$\theta = \frac{1}{4}(2\pi) = \frac{\pi}{2}. \text{ Thus,}$$

$$s = r\theta \Rightarrow s = 2\left(\frac{\pi}{2}\right) = \pi \text{ in.}$$

**19.** Since $\theta = 3(2\pi) = 6\pi$, we have

$$s = r\theta \Rightarrow s = 2(6\pi) = 12\pi \text{ in.}$$

**21.**  $r = 15.2$ cm, $\theta = \frac{3\pi}{4}$

$$s = r\theta \Rightarrow s = 15.2\left(\frac{3\pi}{4}\right) = 11.4\pi \approx 35.8 \text{ cm}$$

**23.** $r = 8.973$ cm, $\theta = 49.06°$

First convert $\theta = 49.06\dfrac{\pi}{180°}$ to radians:

$$\theta = 49.06° = 49.06\left(\frac{\pi}{180}\right) = \frac{49.06\pi}{180}$$

$$s = r\theta \Rightarrow s = 8.973\left(\frac{49.06\pi}{180}\right) \approx 7.683 \text{ cm}$$

**25.** $r = 38.0$ m, $\theta = 21°40'$

First convert $\theta = 21°40'$ to radians:

$$\theta = 21°40' = \left(21 + \tfrac{40}{60}\right)\left(\frac{\pi}{180}\right)$$

$$= \frac{65}{3}\left(\frac{\pi}{180}\right) = \frac{13\pi}{108}$$

$$A = \frac{1}{2}r^2\theta$$

$$A = \frac{1}{2}(38.0)^2\left(\frac{13\pi}{108}\right) \approx 273 \text{ m}^2$$

**27.** The cities are at 28°N and 12°S.
12°S = −12°N, so

$$\theta = 28° - (-12°) = 40° = 40\left(\frac{\pi}{180}\right) = \frac{2\pi}{9}$$

radians

$$s = r\theta = 6400\left(\frac{2\pi}{9}\right) \approx 4500 \text{ km} \quad \text{(rounded to}$$

two significant digits)

**29.** $r = 2$, $s = 1.5$

$$s = r\theta \Rightarrow 1.5 = 2\theta \Rightarrow \theta = \frac{1.5}{2} = \frac{3}{4} \text{ radian}$$

$$A = \frac{1}{2}r^2\theta \Rightarrow$$

$$A = \frac{1}{2}(2)^2\left(\frac{3}{4}\right) = \frac{1}{2}(4)\left(\frac{3}{4}\right) = \frac{3}{2} = 1.5 \text{ sq units}$$

**31. (a)** The hour hand of a clock moves through an angle of $2\pi$ radians in 12 hours, so

$$\omega = \frac{\theta}{t} = \frac{2\pi}{12} = \frac{\pi}{6} \text{ radian per hour}$$

In two hours, the angle would be

$$2\left(\frac{\pi}{6}\right) = \frac{\pi}{3} \text{ radians.}$$

**(b)** The distance $s$ the tip of the hour hand travels during the time period from 1 o'clock to 3 o'clock is the arc length for $\theta = \dfrac{\pi}{3}$ and $r = 6$ in. Thus,

$$s = r\theta = 6\left(\frac{\pi}{3}\right) = 2\pi \text{ in.}$$

**33.** $\tan\dfrac{\pi}{3}$

Converting $\dfrac{\pi}{3}$ to degrees, we have

$$\frac{\pi}{3} = \frac{1}{3}(180°) = 60° \Rightarrow \tan\frac{\pi}{3} = \tan 60° = \sqrt{3}$$

**35.** $\sin\left(-\dfrac{5\pi}{6}\right)$

$-\dfrac{5\pi}{6}$ is coterminal with $-\dfrac{5\pi}{6} + 2\pi = \dfrac{7\pi}{6}$.

Since $\dfrac{7\pi}{6}$ is in quadrant III, the reference angle is $\dfrac{7\pi}{6} - \pi = \dfrac{\pi}{6}$. In quadrant III, the sine is negative. Thus,

$$\sin\left(-\frac{5\pi}{6}\right) = \sin\frac{7\pi}{6} = -\sin\frac{\pi}{6} = -\frac{1}{2}$$

Converting $\dfrac{5\pi}{6}$ to degrees, we have

$$\frac{7\pi}{6}\left(\frac{180°}{\pi}\right) = 210°. \text{ The reference angle is}$$

$210° - 180° = 30°.$ Thus,

$$\sin\left(-\frac{5\pi}{6}\right) = \sin\frac{7\pi}{6} = \sin 210°$$

$$= -\sin 30° = -\frac{1}{2}$$

**37.** $\csc\left(-\dfrac{11\pi}{6}\right)$

$-\dfrac{11\pi}{6}$ is coterminal with

$$-\frac{11\pi}{6} + 2\pi = -\frac{11\pi}{6} + \frac{12\pi}{6} = \frac{\pi}{6}. \text{ Since } \frac{\pi}{6} \text{ is}$$

in quadrant I, we have

$$\csc\left(-\frac{11\pi}{6}\right) = \csc\frac{\pi}{6} = 2. \text{ Converting } \frac{\pi}{6} \text{ to}$$

degrees, we have $\dfrac{\pi}{6} = \dfrac{1}{6}(180°) = 30°.$ Thus,

$$\csc\left(-\frac{11\pi}{6}\right) = \csc\frac{\pi}{6} = \csc 30° = 2.$$

**39.** Since $0 < 1 < \dfrac{\pi}{2}$, sin 1 and cos 1 are both

positive. Thus, tan 1 is positive. Also, since

$\dfrac{\pi}{2} < 2 < \pi$, sin 2 is positive and cos 2 is

negative. Thus, tan 2 is negative. Therefore,

tan 1 > tan 2.

**41.** Snce $\dfrac{\pi}{2} < 2 < \pi$, sin 2 is positive and cos 2 is

negative. Thus, sin 2 > cos 2.

**43.** sin 1.0472 ≈ .8660

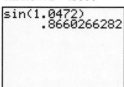

**45.** cos(−.2443) ≈ .9703

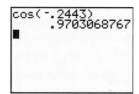

**47.** sec 7.3159 ≈ 1.9513

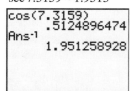

**49.** cos s = .9250 ⇒ s ≈ .3898

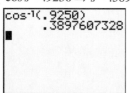

**51.** sin s = .4924 ⇒ s ≈ .5148

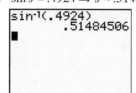

**53.** cot s = .5022 ⇒ s ≈ 1.1054

```
1/.5022
        1.99123855
tan⁻¹(Ans)
        1.105390267
```

**55.** $\left[0, \dfrac{\pi}{2}\right]$, $\cos s = \dfrac{\sqrt{2}}{2}$

Because $\cos s = \dfrac{\sqrt{2}}{2}$, the reference angle for $s$

must be $\dfrac{\pi}{4}$ since $\cos \dfrac{\pi}{4} = \dfrac{\sqrt{2}}{2}$.

For $s$ to be in the interval $\left[0, \dfrac{\pi}{2}\right]$, $s$ must be

the reference angle. Therefore, $s = \dfrac{\pi}{4}$.

**57.** $\left[\pi, \dfrac{3\pi}{2}\right]$, $\sec s = -\dfrac{2\sqrt{3}}{3}$

Because $\sec s = -\dfrac{2\sqrt{3}}{3}$, the reference angle

for $s$ must be $\dfrac{\pi}{6}$ since $\sec \dfrac{\pi}{6} = \dfrac{2\sqrt{3}}{3}$. For $s$ to

be in the interval $\left[\pi, \dfrac{3\pi}{2}\right]$, we must add the

reference angle to $\pi$. Therefore,

$s = \pi + \dfrac{\pi}{6} = \dfrac{7\pi}{6}$.

**59.** $\theta = \dfrac{5\pi}{12}$, $\omega = \dfrac{8\pi}{9}$ radians per sec

$$\omega = \frac{\theta}{t} \Rightarrow \frac{8\pi}{9} = \frac{\frac{5\pi}{12}}{t} \Rightarrow \frac{8\pi}{9} = \frac{5\pi}{12t} \Rightarrow$$

$$96\pi t = 45\pi \Rightarrow t = \frac{45\pi}{96\pi} = \frac{15}{32} \text{ sec}$$

**61.** $t = 8$ sec, $\theta = \dfrac{2\pi}{5}$ radians

$$\omega = \frac{\theta}{t} \Rightarrow \omega = \frac{\frac{2\pi}{5}}{8} = \frac{2\pi}{5}\left(\frac{1}{8}\right) = \frac{\pi}{20} \text{ radians per}$$
sec

**63.** $r = 11.46$ cm, $\omega = 4.283$ radians per sec,
$t = 5.813$ sec
$$s = r\theta = r\omega t = (11.46)(4.283)(5.813)$$
$$\approx 285.3 \text{ cm}$$

**65.** Since $t = 30$ sec and $\theta = \dfrac{5\pi}{6}$, radians we

have $\omega = \dfrac{\theta}{t} \Rightarrow \omega = \dfrac{\frac{5\pi}{6}}{30} = \dfrac{5\pi}{6} \cdot \dfrac{1}{30} = \dfrac{\pi}{36}$ radian

per sec.

**67. (a)** Because alternate interior angles of parallel lines with transversal have the same measure, the measure of angle $ABC$ is equal to the measure of angle $BCD$ (the angle of elevation). Moreover, triangle BAC is a right triangle and we can write

the relation $\sin\theta = \dfrac{h}{d}$.

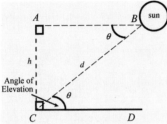

Solving for d we have,

$\sin\theta = \dfrac{h}{d} \Rightarrow d\sin\theta = h \Rightarrow$

$d = h \cdot \dfrac{1}{\sin\theta} \Rightarrow d = h\csc\theta$

**(b)** Since $d = h\csc\theta$, we substitute $2h$ for $d$ and solve.

$2h = h\csc\theta \Rightarrow 2 = \csc\theta \Rightarrow$

$\sin\theta = \dfrac{1}{2} \Rightarrow \theta = \dfrac{\pi}{6}$

$d$ is double $h$ when the sun is $\dfrac{\pi}{6}$ radians (30°) above the horizon.

**(c)** $\csc\dfrac{\pi}{2} = 1$ and $\csc\dfrac{\pi}{3} = \dfrac{2\sqrt{3}}{3} \approx 1.15$

When the sun is lower in the sky $\left(\theta = \dfrac{\pi}{3}\right)$, sunlight is filtered by more atmosphere. There is less ultraviolet light reaching the earth's surface, and therefore, there is less likelihood of becoming sunburned. In this case, sunlight passes through 15% more atmosphere.

## Chapter 3 Test

**1.** $120° = 120\left(\dfrac{\pi}{180}\text{ radian}\right) = \dfrac{2\pi}{3}$ radians

**2.** $-45° = -45\left(\dfrac{\pi}{180}\text{ radian}\right) = -\dfrac{\pi}{4}$ radian

**3.** $5° = 5\left(\dfrac{\pi}{180}\text{ radian}\right) = \dfrac{\pi}{36} \approx .09$ radian

**4.** $\dfrac{3\pi}{4} = \dfrac{3\pi}{4}\left(\dfrac{180°}{\pi}\right) = 135°$

**5.** $-\dfrac{7\pi}{6} = -\dfrac{7\pi}{6}\left(\dfrac{180°}{\pi}\right) = -210°$

**6.** $4 = 4\left(\dfrac{180°}{\pi}\right) \approx 229.18°$

**7.** $r = 150$ cm, $s = 200$ cm

**(a)** $s = r\theta \Rightarrow 200 = 150\theta \Rightarrow \theta = \dfrac{200}{150} = \dfrac{4}{3}$

**(b)** $A = \dfrac{1}{2}r^2\theta$

$A = \dfrac{1}{2}(150)^2\left(\dfrac{4}{3}\right) = \dfrac{1}{2}(22,500)\left(\dfrac{4}{3}\right)$

$= 15,000 \text{ cm}^2$

**8.** $r = \dfrac{1}{2}$ in., $s = 1$ in.

$s = r\theta \Rightarrow 1 = \dfrac{1}{2}\theta \Rightarrow \theta = 2$ radians

For Exercises 9–14, refer to Figure 12 on page 119 of the text.

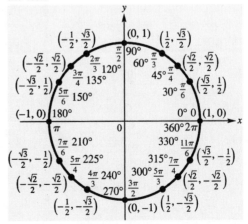

9. $\sin\dfrac{3\pi}{4}$

Since $\dfrac{3\pi}{4}$ is in quadrant II, the reference

angle is $\pi-\dfrac{3\pi}{4}=\dfrac{4\pi}{4}-\dfrac{3\pi}{4}=\dfrac{\pi}{4}$. In quadrant

II, the sine is positive. Thus,

$\sin\dfrac{3\pi}{4}=\sin\dfrac{\pi}{4}=\dfrac{\sqrt{2}}{2}$. Converting $\dfrac{3\pi}{4}$ to

degrees, we have $\dfrac{3\pi}{4}=\dfrac{3}{4}(180°)=135°$. The

reference angle is $180°-135°=45°$. Thus,

$\sin\dfrac{3\pi}{4}=\sin135°=\sin45°=\dfrac{\sqrt{2}}{2}$.

10. $\cos\left(-\dfrac{7\pi}{6}\right)$

$-\dfrac{7\pi}{6}$ is coterminal with

$-\dfrac{7\pi}{6}+2\pi=-\dfrac{7\pi}{6}+\dfrac{12\pi}{6}=\dfrac{5\pi}{6}$. Since $\dfrac{5\pi}{6}$ is

in quadrant II, the reference angle is

$\pi-\dfrac{5\pi}{6}=\dfrac{6\pi}{6}-\dfrac{5\pi}{6}=\dfrac{\pi}{6}$. In quadrant II, the

cosine is negative. Thus,

$\cos\left(-\dfrac{7\pi}{6}\right)=\cos\dfrac{5\pi}{6}=-\cos\dfrac{\pi}{6}=-\dfrac{\sqrt{3}}{2}$.

Converting $\dfrac{5\pi}{6}$ to degrees, we have

$\dfrac{5\pi}{6}=\dfrac{5}{6}(180°)=150°$.

The reference angle is $180°-150°=30°$.

Thus, $\cos\left(-\dfrac{7\pi}{6}\right)=\cos\dfrac{5\pi}{6}=\cos150°$

$=-\cos30°=-\dfrac{\sqrt{3}}{2}$

11. $\tan\dfrac{3\pi}{2}=\tan270°$ is undefined.

12. $\sec\dfrac{8\pi}{3}$

$\dfrac{8\pi}{3}$ is coterminal with $\dfrac{8\pi}{3}-2\pi=\dfrac{2\pi}{3}$.

Since $\dfrac{2\pi}{3}$ is in quadrant II, the reference

angles is $\pi-\dfrac{2\pi}{3}=\dfrac{\pi}{3}$. In quadrant II, the

secant is negative.

Thus, $\sec\dfrac{8\pi}{3}=\sec\dfrac{2\pi}{3}=-\sec\dfrac{\pi}{3}=-2$.

Converting $\dfrac{2\pi}{3}$ to degrees, we have

$\dfrac{2\pi}{3}=\dfrac{2\pi}{3}\cdot\dfrac{180°}{\pi}=120°$. The reference angle

is $180°-120°=60°$. Thus,

$\sec\dfrac{8\pi}{3}=\sec\dfrac{2\pi}{3}=\sec120°=-\sec60°=-2$.

13. $\tan\pi=\tan180°=0$

14. $\cos\dfrac{3\pi}{2}=\cos270°=0$

15. $s=\dfrac{7\pi}{6}$

Since $\dfrac{7\pi}{6}$ is in quadrant III, the reference

angle is $\dfrac{7\pi}{6}-\pi=\dfrac{\pi}{6}$. In quadrant III, the sine

and cosine are negative.

$\sin\dfrac{7\pi}{6}=-\sin\dfrac{\pi}{6}=-\dfrac{1}{2}$

$\cos\dfrac{7\pi}{6}=-\cos\dfrac{\pi}{6}=-\dfrac{\sqrt{3}}{2}$

$\tan\dfrac{7\pi}{6}=\sin\dfrac{\pi}{6}=\dfrac{\sqrt{3}}{3}$

16. For any point $(x, y)$ on the unit circle,

$\sin s=\dfrac{y}{1}=y$ and $\cos s=\dfrac{x}{1}$, both of which

are defined for all values of $x$ and $y$. Thus, the

domains of the sine and cosine functions are

both $(-\infty,\infty)$. $\tan s=\dfrac{y}{x}$ and $\sec x=\dfrac{1}{x}$, so

these functions are not defined for $x=0$. This

occurs at

$s=\ldots,-\dfrac{5\pi}{2},-\dfrac{3\pi}{2},-\dfrac{\pi}{2},\dfrac{\pi}{2},\dfrac{3\pi}{2},\dfrac{5\pi}{2},\ldots$

Therefore, the domains of the tangent and

secant functions are both

$\left\{s\,|\,s\neq(2n+1)\dfrac{\pi}{2},\text{ where }n\text{ is any integer}\right\}$.

$\cot s=\dfrac{x}{y}$ and $\csc s=\dfrac{1}{y}$ are not defined for

$y=0$. This occurs at

$s=\ldots,-3\pi,-2\pi,-\pi,0,\pi,2\pi,3\pi,\ldots$

Therefore, the domains of the cotangent and

cosecant functions are both

$\left\{s\,|\,s\neq n\pi,\text{ where }n\text{ is any integer}\right\}$

**17.** **(a)** $\sin s = .8258 \Rightarrow s \approx .9716$

```
sin-1(.8258)
              .9716
```

**(b)** Since $\cos \dfrac{\pi}{3} = \dfrac{1}{2}$ and $0 \le \dfrac{\pi}{3} \le \dfrac{\pi}{2}$,

$s = \dfrac{\pi}{3}$.

**18.** **(a)** The speed of ray $OP$ is $\omega = \dfrac{\pi}{12}$ radian

per sec. Since $\omega = \dfrac{\theta}{t}$, then in 8 sec,

$\omega = \dfrac{\theta}{t} \Rightarrow \dfrac{\pi}{12} = \dfrac{\theta}{8} \Rightarrow \theta = \dfrac{8\pi}{12} = \dfrac{2\pi}{3}$

radians

**(b)** From part (a), $P$ generates an angle of

$\dfrac{2\pi}{3}$ radians in 8 sec. The distance

traveled by $P$ along the circle is

$s = r\theta \Rightarrow s = 60 \left( \dfrac{2\pi}{3} \right) = 40\pi$ cm

**(c)** $v = \dfrac{s}{t} \Rightarrow \dfrac{40\pi}{8} = 5\pi$ cm per sec.

**19.** $r = 483{,}600{,}000$ mi
In 11.64 years, Jupiter travels $2\pi$ radians.

$\omega = \dfrac{\theta}{t} \Rightarrow \omega = \dfrac{2\pi}{11.64}$

$v = r\omega \Rightarrow v = 483{,}600{,}000 \left( \dfrac{2\pi}{11.64} \right)$

$\approx 261{,}043{,}678.2$ miles per year

Now convert this to miles per second:

$v = \left( \dfrac{261{,}043{,}678.2 \text{ mi}}{\text{yr}} \right) \left( \dfrac{1 \text{ yr}}{365 \text{ days}} \right)$

$\cdot \left( \dfrac{1 \text{ day}}{24 \text{ hr}} \right) \left( \dfrac{1 \text{ hr}}{60 \text{ min}} \right) \left( \dfrac{1 \text{ min}}{60 \text{ sec}} \right)$

$\approx 8.278$ mi per sec

**20.** **(a)** Suppose the person takes a seat at point $A$.

When the person travels $\dfrac{\pi}{2}$ radians, the

person is 50 ft above the ground. When the

person travels $\dfrac{\pi}{6}$ more radians, we can let

$x$ be the additional vertical distance
traveled:

$\sin \dfrac{\pi}{6} = \dfrac{x}{50} \Rightarrow x = 50 \sin \dfrac{\pi}{6} = 50 \left( \dfrac{1}{2} \right) = 25$

Thus, the person traveled an additional 25
ft above the ground, for a total of 75 ft
above the ground.

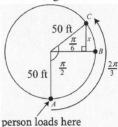

person loads here

**(b)** The Ferris wheel goes $\dfrac{2\pi}{3}$ radians per 30

sec or $\dfrac{2\pi}{90} = \dfrac{\pi}{45}$ radians per second.

# Chapter 4

## Graphs of the Circular Functions

### Section 4.1: Graphs of the Sine and Cosine Functions

**Connections (page 152)**

1. $X = -.4161468$, $Y = .90929743$. $X$ is cos 2, and $Y$ is sin 2.

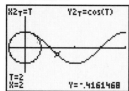 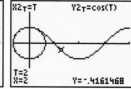

2. $X = 1.9$, $Y = .94630009$; sin 1.9 = .94630009

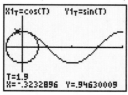

$X = 1.9$, $Y = -.3232896$; cos 1.9 = −.3232896

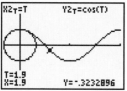

**Exercises**

1. $y = \sin x$

   The graph is a sinusoidal curve with amplitude 1 and period $2\pi$. Since sin 0 = 0, the point (0, 0) is on the graph. This matches with graph G.

3. $y = -\sin x$

   The graph is a sinusoidal curve with amplitude 1 and period $2\pi$. Bcause $a = -1$, the graph is a reflection of $y = \sin x$ in the $x$-axis. This matches with graph E.

5. $y = \sin 2x$

   The graph is a sinusoidal curve with amplitude 1 and period $\pi$. Since $\sin(2 \cdot 0) = \sin 0 = 0$, the point (0, 0) is on the graph. This matches with graph B.

7. $y = 2\sin x$

   The graph is a sinusoidal curve with amplitude 2 and period $2\pi$. Since $2\sin 0 = 2 \cdot 0 = 0$ and $2\sin \pi = 2 \cdot 1 = 2$, the points (0, 0) and $(\pi, 2)$, are on the graph. This matches with graph F.

9. $y = \sin 3x$

   The graph is a sinusoidal curve with amplitude 1 and period $\dfrac{2\pi}{3}$. Since $\sin(3 \cdot 0) = \sin 0 = 0$, the point (0, 0) is on the graph. This matches with graph D.

11. $y = 3\cos x$

    The graph is a sinusoidal curve with amplitude 3 and period $2\pi$. Since $3\cos 0 = 3 \cdot 1 = 3$, the point (0, 3) is on the graph. This matches with graph C.

13. $y = 2\cos x$

    Amplitude: $|2| = 2$

    | $x$ | 0 | $\dfrac{\pi}{2}$ | $\pi$ | $\dfrac{3\pi}{2}$ | $2\pi$ |
    |---|---|---|---|---|---|
    | cos $x$ | 1 | 0 | −1 | 0 | 1 |
    | 2 cos $x$ | 2 | 0 | −2 | 0 | 2 |

    This table gives five values for graphing one period of the function. Repeat this cycle for the interval $[-2\pi, 0]$.

    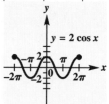

15. $y = \dfrac{2}{3}\sin x$

    Amplitude: $\left|\dfrac{2}{3}\right| = \dfrac{2}{3}$

    | $x$ | 0 | $\dfrac{\pi}{2}$ | $\pi$ | $\dfrac{3\pi}{2}$ | $2\pi$ |
    |---|---|---|---|---|---|
    | sin $x$ | 0 | 1 | 0 | −1 | 0 |
    | $\dfrac{2}{3}\sin x$ | 0 | $\dfrac{2}{3} \approx .7$ | 0 | $-\dfrac{2}{3} \approx -.7$ | 0 |

This table gives five values for graphing one period of $y = \dfrac{2}{3}\sin x$. Repeat this cycle for the interval $[-2\pi, 0]$.

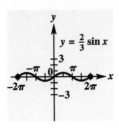

**17.**  $y = -\cos x$

Amplitude: $|-1| = 1$

| $x$ | 0 | $\dfrac{\pi}{2}$ | $\pi$ | $\dfrac{3\pi}{2}$ | $2\pi$ |
|---|---|---|---|---|---|
| $\cos x$ | 1 | 0 | $-1$ | 0 | 1 |
| $-\cos x$ | $-1$ | 0 | 1 | 0 | $-1$ |

This table gives five values for graphing one period of $y = -\cos x$. Repeat this cycle for the interval $[-2\pi, 0]$.

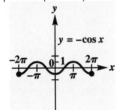

**19.**  $y = -2\sin x$

Amplitude: $|-2| = 2$

| $x$ | 0 | $\dfrac{\pi}{2}$ | $\pi$ | $\dfrac{3\pi}{2}$ | $2\pi$ |
|---|---|---|---|---|---|
| $\sin x$ | 0 | 1 | 0 | $-1$ | 0 |
| $-2\sin x$ | 0 | $-2$ | 0 | 2 | 0 |

This table gives five values for graphing one period of $y = -2\sin x$. Repeat this cycle for the interval $[-2\pi, 0]$.

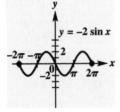

**21.**  $y = \sin(-x)$

Amplitude: 1

| $x$ | 0 | $\dfrac{\pi}{2}$ | $\pi$ | $\dfrac{3\pi}{2}$ | $2\pi$ |
|---|---|---|---|---|---|
| $-x$ | 0 | $-\dfrac{\pi}{2}$ | $-\pi$ | $-\dfrac{3\pi}{2}$ | $-2\pi$ |
| $\sin(-x)$ | 0 | $-1$ | 0 | 1 | 0 |

This table gives five values for graphing one period of $y = \sin(-x)$. Repeat this cycle for the interval $[-2\pi, 0]$.

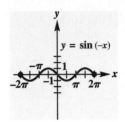

**23.**  $y = \sin\dfrac{1}{2}x$

Period: $\dfrac{2\pi}{\frac{1}{2}} = 4\pi$  and amplitude: $|1| = 1$

Divide the interval $[0, 4\pi]$ into four equal parts to get $x$-values that will yield minimum and maximum points and $x$-intercepts. Then make a table. Repeat this cycle for the interval $[-4\pi, 0]$.

| $x$ | 0 | $\pi$ | $2\pi$ | $3\pi$ | $4\pi$ |
|---|---|---|---|---|---|
| $\dfrac{1}{2}x$ | 0 | $\dfrac{\pi}{2}$ | $\pi$ | $\dfrac{3\pi}{2}$ | $2\pi$ |
| $\sin\dfrac{1}{2}x$ | 0 | 1 | 0 | $-1$ | 0 |

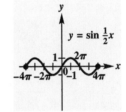

**25.**  $y = \cos\dfrac{3}{4}x$

Period: $\dfrac{2\pi}{\frac{3}{4}} = 2\pi \cdot \dfrac{4}{3} = \dfrac{8\pi}{3}$  and amplitude:

$|1| = 1$

Divide the interval $\left[0, \dfrac{8\pi}{3}\right]$ into four equal parts to get the $x$-values that will yield minimum and maximum points and $x$-intercepts. Then make a table. Repeat this cycle for the interval $\left[-\dfrac{8\pi}{3}, 0\right]$.

| $x$ | 0 | $\dfrac{2\pi}{3}$ | $\dfrac{4\pi}{3}$ | $2\pi$ | $\dfrac{8\pi}{3}$ |
|---|---|---|---|---|---|
| $\dfrac{3}{4}x$ | 0 | $\dfrac{\pi}{2}$ | $\pi$ | $\dfrac{3\pi}{2}$ | $2\pi$ |
| $\cos\dfrac{3}{4}x$ | 1 | 0 | $-1$ | 0 | 1 |

*(continued on next page)*

(*continued from page 73*)

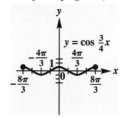

**27.** $y = \sin 3x$

Period: $\dfrac{2\pi}{3}$ and amplitude: $|1| = 1$

Divide the interval $\left[0, \dfrac{2\pi}{3}\right]$ into four equal

parts to get the $x$-values that will yield minimum and maximum points and $x$-intercepts. Then make a table. Repeat this cycle for the interval $\left[-\dfrac{2\pi}{3}, 0\right]$.

| $x$ | 0 | $\dfrac{\pi}{6}$ | $\dfrac{\pi}{3}$ | $\dfrac{\pi}{2}$ | $\dfrac{2\pi}{3}$ |
|---|---|---|---|---|---|
| $3x$ | 0 | $\dfrac{\pi}{2}$ | $\pi$ | $\dfrac{3\pi}{2}$ | $2\pi$ |
| $\sin 3x$ | 0 | 1 | 0 | $-1$ | 0 |

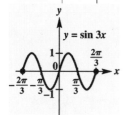

**29.** $y = 2 \sin \dfrac{1}{4} x$

Period: $\dfrac{2\pi}{\frac{1}{4}} = 2\pi \cdot \dfrac{4}{1} = 8\pi$ and amplitude:

$|2| = 2$

Divide the interval $[0, 8\pi]$ into four equal

parts to get the $x$-values that will yield minimum and maximum points and $x$-intercepts. Then make a table. Repeat this cycle for the interval $[-8\pi, 0]$.

| $x$ | 0 | $2\pi$ | $4\pi$ | $6\pi$ | $8\pi$ |
|---|---|---|---|---|---|
| $\dfrac{1}{4}x$ | 0 | $\dfrac{\pi}{2}$ | $\pi$ | $\dfrac{3\pi}{2}$ | $2\pi$ |
| $\sin \dfrac{1}{4}x$ | 0 | 1 | 0 | $-1$ | 0 |
| $2\sin \dfrac{1}{4}x$ | 0 | 2 | 0 | $-2$ | 0 |

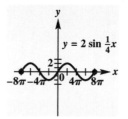

**31.** $y = -2 \cos 3x$

Period: $\dfrac{2\pi}{3}$ and amplitude: $|-2| = 2$

Divide the interval $\left[0, \dfrac{2\pi}{3}\right]$ into four equal

parts to get the $x$-values that will yield minimum and maximum points and $x$-intercepts. Then make a table. Repeat this cycle for the interval $\left[-\dfrac{2\pi}{3}, 0\right]$.

| $x$ | 0 | $\dfrac{\pi}{6}$ | $\dfrac{\pi}{3}$ | $\dfrac{\pi}{2}$ | $\dfrac{2\pi}{3}$ |
|---|---|---|---|---|---|
| $3x$ | 0 | $\dfrac{\pi}{2}$ | $\pi$ | $\dfrac{3\pi}{2}$ | $2\pi$ |
| $\cos 3x$ | 1 | 0 | $-1$ | 0 | 1 |
| $-2\cos 3x$ | $-2$ | 0 | 2 | 0 | $-2$ |

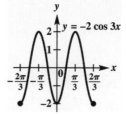

**33.** $y = \cos \pi x$

Period: $\dfrac{2\pi}{\pi} = 2$ and amplitude: $|1| = 1$

Divide the interval $[0, 2]$ into four equal parts to get the $x$-values that will yield minimum and maximum points and $x$-intercepts. Then make a table. Repeat this cycle for the interval $[-2, 0]$.

| $x$ | 0 | $\dfrac{1}{2}$ | 1 | $\dfrac{3}{2}$ | 2 |
|---|---|---|---|---|---|
| $\pi x$ | 0 | $\dfrac{\pi}{2}$ | $\pi$ | $\dfrac{3\pi}{2}$ | $2\pi$ |
| $\cos \pi x$ | 1 | 0 | $-1$ | 0 | 1 |

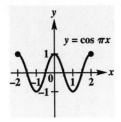

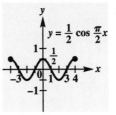

**35.**   $y = -2\sin 2\pi x$

Period: $\dfrac{2\pi}{2\pi} = 1$ and amplitude: $|-2| = 2$

Divide the interval [0, 1] into four equal parts to get the $x$-values that will yield minimum and maximum points and $x$-intercepts. Then make a table. Repeat this cycle for the interval [−1, 0].

| $x$ | 0 | $\dfrac{1}{4}$ | $\dfrac{1}{2}$ | $\dfrac{3}{4}$ | 1 |
|---|---|---|---|---|---|
| $2\pi x$ | 0 | $\dfrac{\pi}{2}$ | $\pi$ | $\dfrac{3\pi}{2}$ | $2\pi$ |
| $\sin 2\pi x$ | 0 | 1 | 0 | −1 | 0 |
| $-2\sin \pi x$ | 0 | −2 | 0 | 2 | 0 |

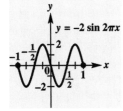

**37.**   $y = \dfrac{1}{2}\cos \dfrac{\pi}{2}x$

Period: $\dfrac{2\pi}{\frac{\pi}{2}} = 2\pi \cdot \dfrac{2}{\pi} = 4$ and amplitude:

$\left|\dfrac{1}{2}\right| = \dfrac{1}{2}$

Divide the interval [0, 4] into four equal parts to get the $x$-values that will yield minimum and maximum points and $x$-intercepts. Then make a table. Repeat this cycle for the interval [−4, 0].

| $x$ | 0 | 1 | 2 | 3 | 4 |
|---|---|---|---|---|---|
| $\dfrac{\pi}{2}x$ | 0 | $\dfrac{\pi}{2}$ | $\pi$ | $\dfrac{3\pi}{2}$ | $2\pi$ |
| $\cos \dfrac{\pi}{2}x$ | 1 | 0 | −1 | 0 | 1 |
| $\dfrac{1}{2}\cos \dfrac{\pi}{2}x$ | $\dfrac{1}{2}$ | 0 | $-\dfrac{1}{2}$ | 0 | $\dfrac{1}{2}$ |

**39.**   $y = \pi \sin \pi x$

Period: $\dfrac{2\pi}{\pi} = 2$ and amplitude: $|\pi| = \pi$

Divide the interval [0, 2] into four equal parts to get the $x$-values that will yield minimum and maximum points and $x$-intercepts. Then make a table. Repeat this cycle for the interval [−2, 0].

| $x$ | 0 | $\dfrac{1}{2}$ | 1 | $\dfrac{3}{2}$ | 2 |
|---|---|---|---|---|---|
| $\pi x$ | 0 | $\dfrac{\pi}{2}$ | $\pi$ | $\dfrac{3\pi}{2}$ | $2\pi$ |
| $\sin \pi x$ | 0 | 1 | 0 | −1 | 0 |
| $\pi \sin \pi x$ | 0 | $\pi$ | 0 | $\pi$ | 0 |

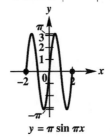

$y = \pi \sin \pi x$

**41.**   The amplitude is $\dfrac{1}{2}\left[2 - (-2)\right] = \dfrac{1}{2}(4) = 2$, so $a = 2$. One complete cycle of the graph is achieved in $\pi$ units, so the period

$\pi = \dfrac{2\pi}{b} \Rightarrow b = \dfrac{2\pi}{\pi} = 2$. Comparing the given graph with the general sine and cosine curves, we see that this graph is a cosine curve. Substituting $a = 2$ and $b = 2$, the function is $y = 2\cos 2x$. Verify by confirming minimum and maximum points and $x$-intercepts from the graph:

$(0, 2) \Rightarrow 2 = 2\cos(2 \cdot 0) = 2\cos 0 = 2 \cdot 1 = 2$

$\left(\dfrac{\pi}{4}, 0\right) \Rightarrow 0 = 2\cos\left(2 \cdot \dfrac{\pi}{4}\right) = 2\cos\dfrac{\pi}{2} = 2 \cdot 0 = 0$

$\left(\dfrac{\pi}{2}, -2\right) \Rightarrow 0 = 2\cos\left(2 \cdot \dfrac{\pi}{2}\right) = 2\cos \pi$

$\qquad\qquad = 2(-1) = -2$

*(continued on next page)*

*(continued from page 75)*

$$\left(\frac{3\pi}{4}, 0\right) \Rightarrow 0 = 2\cos\left(2 \cdot \frac{3\pi}{4}\right)$$
$$= 2\cos\frac{3\pi}{2} = 2 \cdot 0 = 0$$
$$(\pi, 2) \Rightarrow 2 = 2\cos(2\pi) = 2(1) = 2$$

**43.** The amplitude is $\frac{1}{2}\left[3 - (-3)\right] = \frac{1}{2}(6) = 3$, so $a = 3$. One-half of a cycle of the graph is achieved in $2\pi$ units, so the period is

$$2 \cdot 2\pi = 4\pi \text{ and } 4\pi = \frac{2\pi}{b} \Rightarrow b = \frac{2\pi}{4\pi} = \frac{1}{2}.$$

Comparing the given graph with the general sine and cosine curves, we see that this graph is the reflection of the cosine curve in the $x$-axis. Thus, $a = -3$. Substituting $a = -3$ and $b = \frac{1}{2}$, the function is $y = -3\cos\frac{1}{2}x$. Verify by confirming minimum and maximum points and $x$–intercepts from the graph:

$$(0, -3) \Rightarrow -3 = -3\cos\left(\frac{1}{2} \cdot 0\right) = -3\cos 0$$
$$= -3 \cdot 1 = -3$$
$$(\pi, 0) \Rightarrow 0 = -3\cos\left(\frac{1}{2} \cdot \pi\right) = -3\cos\frac{\pi}{2}$$
$$= -3(0) = 0$$
$$(2\pi, 3) \Rightarrow 3 = -3\cos\left(\frac{1}{2} \cdot 2\pi\right) = -3\cos\pi$$
$$= -3(-1) = 3$$

**45.** The amplitude is $\frac{1}{2}\left[3 - (-3)\right] = \frac{1}{2}(6) = 3$, so $a = 3$. One complete cycle of the graph is achieved in $\frac{\pi}{2}$ units, so the period

$$\frac{\pi}{2} = \frac{2\pi}{b} \Rightarrow b = 2\pi \cdot \frac{2}{\pi} = 4. \text{ Comparing the}$$

given graph with the general sine and cosine curves, we see that this graph is a sine curve. Substituting $a = 3$ and $b = 4$, the function is $y = 3\sin 4x$. Verify by confirming minimum and maximum points and $x$–intercepts from the graph:

$$(0, 0) \Rightarrow 0 = 3\sin(4 \cdot 0) = 3\sin 0 = 3 \cdot 0 = 0$$
$$\left(\frac{\pi}{8}, 3\right) \Rightarrow 3 = 3\sin\left(4 \cdot \frac{\pi}{8}\right) = 3\sin\frac{\pi}{2}$$
$$= 3(1) = 3$$
$$\left(\frac{\pi}{4}, 0\right) \Rightarrow 0 = 3\sin\left(2 \cdot \frac{\pi}{4}\right) = 3\sin\frac{\pi}{2}$$
$$= 3(0) = 0$$

$$\left(\frac{3\pi}{8}, -3\right) \Rightarrow -3 = 3\sin\left(4 \cdot \frac{3\pi}{8}\right)$$
$$= 3\sin\frac{3\pi}{2} = 3(-1) = -3$$
$$\left(\frac{\pi}{2}, 0\right) \Rightarrow 0 = 3\sin\left(2 \cdot \frac{\pi}{2}\right) = 3\sin\pi = 3(0) = 0$$

**47.** **(a)** The highest temperature is $80°$; the lowest is $50°$.

**(b)** The amplitude is
$$\frac{1}{2}(80 - 50) = \frac{1}{2}(30) = 15.$$

**(c)** The period is about 35,000 yr.

**(d)** The trend of the temperature now is downward.

**49.** **(a)** The latest time that the animals begin their evening activity is 8:00 P.M., the earliest time is 4:00 P.M. So, $4:00 \leq y \leq$ 8:00. Since there is a difference of 4 hr in these times, the amplitude is
$$\frac{1}{2}(4) = 2 \text{ hr.}$$

**(b)** The length of this period is 1 yr.

**51.** $E = 5\cos 120\pi r$

**(a)** Amplitude: $|5| = 5$ and period:
$$\frac{2\pi}{120\pi} = \frac{1}{60} \text{ sec}$$

**(b)** Since the period is $\frac{1}{60}$, one cycle is completed in $\frac{1}{60}$ sec. Therefore, in 1 sec, 60 cycles are completed.

**(c)** $t = 0$,
$$E = 5\cos 120\pi(0) = 5\cos 0 = 5(1) = 5$$
$t = .03$,
$$E = 5\cos 120\pi(.03) = 5\cos 3.6\pi \approx 1.545$$
$t = .06$,
$$E = 5\cos 120\pi(.06) = 5\cos 7.2\pi \approx -4.045$$
$t = .09$,
$$E = 5\cos 120\pi(.09)$$
$$= 5\cos 10.8\pi 0 \approx -4.045$$
$t = .12$,
$$E = 5\cos 120\pi(.12) = 5\cos 14.4\pi \approx 1.545$$

**(d)**

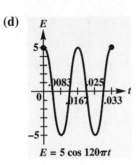

$E = 5\cos 120\pi t$

**53. (a)** The graph has a general upward trend along with small annual oscillations.

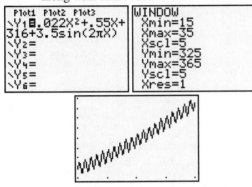

**(b)** The seasonal variations are caused by the term $3.5\sin 2\pi x$. The maximums will occur when $2\pi x = \dfrac{\pi}{2} + 2n\pi$, where $n$ is an integer. Since $x$ cannot be negative, $n$ cannot be negative. This is equivalent to

$$2\pi x = \frac{\pi}{2} + 2n\pi, \; n = 0, 1, 2,\ldots$$

$$2x = \frac{1}{2} + 2n, \; n = 0, 1, 2,\ldots$$

$$x = \frac{1}{4} + n, \; n = 0, 1, 2,\ldots$$

$$x = \frac{4n+1}{4}, \; n = 0, 1, 2,\ldots$$

$$x = \frac{1}{4}, \frac{5}{4}, \frac{9}{4},\ldots$$

Since $x$ is in years, $x = \dfrac{1}{4}$ corresponds to April when the seasonal carbon dioxide levels are maximum.

The minimums will occur when

$2\pi x = \dfrac{3\pi}{2} + 2n\pi$, where $n$ is an integer.

Since $x$ cannot be negative, $n$ cannot be negative. This is equivalent to

$$2\pi x = \frac{3\pi}{2} + 2n\pi, \; n = 0, 1, 2,\ldots$$

$$2x = \frac{3}{2} + 2n, \; n = 0, 1, 2,\ldots$$

$$x = \frac{3}{4} + n, \; n = 0, 1, 2,\ldots$$

$$x = \frac{4n+3}{4}, \; n = 0, 1, 2,\ldots$$

$$x = \frac{3}{4}, \frac{7}{4}, \frac{11}{4},\ldots$$

This is $\dfrac{1}{2}$ yr later, which corresponds to October.

**(c)** Answers will vary.

**55.** $T(x) = 37\sin\left[\dfrac{2\pi}{365}(x-101)\right] + 25$

**(a)** March 15 (day 74)

$$T(74) = 37\sin\left[\frac{2\pi}{365}(74-101)\right] + 25$$
$$\approx 8.4° \approx 8°$$

**(b)** April 5 (day 95)

$$T(95) = 37\sin\left[\frac{2\pi}{365}(95-101)\right] + 25$$
$$\approx 21.1° \approx 21°$$

**(c)** Day 200

$$T(200) = 37\sin\left[\frac{2\pi}{365}(200-101)\right] + 25$$
$$\approx 61.67° \approx 62°$$

**(d)** June 25 is day 176.
$(31 + 28 + 31 + 30 + 31 + 25 = 176)$

$$T(176) = 37\sin\left[\frac{2\pi}{365}(176-101)\right] + 25$$
$$\approx 60.56° \approx 61°$$

**(e)** October 1 is day 274.
$$31 + 28 + 31 + 30 + 31$$
$$+ 30 + 31 + 31 + 30 + 1 = 274$$

$$T(274) = 37\sin\left[\frac{2\pi}{365}(274-101)\right] + 25$$
$$\approx 31.02° \approx 31°$$

**(f)** December 31 is day 365.

$$T(365) = 37\sin\left[\frac{2\pi}{365}(365-101)\right] + 25$$
$$\approx -11.48° \approx -11°$$

**57.** The graph repeats each day, so the period is 24 hours.

**59.** On January 20, low tide was at approximately 6 P.M., with height approximately .2 ft.

**61.** On January 22, high tide was at approximately 2 A.M., with height approximately 2.6 feet.

**63.** $-1 \le y \le 1$
Amplitude: 1

Period: 8 squares $= 8(30°) = 240°$ or $\dfrac{4\pi}{3}$

**65. (a)** No, we can't say that $\sin bx = b \sin x$. If $b$ is not zero, then the period of $y = \sin bx$

is $\dfrac{2\pi}{|b|}$, and the amplitude is 1. The

period of $y = b \sin x$ is $2\pi$, and the amplitude is $|b|$.

**(b)** No, we can't say that $\cos bx = b \cos x$. If $b$ is not zero, then the period of

$y = \cos bx$ is $\dfrac{2\pi}{|b|}$, and the amplitude is

1. The period of $y = b \cos x$ is $2\pi$, and the amplitude is $|b|$.

## Section 4.2: Translations of the Graphs of the Sine and Cosine Functions

**1.** $y = \sin\left(x - \dfrac{\pi}{4}\right)$ is the graph of $y = \sin x$,

shifted to the right $\dfrac{\pi}{4}$ unit. This matches choice D.

**3.** $y = \cos\left(x - \dfrac{\pi}{4}\right)$ is the graph of $y = \cos x$,

shifted to the right $\dfrac{\pi}{4}$ unit. This matches choice H.

**5.** $y = 1 + \sin x$ is the graph of $y = \sin x$, translated vertically 1 unit up. This matches choice B.

**7.** $y = 1 + \cos x$ is the graph of $y = \cos x$, translated vertically 1 unit up. This matches choice F.

**9.** $y = \cos\left(x - \dfrac{\pi}{4}\right)$ is the graph of $y = \cos x$,

shifted to the right $\dfrac{\pi}{4}$ unit. This matches choice C.

**11.** $y = 1 + \sin x$ is the graph of $y = \sin x$, translated vertically 1 unit up. This matches choice A.

**13.** The graph of $y = \sin x + 1$ is the graph of $y = \sin x$ translated vertically 1 unit up, while the graph of $y = \sin(x + 1)$ is the graph of $y = \sin x$ shifted horizontally 1 unit left.

**15.** $y = 3 \sin(2x - 4) = 3 \sin\left[2(x - 2)\right]$

The amplitude $= |3| = 3$, period $= \dfrac{2\pi}{2} = \pi$, and phase shift $= 2$. This matches choice B.

**17.** $y = 4 \sin(3x - 2) = 4 \sin\left[3\left(x - \dfrac{2}{3}\right)\right]$

The amplitude $= |4| = 4$, period $= \dfrac{2\pi}{3}$, and

phase shift $= \dfrac{2}{3}$. This matches choice C.

**19.** If the graph of $y = \cos x$ is translated $\frac{\pi}{2}$ units horizontally to the <u>right</u>, it will coincide with the graph of $y = \sin x$.

**21.** This is a sine curve that has been shifted one unit down, so the equation is $y = -1 + \sin x$. Verify by confirming minimum and maximum points and $x$–intercepts from the graph:
$(0, -1) \Rightarrow -1 = -1 + \sin 0 = -1 + 0 = -1$

$\left(\dfrac{\pi}{2}, 0\right) \Rightarrow 0 = -1 + \sin\dfrac{\pi}{2} = -1 + 1 = 0$

$(\pi, -1) \Rightarrow -1 = -1 + \sin \pi = -1 + 0 = -1$

$\left(\dfrac{3\pi}{2}, -2\right) \Rightarrow -2 = -1 + \sin\dfrac{3\pi}{2} = -1 + (-1) = -2$

$(2\pi, -1) \Rightarrow -1 = 1 + \sin(2\pi) = -1 + 0 = -1$

**23.** The maximum is at $\left(\dfrac{\pi}{3}, 1\right)$, so this is a cosine

curve that has been shifted $\frac{\pi}{3}$ units to the right.

Thus, the equation is $y = \cos\left(x - \frac{\pi}{3}\right)$. Verify by

confirming minimum and maximum points and $x$–intercepts from the graph:

$\left(\dfrac{\pi}{3}, 1\right) \Rightarrow 1 = \cos\left(\dfrac{\pi}{3} - \dfrac{\pi}{3}\right) = \cos 0 = 1$

$\left(\dfrac{4\pi}{3}, -1\right) \Rightarrow -1 = \cos\left(\dfrac{4\pi}{3} - \dfrac{\pi}{3}\right) = \cos \pi = -1$

$\left(\dfrac{7\pi}{3}, 1\right) \Rightarrow 1 = \cos\left(\dfrac{7\pi}{3} - \dfrac{\pi}{3}\right) = \cos 2\pi = 1$

$$\left(\frac{10\pi}{3}, -1\right) \Rightarrow -1 = \cos\left(\frac{10\pi}{3} - \frac{\pi}{3}\right) = \cos 3\pi = -1$$

$$\left(\frac{13\pi}{3}, 1\right) \Rightarrow 1 = \cos\left(\frac{13\pi}{3} - \frac{\pi}{3}\right) = \cos 4\pi = 1$$

**25.**  $y = 2\sin(x - \pi)$

amplitude: $|2| = 2$; period: $\frac{2\pi}{1} = 2\pi$; There

is no vertical translation. The phase shift is $\pi$
units to the right.

**27.**  $y = 4\cos\left(\frac{1}{2}x + \frac{\pi}{2}\right) = 4\cos\frac{1}{2}\left[x - (-\pi)\right]$

amplitude: $|4| = 4$; period:

$\frac{2\pi}{\frac{1}{2}} = 2\pi \cdot \frac{2}{1} = 4\pi$; There is no vertical

translation. The phase shift is $\pi$ units to the
left.

**29.**  $y = 3\cos\frac{\pi}{2}\left(x - \frac{1}{2}\right)$

amplitude: $|3| = 3$; period: $\frac{2\pi}{\frac{\pi}{2}} = 2\pi \cdot \frac{2}{\pi} = 4$;

There is no vertical translation. The phase

shift is $\frac{1}{2}$ unit to the right.

**31.**  $y = 2 - \sin\left(3x - \frac{\pi}{5}\right) = -\sin 3\left(x - \frac{\pi}{15}\right) + 2$

amplitude: $|-1| = 1$; period: $\frac{2\pi}{3}$; The vertical

translation is 2 units up. The phase shift is $\frac{\pi}{15}$

unit to the right

**33.**  $y = \cos\left(x - \frac{\pi}{2}\right)$

*Step 1*: Find the interval whose length is $\frac{2\pi}{b}$.

$$0 \le x - \frac{\pi}{2} \le 2\pi \Rightarrow 0 + \frac{\pi}{2} \le x \le 2\pi + \frac{\pi}{2} \Rightarrow$$
$$\frac{\pi}{2} \le x \le \frac{5\pi}{2}$$

*Step 2*: Divide the period into four equal parts
to get the following *x*-values: $\frac{\pi}{2}$, $\pi$, $\frac{3\pi}{2}$,

$2\pi$, $\frac{5\pi}{2}$

*Step 3*: Evaluate the function for each of the
five *x*-values

| $x$ | $\frac{\pi}{2}$ | $\pi$ | $\frac{3\pi}{2}$ | $2\pi$ | $\frac{5\pi}{2}$ |
|---|---|---|---|---|---|
| $x - \frac{\pi}{2}$ | 0 | $\frac{\pi}{2}$ | $\pi$ | $\frac{3\pi}{2}$ | $2\pi$ |
| $\cos\left(x - \frac{\pi}{2}\right)$ | 1 | 0 | $-1$ | 0 | 1 |

*Steps 4 and 5*: Plot the points found in the
table and join them with a sinusoidal curve.
By graphing an additional period to the right,
we obtain the following graph.

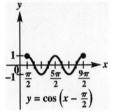

The amplitude is 1. The period is $2\pi$. There is

no vertical translation. The phase shift is $\frac{\pi}{2}$

unit to the right.

**35.**  $y = \sin\left(x + \frac{\pi}{4}\right)$

*Step 1*: Find the interval whose length is $\frac{2\pi}{b}$.

$$0 \le x + \frac{\pi}{4} \le 2\pi \Rightarrow 0 - \frac{\pi}{4} \le x \le 2\pi - \frac{\pi}{4} \Rightarrow$$
$$-\frac{\pi}{4} \le x \le \frac{7\pi}{4}$$

*Step 2*: Divide the period into four equal parts

to get the following *x*-values: $-\frac{\pi}{4}$, $\frac{\pi}{4}$,

$\frac{3\pi}{4}$, $\frac{5\pi}{4}$, $\frac{7\pi}{4}$

*Step 3*: Evaluate the function for each of the
five *x*-values

| $x$ | $-\frac{\pi}{4}$ | $\frac{\pi}{4}$ | $\frac{3\pi}{4}$ | $\frac{5\pi}{4}$ | $\frac{7\pi}{4}$ |
|---|---|---|---|---|---|
| $x + \frac{\pi}{4}$ | 0 | $\frac{\pi}{2}$ | $\pi$ | $\frac{3\pi}{2}$ | $2\pi$ |
| $\sin\left(x + \frac{\pi}{4}\right)$ | 0 | 1 | 0 | $-1$ | 0 |

*Steps 4 and 5*: Plot the points found in the
table and join them with a sinusoidal curve.
By graphing an additional period to the right,
we obtain the following graph.

(*continued on next page*)

(*continued from page 79*)

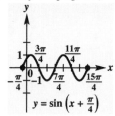

The amplitude is 1. The period is $2\pi$. There is no vertical translation. The phase shift is $\dfrac{\pi}{4}$ unit to the left.

**37.** $y = 2\cos\left(x - \dfrac{\pi}{3}\right)$

*Step 1*: Find the interval whose length is $\dfrac{2\pi}{b}$.

$$0 \le x - \dfrac{\pi}{3} \le 2\pi \Rightarrow 0 + \dfrac{\pi}{3} \le x \le 2\pi + \dfrac{\pi}{3} \Rightarrow$$
$$\dfrac{\pi}{3} \le x \le \dfrac{7\pi}{3}$$

*Step 2*: Divide the period into four equal parts to get the following *x*-values $\dfrac{\pi}{3}, \dfrac{5\pi}{6}, \dfrac{4\pi}{3},$
$\dfrac{11\pi}{6}, \dfrac{7\pi}{3}$

*Step 3*: Evaluate the function for each of the five *x*-values.

| $x$ | $\dfrac{\pi}{3}$ | $\dfrac{5\pi}{6}$ | $\dfrac{4\pi}{3}$ | $\dfrac{11\pi}{6}$ | $\dfrac{7\pi}{3}$ |
|---|---|---|---|---|---|
| $x - \dfrac{\pi}{3}$ | $0$ | $\dfrac{\pi}{2}$ | $\pi$ | $\dfrac{3\pi}{2}$ | $2\pi$ |
| $\cos\left(x - \dfrac{\pi}{3}\right)$ | $1$ | $0$ | $-1$ | $0$ | $1$ |
| $2\cos\left(x - \dfrac{\pi}{3}\right)$ | $2$ | $0$ | $-2$ | $0$ | $2$ |

*Steps 4 and 5*: Plot the points found in the table and join them with a sinusoidal curve. By graphing an additional period to the right, we obtain the following graph.

y

2

$\dfrac{4\pi}{3}$ $\dfrac{10\pi}{3}$

0

$-2$ $\dfrac{\pi}{3}$ $\dfrac{7\pi}{3}$ $\dfrac{13\pi}{3}$

$y = 2\cos\left(x - \dfrac{\pi}{3}\right)$

The amplitude is 2. The period is $2\pi$. There is no vertical translation. The phase shift is $\dfrac{\pi}{3}$ units to the right.

**39.** $y = \dfrac{3}{2}\sin 2\left(x + \dfrac{\pi}{4}\right)$

*Step 1*: Find the interval whose length is $\dfrac{2\pi}{b}$.

$$0 \le 2\left(x + \dfrac{\pi}{4}\right) \le 2\pi \Rightarrow 0 \le x + \dfrac{\pi}{4} \le \pi \Rightarrow$$
$$-\dfrac{\pi}{4} \le x \le \dfrac{3\pi}{4}$$

*Step 2*: Divide the period into four equal parts to get the following *x*-values: $-\dfrac{\pi}{4}, 0, \dfrac{\pi}{4},$
$\dfrac{\pi}{2}, \dfrac{3\pi}{4}$

*Step 3*: Evaluate the function for each of the five *x*-values

| $x$ | $-\dfrac{\pi}{4}$ | $0$ | $\dfrac{\pi}{4}$ | $\dfrac{\pi}{2}$ | $\dfrac{3\pi}{4}$ |
|---|---|---|---|---|---|
| $2\left(x + \dfrac{\pi}{4}\right)$ | $0$ | $\dfrac{\pi}{2}$ | $\pi$ | $\dfrac{3\pi}{2}$ | $2\pi$ |
| $\sin 2\left(x + \dfrac{\pi}{4}\right)$ | $0$ | $1$ | $0$ | $-1$ | $0$ |
| $\dfrac{3}{2}\sin 2\left(x + \dfrac{\pi}{4}\right)$ | $0$ | $\dfrac{3}{2}$ | $0$ | $-\dfrac{3}{2}$ | $0$ |

*Steps 4 and 5*: Plot the points found in the table and join them with a sinusoidal curve.

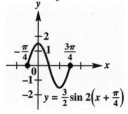

The amplitude is $\dfrac{3}{2}$. The period is $\dfrac{2\pi}{2} = \pi$. There is no vertical translation. The phase shift is $\dfrac{\pi}{4}$ unit to the left.

**41.** $y = -4\sin(2x - \pi) = -4\sin 2\left(x - \dfrac{\pi}{2}\right)$

*Step 1*: Find the interval whose length is $\dfrac{2\pi}{b}$.

$0 \le 2\left(x - \dfrac{\pi}{2}\right) \le 2\pi \Rightarrow 0 \le x - \dfrac{\pi}{2} \le \dfrac{2\pi}{2} \Rightarrow$

$0 \le x - \dfrac{\pi}{2} \le \pi \Rightarrow \dfrac{\pi}{2} \le x \le \dfrac{3\pi}{2}$

*Step 2*: Divide the period into four equal parts to get the following *x*-values: $\dfrac{\pi}{2}$, $\dfrac{3\pi}{4}$, $\pi$,

$\dfrac{5\pi}{4}$, $\dfrac{3\pi}{2}$

*Step 3*: Evaluate the function for each of the five *x*-values

| $x$ | $\dfrac{\pi}{2}$ | $\dfrac{3\pi}{4}$ | $\pi$ | $\dfrac{5\pi}{4}$ | $\dfrac{3\pi}{2}$ |
|---|---|---|---|---|---|
| $2\left(x - \dfrac{\pi}{2}\right)$ | $0$ | $\dfrac{\pi}{2}$ | $\pi$ | $\dfrac{3\pi}{2}$ | $2\pi$ |
| $\sin 2\left(x - \dfrac{\pi}{2}\right)$ | $0$ | $1$ | $0$ | $-1$ | $0$ |
| $-4\sin 2\left(x - \dfrac{\pi}{2}\right)$ | $0$ | $-4$ | $0$ | $4$ | $0$ |

*Steps 4 and 5*: Plot the points found in the table and join them with a sinusoidal curve.

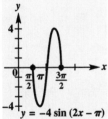

The amplitude is $|-4|$, which is 4. The period is $\dfrac{2\pi}{2}$, which is $\pi$. There is no vertical translation. The phase shift is $\dfrac{\pi}{2}$ units to the right

**43.** $y = \dfrac{1}{2}\cos\left(\dfrac{1}{2}x - \dfrac{\pi}{4}\right) = \dfrac{1}{2}\cos\dfrac{1}{2}\left(x - \dfrac{\pi}{2}\right)$

*Step 1*: Find the interval whose length is $\dfrac{2\pi}{b}$.

$0 \le \dfrac{1}{2}\left(x - \dfrac{\pi}{2}\right) \le 2\pi \Rightarrow 0 \le x - \dfrac{\pi}{2} \le 4\pi \Rightarrow$

$\dfrac{\pi}{2} \le x \le \dfrac{8\pi}{2} + \dfrac{\pi}{2} \Rightarrow \dfrac{\pi}{2} \le x \le \dfrac{9\pi}{2}$

*Step 2*: Divide the period into four equal parts to get the following *x*-values: $\dfrac{\pi}{2}$, $\dfrac{3\pi}{2}$, $\dfrac{5\pi}{2}$,

$\dfrac{7\pi}{2}$, $\dfrac{9\pi}{2}$

*Step 3*: Evaluate the function for each of the five *x*-values.

| $x$ | $\dfrac{\pi}{2}$ | $\dfrac{3\pi}{2}$ | $\dfrac{5\pi}{2}$ | $\dfrac{7\pi}{2}$ | $\dfrac{9\pi}{2}$ |
|---|---|---|---|---|---|
| $\dfrac{1}{2}\left(x - \dfrac{\pi}{2}\right)$ | $0$ | $\dfrac{\pi}{2}$ | $\pi$ | $\dfrac{3\pi}{2}$ | $2\pi$ |
| $\cos\dfrac{1}{2}\left(x - \dfrac{\pi}{2}\right)$ | $1$ | $0$ | $-1$ | $0$ | $1$ |
| $\dfrac{1}{2}\cos\dfrac{1}{2}\left(x - \dfrac{\pi}{2}\right)$ | $\dfrac{1}{2}$ | $0$ | $-\dfrac{1}{2}$ | $0$ | $\dfrac{1}{2}$ |

*Steps 4 and 5*: Plot the points found in the table and join them with a sinusoidal curve.

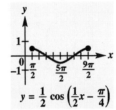

The amplitude is $\dfrac{1}{2}$. The period is $\dfrac{2\pi}{\frac{1}{2}}$,

which is $4\pi$. There is no vertical translation.

The phase shift is $\dfrac{\pi}{2}$ units to the right.

**45.** $y = -3 + 2\sin x$

*Step 1*: The period is $2\pi$.

*Step 2*: Divide the period into four equal parts to get the following *x*-values: $0$, $\dfrac{\pi}{2}$, $\pi$, $\dfrac{3\pi}{2}$,

$2\pi$

Evaluate the function for each of the five *x*-values:

| $x$ | $0$ | $\dfrac{\pi}{2}$ | $\pi$ | $\dfrac{3\pi}{2}$ | $2\pi$ |
|---|---|---|---|---|---|
| $\sin x$ | $0$ | $1$ | $0$ | $-1$ | $0$ |
| $2\sin x$ | $0$ | $2$ | $0$ | $-2$ | $0$ |
| $-3 + 2\sin x$ | $-3$ | $-1$ | $-3$ | $-5$ | $-3$ |

*Steps 4 and 5*: Plot the points found in the table and join them with a sinusoidal curve. By graphing an additional period to the left, we obtain the following graph.

(*continued on next page*)

*(continued from page 81)*

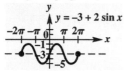

The amplitude is 2. The vertical translation is 3 units down. There is no phase shift.

**47.** $y = -1 - 2\cos 5x$

*Step 1*: Find the interval whose length is $\frac{2\pi}{b}$.

$$0 \le 5x \le 2\pi \Rightarrow 0 \le x \le \frac{2\pi}{5}$$

*Step 2*: Divide the period into four equal parts to get the following $x$-values: $0, \frac{\pi}{10}, \frac{\pi}{5}, \frac{3\pi}{10}, \frac{2\pi}{5}$

*Step 3*: Evaluate the function for each of the five $x$-values.

| $x$ | $0$ | $\frac{\pi}{10}$ | $\frac{\pi}{5}$ | $\frac{3\pi}{10}$ | $\frac{2\pi}{5}$ |
|---|---|---|---|---|---|
| $5x$ | $0$ | $\frac{\pi}{2}$ | $\pi$ | $\frac{3\pi}{2}$ | $2\pi$ |
| $\cos 5x$ | $1$ | $0$ | $-1$ | $0$ | $1$ |
| $-2\cos 5x$ | $-2$ | $0$ | $2$ | $0$ | $-2$ |
| $-1-2\cos 5x$ | $-3$ | $-1$ | $1$ | $-1$ | $-3$ |

*Steps 4 and 5*: Plot the points found in the table and join them with a sinusoidal curve. By graphing an additional period to the left, we obtain the following graph.

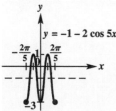

The period is $\frac{2\pi}{5}$. The amplitude is $|-2|$, which is 2. The vertical translation is 1 unit down. There is no phase shift.

**49.** $y = 1 - 2\cos\frac{1}{2}x$

*Step 1*: Find the interval whose length is $\frac{2\pi}{b}$.

$$0 \le \frac{1}{2}x \le 2\pi \Rightarrow 0 \le x \le 4\pi$$

*Step 2*: Divide the period into four equal parts to get the following $x$-values: $0, \pi, 2\pi, 3\pi, 4\pi$

*Step 3*: Evaluate the function for each of the five $x$-values.

| $x$ | $0$ | $\pi$ | $2\pi$ | $3\pi$ | $4\pi$ |
|---|---|---|---|---|---|
| $\frac{1}{2}x$ | $0$ | $\frac{\pi}{2}$ | $\pi$ | $\frac{3\pi}{2}$ | $2\pi$ |
| $\cos\frac{1}{2}x$ | $1$ | $0$ | $-1$ | $0$ | $1$ |
| $-2\cos\frac{1}{2}x$ | $-2$ | $0$ | $2$ | $0$ | $-2$ |
| $1-2\cos\frac{1}{2}x$ | $-1$ | $1$ | $3$ | $1$ | $-1$ |

*Steps 4 and 5*: Plot the points found in the table and join them with a sinusoidal curve. By graphing an additional period to the left, we obtain the following graph.

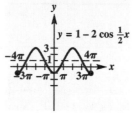

The amplitude is $|-2|$, which is 2. The period is $\frac{2\pi}{\frac{1}{2}}$, which is $4\pi$. The vertical translation is 1 unit up. There is no phase shift.

**51.** $y = -2 + \frac{1}{2}\sin 3x$

*Step 1*: Find the interval whose length is $\frac{2\pi}{b}$.

$$0 \le 3x \le 2\pi \Rightarrow 0 \le x \le \frac{2\pi}{3}$$

*Step 2*: Divide the period into four equal parts to get the following $x$-values: $0, \frac{\pi}{6}, \frac{\pi}{3}, \frac{\pi}{2}, \frac{2\pi}{3}$

*Step 3*: Evaluate the function for each of the five *x*-values.

| $x$ | 0 | $\dfrac{\pi}{6}$ | $\dfrac{\pi}{3}$ | $\dfrac{\pi}{2}$ | $\dfrac{2\pi}{3}$ |
|---|---|---|---|---|---|
| $3x$ | 0 | $\dfrac{\pi}{2}$ | $\pi$ | $\dfrac{3\pi}{2}$ | $2\pi$ |
| $\sin 3x$ | 0 | 1 | 0 | $-1$ | 0 |
| $\dfrac{1}{2}\sin 3x$ | 0 | $\dfrac{1}{2}$ | 0 | $-\dfrac{1}{2}$ | 0 |
| $-2+\dfrac{1}{2}\sin 3x$ | $-2$ | $-\dfrac{3}{2}$ | $-2$ | $-\dfrac{5}{2}$ | $-2$ |

*Steps 4 and 5*: Plot the points found in the table and join them with a sinusoidal curve. By graphing an additional period to the left, we obtain the following graph.

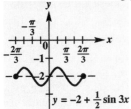

$$y = -2 + \tfrac{1}{2}\sin 3x$$

The amplitude is $\left|\dfrac{1}{2}\right| = \dfrac{1}{2}$. The period is $\dfrac{2\pi}{3}$.

The vertical translation is 2 units down. There is no phase shift.

**53.**   $y = -3 + 2\sin\left(x + \dfrac{\pi}{2}\right)$

*Step 1*: Find the interval whose length is $\dfrac{2\pi}{b}$.

$$0 \le x + \frac{\pi}{2} \le 2\pi \Rightarrow 0 - \frac{\pi}{2} \le x \le 2\pi - \frac{\pi}{2} \Rightarrow$$
$$-\frac{\pi}{2} \le x \le \frac{3\pi}{2}$$

*Step 2*: Divide the period into four equal parts to get the following *x*-values: $-\dfrac{\pi}{2}$, $0$, $\dfrac{\pi}{2}$, $\pi$,

$\dfrac{3\pi}{2}$

*Step 3*: Evaluate the function for each of the five *x*-values

| $x$ | $-\dfrac{\pi}{2}$ | 0 | $\dfrac{\pi}{2}$ | $\pi$ | $\dfrac{3\pi}{2}$ |
|---|---|---|---|---|---|
| $x+\dfrac{\pi}{2}$ | 0 | $\dfrac{\pi}{2}$ | $\pi$ | $\dfrac{3\pi}{2}$ | $2\pi$ |
| $\sin\left(x+\dfrac{\pi}{2}\right)$ | 0 | 1 | 0 | $-1$ | 0 |

| $x$ | $-\dfrac{\pi}{2}$ | 0 | $\dfrac{\pi}{2}$ | $\pi$ | $\dfrac{3\pi}{2}$ |
|---|---|---|---|---|---|
| $2\sin\left(x+\dfrac{\pi}{2}\right)$ | 0 | 2 | 0 | $-2$ | 0 |
| $-3+2\sin\left(x+\dfrac{\pi}{2}\right)$ | $-3$ | $-1$ | $-3$ | $-5$ | $-3$ |

*Steps 4 and 5*: Plot the points found in the table and join them with a sinusoidal curve.

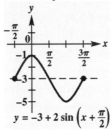

$$y = -3 + 2\sin\left(x+\tfrac{\pi}{2}\right)$$

The amplitude is $\left|2\right|$, which is 2. The period is $2\pi$. The vertical translation is 3 units down.

The phase shift is $\dfrac{\pi}{2}$ units to the left.

**55.**   $y = \dfrac{1}{2} + \sin 2\left(x + \dfrac{\pi}{4}\right)$

*Step 1*: Find the interval whose length is $\dfrac{2\pi}{b}$.

$$0 \le 2\left(x+\frac{\pi}{4}\right) \le 2\pi \Rightarrow 0 \le x + \frac{\pi}{4} \le \frac{2\pi}{2} \Rightarrow$$
$$0 \le x + \frac{\pi}{4} \le \pi \Rightarrow -\frac{\pi}{4} \le x \le \frac{3\pi}{4}$$

*Step 2*: Divide the period into four equal parts to get the following *x*-values: $-\dfrac{\pi}{4}$, $0$, $\dfrac{\pi}{4}$,

$\dfrac{\pi}{2}$, $\dfrac{3\pi}{4}$

*Step 3*: Evaluate the function for each of the five *x*-values.

| $x$ | $-\dfrac{\pi}{4}$ | 0 | $\dfrac{\pi}{4}$ | $\dfrac{\pi}{2}$ | $\dfrac{3\pi}{4}$ |
|---|---|---|---|---|---|
| $2\left(x+\dfrac{\pi}{4}\right)$ | 0 | $\dfrac{\pi}{2}$ | $\pi$ | $\dfrac{3\pi}{2}$ | $2\pi$ |
| $\sin 2\left(x+\dfrac{\pi}{4}\right)$ | 0 | 1 | 0 | $-1$ | 0 |
| $\dfrac{1}{2}+\sin 2\left(x+\dfrac{\pi}{4}\right)$ | $\dfrac{1}{2}$ | $\dfrac{3}{2}$ | $\dfrac{1}{2}$ | $-\dfrac{1}{2}$ | $\dfrac{1}{2}$ |

*(continued on next page)*

(*continued from page 83*)

*Steps 4 and 5*: Plot the points found in the table and join them with a sinusoidal curve.

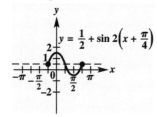

The amplitude is $|1|$, which is 1. The period is $\dfrac{2\pi}{2}$, which is $\pi$. The vertical translation is $\dfrac{1}{2}$ unit up. The phase shift is $\dfrac{\pi}{4}$ units to the left.

**57. (a)** Let January correspond to $x = 1$, February to $x = 2, \dots$, and December of the second year to $x = 24$. Yes, the data appear to outline the graph of a translated sine graph.

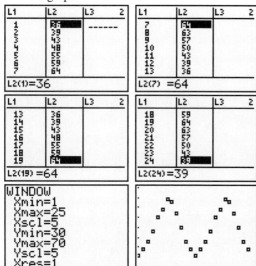

**(b)** The sine graph is vertically centered around the line $y = 50$. This line represents the average yearly temperature in Vancouver of 50°F. (This is also the actual average yearly temperature.)

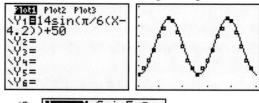

**(c)** The amplitude of the sine graph is approximately 14 since the average monthly high is 64, the average monthly low is 36, and $\dfrac{1}{2}(64 - 36) = \dfrac{1}{2}(28) = 14$.

The period is 12 since the temperature cycles every twelve months. Let $b = \dfrac{2\pi}{12} = \dfrac{\pi}{6}$. One way to determine the phase shift is to use the following technique. The minimum temperature occurs in January. Thus, when $x = 1$,

$b(x - d)$ must equal $\left(-\dfrac{\pi}{2}\right) + 2\pi n$, where

$n$ is an integer, since the sine function is minimum at these values. Solving for $d$, we have

$$\dfrac{\pi}{6}(1 - d) = -\dfrac{\pi}{2} \Rightarrow 1 - d = \dfrac{6}{\pi}\left(-\dfrac{\pi}{2}\right) \Rightarrow$$
$$1 - d = -3 \Rightarrow -d = -4 \Rightarrow d = 4$$

This can be used as a first approximation. Trial and error with a calculator leads to $d = 4.2$

**(d)** Let $f(x) = a \sin b(x - d) + c$. Since the amplitude is 14, let $a = 14$. The period is equal to 1 yr or 12 mo, so $b = \dfrac{\pi}{6}$. The average of the maximum and minimum temperatures is

$\dfrac{1}{2}(64 + 36) = \dfrac{1}{2}(100) = 50$. Thus,

$$f(x) = 14 \sin\left[\dfrac{\pi}{6}(x - 4.2)\right] + 50$$

**(e)** Plotting the data with

$$f(x) = 14 \sin\left[\dfrac{\pi}{6}(x - 4.2)\right] + 50 \text{ on the}$$

same coordinate axes gives a good fit.

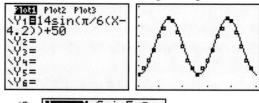

**(f)**

From the sine regression we have

$$y = 13.21\sin(.52x - 2.18) + 49.68$$
$$= 13.21\sin\left[.52(x - 4.19)\right] + 49.68$$

# Chapter 4 Quiz

### (Section 4.1–4.2)

**1.** $y = -4\sin x$

Amplitude: $|-4| = 4$

| $x$ | 0 | $\dfrac{\pi}{2}$ | $\pi$ | $\dfrac{3\pi}{2}$ | $2\pi$ |
|---|---|---|---|---|---|
| $\sin x$ | 0 | 1 | 0 | −1 | 0 |
| $-4\sin x$ | 0 | −4 | 0 | 4 | 0 |

This table gives five values for graphing one period of $y = -4\sin x$. Repeat this cycle for the interval $[-2\pi, 0]$.

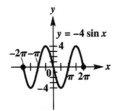

**3.** $y = 3\sin \pi x$

Period: $\dfrac{2\pi}{\pi} = 2$ and amplitude: $|3| = 3$

Divide the interval [0, 2] into four equal parts to get the $x$-values that will yield minimum and maximum points and $x$-intercepts. Make a table. Repeat this cycle for the interval [−2, 0].

| $x$ | 0 | $\dfrac{1}{2}$ | 1 | $\dfrac{3}{2}$ | 2 |
|---|---|---|---|---|---|
| $\pi x$ | 0 | $\dfrac{\pi}{2}$ | $\pi$ | $\dfrac{3\pi}{2}$ | $2\pi$ |
| $\sin \pi x$ | 0 | 1 | 0 | −1 | 0 |
| $3\sin \pi x$ | 0 | 3 | 0 | −3 | 0 |

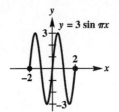

**5.** $y = 2 + \sin(2x - \pi) = 2 + \sin 2\left(x - \dfrac{\pi}{2}\right)$

*Step 1*: Find the interval whose length is $\dfrac{2\pi}{b}$.

$$0 \le 2\left(x - \dfrac{\pi}{2}\right) \le 2\pi \Rightarrow 0 \le x - \dfrac{\pi}{2} \le \dfrac{2\pi}{2} \Rightarrow$$

$$0 \le x - \dfrac{\pi}{2} \le \pi \Rightarrow \dfrac{\pi}{2} \le x \le \dfrac{3\pi}{2}$$

*Step 2*: Divide the period into four equal parts to get the following $x$-values: $\dfrac{\pi}{2}$, $\dfrac{3\pi}{4}$, $\pi$,

$\dfrac{5\pi}{4}$, $\dfrac{3\pi}{2}$

*Step 3*: Evaluate the function for each of the five $x$-values

| $x$ | $\dfrac{\pi}{2}$ | $\dfrac{3\pi}{4}$ | $\pi$ | $\dfrac{5\pi}{4}$ | $\dfrac{3\pi}{2}$ |
|---|---|---|---|---|---|
| $2\left(x - \dfrac{\pi}{2}\right)$ | 0 | $\dfrac{\pi}{2}$ | $\pi$ | $\dfrac{3\pi}{2}$ | $2\pi$ |
| $\sin 2\left(x - \dfrac{\pi}{2}\right)$ | 0 | 1 | 0 | −1 | 0 |
| $2 + \sin 2\left(x - \dfrac{\pi}{2}\right)$ | 2 | 3 | 2 | 1 | 2 |

*Steps 4 and 5*: Plot the points found in the table and join them with a sinusoidal curve.

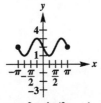

$y = 2 + \sin(2x - \pi)$

The amplitude is $|1|$, which is 1. The period is

$\dfrac{2\pi}{2}$, which is $\pi$.

7.  The amplitude is $\frac{1}{2}\left[1-(-1)\right]=\frac{1}{2}(2)=1$, so

    $a = 1$. One complete cycle of the graph is achieved in $\pi$ units, so the period

    $\pi = \frac{2\pi}{b} \Rightarrow b = \frac{2\pi}{\pi} = 2$. Comparing the given

    graph with the general sine and cosine curves, we see that this graph is a cosine curve. Substituting $a = 1$ and $b = 2$, the function is $y = \cos 2x$. Verify by confirming minimum and maximum points and $x$-intercepts from the graph:

    $(0,1) \Rightarrow 1 = \cos(2 \cdot 0) = \cos 0 = 1$

    $\left(\frac{\pi}{4},0\right) \Rightarrow 0 = \cos\left(2 \cdot \frac{\pi}{4}\right) = \cos\frac{\pi}{2} = 0$

    $\left(\frac{\pi}{2},-1\right) \Rightarrow -1 = \cos\left(2 \cdot \frac{\pi}{2}\right) = \cos\pi = -1$

    $\left(\frac{3\pi}{4},0\right) \Rightarrow 0 = \cos\left(2 \cdot \frac{3\pi}{4}\right) = \cos\frac{3\pi}{2} = 0$

    $(\pi,1) \Rightarrow 1 = \cos(2\pi) = 1$

9.  Refer to the function

    $f(x) = 12\sin\left[\frac{\pi}{6}(x - 3.9)\right] + 72$

    April is represented by $x = 4$.

    $f(4) = 12\sin\left[\frac{\pi}{6}(4 - 3.9)\right] + 72 \approx 73°F$

## Section 4.3: Graphs of the Tangent and Cotangent Functions

1.  $y = -\tan x$
    The graph is the reflection of the graph of $y = \tan x$ about the x-axis. This matches with graph C.

3.  $y = \tan\left(x - \frac{\pi}{4}\right)$

    The graph is the graph of $y = \tan x$ shifted $\frac{\pi}{4}$ units to the right. This matches with graph B.

5.  $y = \cot\left(x + \frac{\pi}{4}\right)$

    The graph is the graph of $y = \cot x$ shifted $\frac{\pi}{4}$ units to the left. This matches with graph F.

7.  $y = \tan 4x$
    *Step 1*: Find the period and locate the vertical

    asymptotes. The period of tangent is $\frac{\pi}{b}$, so

    the period for this function is $\frac{\pi}{4}$. Tangent has

    asymptotes of the form $bx = -\frac{\pi}{2}$ and $bx = \frac{\pi}{2}$.

    Therefore, the asymptotes for $y = \tan 4x$ are

    $4x = -\frac{\pi}{2} \Rightarrow x = -\frac{\pi}{8}$ and $4x = \frac{\pi}{2} \Rightarrow x = \frac{\pi}{8}$.

    *Step 2*: Sketch the two vertical asymptotes found in Step 1.
    *Step 3*: Divide the interval into four equal

    parts: $-\frac{\pi}{8}, -\frac{\pi}{16}, 0, \frac{\pi}{16}, \frac{\pi}{8}$

    *Step 4*: Finding the first-quarter point, midpoint, and third-quarter point, we have

    $\left(-\frac{\pi}{16},-1\right), (0,0), \left(\frac{\pi}{16},1\right)$

    *Step 5*: Join the points with a smooth curve.

    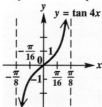

9.  $y = 2\tan x$
    *Step 1*: Find the period and locate the vertical

    asymptotes. The period of tangent is $\frac{\pi}{b}$, so

    the period for this function is $\pi$. Tangent has

    asymptotes of the form $bx = -\frac{\pi}{2}$ and $bx = \frac{\pi}{2}$.

    Therefore, the asymptotes for $y = 2\tan x$ are

    $x = -\frac{\pi}{2}$ and $x = \frac{\pi}{2}$.

    *Step 2*: Sketch the two vertical asymptotes found in Step 1.
    *Step 3*: Divide the interval into four equal

    parts: $-\frac{\pi}{2}, -\frac{\pi}{4}, 0, \frac{\pi}{4}, \frac{\pi}{2}$

    *Step 4*: Finding the first-quarter point, midpoint, and third-quarter point, we have

    $\left(-\frac{\pi}{4},-2\right), (0,0), \left(\frac{\pi}{4},2\right)$

    *Step 5*: Join the points with a smooth curve.
    The graph is "stretched" because $a = 2$ and $|2| > 1$.

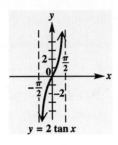

$y = 2\tan x$

**11.**  $y = 2\tan\dfrac{1}{4}x$

*Step 1*: Find the period and locate the vertical

asymptotes. The period of tangent is $\dfrac{\pi}{b}$, so

the period for this function is $4\pi$. Tangent has

asymptotes of the form $bx = -\dfrac{\pi}{2}$ and $bx = \dfrac{\pi}{2}$.

Therefore, the asymptotes for $y = 2\tan\dfrac{1}{4}x$

are

$\dfrac{1}{4}x = -\dfrac{\pi}{2} \Rightarrow x = -2\pi$ and $\dfrac{1}{4}x = \dfrac{\pi}{2} \Rightarrow x = 2\pi$

*Step 2*: Sketch the two vertical asymptotes
found in Step 1.
*Step 3*: Divide the interval into four equal
parts: $-2\pi, -\pi, 0, \pi, 2\pi$
*Step 4*: Finding the first-quarter point,
midpoint, and third-quarter point, we have:
$(-\pi, -2),\ (0,0),\ (\pi, 2)$
*Step 5*: Join the points with a smooth curve.

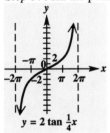

$y = 2\tan\frac{1}{4}x$

**13.**  $y = \cot 3x$

*Step 1*: Find the period and locate the vertical

asymptotes. The period of cotangent is $\dfrac{\pi}{b}$, so

the period for this function is $\dfrac{\pi}{3}$. Cotangent

has asymptotes of the form $bx = 0$ and $bx = \pi$.
The asymptotes for $y = \cot 3x$ are

$3x = 0 \Rightarrow x = 0$ and $3x = \pi \Rightarrow x = \dfrac{\pi}{3}$

*Step 2*: Sketch the two vertical asymptotes
found in Step 1.
*Step 3*: Divide the interval into four equal

parts: $0, \dfrac{\pi}{12}, \dfrac{\pi}{6}, \dfrac{\pi}{4}, \dfrac{\pi}{3}$

*Step 4*: Finding the first-quarter point,
midpoint, and third-quarter point, we have

$\left(\dfrac{\pi}{12}, 1\right),\ \left(\dfrac{\pi}{6}, 0\right),\ \left(\dfrac{\pi}{4}, -1\right)$

*Step 5*: Join the points with a smooth curve.

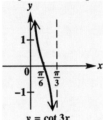

$y = \cot 3x$

**15.**  $y = -2\tan\dfrac{1}{4}x$

*Step 1*: Find the period and locate the vertical

asymptotes. The period of tangent is $\dfrac{\pi}{b}$, so

the period for this function is $4\pi$. Tangent has

asymptotes of the form $bx = -\dfrac{\pi}{2}$ and $bx = \dfrac{\pi}{2}$.

Therefore, the asymptotes for $y = -2\tan\dfrac{1}{4}x$

are

$\dfrac{1}{4}x = -\dfrac{\pi}{2} \Rightarrow x = -2\pi$ and $\dfrac{1}{4}x = \dfrac{\pi}{2} \Rightarrow x = 2\pi$

*Step 2*: Sketch the two vertical asymptotes
found in Step 1.
*Step 3*: Divide the interval into four equal
parts: $-2\pi, -\pi, 0, \pi, 2\pi$
*Step 4*: Finding the first-quarter point,
midpoint, and third-quarter point, we have
$(-\pi, 2),\ (0,0),\ (\pi, -2)$
*Step 5*: Join the points with a smooth curve.

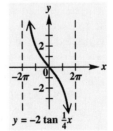

$y = -2\tan\frac{1}{4}x$

**17.** $y = \dfrac{1}{2}\cot 4x$

*Step 1:* Find the period and locate the vertical asymptotes. The period of cotangent is $\dfrac{\pi}{b}$, so the period for this function is $\dfrac{\pi}{4}$. Cotangent has asymptotes of the form $bx = 0$ and $bx = \pi$.

The asymptotes for $y = \dfrac{1}{2}\cot 4x$ are

$$4x = 0 \Rightarrow x = 0 \text{ and } 4x = \pi \Rightarrow x = \dfrac{\pi}{4}$$

*Step 2:* Sketch the two vertical asymptotes found in Step 1.

*Step 3:* Divide the interval into four equal parts: $0, \dfrac{\pi}{16}, \dfrac{\pi}{8}, \dfrac{3\pi}{16}, \dfrac{\pi}{4}$

*Step 4:* Finding the first-quarter point, midpoint, and third-quarter point, we have

$$\left(\dfrac{\pi}{16}, \dfrac{1}{2}\right), \left(\dfrac{\pi}{8}, 0\right), \left(\dfrac{3\pi}{16}, -\dfrac{1}{2}\right)$$

*Step 5:* Join the points with a smooth curve.

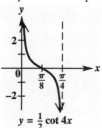

$y = \tfrac{1}{2}\cot 4x$

**19.** $y = \tan(2x - \pi) = \tan 2\left(x - \dfrac{\pi}{2}\right)$

Period: $\dfrac{\pi}{b} = \dfrac{\pi}{2}$

Vertical translation: none

Phase shift (horizontal translation): $\dfrac{\pi}{2}$ units to the right

Because the function is to be graphed over a two-period interval, locate three adjacent vertical asymptotes. Asymptotes of the graph $y = \tan x$ occur at $-\dfrac{\pi}{2}, \dfrac{\pi}{2},$ and $\dfrac{3\pi}{2}$, so use the following equations to locate asymptotes:

$$2\left(x - \dfrac{\pi}{2}\right) = -\dfrac{\pi}{2}, \ 2\left(x - \dfrac{\pi}{2}\right) = \dfrac{\pi}{2}, \text{ and }$$

$$2\left(x - \dfrac{\pi}{2}\right) = \dfrac{3\pi}{2}$$

Solve each of these equations:

$$2\left(x - \dfrac{\pi}{2}\right) = -\dfrac{\pi}{2} \Rightarrow x - \dfrac{\pi}{2} = -\dfrac{\pi}{4} \Rightarrow x = \dfrac{\pi}{4}$$

$$2\left(x - \dfrac{\pi}{2}\right) = \dfrac{\pi}{2} \Rightarrow x - \dfrac{\pi}{2} = \dfrac{\pi}{4} \Rightarrow x = \dfrac{3\pi}{4}$$

$$2\left(x - \dfrac{\pi}{2}\right) = \dfrac{3\pi}{2} \Rightarrow x - \dfrac{\pi}{2} = \dfrac{3\pi}{4} \Rightarrow x = \dfrac{5\pi}{4}$$

Divide the interval $\left(\dfrac{\pi}{4}, \dfrac{3\pi}{4}\right)$ into four equal parts to obtain the following key $x$-values:

first-quarter value: $\dfrac{3\pi}{8}$; middle value: $\dfrac{\pi}{2}$;

third-quarter value: $\dfrac{5\pi}{8}$

Evaluating the given function at these three key $x$-values gives the points

$$\left(\dfrac{3\pi}{8}, -1\right), \left(\dfrac{\pi}{2}, 0\right), \left(\dfrac{5\pi}{8}, 1\right)$$

Connect these points with a smooth curve and continue to graph to approach the asymptote $x = \dfrac{\pi}{4}$ and $x = \dfrac{3\pi}{4}$ to complete one period of the graph. Sketch the identical curve between the asymptotes $x = \dfrac{3\pi}{4}$ and $x = \dfrac{5\pi}{4}$ to complete a second period of the graph.

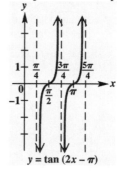

$y = \tan(2x - \pi)$

**21.** $y = \cot\left(3x + \dfrac{\pi}{4}\right) = \cot 3\left(x + \dfrac{\pi}{12}\right)$

Period: $\dfrac{\pi}{b} = \dfrac{\pi}{3}$

Vertical translation: none

Phase shift (horizontal translation): $\dfrac{\pi}{12}$ unit to the left

Because the function is to be graphed over a two-period interval, locate three adjacent vertical asymptotes.

Asymptotes of the graph $y = \cot x$ occur at multiples of $\pi$, so use the following equations to locate asymptotes:

$$3\left(x + \frac{\pi}{12}\right) = 0, \ 3\left(x + \frac{\pi}{12}\right) = \pi, \text{ and}$$

$$3\left(x + \frac{\pi}{12}\right) = 2\pi$$

Solve each of these equations:

$$3\left(x + \frac{\pi}{12}\right) = 0 \Rightarrow x + \frac{\pi}{12} = 0 \Rightarrow x = -\frac{\pi}{12}$$

$$3\left(x + \frac{\pi}{12}\right) = \pi \Rightarrow x + \frac{\pi}{12} = \frac{\pi}{3} \Rightarrow x = \frac{\pi}{3} - \frac{\pi}{12} = \frac{\pi}{4}$$

$$3\left(x + \frac{\pi}{12}\right) = 2\pi \Rightarrow x + \frac{\pi}{12} = \frac{2\pi}{3} \Rightarrow$$

$$x = \frac{2\pi}{3} - \frac{\pi}{12} \Rightarrow x = \frac{7\pi}{12}$$

Divide the interval $\left(\dfrac{\pi}{4}, \dfrac{7\pi}{12}\right)$ into four equal

parts to obtain the following key $x$-values:

first-quarter value: $\dfrac{\pi}{3}$ ; middle value: $\dfrac{5\pi}{12}$ ;

third-quarter value: $\dfrac{\pi}{2}$

Evaluating the given function at these three key $x$-values gives the points.

$$\left(\frac{\pi}{3}, 1\right), \ \left(\frac{5\pi}{12}, 0\right), \ \left(\frac{\pi}{2}, -1\right)$$

Connect these points with a smooth curve and continue to graph to approach the asymptote

$x = \dfrac{\pi}{4}$ and $x = \dfrac{7\pi}{12}$ to complete one period of

the graph. Sketch the identical curve between

the asymptotes $x = -\dfrac{\pi}{12}$ and $x = \dfrac{\pi}{4}$ to

complete a second period of the graph.

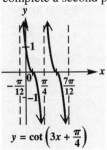

$y = \cot\left(3x + \frac{\pi}{4}\right)$

**23.** $y = 1 + \tan x$
This is the graph of $y = \tan x$ translated vertically 1 unit up.

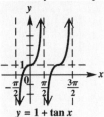

$y = 1 + \tan x$

**25.** $y = 1 - \cot x$
This is the graph of $y = \cot x$ reflected about the $x$-axis and then translated vertically 1 unit up.

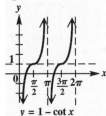

$y = 1 - \cot x$

**27.** $y = -1 + 2 \tan x$
This is the graph of $y = 2 \tan x$ translated vertically 1 unit down.

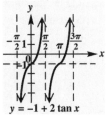

$y = -1 + 2 \tan x$

**29.** $y = -1 + \dfrac{1}{2}\cot(2x - 3\pi) = -1 + \dfrac{1}{2}\cot 2\left(x - \dfrac{3\pi}{2}\right)$

Period: $\dfrac{\pi}{b} = \dfrac{\pi}{2}$.

Vertical translation: 1 unit down

Phase shift (horizontal translation): $\dfrac{3\pi}{2}$ units

to the right

Because the function is to be graphed over a two-period interval, locate three adjacent vertical asymptotes. Asymptotes of the graph $y = \cot x$ occur at multiples of $\pi$, use the following equations to locate asymptotes:

$$2\left(x - \frac{3\pi}{2}\right) = -2\pi, \ 2\left(x - \frac{3\pi}{2}\right) = -\pi, \text{ and}$$

$$2\left(x - \frac{3\pi}{2}\right) = 0$$

*(continued on next page)*

(*continued from page 89*)

Solve each of these equations:

$$2\left(x - \frac{3\pi}{2}\right) = -2\pi \Rightarrow x - \frac{3\pi}{2} = -\pi \Rightarrow$$

$$x = -\pi + \frac{3\pi}{2} = \frac{\pi}{2}$$

$$2\left(x - \frac{3\pi}{2}\right) = -\pi \Rightarrow x - \frac{3\pi}{2} = -\frac{\pi}{2} \Rightarrow$$

$$x = -\frac{\pi}{2} + \frac{3\pi}{2} \Rightarrow x = \frac{2\pi}{2} = \pi$$

$$2\left(x - \frac{3\pi}{2}\right) = 0 \Rightarrow x - \frac{3\pi}{2} = 0 \Rightarrow x = \frac{3\pi}{2}$$

Divide the interval $\left(\frac{\pi}{2}, \pi\right)$ into four equal

parts to obtain the following key $x$-values:

first-quarter value: $\frac{5\pi}{8}$; middle value: $\frac{3\pi}{4}$;

third-quarter value: $\frac{7\pi}{8}$

Evaluating the given function at these three key $x$-values gives the points.

$$\left(\frac{5\pi}{8}, -\frac{1}{2}\right), \left(\frac{3\pi}{4}, -1\right), \left(\frac{7\pi}{8}, -\frac{3}{2}\right)$$

Connect these points with a smooth curve and continue to graph to approach the asymptote

$x = \frac{\pi}{2}$ and $x = \pi$ to complete one period of

the graph. Sketch the identical curve between

the asymptotes $x = \pi$ and $x = \frac{3\pi}{2}$ to

complete a second period of the graph.

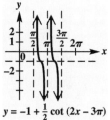

$y = -1 + \frac{1}{2}\cot(2x - 3\pi)$

**31.** $y = 1 - 2\cot 2\left(x + \frac{\pi}{2}\right)$

Period: $\frac{\pi}{b} = \frac{\pi}{2}$

Vertical translation: 1 unit up

Phase shift (horizontal translation): $\frac{\pi}{2}$ unit to

the left

Because the function is to be graphed over a two-period interval, locate three adjacent vertical asymptotes. Asymptotes of the graph $y = \cot x$ occur at multiples of $\pi$, so use the following equations to locate asymptotes.

$$2\left(x + \frac{\pi}{2}\right) = 0, \ 2\left(x + \frac{\pi}{2}\right) = \pi, \text{ and}$$

$$2\left(x + \frac{\pi}{2}\right) = 2\pi$$

Solve each of these equations:

$$2\left(x + \frac{\pi}{2}\right) = 0 \Rightarrow x + \frac{\pi}{2} = 0 \Rightarrow x = 0 - \frac{\pi}{2} = -\frac{\pi}{2}$$

$$2\left(x + \frac{\pi}{2}\right) = \pi \Rightarrow x + \frac{\pi}{2} = \frac{\pi}{2} \Rightarrow$$

$$x = \frac{\pi}{2} - \frac{\pi}{2} \Rightarrow x = 0$$

$$2\left(x + \frac{\pi}{2}\right) = 2\pi \Rightarrow x + \frac{\pi}{2} = \pi \Rightarrow$$

$$x = \pi - \frac{\pi}{2} = \frac{\pi}{2}$$

Divide the interval $\left(0, \frac{\pi}{2}\right)$ into four equal

parts to obtain the following key $x$-values:

first-quarter value: $\frac{\pi}{8}$; middle value: $\frac{\pi}{4}$;

third-quarter value: $\frac{3\pi}{8}$

Evaluating the given function at these three key $x$-values gives the points

$$\left(\frac{\pi}{8}, -1\right), \left(\frac{\pi}{4}, 1\right), \left(\frac{3\pi}{8}, 3\right)$$

Connect these points with a smooth curve and continue to graph to approach the asymptote

$x = 0$ and $x = \frac{\pi}{2}$ to complete one period of

the graph. Sketch the identical curve between

the asymptotes $x = -\frac{\pi}{2}$ and $x = 0$ to

complete a second period of the graph.

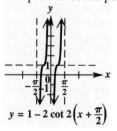

$y = 1 - 2\cot 2\left(x + \frac{\pi}{2}\right)$

**33.** Since the asymptotes are at $-\dfrac{\pi}{2}$, $\dfrac{\pi}{2}$, and

$\dfrac{3\pi}{2}$, this is a tangent function of the form

$y = a \tan x$. The graph passes through the point

$\left(\dfrac{\pi}{4}, -2\right)$. Substituting these values into the

generic equation gives

$y = a \tan x \Rightarrow -2 = a \tan \dfrac{\pi}{4} \Rightarrow -2 = a \cdot 1 \Rightarrow -2 = a$

Thus, the equation of the graph is $y = -2 \tan x$.

**35.** Since the asymptotes occur at $0$, $\dfrac{\pi}{3}$, and $\dfrac{2\pi}{3}$

this is a cotangent function of the form

$y = a \cot bx$. The period of the function is $\dfrac{\pi}{3}$,

so we have $\dfrac{\pi}{b} = \dfrac{\pi}{3} \Rightarrow b = 3$. The graph passes

through the point $\left(\dfrac{\pi}{12}, 1\right)$. Substituting these

values into the generic equation gives

$y = a \cot bx \Rightarrow 1 = a \cot\left(3 \cdot \dfrac{\pi}{12}\right) \Rightarrow$

$1 = a \cot \dfrac{\pi}{4} \Rightarrow 1 = a \cdot 1 \Rightarrow 1 = a$

Thus, the equation of the graph is $y = \cot 3x$.

**37.** True; $\dfrac{\pi}{2}$ is the smallest positive value where

$\cos \dfrac{\pi}{2} = 0$. Since $\tan \dfrac{\pi}{2} = \dfrac{\sin \dfrac{\pi}{2}}{\cos \dfrac{\pi}{2}}$, $\dfrac{\pi}{2}$ is the

smallest positive value where the tangent

function is undefined. Thus, $k = \dfrac{\pi}{2}$ is the

smallest positive value for which $x = k$ is an

asymptote for the tangent function.

**39.** False; $\tan(-x) = \dfrac{\sin(-x)}{\cos(-x)} = \dfrac{-\sin x}{\cos x} = -\tan x$

(since $\sin x$ is odd and $\cos x$ is even) for all $x$

in the domain. Moreover, if $\tan(-x) = \tan x$,

then the graph would be symmetric about the

$y$-axis, which it is not.

**41.** The function $\tan x$ has a period of $\pi$, so it

repeats four times over the interval $(-2\pi, 2\pi]$.

Since its range is $(-\infty, \infty)$, $\tan x = c$ has four

solutions for every value of $c$.

**43.** $\tan(-x) = \dfrac{\sin(-x)}{\cos(-x)} = \dfrac{-\sin x}{\cos x} = -\tan x$,

$\left\{ x \;\middle|\; x \neq (2n+1)\dfrac{\pi}{4}, \text{ where } n \text{ is any integer} \right\}$.

**45.** $d = 4 \tan 2\pi t$

(a) $d = 4 \tan 2\pi(0) = 4 \tan 0 \approx 4(0) = 0$ m

(b) $d = 4 \tan 2\pi(.4) = 4 \tan .8\pi$
$\approx 4(-.7265) \approx -2.9$ m

(c) $d = 4 \tan 2\pi(.8) = 4 \tan 1.6\pi$
$\approx 4(-3.0777) \approx -12.3$ m

(d) $d = 4 \tan 2\pi(1.2) = 4 \tan 2.4\pi$
$\approx 4(3.0777) \approx 12.3$ m

(e) $t = .25$ leads to $\tan \dfrac{\pi}{2}$, which is

undefined.

**47.** $\pi$

**49.** The vertical asymptotes in general occur at

$x = \dfrac{5\pi}{4} + n\pi$, where n is an integer.

**51.** $.3217505544 + \pi \approx 3.463343208$

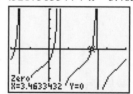

Zero
X=3.4633432  Y=0

# Section 4.4: Graphs of the Secant and Cosecant Functions

**1.** $y = -\csc x$
The graph is the reflection of the graph of
$y = \csc x$ about the $x$-axis. This matches with
graph B.

**3.** $y = \sec\left(x - \dfrac{\pi}{2}\right)$

The graph is the graph of $y = \sec x$ shifted $\dfrac{\pi}{2}$

units to the right. This matches with graph D.

**5.** $y = 3\sec\dfrac{1}{4}x$

*Step 1*: Graph the corresponding reciprocal

function $y = 3\cos\dfrac{1}{4}x$. The period is

$\dfrac{2\pi}{\frac{1}{4}} = 2\pi \cdot \dfrac{4}{1} = 8\pi$ and its amplitude is $|3| = 3$.

One period is in the interval $0 \le x \le 8\pi$.
Dividing the interval into four equal parts
gives us the following key points: $(0,1)$,

$(2\pi,0)$, $(4\pi,-1)$, $(6\pi,0)$, $(8\pi,1)$

*Step 2*: The vertical asymptotes of $y = \sec\dfrac{1}{4}x$

are at the $x$-intercepts of $y = \cos\dfrac{1}{4}x$, which

are $x = 2\pi$ and $x = 6\pi$. Continuing this
pattern to the left, we also have a vertical
asymptote of $x = -2\pi$.

*Step 3*: Sketch the graph.

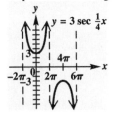

**7.** $y = -\dfrac{1}{2}\csc\left(x + \dfrac{\pi}{2}\right)$

*Step 1*: Graph the corresponding reciprocal

function $y = -\dfrac{1}{2}\sin\left(x + \dfrac{\pi}{2}\right)$. The period is

$2\pi$ and its amplitude is $\left|-\dfrac{1}{2}\right| = \dfrac{1}{2}$. One period

is in the interval $-\dfrac{\pi}{2} \le x \le \dfrac{3\pi}{2}$.

Dividing the interval into four equal parts

gives us the following key points: $\left(-\dfrac{\pi}{2},0\right)$,

$\left(0,-\dfrac{1}{2}\right)$, $\left(\dfrac{\pi}{2},0\right)$, $\left(\pi,\dfrac{1}{2}\right)$, $\left(\dfrac{3\pi}{2},0\right)$

*Step 2*: The vertical asymptotes of

$y = -\dfrac{1}{2}\csc\left(x + \dfrac{\pi}{2}\right)$ are at the $x$-intercepts of

$y = -\dfrac{1}{2}\sin\left(x + \dfrac{\pi}{2}\right)$, which are $x = -\dfrac{\pi}{2}$,

$x = \dfrac{\pi}{2}$, and $x = \dfrac{3\pi}{2}$.

*Step 3*: Sketch the graph.

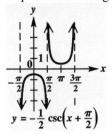

**9.** $y = \csc\left(x - \dfrac{\pi}{4}\right)$

*Step 1*: Graph the corresponding reciprocal

function $y = \sin\left(x - \dfrac{\pi}{4}\right)$ The period is $2\pi$

and its amplitude is $|1| = 1$. One period is in

the interval $\dfrac{\pi}{4} \le x \le \dfrac{9\pi}{4}$. Dividing the

interval into four equal parts gives us the

following key points: $\left(\dfrac{\pi}{4},0\right)$, $\left(\dfrac{3\pi}{4},1\right)$,

$\left(\dfrac{5\pi}{4},0\right)$, $\left(\dfrac{7\pi}{4},-1\right)$, $\left(\dfrac{9\pi}{4},0\right)$

*Step 2*: The vertical asymptotes of

$y = \csc\left(x - \dfrac{\pi}{4}\right)$ are at the $x$-intercepts of

$y = \sin\left(x - \dfrac{\pi}{4}\right)$, which are $x = \dfrac{\pi}{4}$, $x = \dfrac{5\pi}{4}$,

and $x = \dfrac{9\pi}{4}$.

*Step 3*: Sketch the graph.

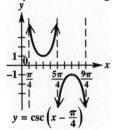

**11.** $y = \sec\left(x + \dfrac{\pi}{4}\right)$

*Step 1:* Graph the corresponding reciprocal

function $y = \cos\left(x + \dfrac{\pi}{4}\right)$. The period is $2\pi$

and its amplitude is $|1| = 1$. One period is in

the interval $-\dfrac{\pi}{4} \le x \le \dfrac{7\pi}{4}$.

Dividing the interval into four equal parts gives us the following key points: $\left(-\dfrac{\pi}{4},1\right)$, $\left(\dfrac{\pi}{4},0\right)$, $\left(\dfrac{3\pi}{4},-1\right)$, $\left(\dfrac{5\pi}{4},0\right)$, $\left(\dfrac{7\pi}{4},1\right)$

*Step 2*: The vertical asymptotes of $y=\sec\left(x+\dfrac{\pi}{4}\right)$ are at the $x$-intercepts of $y=\cos\left(x+\dfrac{\pi}{4}\right)$, which are $x=\dfrac{\pi}{4}$ and $x=\dfrac{5\pi}{4}$. Continuing this pattern to the right, we also have a vertical asymptote of $x=\dfrac{9\pi}{4}$.

*Step 3*: Sketch the graph.

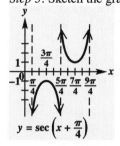

13.  $y=\sec\left(\dfrac{1}{2}x+\dfrac{\pi}{3}\right)=\sec\dfrac{1}{2}\left(x+\dfrac{2\pi}{3}\right)$

*Step 1*: Graph the corresponding reciprocal function $y=\cos\dfrac{1}{2}\left(x+\dfrac{2\pi}{3}\right)$.

The period is $\dfrac{2\pi}{\frac{1}{2}}=2\pi\cdot\dfrac{2}{1}=4\pi$ and its amplitude is $\left|\dfrac{1}{2}\right|=\dfrac{1}{2}$. One period is in the interval $-\dfrac{2\pi}{3}\le x\le\dfrac{10\pi}{3}$. Dividing the interval into four equal parts gives us the following key points: $\left(-\dfrac{2\pi}{3},1\right)$, $\left(\dfrac{\pi}{3},0\right)$, $\left(\dfrac{4\pi}{3},-1\right)$, $\left(\dfrac{7\pi}{3},0\right)$, $\left(\dfrac{10\pi}{3},1\right)$

*Step 2*: The vertical asymptotes of $y=\sec\dfrac{1}{2}\left(x+\dfrac{2\pi}{3}\right)$ are at the $x$-intercepts of $y=\cos\dfrac{1}{2}\left(x+\dfrac{2\pi}{3}\right)$, which are $x=\dfrac{\pi}{3}$ and $x=\dfrac{7\pi}{3}$.

Continuing this pattern to the right, we also have a vertical asymptote of $x=\dfrac{13\pi}{3}$.

*Step 3*: Sketch the graph.

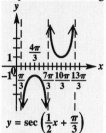

15.  $y=2+3\sec\left(2x-\pi\right)=2+3\sec 2\left(x-\dfrac{\pi}{2}\right)$

*Step 1*: Graph the corresponding reciprocal function $y=2+3\cos 2\left(x-\dfrac{\pi}{2}\right)$. The period is $\pi$ and its amplitude is $|3|=3$. One period is in the interval $\dfrac{\pi}{2}\le x\le\dfrac{3\pi}{2}$. Dividing the interval into four equal parts gives us the following key points: $\left(\dfrac{\pi}{2},5\right)$, $\left(\dfrac{3\pi}{4},2\right)$, $\left(\pi,-1\right)$, $\left(\dfrac{5\pi}{4},2\right)$, $\left(\dfrac{3\pi}{2},5\right)$

*Step 2*: The vertical asymptotes of $y=2+3\sec 2\left(x-\dfrac{\pi}{2}\right)$ are at the $x$-intercepts of $y=3\cos 2\left(x-\dfrac{\pi}{2}\right)$, which are $x=\dfrac{3\pi}{4}$ and $x=\dfrac{5\pi}{4}$. Continuing this pattern to the left, we also have a vertical asymptote of $x=\dfrac{\pi}{4}$.

*Step 3*: Sketch the graph.

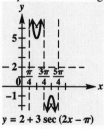

**17.** $y = 1 - \dfrac{1}{2}\csc\left(x - \dfrac{3\pi}{4}\right)$

*Step 1:* Graph the corresponding reciprocal function $y = 1 - \dfrac{1}{2}\sin\left(x - \dfrac{3\pi}{4}\right)$. The period is $2\pi$ and its amplitude is $\dfrac{1}{2}$. One period is in the interval $\dfrac{3\pi}{4} \le x \le \dfrac{11\pi}{4}$. Dividing the interval into four equal parts gives us the following key points: $\left(\dfrac{3\pi}{4}, 1\right)$, $\left(\dfrac{5\pi}{4}, \dfrac{1}{2}\right)$, $\left(\dfrac{7\pi}{4}, 1\right)$, $\left(\dfrac{9\pi}{4}, \dfrac{3}{2}\right)$, $\left(\dfrac{11\pi}{4}, 1\right)$

*Step 2:* The vertical asymptotes of $y = 1 - \dfrac{1}{2}\csc\left(x - \dfrac{3\pi}{4}\right)$ are at the $x$-intercepts of $y = -\dfrac{1}{2}\sin\left(x - \dfrac{3\pi}{4}\right)$, which are $x = \dfrac{3\pi}{4}$, $x = \dfrac{7\pi}{4}$, and $x = \dfrac{11\pi}{4}$.

*Step 3:* Sketch the graph.

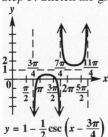

$y = 1 - \dfrac{1}{2}\csc\left(x - \dfrac{3\pi}{4}\right)$

For exercises 19–23, other answers are possible.

**19.** Since the asymptotes occur at multiples of $\pi$, this is a cotangent function of the form $y = a\cot x$. The graph passes through the point $\left(\dfrac{\pi}{4}, -3\right)$. Substituting these values into the generic equation gives

$y = a\cot x \Rightarrow -3 = a\cot\dfrac{\pi}{4} \Rightarrow$

$-3 = a \cdot 1 \Rightarrow -3 = a$

Thus, the equation of the graph is $y = -3\cot x$.

**21.** Since the graph crosses the $y$-axis at $(0, 1)$, this is a secant graph with $a = 1$. The period is

$\left| -\dfrac{\pi}{4} - \dfrac{\pi}{4} \right| = \left| -\dfrac{\pi}{2} \right| = \dfrac{\pi}{2}$. Thus,

$b = \dfrac{2\pi}{\dfrac{\pi}{2}} \Rightarrow b = 4$. The equation of the graph is $y = \sec 4x$.

**23.** This is the graph of $y = \csc x$ translated two units down. Thus, the equation of the graph is $y = -2 + \csc x$.

**25.** True; since $\tan x = \dfrac{\sin x}{\cos x}$ and $\sec x = \dfrac{1}{\cos x}$, the tangent and secant functions will be undefined at the same values.

**27.** True; $\sec(-x) = \dfrac{1}{\cos(-x)} = \dfrac{1}{\cos(x)} = \sec(x)$ (since $\cos x$ is even) for all $x$ in the domain. Moreover, if $\sec(-x) = \sec x$, then the graph would be symmetric about the $y$-axis, which it is.

**29.** None; $\cos x \le 1$ for all $x$, so $\dfrac{1}{\cos x} \ge 1$ and $\sec x \ge 1$. Since $\sec x \ge 1$, $\sec x$ has no values in the interval $(-1, 1)$.

**31.** $\sec(-x) = \dfrac{1}{\cos(-x)} = \dfrac{1}{\cos(x)} = \sec(x)$,

$\left\{ x \,\middle|\, x \ne (2n+1)\dfrac{\pi}{2}, \text{ where } n \text{ is any integer} \right\}$.

**33.** $a = 4\left|\sec 2\pi t\right|$

**(a)** $t = 0$
$a = 4\left|\sec 0\right| = 4\left|1\right| = 4(1) = 4$ m

**(b)** $t = .86$
$a = 4\left|\sec 2\pi(.86)\right| \approx 4\left|1.5688\right|$
$= 4(1.5688) \approx 6.3$ m

**(c)** $t = 1.24$
$a = 4\left|\sec 2\pi(1.24)\right| \approx 4\left|15.9260\right|$
$= 4(15.9260) \approx 63.7$ m

**35.** Graph the functions.

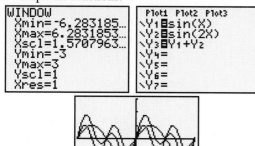

Notice that

$$Y_1\left(\tfrac{\pi}{6}\right) + Y_2\left(\tfrac{\pi}{6}\right) \approx .5 + .8660254 = 1.3660254$$

$$= Y_3\left(\tfrac{\pi}{6}\right) = \left(Y_1 + Y_2\right)\left(\tfrac{\pi}{6}\right)$$

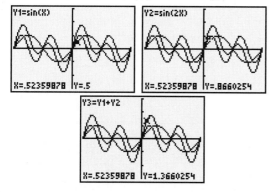

## Summary Exercises on Graphing Circular Functions

**1.** $y = 2\sin \pi x$

Period: $\dfrac{2\pi}{\pi} = 2$ and amplitude: $|2| = 2$

Divide the interval $[0, 2]$ into four equal parts to get the $x$-values that will yield minimum and maximum points and $x$-intercepts. Then make a table.

| $x$ | 0 | $\dfrac{1}{2}$ | 1 | $\dfrac{3}{2}$ | 2 |
|---|---|---|---|---|---|
| $\pi x$ | 0 | $\dfrac{\pi}{2}$ | $\pi$ | $\dfrac{3\pi}{2}$ | $2\pi$ |
| $\sin \pi x$ | 0 | 1 | 0 | $-1$ | 0 |
| $2\sin \pi x$ | 0 | 2 | 0 | $-2$ | 0 |

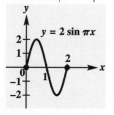

**3.** $y = -2 + .5\cos \dfrac{\pi}{4} x$

*Step 1*: Find the interval whose length is $\dfrac{2\pi}{b}$.

$$0 \le \tfrac{\pi}{4} x \le 2\pi \Rightarrow 0 \le x \le 8$$

*Step 2*: Divide the period into four equal parts to get the following $x$-values: $0, 2, 4, 6, 8$

*Step 3*: Evaluate the function for each of the five $x$-values.

| $x$ | 0 | 2 | 4 | 6 | 8 |
|---|---|---|---|---|---|
| $\dfrac{\pi}{4} x$ | 0 | $\dfrac{\pi}{2}$ | $\pi$ | $\dfrac{3\pi}{2}$ | $2\pi$ |
| $\cos \dfrac{\pi}{4} x$ | 1 | 0 | $-1$ | 0 | 1 |
| $\dfrac{1}{2}\cos \dfrac{\pi}{4} x$ | $\dfrac{1}{2}$ | 0 | $-\dfrac{1}{2}$ | 0 | $\dfrac{1}{2}$ |
| $-2 + \dfrac{1}{2}\cos \dfrac{\pi}{4} x$ | $-\dfrac{3}{2}$ | $-2$ | $-\dfrac{5}{2}$ | $-2$ | $-\dfrac{3}{2}$ |

*Steps 4 and 5*: Plot the points found in the table and join them with a sinusoidal curve. By graphing an additional period to the left, we obtain the following graph.

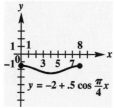

The amplitude is $|.5| = \left|\dfrac{1}{2}\right|$, which is $\dfrac{1}{2}$. The period is 8. The vertical translation is 2 units down. There is no phase shift.

**5.** $y = -4\csc .5x = -4\csc \dfrac{1}{2} x$

*Step 1*: Graph the corresponding reciprocal function $y = -4\sin \dfrac{1}{2} x$ The period is

$$\dfrac{2\pi}{\frac{1}{2}} = 2\pi \cdot \dfrac{2}{1} = 4\pi \text{ and its amplitude is}$$

$$|-4| = 4.$$

One period is in the interval $0 \le x \le 4\pi$. Dividing the interval into four equal parts gives us the following key points: $(0, 0)$, $(\pi, -4)$, $(2\pi, 0)$, $(3\pi, 4)$, $(4\pi, 0)$

*(continued on next page)*

(*continued from page 95*)

*Step 2*: The vertical asymptotes of

$$y = -4\csc\frac{1}{2}x \text{ are at the } x\text{-intercepts of}$$

$$y = -4\sin\frac{1}{2}x, \text{ which are } x = 0, \ x = 2\pi, \text{ and}$$

$x = 4\pi$.

*Step 3*: Sketch the graph.

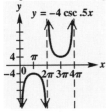

7.  $y = -5\sin\dfrac{x}{3}$

Period: $\dfrac{2\pi}{\frac{1}{3}} = 2\pi \cdot \dfrac{3}{1} = 6\pi$ and amplitude:

$|-5| = 5$

Divide the interval $[0, 6\pi]$ into four equal parts to get the $x$-values that will yield minimum and maximum points and $x$-intercepts. Then make a table. Repeat this cycle for the interval $[-6\pi, 0]$.

| $x$ | 0 | $\dfrac{3\pi}{2}$ | $3\pi$ | $\dfrac{9\pi}{2}$ | $6\pi$ |
|---|---|---|---|---|---|
| $\dfrac{x}{3}$ | 0 | $\dfrac{\pi}{2}$ | $\pi$ | $\dfrac{3\pi}{2}$ | $2\pi$ |
| $\sin\dfrac{x}{3}$ | 0 | 1 | 0 | $-1$ | 0 |
| $-5\sin\dfrac{x}{3}$ | 0 | $-5$ | 0 | 5 | 0 |

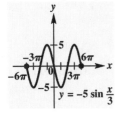

9.  $y = 3 - 4\sin(2.5x + \pi) = 3 - 4\sin\left(\dfrac{5}{2}x + \pi\right)$

$$= 3 - 4\sin\frac{5}{2}\left(x + \frac{2\pi}{5}\right)$$

*Step 1*: Find the interval whose length is $\dfrac{2\pi}{b}$.

$$0 \le \frac{5}{2}\left(x + \frac{2\pi}{5}\right) \le 2\pi \Rightarrow$$

$$0 \le x + \frac{2\pi}{5} \le \frac{4\pi}{5} \Rightarrow -\frac{2\pi}{5} \le x \le \frac{2\pi}{5}$$

*Step 2*: Divide the period into four equal parts to get the following $x$-values: $-\dfrac{2\pi}{5}, \ -\dfrac{\pi}{5}, \ 0,$

$\dfrac{\pi}{5}, \ \dfrac{2\pi}{5}$

*Step 3*: Evaluate the function for each of the five $x$-values

| $x$ | $-\dfrac{2\pi}{5}$ | $-\dfrac{\pi}{5}$ | 0 | $\dfrac{\pi}{5}$ | $\dfrac{2\pi}{5}$ |
|---|---|---|---|---|---|
| $\dfrac{5}{2}\left(x + \dfrac{2\pi}{5}\right)$ | 0 | $\dfrac{\pi}{2}$ | $\pi$ | $\dfrac{3\pi}{2}$ | $2\pi$ |
| $\sin\dfrac{5}{2}\left(x + \dfrac{2\pi}{5}\right)$ | 0 | 1 | 0 | $-1$ | 0 |
| $-4\sin\dfrac{5}{2}\left(x + \dfrac{2\pi}{5}\right)$ | 0 | $-4$ | 0 | 4 | 0 |
| $3 - 4\sin\dfrac{5}{2}\left(x + \dfrac{2\pi}{5}\right)$ | 3 | $-1$ | 3 | 7 | 3 |

*Steps 4 and 5*: Plot the points found in the table and join them with a sinusoidal curve. By graphing an additional period to the right, we obtain the following graph.

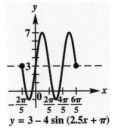

The amplitude is $|-4|$, which is 4. The period

is $\dfrac{2\pi}{\frac{5}{2}}$, which is $\dfrac{4\pi}{5}$. The vertical translation

is 3 units up. The phase shift is $\dfrac{2\pi}{5}$ units to

the left.

## Section 4.5: Harmonic Motion

**1.** $s(0) = 2$ in.; $P = .5$ sec

   **(a)** Given $s(t) = a \cos \omega t$, the period is $\dfrac{2\pi}{\omega}$ and the amplitude is $|a|$.

$$P = .5 \text{ sec} \Rightarrow .5 = \frac{2\pi}{\omega} \Rightarrow$$
$$\frac{1}{2} = \frac{2\pi}{\omega} \Rightarrow \omega = 4\pi$$
$$s(0) = 2 = a \cos\big[\omega(0)\big] \Rightarrow$$
$$2 = a \cos 0 \Rightarrow 2 = a(1) \Rightarrow a = 2$$

Thus, $s(t) = 2 \cos 4\pi t$.

   **(b)** Since
$$s(1) = 2 \cos\big[4\pi(1)\big] = 2 \cos 4\pi = 2(1) = 2,$$
the weight is neither moving upward nor downward. At $t = 1$, the motion of the weight is changing from up to down.

**3.** $s(0) = -3$ in.; $P = .8$ sec

   **(a)** Given $s(t) = a \cos \omega t$, the period is $\dfrac{2\pi}{\omega}$ and the amplitude is $|a|$.

$$P = .8 \text{ sec} \Rightarrow .8 = \frac{2\pi}{\omega} \Rightarrow \frac{4}{5} = \frac{2\pi}{\omega} \Rightarrow$$
$$4\omega = 10\pi \Rightarrow \omega = \frac{10\pi}{4} = 2.5\pi$$
$$s(0) = -3 = a \cos\big[\omega(0)\big] \Rightarrow$$
$$-3 = a \cos 0 \Rightarrow -3 = a(1) \Rightarrow a = -3$$

Thus, $s(t) = -3 \cos 2.5\pi t$.

   **(b)** Since $s(1) = -3 \cos\big[2.5\pi(1)\big]$
$$= -3 \cos \frac{5\pi}{2} = -3(0) = 0$$
the weight is moving upward.

**5.** Since frequency is $\dfrac{\omega}{2\pi}$, we have

$$27.5 = \frac{\omega}{2\pi} \Rightarrow \omega = 55\pi. \text{ Since } s(0) = .21,$$
$$.21 = a \cos\big[\omega(0)\big] \Rightarrow$$
$$.21 = a \cos 0 \Rightarrow .21 = a(1) \Rightarrow a = .21. \text{ Thus,}$$
$$s(t) = .21 \cos 55\pi t.$$

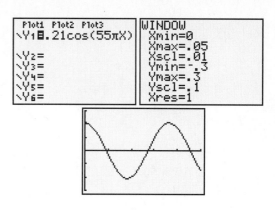

**7.** Since frequency is $\dfrac{\omega}{2\pi}$, we have

$$55 = \frac{\omega}{2\pi} \Rightarrow \omega = 110\pi. \text{ Since } s(0) = .14,$$
$$.14 = a \cos\big[\omega(0)\big] \Rightarrow$$
$$.14 = a \cos 0 \Rightarrow .14 = a(1) \Rightarrow a = .14. \text{ Thus,}$$
$$s(t) = .14 \cos 110\pi t.$$

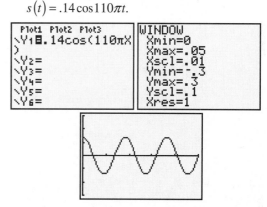

**9. (a)** Since the object is pulled down 4 units, $s(0) = -4$. Thus, we have
$$s(0) = -4 = a \cos\big[\omega(0)\big] \Rightarrow$$
$$-4 = a \cos 0 \Rightarrow -4 = a(1) \Rightarrow a = -4$$

Since the time it takes to complete one oscillation is 3 sec, $P = 3$ sec.

$$P = 3 \text{ sec} \Rightarrow 3 = \frac{2\pi}{\omega} \Rightarrow 3 = \frac{2\pi}{\omega} \Rightarrow$$
$$3\omega = 2\pi \Rightarrow \omega = \frac{2\pi}{3}$$

Therefore, $s(t) = -4 \cos \dfrac{2\pi}{3} t$.

**(b)** $s(1) = -4\cos\left[\dfrac{2\pi}{3}(1.25)\right]$

$= -4\cos\left[\dfrac{2\pi}{3}\left(\dfrac{5}{4}\right)\right] = -4\cos\dfrac{5\pi}{6}$

$= -4\left(-\dfrac{\sqrt{3}}{2}\right) = 2\sqrt{3} \approx 3.46$ units

(rounded to three significant digits)

**(c)** The frequency is the reciprocal of the period, or $\dfrac{1}{3}$ oscillation per second.

**11. (a)** $a = 2, \omega = 2$

$s(t) = a\sin\omega t \Rightarrow s(t) = 2\sin 2t$

amplitude $= |a| = |2| = 2$; period

$= \dfrac{2\pi}{\omega} = \dfrac{2\pi}{2} = \pi$; frequency $= \dfrac{\omega}{2\pi} = \dfrac{1}{\pi}$

rotation per second

**(b)** $a = 2, \omega = 4$

$s(t) = a\sin\omega t \Rightarrow s(t) = 2\sin 4t$

amplitude $= |a| = |2| = 2$; period

$= \dfrac{2\pi}{\omega} = \dfrac{2\pi}{4} = \dfrac{\pi}{2}$; frequency

$= \dfrac{\omega}{2\pi} = \dfrac{4}{2\pi} = \dfrac{2}{\pi}$ rotation per second

**13.** $P = 2\pi\sqrt{\dfrac{L}{32}} \Rightarrow 1 = 2\pi\sqrt{\dfrac{L}{32}} \Rightarrow$

$\dfrac{1}{2\pi} = \sqrt{\dfrac{L}{32}} \Rightarrow \dfrac{1}{(2\pi)^2} = \dfrac{L}{32} \Rightarrow \dfrac{1}{4\pi^2} = \dfrac{L}{32} \Rightarrow$

$\dfrac{32}{4\pi^2} = L \Rightarrow L = \dfrac{8}{\pi^2}$ ft

**15.** $s(t) = -4\cos 8\pi t$

**(a)** The maximum of $s(t) = -4\cos 8\pi t$ is $|-4| = 4$ in.

**(b)** In order for $s(t) = -4\cos 8\pi t$ to reach its maximum, $y = \cos 8\pi t$ needs to be at a minimum. This occurs after

$8\pi t = \pi \Rightarrow t = \dfrac{\pi}{8\pi} = \dfrac{1}{8}$ sec.

**(c)** Since $s(t) = -4\cos 8\pi t$ and

$s(t) = a\cos\omega t$, $\omega = 8\pi$. Therefore,

frequency $= \dfrac{\omega}{2\pi} = \dfrac{8\pi}{2\pi} = 4$ cycles per sec.

The period is the reciprocal of the frequency, or $\dfrac{1}{4}$ sec.

**17.** $s(t) = -5\cos 4\pi t$, $a = |-5| = 5$, $\omega = 4\pi$

**(a)** maximum height = amplitude

$= a = |-5| = 5$ in.

**(b)** frequency

$= \dfrac{\omega}{2\pi} = \dfrac{4\pi}{2\pi} = 2$ cycles per sec; period

$= \dfrac{2\pi}{\omega} = \dfrac{1}{2}$ sec

**19.** $a = -3$

**(a)** We will use a model of the form

$s(t) = a\cos\omega t$ with $a = -3$.

$s(0) = -3\cos\left[\omega(0)\right]$

$= -3\cos 0 = -3(1) = -3$

Using a cosine function rather than a sine function will avoid the need for a phase shift. Since the frequency $= \dfrac{6}{\pi}$ cycles per sec, by definition,

$\dfrac{\omega}{2\pi} = \dfrac{6}{\pi} \Rightarrow \omega\pi = 12\pi \Rightarrow \omega = 12.$

Therefore, a model for the position of the weight at time $t$ seconds is

$s(t) = -3\cos 12t.$

**(b)** The period is the reciprocal of the frequency, or $\dfrac{\pi}{6}$ sec.

**(c)** $s(t) = -5\cos 4\pi t = 5 \Rightarrow \cos 4\pi t = -1 \Rightarrow$

$4\pi t = \pi \Rightarrow t = \dfrac{1}{4}$

The weight first reaches its maximum height after $\dfrac{1}{4}$ sec.

**(d)** Since

$s(1.3) = -5\cos\left[4\pi(1.3)\right] = -5\cos 5.2\pi \approx 4,$

after 1.3 sec, the weight is about 4 in. above the equilibrium position.

For exercise 21, we have

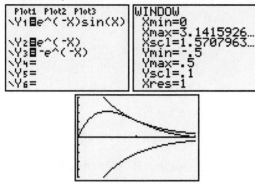

**21.** Since $e^{-t} \neq 0,$ we have

$e^{-t} \sin t = 0 \Rightarrow \sin t = 0 \Rightarrow t = 0,\ \pi.$ The $x$-intercepts of $Y_1$ are the same as these of $\sin x.$

## Chapter 4 Review Exercises

**1.** B; The amplitude is $|4| = 4$ and period is

$\dfrac{2\pi}{2} = \pi.$

**3.** The range of sine and cosine is $[-1, 1]$ and the range of tangent and cotangent is $(-\infty, \infty),$ so those functions can have $y$-value $\dfrac{1}{2}.$

**5..** $y = 2 \sin x$
Amplitude: 2
Period: $2\pi$
Vertical translation: none
Phase shift: none

**7.** $y = -\dfrac{1}{2}\cos 3x$

Amplitude: $\left|-\dfrac{1}{2}\right| = \dfrac{1}{2}$

Period: $\dfrac{2\pi}{3}$

Vertical translation: none
Phase shift: none

**9.** $y = 1 + 2\sin\dfrac{1}{4}x$

Amplitude: $|2| = 2$

Period: $\dfrac{2\pi}{\frac{1}{4}} = 8\pi$

Vertical translation: up 1 unit
Phase shift: none

**11.** $y = 3\cos\left(x + \dfrac{\pi}{2}\right) = 3\cos\left[x - \left(-\dfrac{\pi}{2}\right)\right]$

Amplitude: $|3| = 3$

Period: $2\pi$

Vertical translation: none

Phase shift: $\dfrac{\pi}{2}$ units to the left

**13.** $y = \dfrac{1}{2}\csc\left(2x - \dfrac{\pi}{4}\right) = \dfrac{1}{2}\csc 2\left(x - \dfrac{\pi}{8}\right)$

Amplitude: not applicable

Period: $\dfrac{2\pi}{2} = \pi$

Vertical translation: none

Phase shift: $\dfrac{\pi}{8}$ unit to the right

**15.** $y = \dfrac{1}{3}\tan\left(3x - \dfrac{\pi}{3}\right) = \dfrac{1}{3}\tan 3\left(x - \dfrac{\pi}{9}\right)$

Amplitude: not applicable

Period: $\dfrac{\pi}{3}$

Vertical translation: none

Phase shift: $\dfrac{\pi}{9}$ unit to the right

**17.** The tangent function has a period of $\pi$ and $x$-intercepts at integral multiples of $\pi.$

**19.** The cosine function has a period of $2\pi$ and has the value 0 when $x = \dfrac{\pi}{2}.$

**21.** The cotangent function has a period of $\pi$ and decreases on the interval $(0, \pi).$

**23.** Answers will vary.

**25.** $y = 3\sin x$

Period: $2\pi$ and amplitude: $|3| = 3$

Divide the interval $[0, 2\pi]$ into four equal parts to get $x$-values that will yield minimum and maximum points and $x$-intercepts. Then make a table.

| $x$ | 0 | $\dfrac{\pi}{2}$ | $\pi$ | $\dfrac{3\pi}{2}$ | $2\pi$ |
|---|---|---|---|---|---|
| $\sin x$ | 0 | 1 | 0 | $-1$ | 0 |
| $3\sin x$ | 0 | 3 | 0 | $-3$ | 0 |

This table gives five values for graphing one period of $y = 3\sin x.$

(*continued on next page*)

(*continued from page 99*)

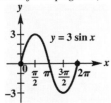

**27.** $y = -\tan x$

This is a reflection of the graph of $y = \tan x$ over the $x$-axis. The period is $\pi$ and vertical asymptotes are $x = -\dfrac{\pi}{2}$ and $x = \dfrac{\pi}{2}$

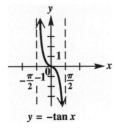

$y = -\tan x$

**29.** $y = 2 + \cot x$

This is the graph of $y = \cot x$ translated up 2 units. The period is $\pi$ and the vertical asymptotes are $x = 0$ and $x = \pi$.

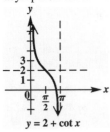

$y = 2 + \cot x$

**31.** $y = \sin 2x$

Period: $\dfrac{2\pi}{2} = \pi$ and amplitude: $|1| = 1$

Divide the interval $[0, \pi]$ into four equal parts to get the $x$-values that will yield minimum and maximum points and $x$-intercepts. Then make a table.

| $x$ | 0 | $\dfrac{\pi}{4}$ | $\dfrac{\pi}{2}$ | $\dfrac{3\pi}{4}$ | $\pi$ |
|---|---|---|---|---|---|
| $2x$ | 0 | $\dfrac{\pi}{2}$ | $\pi$ | $\dfrac{3\pi}{2}$ | $2\pi$ |
| $\sin 2x$ | 0 | 1 | 0 | $-1$ | 0 |

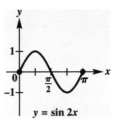

$y = \sin 2x$

**33.** $y = 3 \cos 2x$

Period: $\dfrac{2\pi}{2} = \pi$ and amplitude: $|3| = 3$

Divide the interval $[0, \pi]$ into four equal parts to get the $x$-values that will yield minimum and maximum points and $x$-intercepts. Then make a table.

| $x$ | 0 | $\dfrac{\pi}{4}$ | $\dfrac{\pi}{2}$ | $\dfrac{3\pi}{4}$ | $\pi$ |
|---|---|---|---|---|---|
| $2x$ | 0 | $\dfrac{\pi}{2}$ | $\pi$ | $\dfrac{3\pi}{2}$ | $2\pi$ |
| $\cos 2x$ | 1 | 0 | $-1$ | 0 | 1 |
| $3\cos 2x$ | 3 | 0 | $-3$ | 0 | 3 |

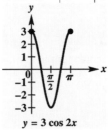

$y = 3 \cos 2x$

**35.** $y = \cos\left(x - \dfrac{\pi}{4}\right)$

*Step 1:* Find the interval whose length is $\dfrac{2\pi}{b}$.

$$0 \le x - \dfrac{\pi}{4} \le 2\pi \Rightarrow 0 + \dfrac{\pi}{4} \le x \le 2\pi + \dfrac{\pi}{4} \Rightarrow$$
$$\dfrac{\pi}{4} \le x \le \dfrac{9\pi}{4}$$

*Step 2:* Divide the period into four equal parts to get the following $x$-values: $\dfrac{\pi}{4}$, $\pi$, $\dfrac{3\pi}{2}$, $2\pi$, $\dfrac{9\pi}{4}$

*Step 3:* Evaluate the function for each of the five $x$-values.

| $x$ | $\dfrac{\pi}{4}$ | $\dfrac{3\pi}{4}$ | $\dfrac{5\pi}{4}$ | $\dfrac{7\pi}{4}$ | $\dfrac{9\pi}{4}$ |
|---|---|---|---|---|---|
| $x - \dfrac{\pi}{4}$ | 0 | $\dfrac{\pi}{2}$ | $\pi$ | $\dfrac{3\pi}{2}$ | $2\pi$ |
| $\cos\left(x - \dfrac{\pi}{4}\right)$ | 1 | 0 | $-1$ | 0 | 1 |

*Steps 4 and 5*: Plot the points found in the table and join them with a sinusoidal curve.

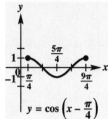

$$y = \cos\left(x - \frac{\pi}{4}\right)$$

The amplitude is 1. The period is $2\pi$. There is no vertical translation. The phase shift is $\frac{\pi}{4}$ units to the right.

**37.** $y = \sec\left(2x + \frac{\pi}{3}\right) = \sec 2\left(x + \frac{\pi}{6}\right)$

*Step 1*: Graph the corresponding reciprocal function $y = \cos 2\left(x + \frac{\pi}{6}\right)$.

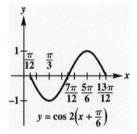

$$y = \cos 2\left(x + \frac{\pi}{6}\right)$$

The period is $\frac{2\pi}{2} = \pi$, and its amplitude is $|1| = 1$. One period is in the interval $\frac{\pi}{12} \le x \le \frac{13\pi}{12}$. Dividing the interval into four equal parts gives the key points $\left(\frac{\pi}{12}, 0\right), \left(\frac{\pi}{3}, -1\right)$, $\left(\frac{7\pi}{12}, 0\right), \left(\frac{5\pi}{6}, 1\right)$, and $\left(\frac{13\pi}{12}, 0\right)$.

*Step 2*: The vertical asymptotes of $y = \sec 2\left(x + \frac{\pi}{6}\right)$ are at the $x$-intercepts of $y = \cos 2\left(x + \frac{\pi}{6}\right)$, which are $x = \frac{\pi}{12}$, $x = \frac{7\pi}{12}$, and $x = \frac{13\pi}{12}$.

*Step 3*: Sketch the graph.

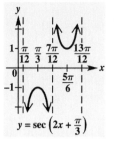

$$y = \sec\left(2x + \frac{\pi}{3}\right)$$

**39.** $y = 1 + 2\cos 3x$

*Step 1*: Find the interval whose length is $\frac{2\pi}{b}$.

$$0 \le 3x \le 2\pi \Rightarrow 0 \le x \le \frac{2\pi}{3}$$

*Step 2*: Divide the period into four equal parts to get the following $x$-values: $0, \frac{\pi}{6}, \frac{\pi}{3}, \frac{\pi}{2}, \frac{2\pi}{3}$

*Step 3*: Evaluate the function for each of the five $x$-values.

| $x$ | $0$ | $\frac{\pi}{6}$ | $\frac{\pi}{3}$ | $\frac{\pi}{2}$ | $\frac{2\pi}{3}$ |
|---|---|---|---|---|---|
| $3x$ | $0$ | $\frac{\pi}{2}$ | $\pi$ | $\frac{3\pi}{2}$ | $2\pi$ |
| $\cos 3x$ | $1$ | $0$ | $-1$ | $0$ | $1$ |
| $2\cos 3x$ | $2$ | $0$ | $-2$ | $0$ | $2$ |
| $1 + 2\cos 3x$ | $3$ | $1$ | $-1$ | $1$ | $3$ |

*Steps 4 and 5*: Plot the points found in the table and join them with a sinusoidal curve.

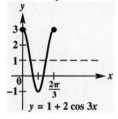

$$y = 1 + 2\cos 3x$$

The period is $\frac{2\pi}{3}$. The amplitude is $|2|$, which is 2. The vertical translation is 1 unit up. There is no phase shift.

**41.** $y = 2\sin \pi x$

Period: $\dfrac{2\pi}{\pi} = 2$ and amplitude: $|1| = 1$

Divide the interval [0, 2] into four equal parts to get the $x$-values that will yield minimum and maximum points and $x$-intercepts. Then make a table.

| $x$ | 0 | $\dfrac{1}{2}$ | 1 | $\dfrac{3}{2}$ | 2 |
|---|---|---|---|---|---|
| $\pi x$ | 0 | $\dfrac{\pi}{2}$ | $\pi$ | $\dfrac{3\pi}{2}$ | $2\pi$ |
| $\sin \pi x$ | 0 | 1 | 0 | $-1$ | 0 |
| $2\sin \pi x$ | 0 | 2 | 0 | $-2$ | 0 |

*Steps 4 and 5*: Plot the point found in the bale and join them with a sinusoidal curve.

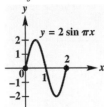

**43.** Answers will vary.

**45. (a)** The shorter leg of the right triangle has length $h_2 - h_1$. Thus, we have

$$\cot \theta = \dfrac{d}{h_2 - h_1} \Rightarrow d = (h_2 - h_1)\cot \theta$$

**(b)** When $h_2 = 55$ and $h_1 = 5$,

$$d = (55 - 5)\cot \theta = 50\cot \theta.$$

The period is $\pi$, but the graph wanted is $d$ for $0 < \theta < \dfrac{\pi}{2}$. The asymptote is the line $\theta = 0$. Also, when

$$\theta = \dfrac{\pi}{4}, d = 50\cot \dfrac{\pi}{4} = 50(1) = 50.$$

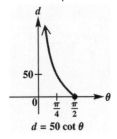

$d = 50 \cot \theta$

**47.** $t = 60 - 30\cos \dfrac{x\pi}{6}$

**(a)** For January, $x = 0$. Thus,

$$t = 60 - 30\cos \dfrac{0 \cdot \pi}{6} = 60 - 30\cos 0$$
$$= 60 - 30(1) = 60 - 30 = 30°$$

**(b)** For April, $x = 3$. Thus,

$$t = 60 - 30\cos \dfrac{3\pi}{6} = 60 - 30\cos \dfrac{\pi}{2}$$
$$= 60 - 30(0) = 60 - 0 = 60°$$

**(c)** For May, $x = 4$. Thus,

$$t = 60 - 30\cos \dfrac{4\pi}{6} = 60 - 30\cos \dfrac{2\pi}{3}$$
$$= 60 - 30\left(-\dfrac{1}{2}\right) = 60 + 15 = 75°$$

**(d)** For June, $x = 5$. Thus,

$$t = 60 - 30\cos \dfrac{5\pi}{6} = 60 - 30\left(-\dfrac{\sqrt{3}}{2}\right)$$
$$= 60 + 15\sqrt{3} \approx 86°$$

**(e)** For August, $x = 7$. Thus,

$$t = 60 - 30\cos \dfrac{7\pi}{6} = 60 - 30\left(-\dfrac{\sqrt{3}}{2}\right)$$
$$= 60 + 15\sqrt{3} \approx 86°$$

**(f)** For October, $x = 9$. Thus,

$$t = 60 - 30\cos \dfrac{9\pi}{6} = 60 - 30\cos \dfrac{3\pi}{2}$$
$$= 60 - 30(0) = 60 - 0 = 60°$$

**49.** $P(x) = 7(1 - \cos 2\pi x)(x + 10) + 100e^{.2x}$

**(a)** January 1, base year $x = 0$
$$P(0) = 7(1 - \cos 0)(10) + 100e^{0}$$
$$= 7(1 - 1)(10) + 100(1)$$
$$= 7(0)(10) + 100 = 0 + 100 = 100$$

**(b)** July 1, base year $x = .5$
$$P(.5) = 7(1 - \cos \pi)(.5 + 10) + 100e^{.2(.5)}$$
$$= 7[1 - (-1)](10.5) + 100e^{.1}$$
$$= 7(2)(10.5) + 100e^{.1}$$
$$= 147 + 100e^{.1} \approx 258$$

**(c)** January 1, following year $x = 1$

$$P(1) = 7(1 - \cos 2\pi)(1 + 10) + 100e^{-2}$$
$$= 7(1 - 1)(1 + 10) + 100e^{-2}$$
$$= 7(0)(11) + 100e^{-2} = 0 + 100e^{-2}$$
$$= 100e^{-2} \approx 122$$

**(d)** July 1, following year $x = 1.5$

$$P(1.5) = 7(1 - \cos 3\pi)(1.5 + 10) + 100e^{-2(1.5)}$$
$$= 7[1 - (-1)](11.5) + 100e^{-3}$$
$$= 7(2)(11.5) + 100e^{-3}$$
$$= 161 + +100e^{-3} \approx 296$$

**51.** $s(t) = 4 \sin \pi t$

$a = 4, \; \omega = \pi$

amplitude $= |a| = 4$ ; period $= \dfrac{2\pi}{\omega} = \dfrac{2\pi}{\pi} = 2$

frequency $= \dfrac{\omega}{2\pi} = \dfrac{\pi}{2\pi} = \dfrac{1}{2}$

**53.** The frequency is the number of cycles in one unit of time.

$$s(1.5) = 4 \sin 1.5\pi = 4 \sin \frac{3\pi}{2} = 4(-1) = -4$$
$$s(2) = 4 \sin 2\pi = 4(0) = 0$$
$$s(3.25) = 4 \sin 3.25\pi = 4 \sin \frac{13\pi}{4} = 4 \sin \frac{5\pi}{4}$$
$$= 4\left(-\frac{\sqrt{2}}{2}\right) = -2\sqrt{2}$$

## Chapter 4 Test

**1. (a)** $y = \sec x$     **(b)** $y = \sin x$

    **(c)** $y = \cos x$     **(d)** $y = \tan x$

    **(e)** $y = \csc x$     **(e)** $y = \cot x$

**2. (a)** The period of this graph is

$\pi = \dfrac{2\pi}{b} \Rightarrow b = \dfrac{2\pi}{\pi} = 2$ , so this is the

graph of $y = \sin 2x$ .

    **(b)** The amplitude of this graph is 2, so this is the graph of $y = 2 \sin x$ .

**3. (a)** The domain of the cosine function is $(-\infty, \infty)$ .

    **(b)** The range of the sine function is $[-1, 1]$ .

    **(c)** The least positive value for which the tangent function is undefined is $\dfrac{\pi}{2}$ .

**(d)** The range of the secant function is $(-\infty, -1] \cup [1, \infty)$ .

**4.** $y = 3 - 6 \sin\left(2x + \dfrac{\pi}{2}\right) = 3 - 6 \sin 2\left(x + \dfrac{\pi}{4}\right)$

$$= 3 - 6 \sin 2\left[x - \left(-\frac{\pi}{4}\right)\right]$$

**(a)** The period is $\dfrac{2\pi}{2} = \pi$ .

**(b)** The amplitude is 6.

**(c)** The range is $[-3, 9]$ .

**(d)** The $y$-intercept occurs when $x = 0$ .

$$-6 \sin\left(2 \cdot 0 + \frac{\pi}{2}\right) + 3 = -6 \sin\left(0 + \frac{\pi}{2}\right) + 3$$
$$= -6 \sin\left(\frac{\pi}{2}\right) + 3$$
$$= -6(1) + 3$$
$$= -6 + 3 = -3$$

**(e)** The phase shift is $\dfrac{\pi}{4}$ unit to the left

$\left(\text{that is, } -\dfrac{\pi}{4}\right)$

**5.** $y = \sin(2x + \pi) = \sin 2\left(x + \dfrac{\pi}{2}\right)$

$$= \sin 2\left[x - \left(-\frac{\pi}{2}\right)\right]$$

*Step 1*: Find the interval whose length is $\dfrac{2\pi}{b}$ .

$$0 \leq 2\left(x + \frac{\pi}{2}\right) \leq 2\pi \Rightarrow 0 \leq x + \frac{\pi}{2} \leq \frac{2\pi}{2} \Rightarrow$$

$$0 \leq x + \frac{\pi}{2} \leq \pi \Rightarrow -\frac{\pi}{2} \leq x \leq \frac{\pi}{2}$$

*Step 2*: Divide the period into four equal parts to get the following $x$-values: $-\dfrac{\pi}{2}$ , $-\dfrac{\pi}{4}$ , 0,

$\dfrac{\pi}{4}$ , $\dfrac{\pi}{2}$

*(continued on next page)*

*(continued from page 103)*

*Step 3*: Evaluate the function for each of the five $x$-values

| $x$ | $-\dfrac{\pi}{2}$ | $-\dfrac{\pi}{4}$ | $0$ | $\dfrac{\pi}{4}$ | $\dfrac{\pi}{2}$ |
|---|---|---|---|---|---|
| $x+\dfrac{\pi}{2}$ | $0$ | $\dfrac{\pi}{4}$ | $\dfrac{\pi}{2}$ | $\dfrac{3\pi}{4}$ | $\pi$ |
| $2\left(x+\dfrac{\pi}{2}\right)$ | $0$ | $\dfrac{\pi}{2}$ | $\pi$ | $\dfrac{3\pi}{2}$ | $2\pi$ |
| $\sin 2\left(x+\dfrac{\pi}{2}\right)$ | $0$ | $1$ | $0$ | $-1$ | $0$ |

*Steps 4 and 5*: Plot the points found in the table and join them with a sinusoidal curve.

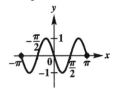

$$y = \sin(2x + \pi)$$

The period is $\pi$. There is no vertical translation. The phase shift is $\dfrac{\pi}{2}$ unit to the right.

6.  $y = -\cos 2x$

Period: $\dfrac{2\pi}{2} = \pi$ and amplitude: $|-1| = 1$

Divide the interval $[0, \pi]$ into four equal parts to get the $x$-values that will yield minimum and maximum points and $x$-intercepts. Then make a table. Repeat this cycle for the interval $[-\pi, 0]$.

| $x$ | $0$ | $\dfrac{\pi}{4}$ | $\dfrac{\pi}{2}$ | $\dfrac{3\pi}{4}$ | $\pi$ |
|---|---|---|---|---|---|
| $2x$ | $0$ | $\dfrac{\pi}{2}$ | $\pi$ | $\dfrac{3\pi}{2}$ | $2\pi$ |
| $\cos 2x$ | $1$ | $0$ | $-1$ | $0$ | $1$ |
| $-\cos 2x$ | $-1$ | $0$ | $1$ | $0$ | $-1$ |

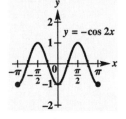

7.  $y = 2 + \cos x$

This is the graph of $y = \cos x$ translated vertically 2 units up.

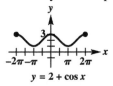

$$y = 2 + \cos x$$

8.  $y = -1 + 2\sin(x + \pi)$

*Step 1*: Find the interval whose length is $\dfrac{2\pi}{b}$.

$0 \le x + \pi \le 2\pi \Rightarrow \Rightarrow -\pi \le x \le \pi$

*Step 2*: Divide the period into four equal parts to get the following $x$-values: $-\pi$, $-\dfrac{\pi}{2}$, $0$,

$\dfrac{\pi}{2}$, $\pi$

*Step 3*: Evaluate the function for each of the five $x$-values

| $x$ | $-\dfrac{\pi}{2}$ | $-\dfrac{\pi}{2}$ | $0$ | $\dfrac{\pi}{2}$ | $\pi$ |
|---|---|---|---|---|---|
| $x+\pi$ | $0$ | $\dfrac{\pi}{2}$ | $\pi$ | $\dfrac{3\pi}{2}$ | $2\pi$ |
| $\sin(x+\pi)$ | $0$ | $1$ | $0$ | $-1$ | $0$ |
| $2\sin(x+\pi)$ | $0$ | $2$ | $0$ | $-2$ | $0$ |
| $-1+2\sin(x+\pi)$ | $-1$ | $1$ | $-1$ | $-3$ | $-1$ |

*Steps 4 and 5*: Plot the points found in the table and join them with a sinusoidal curve. Repeat this cycle for the interval $[-\pi, 0]$.

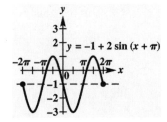

The amplitude is $|2|$, which is 2. The period is $2\pi$. The vertical translation is 1 unit down. The phase shift is $\pi$ units to the left.

**9.** $y = \tan\left(x - \dfrac{\pi}{2}\right)$

Period: $\pi$

Vertical translation: none

Phase shift (horizontal translation): $\dfrac{\pi}{2}$ units to the right

Because the function is to be graphed over a two-period interval, locate three adjacent vertical asymptotes. Because asymptotes of the graph $y = \tan x$ occur at $-\dfrac{\pi}{2}$, and $\dfrac{\pi}{2}$, use the following equations to locate asymptotes: $x - \dfrac{\pi}{2} = -\dfrac{\pi}{2} \Rightarrow x = 0$ and

$x - \dfrac{\pi}{2} = \dfrac{\pi}{2} \Rightarrow x = \pi$. Divide the interval $(0, \pi)$ into four equal parts to obtain the key $x$-values: first-quarter value: $\dfrac{\pi}{4}$;

middle value: $\dfrac{\pi}{2}$; third-quarter value: $\dfrac{3\pi}{4}$

Evaluating the given function at these three key $x$-values gives the points: $\left(\dfrac{\pi}{4}, -1\right)$,

$\left(\dfrac{\pi}{2}, 0\right)$, $\left(\dfrac{3\pi}{4}, 1\right)$. Connect these points with a smooth curve and continue to graph to approach the asymptote $x = 0$ and $x = \pi$ to complete one period of the graph. Repeat this cycle for the interval $[-\pi, 0]$.

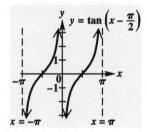

**10.** $y = -2 - \cot\left(x - \dfrac{\pi}{2}\right)$

Period: $\dfrac{\pi}{b} = \dfrac{\pi}{1} = \pi$

Vertical translation: 2 units down

Phase shift (horizontal translation): $\dfrac{\pi}{2}$ units to the right

Because the function is to be graphed over a two-period interval, locate three adjacent vertical asymptotes. Because asymptotes of the graph $y = \cot x$ occur at multiples of $\pi$, use the following equations to locate asymptotes: $x - \dfrac{\pi}{2} = -\pi$, $x - \dfrac{\pi}{2} = 0$, and

$x - \dfrac{\pi}{2} = \pi$. Solve each of these equations:

$x - \dfrac{\pi}{2} = -\pi \Rightarrow x = -\dfrac{\pi}{2}$; $x - \dfrac{\pi}{2} = 0 \Rightarrow x = \dfrac{\pi}{2}$;

$x - \dfrac{\pi}{2} = \pi \Rightarrow x = \dfrac{3\pi}{2}$

Divide the interval $\left(-\dfrac{\pi}{2}, \dfrac{\pi}{2}\right)$ into four equal parts to obtain the following key $x$-values.

first-quarter value: $-\dfrac{\pi}{4}$ middle value: $0$;

third-quarter value: $\dfrac{\pi}{4}$

Evaluating the given function at these three key $x$-values gives the points: $\left(-\dfrac{\pi}{4}, -3\right)$,

$(0, -2)$, $\left(\dfrac{\pi}{4}, -1\right)$

Connect these points with a smooth curve and continue to graph to approach the asymptote $x = -\dfrac{\pi}{2}$ and $x = \dfrac{\pi}{2}$ to complete one period of the graph. Sketch the identical curve between the asymptotes $x = \dfrac{\pi}{2}$ and $x = \dfrac{3\pi}{2}$ to complete a second period of the graph.

**11.** $y = -\csc 2x$

*Step 1*: Graph the corresponding reciprocal function $y = -\sin 2x$ The period is $\frac{2\pi}{2} = \pi$ and its amplitude is $|-1| = 1$. One period is in the interval $0 \le x \le \pi$. Dividing the interval into four equal parts gives us the key points: $(0,0)$, $\left(\frac{\pi}{2},-1\right)$, $\left(\frac{\pi}{2},0\right)$, $\left(\frac{3\pi}{4},1\right)$, $(\pi,0)$

*Step 2*: The vertical asymptotes of $y = -\csc 2x$ are at the x-intercepts of $y = -\sin 2x$, which are $x = 0, x = \frac{\pi}{2}$, and $x = \pi$. Continuing this pattern to the right, we also have a vertical asymptotes of $x = \frac{3\pi}{2}$ and $x = 2\pi$.

*Step 3*: Sketch the graph.

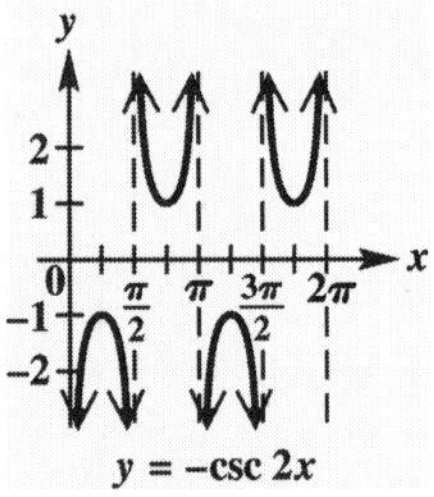


**12.** $y = 3\csc \pi x$

*Step 1*: Graph the corresponding reciprocal function $y = 3\sin \pi x$. The period is $\frac{2\pi}{2} = 2$, and the amplitude is $|3| = 3$. One period is in the interval $0 \le x \le 2$. Dividng the interval into four equal parts give the key points $(0,0)$, $\left(\frac{1}{2},1\right)$, $(1,0)$, $\left(\frac{3}{2},-1\right)$, $(2,0)$.

*Step 2*: The vertical asymptotes of $y = 3\csc \pi x$ are at the $x$-intercepts of $y = 3\sin \pi x$, namely $x = 0$, $x = 1$, and $x = 2$.

*Step 3*: Sketch the graph. Repeat this cycle for the interval [−2, 0].

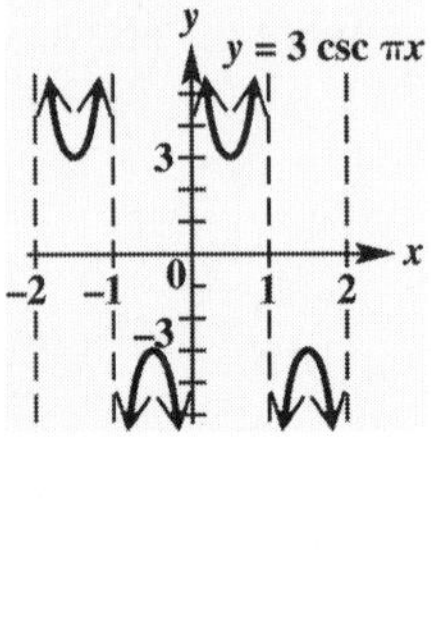


Period: 2.
Amplitude: Not applicable
Phase shift: none
Vertical translation: none

**13.** **(a)** $f(x) = 17.5\sin\left[\frac{\pi}{6}(x-4)\right] + 67.5$

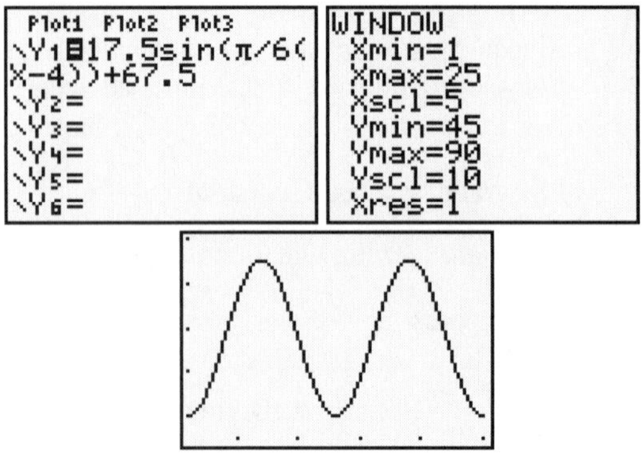


**(b)** Amplitude: 17.5

Period: $\frac{2\pi}{\frac{\pi}{6}} = 2\pi \cdot \frac{6}{\pi} = 12;$

Phase shift: 4 units to the right
Vertical translation: 67.5 units up

**(c)** For the month of December, $x = 12$.

$$f(12) = 17.5\sin\left[\frac{\pi}{6}(12-4)\right] + 67.5$$
$$= 17.5\sin\left(\frac{4\pi}{3}\right) + 67.5$$
$$= 17.5\left(-\frac{\sqrt{3}}{2}\right) + 67.5 \approx 52°$$

**(d)** A minimum of 50 occurring at $x = 13 = 12 + 1$ implies 50°F in January.

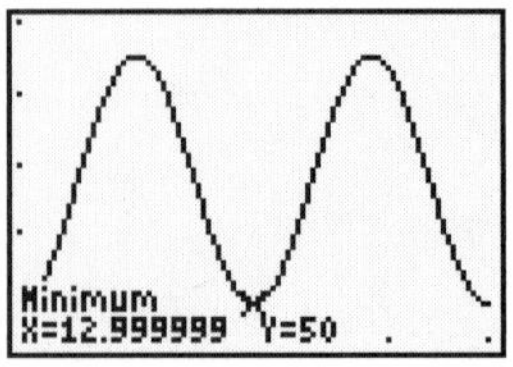


A maximum of 85 occurring at $x = 7$ and $x = 19 = 12 + 7$ implies 85°F in July.

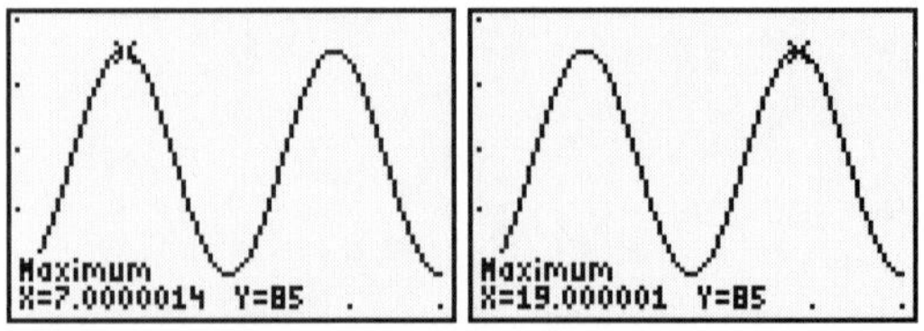


**(e)** Approximately 67.5° would be an average yearly temperature. This is the vertical translation.

**14.** $s(t) = -4\cos 8\pi t$, $a = |-4| = 4$, $\omega = 8\pi$

    **(a)** maximum height = amplitude
$$= a = |-4| = 4 \text{ in.}$$

    **(b)** $s(t) = -4\cos 8\pi t = 4 \Rightarrow \cos 8\pi t = -1 \Rightarrow$
$$8\pi t = \pi \Rightarrow t = \frac{1}{8}$$

    The weight first reaches its maximum
height after $\dfrac{1}{8}$ sec.

    **(c)** frequency $= \dfrac{\omega}{2\pi} = \dfrac{8\pi}{2\pi} = 4$ cycles per sec;

    period $= \dfrac{2\pi}{\omega} = \dfrac{2\pi}{8\pi} = \dfrac{1}{4}$ sec

**15.** The functions $y = \sin x$ and $y = \cos x$ both have all real numbers as their domains. The

functions $f(x) = \tan x = \dfrac{\sin x}{\cos x}$ and

$f(x) = \sec x = \dfrac{1}{\cos x}$ both have cos $x$ in their

denominators. Therefore, both the tangent and secnt functions have the same restrictions on their domains. Similarly, $f(x) = \cot x = \dfrac{\cos x}{\sin x}$

and $f(x) = \csc x = \dfrac{1}{\sin x}$ both have sin $x$ in

their denominators, and so have the same restrictions on their domains.

# Chapter 5

## Trigonometric Identities And Equations

### Section 5.1: Fundamental Identities

1. By a negative-angle identity,
   $\tan(-\theta) = -\tan\theta.$ Thus, if $\tan\theta = 2.6,$ then
   $\tan(-\theta) = \underline{-2.6}.$

3. By a reciprocal identity, $\cot\theta = \dfrac{1}{\tan\theta}.$ Thus,
   if $\tan\theta = 1.6,$ then $\cot\theta = \underline{.625}.$

5. By a negative angle identity $\sin(-\theta) = -\sin\theta.$
   Thus, $-\sin(-\theta) = -(-\sin\theta) = \sin\theta.$ So, if
   $\sin\theta = \dfrac{2}{3},$ then $-\sin(-\theta) = \underline{\dfrac{2}{3}}.$

7. $\cos\theta = \dfrac{3}{4},$ $\theta$ is in quadrant I.
   An identity that relates sine and cosine is
   $\sin^2 s + \cos^2 s = 1.$

   $$\sin^2\theta + \cos^2\theta = 1 \Rightarrow \sin^2\theta + \left(\dfrac{3}{4}\right)^2 = 1 \Rightarrow$$

   $$\sin^2\theta = 1 - \dfrac{9}{16} = \dfrac{7}{16} \Rightarrow \sin\theta = \pm\dfrac{\sqrt{7}}{4}$$

   Since $\theta$ is in quadrant I, $\sin\theta = \dfrac{\sqrt{7}}{4}.$

9. $\cos(-\theta) = \dfrac{\sqrt{5}}{5},$ $\tan\theta < 0$

   Since $\cos(-\theta) = \dfrac{\sqrt{5}}{5},$ we have $\cos\theta = \dfrac{\sqrt{5}}{5}$
   by a negative angle identity. An identity that
   relates sine and cosine is $\sin^2\theta + \cos^2\theta = 1.$

   $$\sin^2\theta + \cos^2\theta = 1 \Rightarrow \sin^2\theta + \left(\dfrac{\sqrt{5}}{5}\right)^2 = 1 \Rightarrow$$

   $$\sin^2\theta + \dfrac{5}{25} = 1 \Rightarrow \sin^2\theta + \dfrac{1}{5} = 1$$

   $$\sin^2\theta = 1 - \dfrac{1}{5} = \dfrac{4}{5} \Rightarrow$$

   $$\sin\theta = \pm\dfrac{2}{\sqrt{5}} = \pm\dfrac{2}{\sqrt{5}} \cdot \dfrac{\sqrt{5}}{\sqrt{5}} = \pm\dfrac{2\sqrt{5}}{5}$$

Since $\tan\theta < 0$ and $\cos\theta > 0,$ $\theta$ is in
quadrant IV, so $\sin\theta < 0.$ Thus,

$$\sin\theta = -\dfrac{2\sqrt{5}}{5}.$$

11. $\sec\theta = \dfrac{11}{4},$ $\tan\theta < 0$

    Since $\cos\theta = \dfrac{1}{\sec\theta},$ $\cos\theta = \dfrac{1}{\dfrac{11}{4}} = \dfrac{4}{11}.$ Use the

    identity $\sin^2\theta + \cos^2\theta = 1,$ to obtain

    $$\sin^2\theta + \cos^2\theta = 1 \Rightarrow \sin^2\theta + \left(\dfrac{4}{11}\right)^2 = 1 \Rightarrow$$

    $$\sin^2\theta + \dfrac{16}{121} = 1 \Rightarrow \sin^2\theta = 1 - \dfrac{16}{121} \Rightarrow$$

    $$\sin^2\theta = \dfrac{105}{121} \Rightarrow \sin\theta = \pm\dfrac{\sqrt{105}}{11}$$

    Since $\tan\theta < 0$ and $\sec\theta > 0,$ $\theta$ is in
    quadrant IV, so $\sin\theta < 0.$ Thus,

    $$\sin\theta = -\dfrac{\sqrt{105}}{11}.$$

13. The quadrants are given so that one can
    determine which sign $(+ \text{ or } -)$ $\sin\theta$ will take.

    Since $\sin\theta = \dfrac{1}{\csc\theta},$ the sign of $\sin\theta$ will be
    the same as $\csc\theta.$

15. $\sin(-x) = \underline{-\sin x}$

17. $\cos(-x) = \underline{\cos x}$

19. $\tan(-x) = \underline{-\tan x}$

21. This is the graph of $f(x) = \sec x.$ It is
    symmetric about the $y$-axis. Moreover, since

    $$f(-x) = \sec(-x) = \dfrac{1}{\cos(-x)} = \dfrac{1}{\cos x} = \sec x = f(x),$$

    $$f(-x) = f(x).$$

**23.** This is the graph of $f(x) = \cot x$. It is symmetric about the origin. Moreover, since

$$f(-x) = \cot(-x) = \frac{\cos(-x)}{\sin(-x)} = \frac{\cos x}{-\sin x}$$

$$= -\frac{\cos x}{\sin x} = -\cot x = -f(x)$$

$$f(-x) = -f(x).$$

**25.** $\sin\theta = \dfrac{2}{3}, \theta$ in quadrant II

Since $\theta$ is in quadrant II, the sine and cosecant function values are positive. The cosine, tangent, cotangent, and secant function values are negative.

$$\sin^2\theta + \cos^2\theta = 1 \Rightarrow$$

$$\cos^2\theta = 1 - \sin^2\theta = 1 - \left(\frac{2}{3}\right)^2 = 1 - \frac{4}{9} = \frac{5}{9} \Rightarrow$$

$$\cos\theta = -\frac{\sqrt{5}}{3}, \text{ since } \cos\theta < 0$$

$$\tan\theta = \frac{\sin\theta}{\cos\theta} = \frac{\frac{2}{3}}{-\frac{\sqrt{5}}{3}} = -\frac{2}{\sqrt{5}}$$

$$= -\frac{2}{\sqrt{5}} \cdot \frac{\sqrt{5}}{\sqrt{5}} = -\frac{2\sqrt{5}}{5}$$

$$\cot\theta = \frac{1}{\tan\theta} = \frac{1}{-\frac{2}{\sqrt{5}}} = -\frac{\sqrt{5}}{2}$$

$$\sec\theta = \frac{1}{\cos\theta} = \frac{1}{-\frac{\sqrt{5}}{3}} = -\frac{3}{\sqrt{5}}$$

$$= -\frac{3}{\sqrt{5}} \cdot \frac{\sqrt{5}}{\sqrt{5}} = -\frac{3\sqrt{5}}{5}$$

$$\csc\theta = \frac{1}{\sin\theta} = \frac{1}{\frac{2}{3}} = \frac{3}{2}$$

**27.** $\tan\theta = -\dfrac{1}{4}, \theta$ in quadrant IV

Since $\theta$ is in quadrant IV, the cosine and secant function values are positive. The sine, tangent, cotangent, and cosecant function values are negative.

$$\cot\theta = \frac{1}{\tan\theta} = \frac{1}{-\frac{1}{4}} = -4$$

$$\sec^2\theta = 1 + \tan^2\theta = 1 + \left(-\frac{1}{4}\right)^2$$

$$= 1 + \frac{1}{16} = \frac{17}{16} \Rightarrow$$

$$\sec\theta = \frac{\sqrt{17}}{4}, \text{ since } \sec\theta > 0$$

$$\cos\theta = \frac{1}{\sec\theta} = \frac{1}{\frac{\sqrt{17}}{4}} = \frac{4}{\sqrt{17}}$$

$$= \frac{4}{\sqrt{17}} \cdot \frac{\sqrt{17}}{\sqrt{17}} = \frac{4\sqrt{17}}{17}$$

$$\sin^2\theta + \cos^2\theta = 1 \Rightarrow$$

$$\sin^2\theta = 1 - \cos^2\theta = 1 - \left(\frac{4}{\sqrt{17}}\right)^2$$

$$\sin^2\theta = 1 - \frac{16}{17} = \frac{1}{17} \Rightarrow$$

$$\sin\theta = -\frac{1}{\sqrt{17}} = -\frac{1}{\sqrt{17}} \cdot \frac{\sqrt{17}}{\sqrt{17}} = -\frac{\sqrt{17}}{17},$$

since $\sin\theta < 0$

$$\csc\theta = \frac{1}{\sin\theta} = \frac{1}{-\frac{1}{\sqrt{17}}} = -\sqrt{17}$$

**29.** $\cot\theta = \dfrac{4}{3}, \sin\theta > 0$

Since $\cot\theta > 0$ and $\sin\theta > 0, \theta$ is in quadrant I, so all the function values are positive.

$$\tan = \frac{1}{\cot\theta} = \frac{1}{\frac{4}{3}} = \frac{3}{4}$$

$$\sec^2\theta = 1 + \tan^2\theta = 1 + \left(\frac{3}{4}\right)^2 = 1 + \frac{9}{16} = \frac{25}{16} \Rightarrow$$

$$\sec\theta = \frac{5}{4}, \text{ since } \sec\theta > 0$$

$$\cos\theta = \frac{1}{\sec\theta} = \frac{1}{\frac{5}{4}} = \frac{4}{5}$$

$$\sin^2\theta = 1 - \cos^2\theta = 1 - \left(\frac{4}{5}\right)^2 = 1 - \frac{16}{25} = \frac{9}{25} \Rightarrow$$

$$\sin\theta = \frac{3}{5}, \text{ since } \sin\theta > 0$$

$$\csc\theta = \frac{1}{\sin\theta} = \frac{1}{\frac{3}{5}} = \frac{5}{3}$$

**31.** $\sec\theta = \dfrac{4}{3}, \sin\theta < 0$

Since $\sec\theta > 0$ and $\sin\theta < 0, \theta$ is in quadrant IV. Since $\theta$ is in quadrant IV, the cosine function value is positive. The tangent, cotangent, and cosecant function values are negative.

$$\cos\theta = \frac{1}{\sec\theta} = \frac{1}{\frac{4}{3}} = \frac{3}{4}$$

*(continued on next page)*

*(continued from page 109)*

$$\sin^2\theta = 1 - \cos^2\theta = 1 - \left(\frac{3}{4}\right)^2 = 1 - \frac{9}{16} = \frac{7}{16} \Rightarrow$$

$$\sin\theta = -\frac{\sqrt{7}}{4}, \text{ since } \sin\theta < 0$$

$$\tan\theta = \frac{\sin\theta}{\cos\theta} = \frac{-\frac{\sqrt{7}}{4}}{\frac{3}{4}} = -\frac{\sqrt{7}}{3}$$

$$\cot\theta = \frac{1}{\tan\theta} = -\frac{1}{\frac{\sqrt{7}}{3}} = -\frac{3}{\sqrt{7}} \cdot \frac{\sqrt{7}}{\sqrt{7}} = -\frac{3\sqrt{7}}{7}$$

$$\csc\theta = \frac{1}{\sin\theta} = \frac{1}{-\frac{\sqrt{7}}{4}} = -\frac{4}{\sqrt{7}} \cdot \frac{\sqrt{7}}{\sqrt{7}} = -\frac{4\sqrt{7}}{7}$$

**33.** Since $\frac{\cos x}{\sin x} = \cot x$, choose expression B.

**35.** Since $\cos(-x) = \cos x$, choose expression E.

**37.** Since $1 = \sin^2 x + \cos^2 x$, choose expression A.

**39.** Since $\sec^2 x - 1 = \tan^2 x = \frac{\sin^2 x}{\cos^2 x}$, choose expression A.

**41.** Since $1 + \sin^2 x = \left(\csc^2 x - \cot^2 x\right) + \sin^2 x$, choose expression D.

**43.** It is incorrect to state $1 + \cot^2 = \csc^2$. Cotangent and cosecant are functions of some variable such as $\theta, x, \text{ or } t$. An acceptable statement would be $1 + \cot^2\theta = \csc^2\theta$.

**45.** Find $\sin\theta$ if $\cos\theta = \frac{x}{x+1}$.

$\sin^2\theta + \cos^2\theta = 1$ and $\cos\theta = \frac{x}{x+1}$, so

$$\sin^2\theta = 1 - \cos^2\theta = 1 - \left(\frac{x}{x+1}\right)^2$$

$$= 1 - \frac{x^2}{(x+1)^2} = \frac{(x+1)^2 - x^2}{(x+1)^2}$$

$$= \frac{x^2 + 2x + 1 - x^2}{(x+1)^2} = \frac{2x+1}{(x+1)^2}$$

Thus, $\sin\theta = \frac{\pm\sqrt{2x+1}}{x+1}$.

**47.** $\sin^2 x + \cos^2 x = 1 \Rightarrow \sin^2 x = 1 - \cos^2 x \Rightarrow$
$\sin x = \pm\sqrt{1 - \cos^2 x}$

**49.** $\tan^2 x + 1 = \sec^2 x \Rightarrow \tan^2 x = \sec^2 x - 1 \Rightarrow$
$\tan x = \pm\sqrt{\sec^2 x - 1}$

**51.** $\csc x = \frac{1}{\sin x} \Rightarrow$

$$\csc x = \frac{1}{\pm\sqrt{1 - \cos^2 x}}$$

$$= \frac{\pm 1}{\sqrt{1 - \cos^2 x}} \cdot \frac{\sqrt{1 - \cos^2 x}}{\sqrt{1 - \cos^2 x}}$$

$$= \frac{\pm\sqrt{1 - \cos^2 x}}{1 - \cos^2 x}$$

**53.** $\cot\theta\sin\theta = \frac{\cos\theta}{\sin\theta} \cdot \sin\theta = \cos\theta$

**55.** $\cos\theta\csc\theta = \cos\theta \cdot \frac{1}{\sin\theta} = \frac{\cos\theta}{\sin\theta} = \cot\theta$

**57.** $\sin^2\theta\left(\csc^2\theta - 1\right) = \sin^2\theta\left(\frac{1}{\sin^2\theta} - 1\right)$

$$= \frac{\sin^2\theta}{\sin^2\theta} - \sin^2\theta$$

$$= 1 - \sin^2\theta = \cos^2\theta$$

**59.** $\left(1 - \cos\theta\right)\left(1 + \sec\theta\right)$
$\quad = 1 + \sec\theta - \cos\theta - \cos\theta\sec\theta$
$\quad = 1 + \sec\theta - \cos\theta - \cos\theta\left(\frac{1}{\cos\theta}\right)$
$\quad = 1 + \sec\theta - \cos\theta - 1 = \sec\theta - \cos\theta$

**61.** $\dfrac{\cos^2\theta - \sin^2\theta}{\sin\theta\cos\theta} = \dfrac{\cos^2\theta}{\sin\theta\cos\theta} - \dfrac{\sin^2\theta}{\sin\theta\cos\theta}$

$$= \frac{\cos\theta}{\sin\theta} - \frac{\sin\theta}{\cos\theta} = \cot\theta - \tan\theta$$

**63.** $\sec\theta - \cos\theta = \dfrac{1}{\cos\theta} - \cos\theta = \dfrac{1}{\cos\theta} - \dfrac{\cos^2\theta}{\cos\theta}$

$$= \frac{1 - \cos^2\theta}{\cos\theta} = \frac{\sin^2\theta}{\cos\theta}$$

$$= \frac{\sin\theta}{\cos\theta} \cdot \sin\theta = \tan\theta\sin\theta$$

**65.** $\sin\theta\left(\csc\theta - \sin\theta\right) = \sin\theta\csc\theta - \sin^2\theta$

$$= \sin\theta \cdot \frac{1}{\sin\theta} - \sin^2\theta$$

$$= 1 - \sin^2\theta = \cos^2\theta$$

**67.** $\sin^2\theta + \tan^2\theta + \cos^2\theta$
$\quad = \left(\sin^2\theta + \cos^2\theta\right) + \tan^2\theta$
$\quad = 1 + \tan^2\theta = \sec^2\theta$

**69.** Since $\cos x = \dfrac{1}{5}$, which is positive, $x$ is in quadrant I or quadrant IV.

$$\sin x = \pm\sqrt{1-\cos^2 x} = \pm\sqrt{1-\left(\dfrac{1}{5}\right)^2} = \pm\sqrt{\dfrac{24}{25}}$$

$$= \pm\dfrac{\sqrt{24}}{5} = \pm\dfrac{2\sqrt{6}}{5}$$

$$\tan x = \dfrac{\sin x}{\cos x} = \dfrac{\pm\frac{2\sqrt{6}}{5}}{\frac{1}{5}} = \pm 2\sqrt{6}$$

$$\sec x = \dfrac{1}{\cos x} = \dfrac{1}{\frac{1}{5}} = 5$$

Quadrant I:

$$\dfrac{\sec x - \tan x}{\sin x} = \dfrac{5-2\sqrt{6}}{\frac{2\sqrt{6}}{5}} = \dfrac{25-10\sqrt{6}}{2\sqrt{6}}$$

$$= \dfrac{25-10\sqrt{6}}{2\sqrt{6}} \cdot \dfrac{\sqrt{6}}{\sqrt{6}} = \dfrac{25\sqrt{6}-60}{12}$$

Quadrant IV:

$$\dfrac{\sec x - \tan x}{\sin x} = \dfrac{5-\left(-2\sqrt{6}\right)}{-\frac{2\sqrt{5}}{5}} = \dfrac{25+10\sqrt{6}}{-2\sqrt{6}}$$

$$= \dfrac{25+10\sqrt{6}}{-2\sqrt{6}} \cdot \dfrac{-\sqrt{6}}{-\sqrt{6}} = \dfrac{-25\sqrt{6}-60}{12}$$

**71.** $y = \sin(-2x) \Rightarrow y = -\sin(2x)$

**73.** $y = \cos(-4x) \Rightarrow y = \cos(4x)$

**75.** **(a)** $y = \sin(-4x) \Rightarrow y = -\sin(4x)$

**(b)** $y = \cos(-2x) \Rightarrow y = \cos(2x)$

**(c)** $y = -5\sin(-3x) \Rightarrow y = -5\left[-\sin(3x)\right] \Rightarrow$
$y = 5\sin(3x)$

In Exercises 77–81, the functions are graphed in the following window.

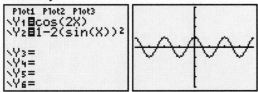

**77.** The equation $\cos 2x = 1 - 2\sin^2 x$ is an identity.

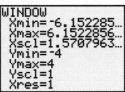

**79.** The equation $\sin x = \sqrt{1-\cos^2 x}$ is not an identity.

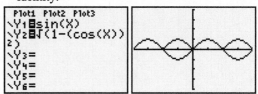

**81.** Does $\cos(x-y) = \cos x - \cos y$ ? If it does, then the graphs of $Y_1 = \cos(x-y)$ and $Y_2 = \cos x - \cos y$ will overlap for specific values of y. We will graph 3 cases:

$$y = \dfrac{\pi}{4}, \; y = \dfrac{\pi}{2}, \text{ and } y = \pi.$$

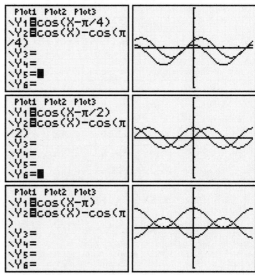

The equation $\cos(x-y) = \cos x - \cos y$ is not an identity.

# Section 5.2: Verifying Trigonometric Identities

**1.** $\cot\theta + \dfrac{1}{\cot\theta} = \cot\theta + \tan\theta$

$$= \dfrac{\cos\theta}{\sin\theta} + \dfrac{\sin\theta}{\cos\theta}$$

$$= \dfrac{\cos^2\theta + \sin^2\theta}{\sin\theta\cos\theta}$$

$$= \dfrac{1}{\sin\theta\cos\theta} \text{ or } \csc\theta\sec\theta$$

**3.** $\tan s(\cot s + \csc s) = \dfrac{\sin s}{\cos s}\left(\dfrac{\cos s}{\sin s} + \dfrac{1}{\sin s}\right)$

$$= 1 + \dfrac{1}{\cos s}$$

$$= 1 + \sec s$$

**5.** $\dfrac{1}{\csc^2 \theta} + \dfrac{1}{\sec^2 \theta} = \sin^2 \theta + \cos^2 \theta = 1$

**7.** $\dfrac{\cos x}{\sec x} + \dfrac{\sin x}{\csc x} = \dfrac{\cos x}{\dfrac{1}{\cos x}} + \dfrac{\sin x}{\dfrac{1}{\sin x}}$

$\qquad = \cos x \left( \dfrac{\cos x}{1} \right) + \sin x \left( \dfrac{\sin x}{1} \right)$

$\qquad = \cos^2 x + \sin^2 x = 1$

**9.** $(1 + \sin t)^2 + \cos^2 t = 1 + 2\sin t + \sin^2 t + \cos^2 t$

$\qquad = 1 + 2\sin t + \left( \sin^2 t + \cos^2 t \right)$

$\qquad = 1 + 2\sin t + 1 = 2 + 2\sin t$

**11.** $\dfrac{1}{1 + \cos x} - \dfrac{1}{1 - \cos x}$

$= \dfrac{1 - \cos x}{(1 + \cos x)(1 - \cos x)} - \dfrac{1 + \cos x}{(1 + \cos x)(1 - \cos x)}$

$= \dfrac{(1 - \cos x) - (1 + \cos x)}{(1 + \cos x)(1 - \cos x)}$

$= \dfrac{1 - \cos x - 1 - \cos x}{1 - \cos^2 x} = -\dfrac{2\cos x}{\sin^2 x}$ or

$-\dfrac{2\cos x}{\sin^2 x} = -\dfrac{2\cos x}{\sin x \sin x} = -2 \left( \dfrac{\cos x}{\sin x} \right) \left( \dfrac{1}{\sin x} \right)$

$\qquad\qquad = -2 \cot x \csc x$

**13.** $\sin^2 \theta - 1 = (\sin \theta + 1)(\sin \theta - 1)$

**15.** $(\sin x + 1)^2 - (\sin x - 1)^2$

$= \left[ (\sin x + 1) + (\sin x - 1) \right]$

$\qquad\qquad \cdot \left[ (\sin x + 1) - (\sin x - 1) \right]$

$= (\sin x + 1 + \sin x - 1)(\sin x + 1 - \sin x + 1)$

$= (2\sin x)(2) = 4\sin x$

**17.** $2\sin^2 x + 3\sin x + 1$

Let $a = \sin x$.

$2\sin^2 x + 3\sin x + 1 = 2a^2 + 3a + 1$

$\qquad\qquad = (2a + 1)(a + 1)$

$\qquad\qquad = (2\sin x + 1)(\sin x + 1)$

**19.** $\cos^4 x + 2\cos^2 x + 1$

Let $\cos^2 x = a$.

$\cos^4 x + 2\cos^2 x + 1 = a^2 + 2a + 1$

$\qquad\qquad = (a + 1)^2 = \left( \cos^2 x + 1 \right)^2$

**21.** $\sin^3 x - \cos^3 x$

Let $\sin x = a$ and $\cos x = b$.

$\sin^3 x - \cos^3 x$

$= a^3 - b^3 = (a - b)\left( a^2 + ab + b^2 \right)$

$= (\sin x - \cos x)\left( \sin^2 x + \sin x \cos x + \cos^2 x \right)$

$= (\sin x - \cos x)\left[ \left( \sin^2 x + \cos^2 x \right) + \sin x \cos x \right]$

$= (\sin x - \cos x)(1 + \sin x \cos x)$

**23.** $\tan \theta \cos \theta = \dfrac{\sin \theta}{\cos \theta} \cos \theta = \sin \theta$

**25.** $\sec r \cos r = \dfrac{1}{\cos r} \cdot \cos r = 1$

**27.** $\dfrac{\sin \beta \tan \beta}{\cos \beta} = \tan \beta \tan \beta = \tan^2 \beta$

**29.** $\sec^2 x - 1 = \dfrac{1}{\cos^2 x} - 1 = \dfrac{1}{\cos^2 x} - \dfrac{\cos^2 x}{\cos^2 x}$

$\qquad = \dfrac{1 - \cos^2 x}{\cos^2 x} = \dfrac{\sin^2 x}{\cos^2 x} = \tan^2 x$

**31.** $\dfrac{\sin^2 x}{\cos^2 x} + \sin x \csc x = \tan^2 x + \sin x \cdot \dfrac{1}{\sin x}$

$\qquad\qquad = \tan^2 x + 1 = \sec^2 x$

**33.** $1 - \dfrac{1}{\csc^2 x} = 1 - \sin^2 x = \cos^2 x$

**35.** Verify $\dfrac{\cot \theta}{\csc \theta} = \cos \theta$.

$\dfrac{\cot \theta}{\csc \theta} = \dfrac{\dfrac{\cos \theta}{\sin \theta}}{\dfrac{1}{\sin \theta}} = \dfrac{\cos \theta}{\sin \theta} \cdot \dfrac{\sin \theta}{1} = \cos \theta$

**37.** Verify $\dfrac{1 - \sin^2 \beta}{\cos \beta} = \cos \beta$.

$\dfrac{1 - \sin^2 \beta}{\cos \beta} = \dfrac{\cos^2 \beta}{\cos \beta} = \cos \beta$

**39.** Verify $\cos^2\theta(\tan^2\theta+1)=1$.

$$\cos^2\theta(\tan^2\theta+1)=\cos^2\theta\left(\frac{\sin^2\theta}{\cos^2\theta}+1\right)$$

$$=\cos^2\theta\left(\frac{\sin^2\theta}{\cos^2\theta}+\frac{\cos^2\theta}{\cos^2\theta}\right)$$

$$=\cos^2\theta\left(\frac{\sin^2\theta+\cos^2\theta}{\cos^2\theta}\right)$$

$$=\cos^2\theta\left(\frac{1}{\cos^2\theta}\right)=1$$

**41.** Verify $\cot s+\tan s=\sec s\csc s$.

$$\cot s+\tan s=\frac{\cos s}{\sin s}+\frac{\sin s}{\cos s}$$

$$=\frac{\cos^2 s}{\sin s\cos s}+\frac{\sin^2 s}{\sin s\cos s}$$

$$=\frac{\cos^2 s+\sin^2 s}{\cos s\sin s}=\frac{1}{\cos s\sin s}$$

$$=\frac{1}{\cos s}\cdot\frac{1}{\sin s}=\sec s\csc s$$

**43.** Verify $\dfrac{\cos\alpha}{\sec\alpha}+\dfrac{\sin\alpha}{\csc\alpha}=\sec^2\alpha-\tan^2\alpha$.

Working with the left side, we have

$$\frac{\cos\alpha}{\sec\alpha}+\frac{\sin\alpha}{\csc\alpha}=\frac{\cos\alpha}{\dfrac{1}{\cos\alpha}}+\frac{\sin\alpha}{\dfrac{1}{\sin\alpha}}$$

$$=\cos^2\alpha+\sin^2\alpha=1$$

Working with the right side, we have

$$\sec^2\alpha-\tan^2\alpha=1.$$

Since $\dfrac{\cos\alpha}{\sec\alpha}+\dfrac{\sin\alpha}{\csc\alpha}=1=\sec^2\alpha-\tan^2\alpha$,

the statement has been verified.

**45.** Verify $\sin^4\theta-\cos^4\theta=2\sin^2\theta-1$.

$$\sin^4\theta-\cos^4\theta$$

$$=\left(\sin^2\theta+\cos^2\theta\right)\left(\sin^2\theta-\cos^2\theta\right)$$

$$=1\cdot\left(\sin^2\theta-\cos^2\theta\right)=\sin^2\theta-\cos^2\theta$$

$$=\sin^2\theta-\left(1-\sin^2\theta\right)$$

$$=\sin^2\theta-1+\sin^2\theta=2\sin^2\theta-1$$

**47.** Verify $\dfrac{1-\cos x}{1+\cos x}=(\cot x-\csc x)^2$.

Work with the left side.

$$\frac{1-\cos x}{1+\cos x}=\frac{(1-\cos x)(1-\cos x)}{(1+\cos x)(1-\cos x)}$$

$$=\frac{1-2\cos x+\cos^2 x}{1-\cos^2 x}$$

$$=\frac{1-2\cos x+\cos^2 x}{\sin^2 x}$$

Work with the right side.

$$(\cot x-\csc x)^2=\left(\frac{\cos x}{\sin x}-\frac{1}{\sin x}\right)^2$$

$$=\left(\frac{\cos x-1}{\sin x}\right)^2$$

$$=\frac{\cos^2 x-2\cos x+1}{\sin^2 x}$$

$$\frac{1-\cos x}{1+\cos x}=\frac{\cos^2 x-2\cos x+1}{\sin^2 x}$$

$$=(\cot x-\csc x)^2$$

Thus, the statement has been verified.

**49.** Verify $\dfrac{\cos\theta+1}{\tan^2\theta}=\dfrac{\cos\theta}{\sec\theta-1}$.

Work with the left side.

$$\frac{\cos\theta+1}{\tan^2\theta}=\frac{\cos\theta+1}{\sec^2\theta-1}=\frac{\cos\theta+1}{\dfrac{1}{\cos^2\theta}-1}$$

$$=\frac{(\cos\theta+1)\cos^2\theta}{\left(\dfrac{1}{\cos^2\theta}-1\right)\cos^2\theta}$$

$$=\frac{\cos^2\theta(\cos\theta+1)}{1-\cos^2\theta}$$

$$=\frac{\cos^2\theta(\cos\theta+1)}{(1+\cos\theta)(1-\cos\theta)}=\frac{\cos^2\theta}{1-\cos\theta}$$

Now work with the right side.

$$\frac{\cos\theta}{\sec\theta-1}=\frac{\cos\theta}{\dfrac{1}{\cos\theta}-1}$$

$$=\frac{\cos\theta}{\dfrac{1}{\cos\theta}-1}\cdot\frac{\cos\theta}{\cos\theta}=\frac{\cos^2\theta}{1-\cos\theta}$$

$$\frac{\cos\theta+1}{\tan^2\theta}=\frac{\cos^2\theta}{1-\cos\theta}=\frac{\cos\theta}{\sec\theta-1}$$

Thus, the statement has been verified.

**51.** Verify $\dfrac{1}{1-\sin\theta}+\dfrac{1}{1+\sin\theta}=2\sec^2\theta.$

$$\dfrac{1}{1-\sin\theta}+\dfrac{1}{1+\sin\theta}$$

$$=\dfrac{1+\sin\theta}{(1+\sin\theta)(1-\sin\theta)}+\dfrac{1-\sin\theta}{(1+\sin\theta)(1-\sin\theta)}$$

$$=\dfrac{(1+\sin\theta)+(1-\sin\theta)}{(1+\sin\theta)(1-\sin\theta)}$$

$$=\dfrac{1+\sin\theta+1-\sin\theta}{(1+\sin\theta)(1-\sin\theta)}$$

$$=\dfrac{2}{1-\sin^2\theta}=\dfrac{2}{\cos^2\theta}=2\sec^2\theta$$

**53.** Verify $\dfrac{\cot\alpha+1}{\cot\alpha-1}=\dfrac{1+\tan\alpha}{1-\tan\alpha}.$

$$\dfrac{\cot\alpha+1}{\cot\alpha-1}=\dfrac{\dfrac{\cos\alpha}{\sin\alpha}+1}{\dfrac{\cos\alpha}{\sin\alpha}-1}=\dfrac{\dfrac{\cos\alpha}{\sin\alpha}+1}{\dfrac{\cos\alpha}{\sin\alpha}-1}\cdot\dfrac{\sin\alpha}{\sin\alpha}$$

$$=\dfrac{\cos\alpha+\sin\alpha}{\cos\alpha-\sin\alpha}$$

$$=\dfrac{\cos\alpha+\sin\alpha}{\cos\alpha-\sin\alpha}\cdot\dfrac{\dfrac{1}{\cos\alpha}}{\dfrac{1}{\cos\alpha}}$$

$$=\dfrac{\dfrac{\cos\alpha}{\cos\alpha}+\dfrac{\sin\alpha}{\cos\alpha}}{\dfrac{\cos\alpha}{\cos\alpha}-\dfrac{\sin\alpha}{\cos\alpha}}=\dfrac{1+\tan\alpha}{1-\tan\alpha}$$

**55.** Verify $\sec^4 x-\sec^2 x=\tan^4 x+\tan^2 x.$
Simplify the left side.

$$\sec^4 x-\sec^2 x=\sec^2 x\left(\sec^2 x-1\right)$$

$$=\sec^2 x\tan^2 x=\tan^2 x\sec^2 x$$

Simplify the right side.

$$\tan^4 x+\tan^2 x=\tan^2 x\left(\tan^2 x+1\right)$$

$$=\tan^2 x\sec^2 x$$

$$\sec^4 x-\sec^2 x=\tan^2 x\sec^2 x=\tan^4 x+\tan^2 x$$

Thus, the statement has been verified.

**57.** Verify $\dfrac{\sec^4 s-\tan^4 s}{\sec^2 s+\tan^2 s}=\sec^2 s-\tan^2 s.$

$$\dfrac{\sec^4 s-\tan^4 s}{\sec^2 s+\tan^2 s}$$

$$=\dfrac{\left(\sec^2 s+\tan^2 s\right)\left(\sec^2 s-\tan^2 s\right)}{\sec^2 s+\tan^2 s}$$

$$=\sec^2 s-\tan^2 s$$

**59.** Verify $\dfrac{\tan^2 t-1}{\sec^2 t}=\dfrac{\tan t-\cot t}{\tan t+\cot t}.$
Simplify the right side

$$\dfrac{\tan t-\cot t}{\tan t+\cot t}=\dfrac{\tan t-\dfrac{1}{\tan t}}{\tan t+\dfrac{1}{\tan t}}$$

$$=\dfrac{\tan t-\dfrac{1}{\tan t}}{\tan t+\dfrac{1}{\tan t}}\cdot\dfrac{\tan t}{\tan t}$$

$$=\dfrac{\tan^2 t-1}{\tan^2 t+1}=\dfrac{\tan^2 t-1}{\sec^2 t}$$

**61.** Verify

$$\left(1-\cos^2\alpha\right)\left(1+\cos^2\alpha\right)=2\sin^2\alpha-\sin^4\alpha.$$

$$\left(1-\cos^2\alpha\right)\left(1+\cos^2\alpha\right)=\sin^2\alpha\left(1+\cos^2\alpha\right)$$

$$=\sin^2\alpha\left(2-\sin^2\alpha\right)$$

$$=2\sin^2\alpha-\sin^4\alpha$$

**63.** Verify $\sin^2\alpha\sec^2\alpha+\sin^2\alpha\csc^2\alpha=\sec^2\alpha.$

$$\sin^2\alpha\sec^2\alpha+\sin^2\alpha\csc^2\alpha$$

$$=\sin^2\alpha\cdot\dfrac{1}{\cos^2\alpha}+\sin^2\alpha\cdot\dfrac{1}{\sin^2\alpha}$$

$$=\dfrac{\sin^2\alpha}{\cos^2\alpha}+1=\tan^2\alpha+1=\sec^2\alpha$$

**65.** Verify $\dfrac{\tan s}{1+\cos s}+\dfrac{\sin s}{1-\cos s}=\cot s+\sec s\csc s.$

$$\dfrac{\tan s}{1+\cos s}+\dfrac{\sin s}{1-\cos s}$$

$$=\dfrac{\tan s(1-\cos s)}{(1+\cos s)(1-\cos s)}+\dfrac{\sin s(1+\cos s)}{(1+\cos s)(1-\cos s)}$$

$$=\dfrac{\tan s(1-\cos s)+\sin s(1+\cos s)}{(1+\cos s)(1-\cos s)}$$

$$=\dfrac{\tan s-\sin s+\sin s+\sin s\cos s}{1-\cos^2 s}$$

$$=\dfrac{\tan s+\sin s\cos s}{\sin^2 s}=\dfrac{\tan s}{\sin^2 s}+\dfrac{\sin s\cos s}{\sin^2 s}$$

$$=\tan s\cdot\dfrac{1}{\sin^2 s}+\dfrac{\cos s}{\sin s}=\dfrac{\sin s}{\cos s}\cdot\dfrac{1}{\sin^2 s}+\cot s$$

$$=\dfrac{1}{\cos s}\cdot\dfrac{1}{\sin s}+\cot s=\sec s\csc s+\cot s$$

**67.** Verify

$$\frac{1-\sin\theta}{1+\sin\theta} = \sec^2\theta - 2\sec\theta\tan\theta + \tan^2\theta \ .$$

Simplify the right side

$$\sec^2\theta - 2\sec\theta\tan\theta + \tan^2\theta$$

$$= \frac{1}{\cos^2\theta} - 2\cdot\frac{1}{\cos\theta}\cdot\frac{\sin\theta}{\cos\theta} + \frac{\sin^2\theta}{\cos^2\theta}$$

$$= \frac{1-2\sin\theta+\sin^2\theta}{\cos^2\theta} = \frac{(1-\sin\theta)^2}{1-\sin^2\theta}$$

$$= \frac{(1-\sin\theta)^2}{(1+\sin\theta)(1-\sin\theta)} = \frac{1-\sin\theta}{1+\sin\theta}$$

**69.** Verify

$$\frac{\sin\theta}{1-\cos\theta} - \frac{\sin\theta\cos\theta}{1+\cos\theta} = \csc\theta\left(1+\cos^2\theta\right) \ .$$

$$\frac{\sin\theta}{1-\cos\theta} - \frac{\sin\theta\cos\theta}{1+\cos\theta}$$

$$= \frac{\sin\theta(1+\cos\theta)}{(1+\cos\theta)(1-\cos\theta)} - \frac{\sin\theta\cos\theta(1-\cos\theta)}{(1+\cos\theta)(1-\cos\theta)}$$

$$= \frac{\sin\theta(1+\cos\theta)-\sin\theta\cos\theta(1-\cos\theta)}{(1+\cos\theta)(1-\cos\theta)}$$

$$= \frac{\sin\theta+\sin\theta\cos\theta-\sin\theta\cos\theta+\sin\theta\cos^2\theta}{1-\cos^2\theta}$$

$$= \frac{\sin\theta+\sin\theta\cos^2\theta}{\sin^2\theta} = \frac{1+\cos^2\theta}{\sin\theta}$$

$$= \frac{1}{\sin\theta}\left(1+\cos^2\theta\right) = \csc\theta\left(1+\cos^2\theta\right)$$

**71.** Verify $\dfrac{1+\cos x}{1-\cos x} - \dfrac{1-\cos x}{1+\cos x} = 4\cot x\csc x$ .

$$\frac{1+\cos x}{1-\cos x} - \frac{1-\cos x}{1+\cos x}$$

$$= \frac{(1+\cos x)^2}{(1+\cos x)(1-\cos x)} - \frac{(1-\cos x)^2}{(1+\cos x)(1-\cos x)}$$

$$= \frac{1+2\cos x+\cos^2 x}{(1+\cos x)(1-\cos x)} - \frac{1-2\cos x+\cos^2 x}{(1+\cos x)(1-\cos x)}$$

$$= \frac{1+2\cos x+\cos^2 x-1+2\cos x-\cos^2 x}{(1+\cos x)(1-\cos x)}$$

$$= \frac{4\cos x}{1-\cos^2 x} = \frac{4\cos x}{\sin^2 x} = 4\cdot\frac{\cos x}{\sin x}\cdot\frac{1}{\sin x}$$

$$= 4\cot x\csc x$$

**73.** Verify $\dfrac{1+\sin\theta}{1-\sin\theta} - \dfrac{1-\sin\theta}{1+\sin\theta} = 4\tan\theta\sec\theta$

$$\frac{1+\sin\theta}{1-\sin\theta} - \frac{1-\sin\theta}{1+\sin\theta}$$

$$= \frac{(1+\sin\theta)^2 - (1-\sin\theta)^2}{(1-\sin\theta)(1+\sin\theta)}$$

$$= \frac{\left(1+2\sin\theta+\sin^2\theta\right) - \left(1-2\sin\theta+\sin^2\theta\right)}{1-\sin^2\theta}$$

$$= \frac{4\sin\theta}{\cos^2\theta} = 4\frac{\sin\theta}{\cos\theta}\cdot\frac{1}{\cos\theta} = 4\tan\theta\sec\theta$$

**75.** Verify $\left(2\sin x+\cos y\right)^2 + \left(2\cos x-\sin y\right)^2 = 5$

$$\left(2\sin x+\cos x\right)^2 + \left(2\cos x-\sin x\right)^2 = \left(4\sin^2 x+4\sin x\cos x+\cos^2 x\right) + \left(4\cos^2 x-4\sin x\cos x+\sin^2 x\right)$$

$$= 4\left(\sin^2 x+\cos^2 x\right) + \left(\cos^2 x+\sin^2 x\right) = 4+1 = 5$$

**77.** Verify $\sec x - \cos x + \csc x - \sin x - \sin x\tan x = \cos x\cot x$

$$\sec x - \cos x + \csc x - \sin x - \sin x\tan x = \frac{1}{\cos x} - \cos x + \frac{1}{\sin x} - \sin x - \sin x\left(\frac{\sin x}{\cos x}\right)$$

$$= \left(\frac{1}{\cos x} - \cos x\right) + \left(\frac{1}{\sin x} - \sin x\right) - \frac{\sin^2 x}{\cos x}$$

$$= \frac{1-\cos^2 x}{\cos x} + \frac{1-\sin^2 x}{\sin x} - \frac{\sin^2 x}{\cos x}$$

$$= \left(\frac{1-\cos^2 x}{\cos x} - \frac{\sin^2 x}{\cos x}\right) + \frac{1-\sin^2 x}{\sin x} = \frac{1-\cos^2 x-\sin^2 x}{\cos x} + \frac{\cos^2 x}{\sin x}$$

$$= \frac{1-\left(\cos^2 x+\sin^2 x\right)}{\cos x} + \frac{\cos^2 x}{\sin x} = \frac{1-1}{\cos x} + \cos x\cdot\frac{\cos x}{\sin x} = \cos x\cot x$$

In Exercises 79–85, the functions are graphed in the following window.

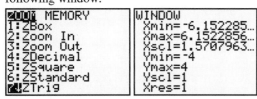

**79.** $(\sec\theta + \tan\theta)(1 - \sin\theta)$ appears to be equivalent to $\cos\theta$.

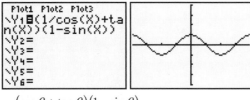

$$(\sec\theta + \tan\theta)(1 - \sin\theta)$$

$$= \left(\frac{1}{\cos\theta} + \frac{\sin\theta}{\cos\theta}\right)(1 - \sin\theta)$$

$$= \left(\frac{1 + \sin\theta}{\cos\theta}\right)(1 - \sin\theta)$$

$$= \frac{(1 + \sin\theta)(1 - \sin\theta)}{\cos\theta}$$

$$= \frac{1 - \sin^2\theta}{\cos\theta} = \frac{\cos^2\theta}{\cos\theta} = \cos\theta$$

**81.** $\dfrac{\cos\theta + 1}{\sin\theta + \tan\theta}$ appears to be equivalent to $\cot\theta$.

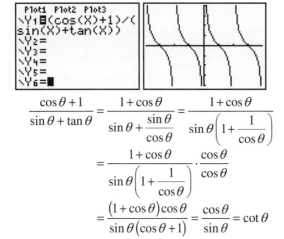

$$\frac{\cos\theta + 1}{\sin\theta + \tan\theta} = \frac{1 + \cos\theta}{\sin\theta + \dfrac{\sin\theta}{\cos\theta}} = \frac{1 + \cos\theta}{\sin\theta\left(1 + \dfrac{1}{\cos\theta}\right)}$$

$$= \frac{1 + \cos\theta}{\sin\theta\left(1 + \dfrac{1}{\cos\theta}\right)} \cdot \frac{\cos\theta}{\cos\theta}$$

$$= \frac{(1 + \cos\theta)\cos\theta}{\sin\theta(\cos\theta + 1)} = \frac{\cos\theta}{\sin\theta} = \cot\theta$$

**83.** Is $\dfrac{2 + 5\cos x}{\sin x} = 2\csc x + 5\cot x$ an identity?

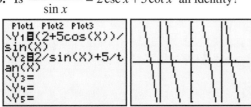

The graphs of $y = \dfrac{2 + 5\cos x}{\sin x}$ and

$y = 2\csc x + 5\cot x$ appear to be the same. The given equation may be an identity. Since

$$\frac{2 + 5\cos x}{\sin x} = \frac{2}{\sin x} + \frac{5\cos x}{\sin x} = 2\csc x + 5\cot x,$$

the given statement is an identity.

**85.** Is $\dfrac{\tan x - \cot x}{\tan x + \cot x} = 2\sin^2 x$ an identity?

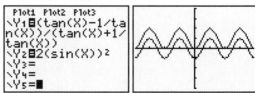

The graphs of

$y = \dfrac{\tan x - \cot x}{\tan x + \cot x}$ and $y = 2\sin^2 x$ are not the same. The given statement is not an identity.

**87.** Show that $\sin(\csc s) = 1$ is not an identity.

We need to find only one value for which the statement is false. Let $s = 2$. Use a calculator to find that $\sin(\csc 2) \approx .891094$, which is not equal to 1. $\sin(\csc s) = 1$ does not hold true for *all* real numbers $s$. Thus, it is not an identity.

**89.** Show that $\csc t = \sqrt{1 + \cot^2 t}$ is not an identity.

Let $t = \dfrac{\pi}{4}$. We have $\csc\dfrac{\pi}{4} = \sqrt{2}$ and

$$\sqrt{1 + \cot^2\frac{\pi}{4}} = \sqrt{1 + 1^2} = \sqrt{1 + 1} = \sqrt{2}.$$ But let

$t = -\dfrac{\pi}{4}$. We have $\csc\left(-\dfrac{\pi}{4}\right) = -\sqrt{2}$ and

$$\sqrt{1 + \cot^2\left(-\frac{\pi}{4}\right)} = \sqrt{1 + (-1)^2} = \sqrt{1 + 1} = \sqrt{2}.$$

$\csc t = \sqrt{1 + \cot^2 t}$ does not hold true for *all* real numbers $s$. Thus, it is not an identity.

**91.** $\sin x = \sqrt{1 - \cos^2 x}$ is a true statement when $\sin x \geq 0$.

**93.** (a)  $I = k\cos^2\theta = k\left(1 - \sin^2\theta\right)$

(b)  For $\theta = 2\pi n$ for all integers $n$, $\cos^2\theta = 1$, its maximum value and $I$ attains a maximum value of $k$.

**95. (a)** The sum of $L$ and $C$ equals 3.

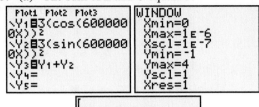

**(b)** Let $Y_1 = L(t), Y_2 = C(t),$ and $Y_3 = E(t)$

$Y_3 = 3$ for all inputs.

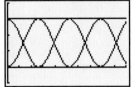

**(c)** $E(t) = L(t) + C(t)$

$= 3\cos^2(6,000,000t) + 3\sin^2(6,000,000t)$

$= 3\left[\cos^2(6,000,000t) + \sin^2(6,000,000t)\right]$

$= 3 \cdot 1 = 3$

# Section 5.3: Sum and Difference Identities for Cosine

**1.** Since $\cos(x+y) = \cos x \cos y - \sin x \sin y,$ the correct choice is F.

**3.** Since $\cos(90° - x) = \sin x,$ the correct choice is E.

**5.** $\cos 75° = \cos(30° + 45°)$

$= \cos 30° \cos 45° - \sin 30° \sin 45°$

$= \dfrac{\sqrt{3}}{2} \cdot \dfrac{\sqrt{2}}{2} - \dfrac{1}{2} \cdot \dfrac{\sqrt{2}}{2}$

$= \dfrac{\sqrt{6}}{4} - \dfrac{\sqrt{2}}{4} = \dfrac{\sqrt{6} - \sqrt{2}}{4}$

**7.** $\cos 105° = \cos(60° + 45°)$

$= \cos 60° \cos 45° - \sin 60° \sin 45°$

$= \dfrac{1}{2} \cdot \dfrac{\sqrt{2}}{2} - \dfrac{\sqrt{3}}{2} \cdot \dfrac{\sqrt{2}}{2}$

$= \dfrac{\sqrt{2}}{4} - \dfrac{\sqrt{6}}{4} = \dfrac{\sqrt{2} - \sqrt{6}}{4}$

**9.** $\cos\left(\dfrac{7\pi}{12}\right) = \cos\left(\dfrac{4\pi}{12} + \dfrac{3\pi}{12}\right) = \cos\left(\dfrac{\pi}{3} + \dfrac{\pi}{4}\right)$

$= \cos\dfrac{\pi}{3}\cos\dfrac{\pi}{4} - \sin\dfrac{\pi}{3}\sin\dfrac{\pi}{4}$

$= \dfrac{1}{2} \cdot \dfrac{\sqrt{2}}{2} - \dfrac{\sqrt{3}}{2} \cdot \dfrac{\sqrt{2}}{2}$

$= \dfrac{\sqrt{2}}{4} - \dfrac{\sqrt{6}}{4} = \dfrac{\sqrt{2} - \sqrt{6}}{4}$

**11.** $\cos 40° \cos 50° - \sin 40° \sin 50°$

$= \cos(40° + 50°) = \cos 90° = 0$

**13.** The answer to exercise 11 is 0. Using a calculator to evaluate $\cos 40° \cos 50° - \sin 40° \sin 50°$ also gives a value of 0. (Make sure that your calculator is in DEGREE mode.)

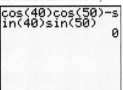

**15.** $\tan 87° = \cot(90° - 87°) = \cot 3°$

**17.** $\cos\dfrac{\pi}{12} = \sin\left(\dfrac{\pi}{2} - \dfrac{\pi}{12}\right) = \sin\dfrac{5\pi}{12}$

**19.** $\csc(-14°24') = \sec\left[90° - (-14°24')\right]$

$= \sec(90° + 14°24')$

$= \sec(104°24')$

**21.** $\sin\dfrac{5\pi}{8} = \cos\left(\dfrac{\pi}{2} - \dfrac{5\pi}{8}\right)$

$= \cos\left(\dfrac{4\pi}{8} - \dfrac{5\pi}{8}\right) = \cos\left(-\dfrac{\pi}{8}\right)$

**23.** $\sec 146°42' = \csc(90° - 146°42')$

$= \csc\left[-(146°42' - 90°)\right]$

$= \csc(-56°42')$

**25.** $\cot 176.9814° = \tan(90° - 176.9814°)$

$= \tan(-86.9814°)$

**27.** Since $\dfrac{\pi}{6} = \dfrac{\pi}{2} - \dfrac{\pi}{3},$ $\cot\dfrac{\pi}{3} = \tan\dfrac{\pi}{6}.$

$\cot\dfrac{\pi}{3} = \underline{\tan\dfrac{\pi}{6}}$

**29.** Since $90° - 57° = 33°,$ $\sin 57° = \cos 33°.$

$\sin 57° = \underline{\cos 33°}$

**31.** Since $90° - 70° = 20°,$ and $\sin x = \dfrac{1}{\csc x},$

$$\cos 70° = \sin 20° = \frac{1}{\csc 20°}.$$

$$\cos 70° = \frac{1}{\underline{\csc 20°}}$$

For exercises 33–37, other answers are possible.

**33.** $\tan\theta = \cot\left(45° + 2\theta\right)$

Since $\tan\theta = \cot\left(90° - \theta\right),$

$$90° - \theta = 45° + 2\theta \Rightarrow 90° = 45° + 3\theta \Rightarrow$$
$$3\theta = 45° \Rightarrow \theta = 15°$$

**35.** $\sec\theta = \csc\left(\dfrac{\theta}{2} + 20°\right)$

By a cofunction identity, $\sec\theta = \csc\left(90° - \theta\right).$

Thus, $\csc\left(\dfrac{\theta}{2} + 20°\right) = \csc\left(90° - \theta\right).$ So,

$$\csc\left(\frac{\theta}{2} + 20°\right) = \csc\left(90° - \theta\right) \Rightarrow$$

$$\frac{\theta}{2} + 20° = 90° - \theta \Rightarrow \frac{3\theta}{2} + 20° = 90° \Rightarrow$$

$$\frac{3\theta}{2} = 70° \Rightarrow \theta = \frac{140°}{3}$$

**37.** $\sin\left(3\theta - 15°\right) = \cos\left(\theta + 25°\right)$

Since $\sin\theta = \cos\left(90° - \theta\right),$ we have

$$\sin\left(3\theta - 15°\right) = \cos\left[90° - \left(3\theta - 15°\right)\right]$$
$$= \cos\left(90° - 3\theta + 15°\right)$$
$$= \cos\left(105° - 3\theta\right)$$

Solve $\cos\left(105° - 3\theta\right) = \cos\left(\theta + 25°\right).$

$$\cos\left(105° - 3\theta\right) = \cos\left(\theta + 25°\right)$$
$$105° - 3\theta = \theta + 25°$$
$$105° = 4\theta + 25°$$
$$4\theta = 80° \Rightarrow \theta = 20°$$

**39.** $\cos\left(0° - \theta\right) = \cos 0°\cos\theta + \sin 0°\sin\theta$
$$= (1)\cos\theta + (0)\sin\theta = \cos\theta$$

**41.** $\cos\left(180° - \theta\right) = \cos 180°\cos\theta + \sin 180°\sin\theta$
$$= (-1)\cos\theta + (0)\sin\theta$$
$$= -\cos\theta + 0 = -\cos\theta$$

**43.** $\cos\left(0° + \theta\right) = \cos 0°\cos\theta - \sin 0°\sin\theta$
$$= (1)\cos\theta - (0)\sin\theta = \cos\theta$$

**45.** $\cos\left(180° + \theta\right) = \cos 180°\cos\theta - \sin 180°\sin\theta$
$$= (-1)\cos\theta - (0)\sin\theta$$
$$= -\cos\theta - 0 = -\cos\theta$$

**47.** $\cos s = -\dfrac{1}{5},\ \sin t = \dfrac{3}{5},\ s$ and $t$ are in quadrant II.

$$\cos s = \frac{x}{r} \Rightarrow \cos s = -\frac{1}{5} = \frac{-1}{5} \Rightarrow x = -1, r = 5.$$

Substituting into the Pythagorean theorem, we have $(-1)^2 + y^2 = 5^2 \Rightarrow y^2 = 24 \Rightarrow y = \sqrt{24},$

since $\sin x > 0.$ Thus, $\sin s = \dfrac{y}{r} = \dfrac{\sqrt{24}}{5}.$

We will use a Pythagorean identity to find the value of $\cos t.$

$$\cos t = -\sqrt{1 - \sin^2 t} = -\sqrt{1 - \left(\frac{3}{5}\right)^2}$$

$$= -\sqrt{1 - \frac{9}{25}} = -\sqrt{\frac{16}{25}} = -\frac{4}{5}$$

$$\cos(s + t) = \cos s\cos t - \sin s\sin t$$

$$= \left(-\frac{1}{5}\right)\left(-\frac{4}{5}\right) - \left(\frac{\sqrt{24}}{5}\right)\left(\frac{3}{5}\right)$$

$$= \frac{4}{25} - \frac{3\sqrt{24}}{25} = \frac{4}{25} - \frac{6\sqrt{6}}{25} = \frac{4 - 6\sqrt{6}}{25}$$

$$\cos(s - t) = \cos s\cos t + \sin s\sin t$$

$$= \left(-\frac{1}{5}\right)\left(-\frac{4}{5}\right) + \left(\frac{\sqrt{24}}{5}\right)\left(\frac{3}{5}\right)$$

$$= \frac{4}{25} + \frac{3\sqrt{24}}{25} = \frac{4}{25} + \frac{6\sqrt{6}}{25} = \frac{4 + 6\sqrt{6}}{25}$$

**49.** $\sin s = \dfrac{3}{5}$ and $\sin t = -\dfrac{12}{13},\ s$ is in quadrant I and $t$ is in quadrant III.

$$\sin s = \frac{y}{r} = \frac{3}{5} \Rightarrow y = 3, r = 5.$$ Substituting into the Pythagorean theorem, we have

$$x^2 + 3^2 = 5^2 \Rightarrow x^2 = 16 \Rightarrow x = 4,\ \text{since}$$

$\cos s > 0.$ Thus, $\cos s = \dfrac{x}{r} = \dfrac{4}{5}.$ We will use a Pythagorean identity to find the value of $\cos t.$

$$\cos t = -\sqrt{1 - \left(-\frac{12}{13}\right)^2} = -\sqrt{1 - \frac{144}{169}}$$

$$= -\sqrt{\frac{25}{169}} = -\frac{5}{13}$$

$$\cos(s + t) = \cos s\cos t - \sin s\sin t$$

$$= \left(\frac{4}{5}\right)\left(-\frac{5}{13}\right) - \left(\frac{3}{5}\right)\left(-\frac{12}{13}\right)$$

$$= -\frac{20}{65} + \frac{36}{65} = \frac{16}{65}$$

$\cos(s-t) = \cos s \cos t + \sin s \sin t$

$$= \left(\frac{4}{5}\right)\left(-\frac{5}{13}\right) + \left(\frac{3}{5}\right)\left(-\frac{12}{13}\right)$$

$$= -\frac{20}{65} - \frac{36}{65} = -\frac{56}{65}$$

**51.** $\sin s = \dfrac{\sqrt{5}}{7}$ and $\sin t = \dfrac{\sqrt{6}}{8}$, $s$ and $t$ are in quadrant I.

$\sin s = \dfrac{y}{r} = \dfrac{\sqrt{5}}{7} \Rightarrow y = \sqrt{5}, r = 7$. Substituting into the Pythagorean theorem, we have

$x^2 + \left(\sqrt{5}\right)^2 = 7^2 \Rightarrow x^2 = 44 \Rightarrow x = \sqrt{44}$, since

$\cos s > 0$. Thus, $\cos s = \dfrac{x}{r} = \dfrac{\sqrt{44}}{7}$.

We will use a Pythagorean identity to find the value of $\cos t$.

$$\cos t = \sqrt{1 - \left(\frac{\sqrt{6}}{8}\right)^2} = \sqrt{1 - \frac{6}{64}}$$

$$= \sqrt{\frac{58}{64}} = \frac{\sqrt{58}}{8}$$

$\cos(s+t) = \cos s \cos t - \sin s \sin t$

$$= \left(\frac{\sqrt{44}}{7}\right)\left(\frac{\sqrt{58}}{8}\right) - \left(\frac{\sqrt{5}}{7}\right)\left(\frac{\sqrt{6}}{8}\right)$$

$$= \frac{2\sqrt{638}}{56} - \frac{\sqrt{30}}{56} = \frac{2\sqrt{638} - \sqrt{30}}{56}$$

$\cos(s+t) = \cos s \cos t - \sin s \sin t$

$$= \left(\frac{\sqrt{44}}{7}\right)\left(\frac{\sqrt{58}}{8}\right) + \left(\frac{\sqrt{5}}{7}\right)\left(\frac{\sqrt{6}}{8}\right)$$

$$= \frac{2\sqrt{638}}{56} + \frac{\sqrt{30}}{56} = \frac{2\sqrt{638} + \sqrt{30}}{56}$$

**53.** True or false: $\cos 42° = \cos(30° + 12°)$

$42° = 30° + 12°$. Thus, the given statement is true.

**55.** True or false:
$\cos 74° = \cos 60° \cos 14° + \sin 60° \sin 14°$

$\cos 74° = \cos(60° + 14°)$

$\qquad = \cos 60° \cos 14° - \sin 60° \sin 14°$

$\qquad \neq \cos 60° \cos 14° + \sin 60° \sin 14°$

Thus, the given statement is false.

**57.** True or false:
$$\cos\frac{\pi}{3} = \cos\frac{\pi}{12}\cos\frac{\pi}{4} - \sin\frac{\pi}{12}\sin\frac{\pi}{4}.$$

$$\cos\frac{\pi}{3} = \cos\left(\frac{\pi}{12} + \frac{\pi}{4}\right)$$

$$= \cos\frac{\pi}{12}\cos\frac{\pi}{4} - \sin\frac{\pi}{12}\sin\frac{\pi}{4}$$

Thus, the given statement is true.

**59.** True or false:
$\cos 70° \cos 20° - \sin 70° \sin 20° = 0$.

$\cos 70° \cos 20° - \sin 70° \sin 20°$

$\qquad = \cos(70° + 20°) = \cos 90° = 0$

Thus, the given statement is true.

**61.** True or false: $\tan\left(\theta - \dfrac{\pi}{2}\right) = \cot\theta$.

$$\tan\left(\theta - \frac{\pi}{2}\right) = -\tan\left[-\left(\theta - \frac{\pi}{2}\right)\right]$$

$$= -\tan\left(\frac{\pi}{2} - \theta\right)$$

$$= -\cot\theta \neq \cot\theta$$

Thus, the given statement is false.

**63.** Verify $\cos\left(\dfrac{\pi}{2} + x\right) = -\sin x$.

$$\cos\left(\frac{\pi}{2} + x\right) = \cos\frac{\pi}{2}\cos x - \sin\frac{\pi}{2}\sin x$$

$$= (0)\cos x - (1)\sin x = -\sin x$$

**65.** Verify $\cos 2x = \cos^2 x - \sin^2 x$.

$\cos 2x = \cos(x + x)$

$\qquad = \cos x \cos x - \sin x \sin x$

$\qquad = \cos^2 x - \sin^2 x$

**67.** $\cos 195° = \cos(180° + 15°)$

$\qquad = \cos 180° \cos 15° - \sin 180° \sin 15°$

$\qquad = (-1)\cos 15° - (0)\sin 15°$

$\qquad = -\cos 15° - 0 = -\cos 15°$

**69.** $\cos 195° = -\cos 15° = \dfrac{-\sqrt{6} - \sqrt{2}}{4}$

**71. (a)** Since there are 60 cycles per sec, the number of cycles in .05 sec is given by (.05 sec)(60 cycles per sec) = 3 cycles

**(b)** Since $V = 163\sin\omega t$ and the maximum value of $\sin\omega t$ is 1, the maximum voltage is 163. Since $V = 163\sin\omega t$ and the minimum value of $\sin\omega t$ is $-1$, the minimum voltage is $-163$. Therefore, the voltage is not always equal to 115.

## Section 5.4 Sum and Difference Identities for Sine and Tangent

1. $\sin 15° = \sin(45° - 30°)$
$= \sin 45° \cos 30° - \cos 45° \sin 30°$
$= \dfrac{\sqrt{2}}{2} \cdot \dfrac{\sqrt{3}}{2} - \dfrac{\sqrt{2}}{2} \cdot \dfrac{1}{2}$
$= \dfrac{\sqrt{6}}{4} - \dfrac{\sqrt{2}}{4} = \dfrac{\sqrt{6} - \sqrt{2}}{4}$

   Thus, the correct choice is C.

3. $\tan 15° = \tan(60° - 45°)$
$= \dfrac{\tan 60° - \tan 45°}{1 + \tan 60° \tan 45°}$
$= \dfrac{\sqrt{3} - 1}{1 + \sqrt{3}(1)} = \dfrac{\sqrt{3} - 1}{1 + \sqrt{3}} \cdot \dfrac{1 - \sqrt{3}}{1 - \sqrt{3}}$
$= \dfrac{\sqrt{3} - 3 - 1 + \sqrt{3}}{1 - 3} = \dfrac{-4 + 2\sqrt{3}}{-2} = 2 - \sqrt{3}$

   Thus, the correct choice is E.

5. $\sin(-105°) = \sin(45° - 150°)$
$= \sin 45° \cos 150° - \cos 45° \sin 150°$
$= \dfrac{\sqrt{2}}{2}\left(-\dfrac{\sqrt{3}}{2}\right) - \dfrac{\sqrt{2}}{2} \cdot \dfrac{1}{2}$
$= -\dfrac{\sqrt{6}}{4} - \dfrac{\sqrt{2}}{4} = \dfrac{-\sqrt{6} - \sqrt{2}}{4}$

   Thus, the correct choice is B.

7. Answers will vary.

9. $\sin \dfrac{5\pi}{12} = \sin\left(\dfrac{\pi}{4} + \dfrac{\pi}{6}\right)$
$= \sin \dfrac{\pi}{4} \cos \dfrac{\pi}{6} + \cos \dfrac{\pi}{4} \sin \dfrac{\pi}{6}$
$= \dfrac{\sqrt{2}}{2} \cdot \dfrac{\sqrt{3}}{2} + \dfrac{\sqrt{2}}{2} \cdot \dfrac{1}{2}$
$= \dfrac{\sqrt{6}}{4} + \dfrac{\sqrt{2}}{4} = \dfrac{\sqrt{6} + \sqrt{2}}{4}$

11. $\tan \dfrac{\pi}{12} = \tan\left(\dfrac{\pi}{4} - \dfrac{\pi}{6}\right) = \dfrac{\tan \dfrac{\pi}{4} - \tan \dfrac{\pi}{6}}{1 + \tan \dfrac{\pi}{4} \tan \dfrac{\pi}{6}}$
$= \dfrac{1 - \dfrac{\sqrt{3}}{3}}{1 + \dfrac{\sqrt{3}}{3}} = \dfrac{1 - \dfrac{\sqrt{3}}{3}}{1 + \dfrac{\sqrt{3}}{3}} \cdot \dfrac{3}{3} = \dfrac{3 - \sqrt{3}}{3 + \sqrt{3}}$

$= \dfrac{3 - \sqrt{3}}{3 + \sqrt{3}} \cdot \dfrac{3 - \sqrt{3}}{3 - \sqrt{3}} = \dfrac{\left(3 - \sqrt{3}\right)^2}{3^2 - \left(\sqrt{3}\right)^2}$
$= \dfrac{9 - 6\sqrt{3} + 3}{9 - 3} = \dfrac{12 - 6\sqrt{3}}{6} = 2 - \sqrt{3}$

12. $\sin \dfrac{\pi}{12} = \sin\left(\dfrac{\pi}{4} - \dfrac{\pi}{6}\right)$
$= \sin \dfrac{\pi}{4} \cos \dfrac{\pi}{6} - \cos \dfrac{\pi}{4} \sin \dfrac{\pi}{6}$
$= \dfrac{\sqrt{2}}{2} \cdot \dfrac{\sqrt{3}}{2} - \dfrac{\sqrt{2}}{2} \cdot \dfrac{1}{2}$
$= \dfrac{\sqrt{6}}{4} - \dfrac{\sqrt{2}}{4} = \dfrac{\sqrt{6} - \sqrt{2}}{4}$

13. $\sin\left(-\dfrac{7\pi}{12}\right) = \sin\left(-\dfrac{\pi}{3} - \dfrac{\pi}{4}\right)$
$= \sin\left(-\dfrac{\pi}{3}\right)\cos \dfrac{\pi}{4} - \cos\left(-\dfrac{\pi}{3}\right)\sin \dfrac{\pi}{4}$
$= -\sin \dfrac{\pi}{3} \cos \dfrac{\pi}{4} - \cos \dfrac{\pi}{3} \sin \dfrac{\pi}{4}$
$= -\dfrac{\sqrt{3}}{2} \cdot \dfrac{\sqrt{2}}{2} - \dfrac{1}{2} \cdot \dfrac{\sqrt{2}}{2}$
$= -\dfrac{\sqrt{6}}{4} - \dfrac{\sqrt{2}}{4} = \dfrac{-\sqrt{6} - \sqrt{2}}{4}$

15. $\sin 76° \cos 31° - \cos 76° \sin 31° = \sin(76° - 31°)$
$= \sin 45° = \dfrac{\sqrt{2}}{2}$

17. $\dfrac{\tan 80° + \tan 55°}{1 - \tan 80° \tan 55°} = \tan(80° + 55°)$
$= \tan 135° = -1$

19. $\dfrac{\tan 100° + \tan 80°}{1 - \tan 100° \tan 80°} = \tan(100° + 80°)$
$= \tan 180° = 0$

21. $\sin \dfrac{\pi}{5} \cos \dfrac{3\pi}{10} + \cos \dfrac{\pi}{5} \sin \dfrac{3\pi}{10}$
$= \sin\left(\dfrac{\pi}{5} + \dfrac{3\pi}{10}\right) = \sin\left(\dfrac{\pi}{2}\right) = 1$

23. $\cos(30° + \theta) = \cos 30° \cos \theta - \sin 30° \sin \theta$
$= \dfrac{\sqrt{3}}{2} \cos \theta - \dfrac{1}{2} \sin \theta$
$= \dfrac{1}{2}\left(\sqrt{3} \cos \theta - \sin \theta\right)$
$= \dfrac{\sqrt{3} \cos \theta - \sin \theta}{2}$

**25.** $\cos\left(60°+\theta\right)=\cos 60°\cos\theta-\sin 60°\sin\theta$

$$=\frac{1}{2}\cos\theta-\frac{\sqrt{3}}{2}\sin\theta$$

$$=\frac{1}{2}\left(\cos\theta-\sqrt{3}\sin\theta\right)$$

$$=\frac{\cos\theta-\sqrt{3}\sin\theta}{2}$$

**27.** $\cos\left(\dfrac{3\pi}{4}-x\right)=\cos\dfrac{3\pi}{4}\cos x+\sin\dfrac{3\pi}{4}\sin x$

$$=\left(-\frac{\sqrt{2}}{2}\right)\cos x+\left(\frac{\sqrt{2}}{2}\right)\sin x$$

$$=\frac{\sqrt{2}}{2}\left(-\cos x+\sin x\right)$$

$$=\frac{\sqrt{2}\left(\sin x-\cos x\right)}{2}$$

**29.** $\tan\left(\theta+30°\right)=\dfrac{\tan\theta+\tan 30°}{1-\tan\theta\tan 30°}$

$$=\frac{\tan\theta+\dfrac{1}{\sqrt{3}}}{1-\left(\dfrac{1}{\sqrt{3}}\right)\tan\theta}$$

$$=\frac{\sqrt{3}\tan\theta+1}{\sqrt{3}-\tan\theta}$$

**31.** $\sin\left(\dfrac{\pi}{4}+x\right)=\sin\dfrac{\pi}{4}\cos x+\cos\dfrac{\pi}{4}\sin x$

$$=\frac{\sqrt{2}}{2}\cos x+\frac{\sqrt{2}}{2}\sin x$$

$$=\frac{\sqrt{2}\left(\cos x+\sin x\right)}{2}$$

**33.** $\sin\left(270°-\theta\right)=\sin 270°\cos\theta-\cos 270°\sin\theta$

$$=\left(-1\right)\left(\cos\theta\right)-\left(0\right)\left(\sin\theta\right)$$

$$=-\cos\theta$$

**35.** $\tan\left(360°-\theta\right)=\dfrac{\tan 360°-\tan\theta}{1+\tan 360°\tan\theta}$

$$=\frac{0-\tan\theta}{1+0\cdot\tan\theta}=-\tan\theta$$

**37.** $\tan\left(\pi-\theta\right)=\dfrac{\tan\pi-\tan\theta}{1+\tan\pi\tan\theta}$

$$=\frac{0-\tan\theta}{1+0\cdot\tan\theta}=-\tan\theta$$

**39.** Answers will vary.

**41.** $\cos s=\dfrac{3}{5}$, $\sin t=\dfrac{5}{13}$, and $s$ and $t$ are in quadrant I.

First find the values of sin $s$, tan $s$, cos $t$, and tan $t$. Because $s$ and $t$ are both in quadrant I, the values of sin $s$ and cos $t$, tan $s$, and tan $t$ will be positive.

$$\sin s=\sqrt{1-\left(\frac{3}{5}\right)^2}=\sqrt{1-\frac{9}{25}}=\sqrt{\frac{16}{25}}=\frac{4}{5}$$

$$\cos t=\sqrt{1-\left(\frac{5}{13}\right)^2}=\sqrt{1-\frac{25}{169}}=\sqrt{\frac{144}{169}}=\frac{12}{13}$$

$$\tan s=\frac{\sin s}{\cos s}=\frac{\frac{4}{5}}{\frac{3}{5}}=\frac{4}{3};\ \tan t=\frac{\sin t}{\cos t}=\frac{\frac{5}{13}}{\frac{12}{13}}=\frac{5}{12}$$

**(a)** $\sin\left(s+t\right)=\sin s\cos t+\cos s\sin t$

$$=\left(\frac{4}{5}\right)\left(\frac{12}{13}\right)+\left(\frac{3}{5}\right)\left(\frac{5}{13}\right)$$

$$=\frac{48}{65}+\frac{15}{65}=\frac{63}{65}$$

**(b)** $\tan\left(s+t\right)=\dfrac{\tan s+\tan t}{1-\tan s\tan t}=\dfrac{\frac{4}{3}+\frac{5}{12}}{1-\left(\frac{4}{3}\right)\left(\frac{5}{12}\right)}$

$$=\frac{48+15}{36-20}=\frac{63}{16}$$

**(c)** From parts (a) and (b), sin $(s+t)>0$ and tan $(s+t)>0$. The only quadrant in which the values of both the sine and the tangent are positive is quadrant I, so $s+t$ is in quadrant I.

**43.** $\sin s=\dfrac{2}{3}$ and $\sin t=-\dfrac{1}{3}$, $s$ is in quadrant II and $t$ is in quadrant IV.

First find the values of cos $s$, cos $t$, tan $s$, and tan $t$. Because $s$ is in quadrant II and $t$ is in quadrant IV, the values of cos $s$, tan $s$, and tan $t$ will be negative, while cos $t$ will be positive.

$$\cos s=-\sqrt{1-\left(\frac{2}{3}\right)^2}=-\sqrt{1-\frac{4}{9}}=-\sqrt{\frac{5}{9}}=-\frac{\sqrt{5}}{3}$$

$$\cos t=\sqrt{1-\left(-\frac{1}{3}\right)^2}=\sqrt{1-\frac{1}{9}}=\sqrt{\frac{8}{9}}$$

$$=\frac{\sqrt{8}}{3}=\frac{2\sqrt{2}}{3}$$

$$\tan s=\frac{\sin s}{\cos s}=\frac{\frac{2}{3}}{-\frac{\sqrt{5}}{3}}=-\frac{2}{\sqrt{5}}=-\frac{2\sqrt{5}}{5}$$

$$\tan t=\frac{\sin t}{\cos t}=\frac{-\frac{1}{3}}{\frac{2\sqrt{2}}{3}}=-\frac{1}{2\sqrt{2}}=-\frac{\sqrt{2}}{4}$$

(*continued on next page*)

*(continued from page 121)*

(a) $\sin(s+t) = \sin s \cos t + \cos s \sin t$

$$= \left(\frac{2}{3}\right)\left(\frac{2\sqrt{2}}{3}\right) + \left(-\frac{\sqrt{5}}{3}\right)\left(-\frac{1}{3}\right)$$

$$= \frac{4\sqrt{2}}{9} + \frac{\sqrt{5}}{9} = \frac{4\sqrt{2}+\sqrt{5}}{9}$$

(b) Different forms of $\tan(s+t)$ will be obtained depending on whether $\tan s$ and $\tan t$ are written with rationalized denominators.

$$\tan(s+t) = \frac{-\frac{2\sqrt{5}}{5} + \left(-\frac{\sqrt{2}}{4}\right)}{1 - \left(-\frac{2\sqrt{5}}{5}\right)\left(-\frac{\sqrt{2}}{4}\right)} = \frac{-8\sqrt{5}-5\sqrt{2}}{20-2\sqrt{10}}$$

or

$$\tan(s+t) = \frac{-\frac{2}{\sqrt{5}} + \left(-\frac{1}{2\sqrt{2}}\right)}{1 - \left(-\frac{2}{\sqrt{5}}\right)\left(-\frac{1}{2\sqrt{2}}\right)}$$

$$= \frac{-4\sqrt{2}-\sqrt{5}}{2\sqrt{10}-2} = \frac{4\sqrt{2}+\sqrt{5}}{2-2\sqrt{10}}$$

(c) To find the quadrant of $s+t$, notice from the preceding that $\sin(s+t) = \frac{4\sqrt{2}+\sqrt{5}}{9} > 0$

and $\tan(s+t) = \frac{-8\sqrt{5}-5\sqrt{2}}{20-2\sqrt{10}} \approx -1.8 < 0$.

The only quadrant in which the values of sine are positive and tangent is negative is quadrant II. Therefore, $s+t$ is in quadrant II.

**45.** $\cos s = -\dfrac{8}{17}$ and $\cos t = -\dfrac{3}{5}$, $s$ and $t$ are in quadrant III

First find the values of $\sin s$, $\sin t$, $\tan s$, and $\tan t$. Because $s$ and $t$ are both in quadrant III, the values of $\sin s$ and $\sin t$ will be negative, while $\tan s$ and $\tan t$ will be positive.

$$\sin s = -\sqrt{1-\cos^2 s} = -\sqrt{1-\left(-\frac{8}{17}\right)^2}$$

$$= -\sqrt{1-\frac{64}{289}} = -\sqrt{\frac{225}{289}} = -\frac{15}{17}$$

$$\sin t = -\sqrt{1-\cos^2 t} = -\sqrt{1-\left(-\frac{3}{5}\right)^2}$$

$$= -\sqrt{1-\frac{9}{25}} = -\sqrt{\frac{16}{25}} = -\frac{4}{5}$$

$$\tan s = \frac{\sin s}{\cos s} = \frac{-\frac{15}{17}}{-\frac{8}{17}} = \frac{15}{8}$$

$$\tan t = \frac{\sin t}{\cos t} = \frac{-\frac{4}{5}}{-\frac{3}{5}} = \frac{4}{3}$$

(a) $\sin(s+t) = \sin s \cos t + \cos s \sin t$

$$= \left(-\frac{15}{17}\right)\left(-\frac{3}{5}\right) + \left(-\frac{8}{17}\right)\left(-\frac{4}{5}\right)$$

$$= \frac{45}{85} + \frac{32}{85} = \frac{77}{85}$$

(b) $\tan(s+t) = \dfrac{\frac{15}{8} + \frac{4}{3}}{1 - \left(\frac{15}{8}\right)\left(\frac{4}{3}\right)} = \dfrac{45+32}{24-60}$

$$= \frac{77}{-36} = -\frac{77}{36}$$

(c) From parts (a) and (b), $\sin(s+t) > 0$ and $\tan(s+t) < 0$. The only quadrant in which the value of the sine is positive and the value of the tangent is negative is quadrant II, so $s+t$ is in quadrant II.

**47.** $\sin 165° = \sin(180° - 15°)$

$$= \sin 180° \cos 15° - \cos 180° \sin 15°$$

$$= (0)\cos 15° - (-1)\sin 15° = 0 + \sin 15°$$

$$= \sin 15°$$

Now use a difference identity to find $\sin 15°$.

$\sin 15° = \sin(45° - 30°)$

$$= \sin 45° \cos 30° - \cos 45° \sin 30°$$

$$= \frac{\sqrt{2}}{2} \cdot \frac{\sqrt{3}}{2} - \frac{\sqrt{2}}{2} \cdot \frac{1}{2}$$

$$= \frac{\sqrt{6}}{4} - \frac{\sqrt{2}}{4} = \frac{\sqrt{6}-\sqrt{2}}{4}$$

**49.** $\sin 255° = \sin(270° - 15°)$

$$= \sin 270° \cos 15° - \cos 270° \sin 15°$$

$$= (-1)\cos 15° - (0)\sin 15°$$

$$= -\cos 15° - 0 = -\cos 15°$$

Now use a difference identity to find $\cos 15°$.

$\cos 15° = \cos(45° - 30°)$

$$= (\cos 45° \cos 30° + \sin 45° \sin 30°)$$

$$= \frac{\sqrt{2}}{2} \cdot \frac{\sqrt{3}}{2} + \frac{\sqrt{2}}{2} \cdot \frac{1}{2}$$

$$= \frac{\sqrt{6}}{4} + \frac{\sqrt{2}}{4} = \frac{\sqrt{6}+\sqrt{2}}{4}$$

Thus, $\sin 255° = -\cos 15° = -\left(\dfrac{\sqrt{6}+\sqrt{2}}{4}\right)$

$$= \frac{-\sqrt{6}-\sqrt{2}}{4}$$

**51.**  $\tan\dfrac{11\pi}{12} = \tan\left(\pi - \dfrac{\pi}{12}\right)$

$= \dfrac{\tan\pi - \tan\frac{\pi}{12}}{1 + \tan\pi\tan\frac{\pi}{12}} = -\tan\dfrac{\pi}{12}$

Now use a difference identity to find  $\tan\dfrac{\pi}{12}$.

$\tan\dfrac{\pi}{12} = \tan\left(\dfrac{\pi}{4} - \dfrac{\pi}{6}\right) = \dfrac{\tan\frac{\pi}{4} - \tan\frac{\pi}{6}}{1 + \tan\frac{\pi}{4}\tan\frac{\pi}{6}}$

$= \dfrac{1 - \frac{\sqrt{3}}{3}}{1 + 1\cdot\frac{\sqrt{3}}{3}} = \dfrac{1 - \frac{\sqrt{3}}{3}}{1 + \frac{\sqrt{3}}{3}} = \dfrac{\frac{3 - \sqrt{3}}{3}}{\frac{3 + \sqrt{3}}{3}}$

$= \dfrac{3 - \sqrt{3}}{3 + \sqrt{3}} = \dfrac{3 - \sqrt{3}}{3 + \sqrt{3}}\cdot\dfrac{3 - \sqrt{3}}{3 - \sqrt{3}}$

$= \dfrac{9 - 6\sqrt{3} + 3}{9 - 3} = \dfrac{12 - 6\sqrt{3}}{6}$

$= \dfrac{6\left(2 - \sqrt{3}\right)}{6} = 2 - \sqrt{3}$

Thus,

$\tan\dfrac{11\pi}{12} = -\tan\dfrac{\pi}{12} = -\left(2 - \sqrt{3}\right) = -2 + \sqrt{3}.$

**53.**  $\sin\left(\dfrac{\pi}{2} + \theta\right)$ appears to be equivalent to

$\cos\theta.$

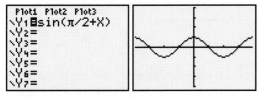

$\sin\left(\dfrac{\pi}{2} + x\right) = \sin\dfrac{\pi}{2}\cos x + \sin x\cos\dfrac{\pi}{2}$

$= 1\cdot\cos x + \sin x\cdot 0$

$= \cos x + 0 = \cos x$

**55.**  $\tan\left(\dfrac{\pi}{2} + \theta\right)$ appears to be equivalent to

$-\cot\theta.$

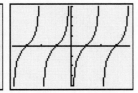

$\tan\left(\dfrac{\pi}{2} + \theta\right) = \dfrac{\sin\left(\dfrac{\pi}{2} + \theta\right)}{\cos\left(\dfrac{\pi}{2} + \theta\right)}$

$= \dfrac{\sin\dfrac{\pi}{2}\cos\theta + \cos\dfrac{\pi}{2}\sin\theta}{\cos\dfrac{\pi}{2}\cos\theta - \sin\dfrac{\pi}{2}\sin\theta}$

$= \dfrac{1\cdot\cos\theta + 0\cdot\sin\theta}{0\cdot\cos\theta - 1\cdot\sin\theta} = \dfrac{\cos\theta}{-\sin\theta}$

$= -\cot\theta$

**57.**  Verify  $\sin 2x = 2\sin x\cos x$  is an identity.

$\sin 2x = \sin(x + x) = \sin x\cos x + \cos x\sin x$

$= 2\sin x\cos x$

**59.**  Verify  $\sin(210° + x) - \cos(210° + x) = 0$  is an

identity.

$\sin(210° + x) - \cos(120° + x)$

$= (\sin 210°\cos x + \cos 210°\sin x)$

$\qquad - (\cos 120°\cos x - \sin 120°\sin x)$

$= \left(-\dfrac{1}{2}\cos x - \dfrac{\sqrt{3}}{2}\sin x\right)$

$\qquad - \left(-\dfrac{1}{2}\cos x - \dfrac{\sqrt{3}}{2}\sin x\right) = 0$

**61.**  Verify  $\dfrac{\cos(\alpha - \beta)}{\cos\alpha\sin\beta} = \tan\alpha + \cot\beta$  is an

identity.

$\dfrac{\cos(\alpha - \beta)}{\cos\alpha\sin\beta} = \dfrac{\cos\alpha\cos\beta + \sin\alpha\sin\beta}{\cos\alpha\sin\beta}$

$= \dfrac{\cos\alpha\cos\beta}{\cos\alpha\sin\beta} + \dfrac{\sin\alpha\sin\beta}{\cos\alpha\sin\beta}$

$= \dfrac{\cos\beta}{\sin\beta} + \dfrac{\sin\alpha}{\cos\alpha} = \cot\beta + \tan\alpha$

**63.** Verify that $\dfrac{\sin(x-y)}{\sin(x+y)} = \dfrac{\tan x - \tan y}{\tan x + \tan y}$ is an identity

$$\dfrac{\sin(x-y)}{\sin(x+y)} = \dfrac{\sin x \cos y - \cos x \sin y}{\sin x \cos y + \cos x \sin y} = \dfrac{\dfrac{\sin x \cos y}{\cos x \cos y} - \dfrac{\cos x \sin y}{\cos x \cos y}}{\dfrac{\sin x \cos y}{\cos x \cos y} + \dfrac{\cos x \sin y}{\cos x \cos y}} = \dfrac{\dfrac{\sin x}{\cos x} \cdot \dfrac{\cos y}{\cos y} - \dfrac{\cos x}{\cos x} \cdot \dfrac{\sin y}{\cos y}}{\dfrac{\sin x}{\cos x} \cdot \dfrac{\cos y}{\cos y} + \dfrac{\cos x}{\cos x} \cdot \dfrac{\sin y}{\cos y}}$$

$$= \dfrac{\dfrac{\sin x}{\cos x} \cdot 1 - 1 \cdot \dfrac{\sin y}{\cos y}}{\dfrac{\sin x}{\cos x} \cdot 1 + 1 \cdot \dfrac{\sin y}{\cos y}} = \dfrac{\dfrac{\sin x}{\cos x} - \dfrac{\sin y}{\cos y}}{\dfrac{\sin x}{\cos x} + \dfrac{\sin y}{\cos y}} = \dfrac{\tan x - \tan y}{\tan x + \tan y}$$

**65.** Verify $\dfrac{\sin(s-t)}{\sin t} + \dfrac{\cos(s-t)}{\cos t} = \dfrac{\sin s}{\sin t \cos t}$ is an identity.

$$\dfrac{\sin(s-t)}{\sin t} + \dfrac{\cos(s-t)}{\cos t} = \dfrac{\sin s \cos t - \sin t \cos s}{\sin t} + \dfrac{\cos s \cos t + \sin t \sin s}{\cos t}$$

$$= \dfrac{\sin s \cos^2 t - \sin t \cos t \cos s}{\sin t \cos t} + \dfrac{\sin t \cos t \cos s + \sin^2 t \sin s}{\sin t \cos t} = \dfrac{\sin s \cos^2 t + \sin s \sin^2 t}{\sin t \cos t}$$

$$= \dfrac{\sin s \left(\cos^2 t + \sin^2 t\right)}{\sin t \cos t} = \dfrac{\sin s}{\sin t \cos t}$$

**67.** $\angle\beta$ and $\angle ABC$ are supplementary, so $m\angle ABC = 180° - \beta$.

**69.** $\tan\theta = \tan(\beta - \alpha) = \dfrac{\tan\beta - \tan\alpha}{1 + \tan\alpha \tan\beta}$

**71.** $x + y = 9,\ 2x + y = -1$

Convert each equation to slope-intercept form to find the slopes:

$x + y = 9 \Rightarrow y = -x + 9 \Rightarrow m_1 = -1$

$2x + y = -1 \Rightarrow y = -2x - 1 \Rightarrow m_2 = -2$

$\tan\theta = \dfrac{m_2 - m_1}{1 + m_1 m_2} = \dfrac{-2 - (-1)}{1 + (-1)(-2)} = -\dfrac{1}{3} \Rightarrow$

$\theta = \tan^{-1}\left(-\dfrac{1}{3}\right) \approx -18.4°$

Note that the angle must be positive, so $\theta \approx 18.4°$

**73.** **(a)**   $F = \dfrac{.6W \sin(\theta + 90°)}{\sin 12°}$

$= \dfrac{.6(170)\sin(30 + 90)°}{\sin 12°}$

$= \dfrac{102 \sin 120°}{\sin 12°} \approx 425\ \text{lb}$

(This is a good reason why people frequently have back problems.)

**(b)**   $F = \dfrac{.6W \sin(\theta + 90°)}{\sin 12°}$

$= \dfrac{.6W\left(\sin\theta \cos 90° + \sin 90° \cos\theta\right)}{\sin 12°}$

$= \dfrac{.6W\left(\sin\theta \cdot 0 + 1 \cdot \cos\theta\right)}{\sin 12°}$

$= \dfrac{.6W\left(0 + \cos\theta\right)}{\sin 12°}$

$= \dfrac{.6}{\sin 12°}W \cos\theta \approx 2.9W \cos\theta$

**(c)**   $F$ will be maximum when $\cos\theta = 1$ or $\theta = 0°$. ($\theta = 0°$ corresponds to the back being horizontal which gives a maximum force on the back muscles. This agrees with intuition since stress on the back increases as one bends farther until the back is parallel with the ground.)

**75.**   $E = 20 \sin\left(\dfrac{\pi t}{4} - \dfrac{\pi}{2}\right)$

$= 20\left(\sin\dfrac{\pi t}{4}\cos\dfrac{\pi}{2} - \cos\dfrac{\pi t}{4}\sin\dfrac{\pi}{2}\right)$

$= 20\left(\sin\dfrac{\pi t}{4}(0) - \cos\dfrac{\pi t}{4}(1)\right)$

$= 20\left(0 - \cos\dfrac{\pi t}{4}\right) = -20\cos\dfrac{\pi t}{4}$

## Chapter 5 Quiz
**(Sections 5.1–5.4)**

**1.** $\sin\theta = -\dfrac{7}{25}$, $\theta$ is in quadrant IV

In quadrant IV, the cosine and secant function values are positive. The tangent, cotangent, and cosecant function values are negative.

$$\cos\theta = \sqrt{1-\sin^2\theta} = \sqrt{1-\left(-\dfrac{7}{25}\right)^2}$$

$$= \sqrt{1-\dfrac{49}{625}} = \sqrt{\dfrac{576}{625}} = \dfrac{24}{25}$$

$$\tan\theta = \dfrac{\sin\theta}{\cos\theta} = \dfrac{-\frac{7}{25}}{\frac{24}{25}} = -\dfrac{7}{24}$$

$$\cot\theta = \dfrac{1}{\tan\theta} = \dfrac{1}{-\frac{7}{24}} = -\dfrac{24}{7}$$

$$\sec\theta = \dfrac{1}{\cos\theta} = \dfrac{1}{\frac{24}{25}} = \dfrac{25}{24}$$

$$\csc\theta = \dfrac{1}{\sin\theta} = \dfrac{1}{-\frac{7}{25}} = -\dfrac{25}{7}$$

**3.** $\sin\left(-\dfrac{7\pi}{12}\right) = -\sin\left(\dfrac{7\pi}{12}\right) = -\sin\left(\dfrac{\pi}{3}+\dfrac{\pi}{4}\right)$

$$= -\left(\sin\dfrac{\pi}{3}\cos\dfrac{\pi}{4} + \cos\dfrac{\pi}{3}\sin\dfrac{\pi}{4}\right)$$

$$= -\left[\dfrac{\sqrt3}{2}\left(\dfrac{\sqrt2}{2}\right) + \dfrac{1}{2}\left(\dfrac{\sqrt2}{2}\right)\right]$$

$$= -\left(\dfrac{\sqrt6}{4}+\dfrac{\sqrt2}{4}\right) = -\left(\dfrac{\sqrt6+\sqrt2}{4}\right)$$

$$= \dfrac{-\sqrt6-\sqrt2}{4}$$

**5.** $\cos A = \dfrac{3}{5}, \sin B = -\dfrac{5}{13}, 0 < A < \dfrac{\pi}{2}$, and $\pi < B < \dfrac{3\pi}{2}$

$\cos A = \dfrac{3}{5} = \dfrac{y}{r} \Rightarrow y = 3, r = 5$. Substituting into the Pythagorean theorem, we have $x^2 + 3^3 = 5^2 \Rightarrow x = 4$, since $\sin A > 0$. Thus, $\sin A = \dfrac{4}{5}$.

We will use a Pythagorean identity to find the value of $\cos B$.

$$\cos B = -\sqrt{1-\left(-\dfrac{5}{13}\right)^2} = -\sqrt{1-\dfrac{25}{169}}$$

$$= -\sqrt{\dfrac{144}{169}} = -\dfrac{12}{13}$$

Note that $\cos B$ is negative because $B$ is in quadrant III.

**(a)** $\cos(A+B) = \cos A\cos B - \sin A\sin B$

$$= \dfrac{3}{5}\left(-\dfrac{12}{13}\right) - \dfrac{4}{5}\left(-\dfrac{5}{13}\right)$$

$$= -\dfrac{36}{65} + \dfrac{20}{65} = -\dfrac{16}{65}$$

**(b)** $\sin(A+B) = \sin A\cos B + \cos A\sin B$

$$= \dfrac{4}{5}\left(-\dfrac{12}{13}\right) + \dfrac{3}{5}\left(-\dfrac{5}{13}\right)$$

$$= -\dfrac{48}{65} - \dfrac{15}{65} = -\dfrac{63}{65}$$

**(c)** Both $\cos(A+B)$ and $\sin(A+B)$ are negative. Thus $(A+B)$ is in quadrant III.

**7.** Verify $\dfrac{1+\sin\theta}{\cot^2\theta} = \dfrac{\sin\theta}{\csc\theta-1}$ is an identity.

Working with the right side, we have

$$\dfrac{\sin\theta}{\csc\theta-1} = \dfrac{\sin\theta}{\csc\theta-1}\cdot\dfrac{\csc\theta+1}{\csc\theta+1}$$

$$= \dfrac{\sin\theta\csc\theta+\sin\theta}{\csc^2\theta-1} = \dfrac{1+\sin\theta}{\cot^2\theta}$$

**9.** Verify $\sin\left(\dfrac{\pi}{3}+\theta\right) - \sin\left(\dfrac{\pi}{3}-\theta\right) = \sin\theta$ is an identity.

$$\sin\left(\dfrac{\pi}{3}+\theta\right) - \sin\left(\dfrac{\pi}{3}-\theta\right) = \left(\sin\dfrac{\pi}{3}\cos\theta + \cos\dfrac{\pi}{3}\sin\theta\right) - \left(\sin\dfrac{\pi}{3}\cos\theta - \cos\dfrac{\pi}{3}\sin\theta\right)$$

$$= 2\cos\dfrac{\pi}{3}\sin\theta = 2\left(\dfrac{1}{2}\right)\sin\theta = \sin\theta$$

## Section 5.5: Double-Angle Identities

1. C. $2\cos^2 15° - 1 = \cos(2 \cdot 15°) = \cos 30° = \dfrac{\sqrt{3}}{2}$

3. B. $2\sin 22.5° \cos 22.5° = \sin(2 \cdot 22.5°)$
$$= \sin 45° = \dfrac{\sqrt{2}}{2}$$

5. C. $2\sin\dfrac{\pi}{3}\cos\dfrac{\pi}{3} = \sin\left(2\cdot\dfrac{\pi}{3}\right) = \sin\dfrac{2\pi}{3} = \dfrac{\sqrt{3}}{2}$

7. $\cos 2\theta = \dfrac{3}{5}, \theta$ is in quadrant I.

   $\cos 2\theta = 2\cos^2\theta - 1 \Rightarrow \dfrac{3}{5} = 2\cos^2\theta - 1 \Rightarrow$

   $2\cos^2\theta = \dfrac{3}{5} + 1 = \dfrac{8}{5} \Rightarrow \cos^2\theta = \dfrac{8}{10} = \dfrac{4}{5}$

   Since $\theta$ is in quadrant I, $\cos\theta > 0$. Thus,

   $\cos\theta = \sqrt{\dfrac{4}{5}} = \dfrac{2}{\sqrt{5}} = \dfrac{2\sqrt{5}}{5}$.

   Since $\theta$ is in quadrant I, $\sin\theta > 0$.

   $\sin\theta = \sqrt{1 - \cos^2\theta} = \sqrt{1 - \left(\dfrac{2\sqrt{5}}{5}\right)^2} = \sqrt{1 - \dfrac{20}{25}}$

   $= \sqrt{\dfrac{5}{25}} = \sqrt{\dfrac{1}{5}} = \dfrac{1}{\sqrt{5}} = \dfrac{1}{\sqrt{5}} \cdot \dfrac{\sqrt{5}}{\sqrt{5}} = \dfrac{\sqrt{5}}{5}$

9. $\cos 2\theta = -\dfrac{5}{12}, \dfrac{\pi}{2} < \theta < \pi$

   $\cos 2\theta = 2\cos^2\theta - 1 \Rightarrow$

   $2\cos^2\theta = \cos 2\theta + 1 = -\dfrac{5}{12} + 1 = \dfrac{7}{12} \Rightarrow$

   $\cos^2\theta = \dfrac{7}{24}$

   Since $\dfrac{\pi}{2} < \theta < \pi, \cos\theta < 0$. Thus,

   $\cos\theta = -\sqrt{\dfrac{7}{24}} = -\dfrac{\sqrt{7}}{\sqrt{24}} = -\dfrac{\sqrt{7}}{2\sqrt{6}}$

   $= -\dfrac{\sqrt{7}}{2\sqrt{6}} \cdot \dfrac{\sqrt{6}}{\sqrt{6}} = -\dfrac{\sqrt{42}}{12}$

   $\sin^2\theta = 1 - \cos^2\theta = 1 - \left(-\sqrt{\dfrac{7}{24}}\right)^2$

   $= 1 - \dfrac{7}{24} = \dfrac{17}{24}$

Since $\dfrac{\pi}{2} < \theta < \pi, \sin\theta > 0$. Thus,

$\sin\theta = \sqrt{\dfrac{17}{24}} = \dfrac{\sqrt{17}}{\sqrt{24}} = \dfrac{\sqrt{17}}{2\sqrt{6}}$

$= \dfrac{\sqrt{17}}{2\sqrt{6}} \cdot \dfrac{\sqrt{6}}{\sqrt{6}} = \dfrac{\sqrt{102}}{12}$

11. $\sin\theta = \dfrac{2}{5}, \cos\theta < 0$

    $\cos 2\theta = 1 - 2\sin^2\theta \Rightarrow$

    $\cos 2\theta = 1 - 2\left(\dfrac{2}{5}\right)^2 1 - 2 \cdot \dfrac{4}{25} = 1 - \dfrac{8}{25} = \dfrac{17}{25}$

    $\cos^2 2\theta + \sin^2 2\theta = 1 \Rightarrow$

    $\sin^2 2\theta = 1 - \cos^2 2\theta \Rightarrow$

    $\sin^2 2\theta = 1 - \left(\dfrac{17}{25}\right)^2 = 1 - \dfrac{289}{625} = \dfrac{336}{625}$

    Since $\cos\theta < 0, \sin 2\theta < 0$ because
    $\sin 2\theta = 2\sin\theta\cos\theta < 0$ and $\sin\theta > 0$.

    $\sin 2\theta = -\sqrt{\dfrac{336}{625}} = -\dfrac{\sqrt{336}}{25} = -\dfrac{4\sqrt{21}}{25}$

13. $\tan x = 2, \cos x > 0$

    $\tan 2x = \dfrac{2\tan x}{1 - \tan^2 x} = \dfrac{2(2)}{1 - 2^2} = \dfrac{4}{1 - 4} = -\dfrac{4}{3}$

    Since both $\tan x$ and $\cos x$ are positive, $x$ must be in quadrant I. Since $0° < x < 90°$, then $0° < 2x < 180°$. Thus, $2x$ must be in either quadrant I or quadrant II. However, $\tan 2x < 0$, so $2x$ is in quadrant II, and $\sec 2x$ is negative.

    $\sec^2 2x = 1 + \tan^2 2x = 1 + \left(-\dfrac{4}{3}\right)^2 = 1 + \dfrac{16}{9} = \dfrac{25}{9}$

    $\sec 2x = -\dfrac{5}{3} \Rightarrow \cos 2x = \dfrac{1}{\sec 2x} = -\dfrac{3}{5}$

    $\cos^2 2x + \sin^2 2x = 1 \Rightarrow \sin^2 2x = 1 - \cos^2 2x \Rightarrow$

    $\sin^2 2x = 1 - \left(-\dfrac{3}{5}\right)^2 \Rightarrow \sin^2 2x = 1 - \dfrac{9}{25} = \dfrac{16}{25}$

    In quadrants I and II, $\sin 2x > 0$. Thus, we
    have $\sin 2x = \sqrt{\dfrac{16}{25}} = \dfrac{4}{5}$.

15. $\sin\theta = -\dfrac{\sqrt{5}}{7}, \cos\theta > 0$

    $\cos^2\theta = 1 - \sin^2\theta = 1 - \left(-\dfrac{\sqrt{5}}{7}\right)^2 = 1 - \dfrac{5}{49} = \dfrac{44}{49}$

    Since $\cos\theta > 0$, $\cos\theta = \sqrt{\dfrac{44}{49}} = \dfrac{\sqrt{44}}{7} = \dfrac{2\sqrt{11}}{7}$.

$$\cos 2\theta = 1 - 2\sin^2\theta = 1 - 2\left(-\frac{\sqrt{5}}{7}\right)^2$$

$$= 1 - 2 \cdot \frac{5}{49} = 1 - \frac{10}{49} = \frac{39}{49}$$

$$\sin 2\theta = 2\sin\theta\cos\theta = 2\left(-\frac{\sqrt{5}}{7}\right)\left(\frac{2\sqrt{11}}{7}\right)$$

$$= -\frac{4\sqrt{55}}{49}$$

**17.**  $\cos^2 15° - \sin^2 15° = \cos\left[2(15°)\right]$

$$= \cos 30° = \frac{\sqrt{3}}{2}$$

**19.**  $1 - 2\sin^2 15° = \cos\left[2(15°)\right] = \cos 30° = \frac{\sqrt{3}}{2}$

**21.**  $2\cos^2 67\frac{1}{2}° - 1 = \cos^2 67\frac{1}{2}° - \sin^2 67\frac{1}{2}°$

$$= \cos 2\left(67\frac{1}{2}°\right) = \cos 135°$$

$$= -\frac{\sqrt{2}}{2}$$

**23.**  $\dfrac{\tan 51°}{1 - \tan^2 51°}$

Since $\dfrac{2\tan A}{1 - \tan^2 A} = \tan 2A$ , we have

$$\frac{1}{2}\left(\frac{2\tan A}{1 - \tan^2 A}\right) = \frac{1}{2}\tan 2A \Rightarrow$$

$$\frac{\tan A}{1 - \tan^2 A} = \frac{1}{2}\tan 2A \Rightarrow$$

$$\frac{\tan 51°}{1 - \tan^2 51°} = \frac{1}{2}\tan\left[2(51°)\right] = \frac{1}{2}\tan 102°$$

**25.**  $\dfrac{1}{4} - \dfrac{1}{2}\sin^2 47.1° = \dfrac{1}{4}\left(1 - 2\sin^2 47.1°\right)$

$$= \frac{1}{4}\cos\left[2(47.1°)\right]$$

$$= \frac{1}{4}\cos 94.2°$$

**27.**  $\sin^2\dfrac{2\pi}{5} - \cos^2\dfrac{2\pi}{5} = -\left(\cos^2\dfrac{2\pi}{5} - \sin^2\dfrac{2\pi}{5}\right)$

$$\cos^2 A - \sin^2 A = \cos 2A \Rightarrow$$

$$-\left(\cos^2\frac{2\pi}{5} - \sin^2\frac{2\pi}{5}\right) = -\cos\left(2 \cdot \frac{2\pi}{5}\right)$$

$$= -\cos\frac{4\pi}{5}$$

**29.**  $\sin 4x = \sin\left[2(2x)\right] = 2\sin 2x\cos 2x$

$$= 2\left(2\sin x\cos x\right)\left(\cos^2 x - \sin^2 x\right)$$

$$= 4\sin x\cos^3 x - 4\sin^3 x\cos x$$

**31.**  $\tan 3x = \tan\left(2x + x\right) = \dfrac{\tan 2x + \tan x}{1 - \tan 2x\tan x}$

$$= \frac{\dfrac{2\tan x}{1 - \tan^2 x} + \tan x}{1 - \dfrac{2\tan x}{1 - \tan^2 x} \cdot \tan x}$$

$$= \frac{\dfrac{2\tan x}{1 - \tan^2 x} + \dfrac{\left(1 - \tan^2 x\right)\tan x}{1 - \tan^2 x}}{\dfrac{1 - \tan^2 x}{1 - \tan^2 x} - \dfrac{2\tan^2 x}{1 - \tan^2 x}}$$

$$= \frac{\dfrac{2\tan x}{1 - \tan^2 x} + \dfrac{\tan x - \tan^3 x}{1 - \tan^2 x}}{\dfrac{1 - \tan^2 x}{1 - \tan^2 x} - \dfrac{2\tan^2 x}{1 - \tan^2 x}} \cdot \frac{1 - \tan^2 x}{1 - \tan^2 x}$$

$$= \frac{2\tan x + \tan x - \tan^3 x}{1 - \tan^2 x - 2\tan^2 x}$$

$$= \frac{3\tan x - \tan^3 x}{1 - 3\tan^2 x}$$

Exercises 33–36 are graphed in the following window:

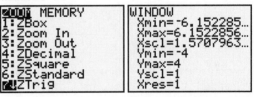

**33.**  $\cos^4 x - \sin^4 x$ appears to be equivalent to $\cos 2x$.

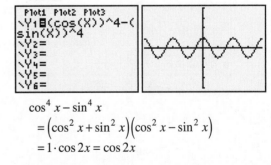

$$\cos^4 x - \sin^4 x$$

$$= \left(\cos^2 x + \sin^2 x\right)\left(\cos^2 x - \sin^2 x\right)$$

$$= 1 \cdot \cos 2x = \cos 2x$$

**35.**  $\dfrac{2\tan x}{2 - \sec^2 x}$ appears to be equivalent to $\tan 2x$.

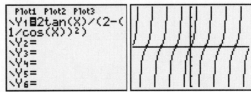

*(continued on next page)*

*(continued from page 127)*

$$\frac{2\tan x}{2-\sec^2 x} = \frac{2\tan x}{1-\left(\sec^2 x - 1\right)}$$
$$= \frac{2\tan x}{1-\tan^2 x} = \tan 2x$$

**37.** Verify $\left(\sin x + \cos x\right)^2 = \sin 2x + 1$ is an identity.

$$\left(\sin x + \cos x\right)^2 = \sin^2 x + 2\sin x\cos x + \cos^2 x$$
$$= \left(\sin^2 x + \cos^2 x\right) + 2\sin x\cos x$$
$$= 1 + \sin 2x$$

**39.** Verify $\tan 8\theta - \tan 8\theta \tan^2 4\theta = 2\tan 4\theta$ is an identity.

$$\tan 8\theta - \tan 8\theta \tan^2 4\theta$$
$$= \tan 8\theta\left(1 - \tan^2 4\theta\right)$$
$$= \frac{2\tan 4\theta}{1-\tan^2 4\theta}\left(1 - \tan^2 4\theta\right) = 2\tan 4\theta$$

**41.** Verify $\cos 2\theta = \frac{2-\sec^2 \theta}{\sec^2 \theta}$ is an identity.

Working with the right side, we have

$$\frac{2-\sec^2 \theta}{\sec^2 \theta} = \frac{2 - \dfrac{1}{\cos^2 \theta}}{\dfrac{1}{\cos^2 \theta}} = \frac{2 - \dfrac{1}{\cos^2 \theta}}{\dfrac{1}{\cos^2 \theta}} \cdot \frac{\cos^2 \theta}{\cos^2 \theta}$$
$$= \frac{2\cos^2 \theta - 1}{1} = \cos 2\theta$$

**43.** Verify that $\sin 4x = 4\sin x\cos x\cos 2x$ is an identity.

$$\sin 4x = \sin 2\left(2x\right) = 2\sin 2x\cos 2x$$
$$= 2\left(2\sin x\cos x\right)\cos 2x$$
$$= 4\sin x\cos x\cos 2x$$

**45.** Verify $\frac{2\cos 2\theta}{\sin 2\theta} = \cot \theta - \tan \theta$ is an identity.

Work with the right side.

$$\cot \theta - \tan \theta = \frac{\cos \theta}{\sin \theta} - \frac{\sin \theta}{\cos \theta}$$
$$= \frac{\cos \theta}{\sin \theta}\cdot\frac{\cos \theta}{\cos \theta} - \frac{\sin \theta}{\cos \theta}\cdot\frac{\sin \theta}{\sin \theta}$$
$$= \frac{\cos^2 \theta - \sin^2 \theta}{\sin \theta \cos \theta}$$
$$= \frac{2\left(\cos^2 \theta - \sin^2 \theta\right)}{2\sin \theta \cos \theta} = \frac{2\cos 2\theta}{\sin 2\theta}$$

**47.** Verify $\tan x + \cot x = 2\csc 2x$ is an identity.

$$\tan x + \cot x = \frac{\sin x}{\cos x}\cdot\frac{\sin x}{\sin x} + \frac{\cos x}{\sin x}\cdot\frac{\cos x}{\cos x}$$
$$= \frac{\sin^2 x + \cos^2 x}{\cos x\sin x} = \frac{1}{\cos x\sin x}$$
$$= \frac{2}{2\cos x\sin x} = \frac{2}{\sin 2x} = 2\csc 2x$$

**49.** Verify $1 + \tan x\tan 2x = \sec 2x$ is an identity.

$$1 + \tan x\tan 2x = 1 + \tan x\left(\frac{2\tan x}{1-\tan^2 x}\right)$$
$$= 1 + \frac{2\tan^2 x}{1-\tan^2 x}$$
$$= \frac{\left(1-\tan^2 x\right) + 2\tan^2 x}{1-\tan^2 x}$$
$$= \frac{1+\tan^2 x}{1-\tan^2 x} = \frac{1+\dfrac{\sin^2 x}{\cos^2 x}}{1-\dfrac{\sin^2 x}{\cos^2 x}}$$
$$= \frac{1+\dfrac{\sin^2 x}{\cos^2 x}}{1-\dfrac{\sin^2 x}{\cos^2 x}} \cdot \frac{\cos^2 x}{\cos^2 x}$$
$$= \frac{\cos^2 x + \sin^2 x}{\cos^2 x - \sin^2 x}$$
$$= \frac{1}{\cos 2x} = \sec 2x$$

**51.** Verify $\sin 2A\cos 2A = \sin 2A - 4\sin^3 A\cos A$ is an identity.

$$\sin 2A\cos 2A = \left(2\sin A\cos A\right)\left(1 - 2\sin^2 A\right)$$
$$= 2\sin A\cos A - 4\sin^3 A\cos A$$
$$= \sin 2A - 4\sin^3 A\cos A$$

**53.** Verify $\tan(\theta - 45°) + \tan(\theta + 45°) = 2\tan 2\theta$ is an identity.

$$\tan(\theta - 45°) + \tan(\theta + 45°) = \frac{\tan\theta - \tan 45°}{1 + \tan\theta\tan 45°} + \frac{\tan\theta + \tan 45°}{1 - \tan\theta\tan 45°} = \frac{\tan\theta - 1}{1 + \tan\theta} + \frac{\tan\theta + 1}{1 - \tan\theta} = \frac{\tan\theta - 1}{\tan\theta + 1} - \frac{\tan\theta + 1}{\tan\theta - 1}$$

$$= \frac{(\tan\theta - 1)^2 - (\tan\theta + 1)^2}{(\tan\theta + 1)(\tan\theta - 1)} = \frac{(\tan^2\theta - 2\tan\theta + 1) - (\tan^2\theta + 2\tan\theta + 1)}{\tan^2\theta - 1}$$

$$= \frac{-4\tan\theta}{\tan^2\theta - 1} = \frac{4\tan\theta}{1 - \tan^2\theta} = \frac{2(2\tan\theta)}{1 - \tan^2\theta} = 2\tan 2\theta$$

**55.** $2\sin 58° \cos 102° = 2\left(\frac{1}{2}\left[\sin(58° + 102°) + \sin(58° - 102°)\right]\right) = \sin 160° + \sin(-44°) = \sin 160° - \sin 44°$

**57.** $2\sin\dfrac{\pi}{6}\cos\dfrac{\pi}{3} = 2\cdot\dfrac{1}{2}\left[\sin\left(\dfrac{\pi}{6} + \dfrac{\pi}{3}\right) + \sin\left(\dfrac{\pi}{6} - \dfrac{\pi}{3}\right)\right] = \sin\dfrac{\pi}{2} + \sin\left(-\dfrac{\pi}{6}\right) = \sin\dfrac{\pi}{2} - \sin\dfrac{\pi}{6}$

**59.** $6\sin 4x\sin 5x = 6\cdot\dfrac{1}{2}\left[\cos(4x - 5x) - \cos(4x + 5x)\right] = 3\cos(-x) - 3\cos 9x = 3\cos x - 3\cos 9x$

**61.** $\cos 4x - \cos 2x = -2\sin\left(\dfrac{4x + 2x}{2}\right)\sin\left(\dfrac{4x - 2x}{2}\right) = -2\sin\dfrac{6x}{2}\sin\dfrac{2x}{2} = -2\sin 3x\sin x$

**63.** $\sin 25° + \sin(-48°) = 2\sin\left(\dfrac{25° + (-48°)}{2}\right)\cos\left(\dfrac{25° - (-48°)}{2}\right) = 2\sin\dfrac{-23°}{2}\cos\dfrac{73°}{2}$

$$= 2\sin(-11.5°)\cos 36.5° = -2\sin 11.5°\cos 36.5°$$

**65.** $\cos 4x + \cos 8x = 2\cos\left(\dfrac{4x + 8x}{2}\right)\cos\left(\dfrac{4x - 8x}{2}\right) = 2\cos\dfrac{12x}{2}\cos\dfrac{-4x}{2} = 2\cos 6x\cos(-2x) = 2\cos 6x\cos 2x$

**67.** From Example 6, $W = \dfrac{(163\sin 120\pi t)^2}{15} \Rightarrow W \approx 1771.3(\sin 120\pi t)^2$. Thus, we have

$$1771.3(\sin 120\pi t)^2 = 1771.3\sin 120\pi t\cdot\sin 120\pi t$$

$$= (1771.3)\left(\frac{1}{2}\right)\left[\cos(120\pi t - 120\pi t) - \cos(120\pi t + 120\pi t)\right]$$

$$= 885.6(\cos 0 - \cos 240\pi t) = 885.6(1 - \cos 240\pi t) = -885.6\cos 240\pi t + 885.6$$

If we compare this to $W = a\cos(\omega t) + c$, then $a = -885.6$, $c = 885.6$, and $\omega = 240\pi$.

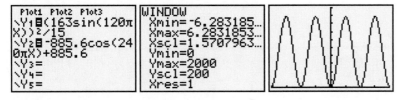

## Section 5.6 Half-Angle Identities

**1.** 195° lies in quadrant III, and the sine is negative in quadrant III, so choose the negative square root.

**3.** 225° lies in quadrant III, and the tangent is positive in quadrant III, so choose the positive square root.

**5.** C. $\sin 15° = \sin\dfrac{30°}{2} = \sqrt{\dfrac{1 - \cos 30°}{2}} = \sqrt{\dfrac{1 - \frac{\sqrt{3}}{2}}{2}}$

$$= \sqrt{\dfrac{2 - \sqrt{3}}{2\cdot 2}} = \sqrt{\dfrac{2 - \sqrt{3}}{4}} = \dfrac{\sqrt{2 - \sqrt{3}}}{2}$$

**7.** D. $\cos\dfrac{\pi}{8} = \cos\dfrac{\frac{\pi}{4}}{2} = \sqrt{\dfrac{1+\cos\frac{\pi}{4}}{2}} = \sqrt{\dfrac{1+\frac{\sqrt{2}}{2}}{2}}$

$\qquad = \sqrt{\dfrac{2+\sqrt{2}}{2\cdot 2}} = \dfrac{\sqrt{2+\sqrt{2}}}{2}$

**9.** F. $\tan 67.5° = \tan\dfrac{135°}{2} = \dfrac{1-\cos 135°}{\sin 135°}$

$\qquad = \dfrac{1-\left(-\frac{\sqrt{2}}{2}\right)}{\frac{\sqrt{2}}{2}} = \dfrac{1-\left(-\frac{\sqrt{2}}{2}\right)}{\frac{\sqrt{2}}{2}}\cdot\dfrac{2}{2}$

$\qquad = \dfrac{2+\sqrt{2}}{\sqrt{2}} = \dfrac{2+\sqrt{2}}{\sqrt{2}}\cdot\dfrac{\sqrt{2}}{\sqrt{2}}$

$\qquad = \dfrac{2\sqrt{2}+2}{2} = \sqrt{2}+1$

**11.** $\sin 67.5° = \sin\left(\dfrac{135°}{2}\right)$

Since $67.5°$ is in quadrant I, $\sin 67.5° > 0$.

$\sin 67.5° = \sqrt{\dfrac{1-\cos 135°}{2}} = \sqrt{\dfrac{1-(-\cos 45°)}{2}}$

$\qquad = \sqrt{\dfrac{1+\frac{\sqrt{2}}{2}}{2}} = \sqrt{\dfrac{1+\frac{\sqrt{2}}{2}}{2}\cdot\dfrac{2}{2}}$

$\qquad = \dfrac{\sqrt{2+\sqrt{2}}}{\sqrt{4}} = \dfrac{\sqrt{2+\sqrt{2}}}{2}$

**13.** $\cos 195° = \cos\left(\dfrac{390°}{2}\right)$

Since $195°$ is in quadrant III, $\cos 195° < 0$.

$\cos 195° = -\sqrt{\dfrac{1+\cos 390°}{2}} = -\sqrt{\dfrac{1+\cos 30°}{2}}$

$\qquad = -\sqrt{\dfrac{1+\frac{\sqrt{3}}{2}}{2}} = -\sqrt{\dfrac{2+\sqrt{3}}{4}} = \dfrac{-\sqrt{2+\sqrt{3}}}{2}$

**15.** $\cos 165° = \cos\left(\dfrac{330°}{2}\right)$

Since $165°$ is in quadrant II, $\cos 165° < 0$.

$\cos 165° = -\sqrt{\dfrac{1+\cos 330°}{2}} = -\sqrt{\dfrac{1+\cos 30°}{2}}$

$\qquad = -\sqrt{\dfrac{1+\frac{\sqrt{3}}{2}}{2}} = -\sqrt{\dfrac{2+\sqrt{3}}{4}}$

$\qquad = \dfrac{-\sqrt{2+\sqrt{3}}}{2}$

**17.** To find $\sin 7.5°$, you could use the half-angle formulas for sine and cosine as follows:

$\sin 7.5° = \sqrt{\dfrac{1-\cos 15°}{2}}$ and

$\cos 15° = \sqrt{\dfrac{1+\cos 30°}{2}} = \sqrt{\dfrac{1+\frac{\sqrt{3}}{2}}{2}}$

$\qquad = \sqrt{\dfrac{2+\sqrt{3}}{4}} = \dfrac{\sqrt{2+\sqrt{3}}}{2}$

Thus,

$\sin 7.5° = \sqrt{\dfrac{1-\cos 15°}{2}} = \sqrt{\dfrac{1-\frac{\sqrt{2+\sqrt{3}}}{2}}{2}}$

$\qquad = \sqrt{\dfrac{2-\sqrt{2+\sqrt{3}}}{4}} = \dfrac{\sqrt{2-\sqrt{2+\sqrt{3}}}}{2}$

**19.** Find $\cos\dfrac{x}{2}$, given $\cos x = \dfrac{1}{4}$, with $0 < x < \dfrac{\pi}{2}$.

Since $0 < x < \dfrac{\pi}{2} \Rightarrow 0 < \dfrac{x}{2} < \dfrac{\pi}{4}$, $\cos\dfrac{x}{2} > 0$.

$\cos\dfrac{x}{2} = \sqrt{\dfrac{1+\cos x}{2}} = \sqrt{\dfrac{1+\frac{1}{4}}{2}} = \sqrt{\dfrac{4+1}{8}} = \sqrt{\dfrac{5}{8}}$

$\qquad = \dfrac{\sqrt{5}}{\sqrt{8}} = \dfrac{\sqrt{5}}{2\sqrt{2}} = \dfrac{\sqrt{5}}{2\sqrt{2}}\cdot\dfrac{\sqrt{2}}{\sqrt{2}} = \dfrac{\sqrt{10}}{4}$

**21.** Find $\tan\dfrac{\theta}{2}$, given $\sin\theta = \dfrac{3}{5}$, with $90° < \theta < 180°$.

To find $\tan\dfrac{\theta}{2}$, we need the values of $\sin\theta$

and $\cos\theta$. We know $\sin\theta = \dfrac{3}{5}$.

$\cos\theta = \pm\sqrt{1-\left(\dfrac{3}{5}\right)^2} = \pm\sqrt{1-\dfrac{9}{25}} = \pm\sqrt{\dfrac{16}{25}} = \pm\dfrac{4}{5}$

Since $90° < \theta < 180°$ ($\theta$ is in quadrant II),

$\cos\theta < 0$. Thus, $\cos\theta = -\dfrac{4}{5}$.

$\tan\dfrac{\theta}{2} = \dfrac{\sin\theta}{1+\cos\theta} = \dfrac{\frac{3}{5}}{1+\left(-\frac{4}{5}\right)} = \dfrac{\frac{3}{5}}{1+\left(-\frac{4}{5}\right)}\cdot\dfrac{5}{5}$

$\qquad = \dfrac{3}{5-4} = \dfrac{3}{1} = 3$

**23.** Find $\sin\dfrac{x}{2}$, given $\tan x = 2$, with $0 < x < \dfrac{\pi}{2}$.

Since $x$ is in quadrant I, $\sec\ x > 0$.

$\sec^2 x = \tan^2 x + 1 \Rightarrow$

$\sec^2 x = 2^2 + 1 = 4 + 1 = 5 \Rightarrow \sec x = \sqrt{5}$

$\cos x = \dfrac{1}{\sec x} = \dfrac{1}{\sqrt{5}} = \dfrac{1}{\sqrt{5}}\cdot\dfrac{\sqrt{5}}{\sqrt{5}} = \dfrac{\sqrt{5}}{5}$

Since $0 < x < \dfrac{\pi}{2} \Rightarrow 0 < \dfrac{x}{2} < \dfrac{\pi}{4}$, $\dfrac{x}{2}$ is in

quadrant I. Thus, $\sin\dfrac{x}{2} > 0$.

$\sin\dfrac{x}{2} = \sqrt{\dfrac{1 - \cos x}{2}} = \sqrt{\dfrac{1 - \frac{\sqrt{5}}{5}}{2}} = \dfrac{\sqrt{50 - 10\sqrt{5}}}{10}$

**25.** Find $\tan\dfrac{\theta}{2}$, given $\tan\theta = \dfrac{\sqrt{7}}{3}$, with

$180° < \theta < 270°$.

$\sec^2\theta = \tan^2\theta + 1 \Rightarrow$

$\sec^2\theta = \left(\dfrac{\sqrt{7}}{3}\right)^2 + 1 = \dfrac{7}{9} + 1 = \dfrac{16}{9}$

Since $\theta$ is in quadrant III, $\sec\ \theta < 0$ and $\sin\theta < 0$.

$\sec\theta = -\sqrt{\dfrac{16}{9}} = -\dfrac{4}{3}$ and

$\cos\theta = \dfrac{1}{\sec\theta} = \dfrac{1}{-\frac{4}{3}} = -\dfrac{3}{4}$

$\sin\theta = -\sqrt{1 - \cos^2\theta}$

$= -\sqrt{1 - \left(-\dfrac{3}{4}\right)^2} - \sqrt{1 - \dfrac{9}{16}} = -\dfrac{\sqrt{7}}{4}$

$\tan\dfrac{\theta}{2} = \dfrac{\sin\theta}{1 + \cos\theta} = \dfrac{-\frac{\sqrt{7}}{4}}{1 + \left(-\frac{3}{4}\right)} = \dfrac{-\sqrt{7}}{4 - 3} = -\sqrt{7}$

**27.** Find $\sin\theta$, given $\cos 2\theta = \dfrac{3}{5}$, $\theta$ is in

quadrant I. Since $\theta$ is in quadrant I, $\sin\theta > 0$.

$\sin\theta = \sqrt{\dfrac{1 - \cos 2\theta}{2}} \Rightarrow$

$\sin\theta = \sqrt{\dfrac{1 - \frac{3}{5}}{2}} = \sqrt{\dfrac{\frac{2}{5}}{2}} = \sqrt{\dfrac{2}{10}} = \sqrt{\dfrac{1}{5}}$

$= \dfrac{1}{\sqrt{5}}\cdot\dfrac{\sqrt{5}}{\sqrt{5}} = \dfrac{\sqrt{5}}{5}$

**29.** Find $\cos x$, given $\cos 2x = -\dfrac{5}{12}, \dfrac{\pi}{2} < x < \pi$.

Since $\dfrac{\pi}{2} < x < \pi, \cos x < 0$.

$\cos x = -\sqrt{\dfrac{1 + \cos 2x}{2}} \Rightarrow$

$\cos x = -\sqrt{\dfrac{1 + \left(-\frac{5}{12}\right)}{2}} = -\sqrt{\dfrac{\frac{7}{12}}{2}} = -\sqrt{\dfrac{7}{24}}$

$= -\dfrac{\sqrt{7}}{2\sqrt{6}}\cdot\dfrac{\sqrt{6}}{\sqrt{6}} = -\dfrac{\sqrt{42}}{12}$

**31.** $\cos x \approx .9682$ and $\sin x \approx .25$

$\tan\dfrac{x}{2} \approx \dfrac{\sin x}{1 + \cos x} = \dfrac{.25}{1 + .9682} = \dfrac{.25}{1.9682} \approx .1270$

(.127 to three significant digits)

**33.** $\sqrt{\dfrac{1 - \cos 40°}{2}} = \sin\dfrac{40°}{2} = \sin 20°$

**35.** $\sqrt{\dfrac{1 - \cos 147°}{1 + \cos 147°}} = \tan\dfrac{147°}{2} = \tan 73.5°$

**37.** $\dfrac{1 - \cos 59.74°}{\sin 59.74°} = \tan\dfrac{59.74°}{2} = \tan 29.87°$

**39.** $\pm\sqrt{\dfrac{1 + \cos 18x}{2}} = \cos\dfrac{18x}{2} = \cos 9x$

**41.** $\pm\sqrt{\dfrac{1 - \cos 8\theta}{1 + \cos 8\theta}} = \tan\dfrac{8\theta}{2} = \tan 4\theta$

**43.** $\pm\sqrt{\dfrac{1 + \cos\frac{x}{4}}{2}} = \cos\dfrac{\frac{x}{4}}{2} = \cos\dfrac{x}{8}$

Exercises 45–47 are graphed in the following window:

**45.** $\dfrac{\sin x}{1 + \cos x}$ appears to be equivalent to $\tan\dfrac{x}{2}$.

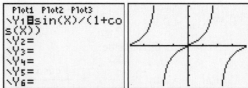

*(continued on next page)*

*(continued from page 131)*

$$\frac{\sin x}{1+\cos x} = \frac{\sin 2\left(\frac{x}{2}\right)}{1+\cos 2\left(\frac{x}{2}\right)} = \frac{2\sin\left(\frac{x}{2}\right)\cos\left(\frac{x}{2}\right)}{1+\left[2\cos^2\left(\frac{x}{2}\right)-1\right]}$$

$$= \frac{2\sin\left(\frac{x}{2}\right)\cos\left(\frac{x}{2}\right)}{2\cos^2\left(\frac{x}{2}\right)} = \frac{\sin\left(\frac{x}{2}\right)}{\cos\left(\frac{x}{2}\right)}$$

$$= \tan\left(\frac{x}{2}\right)$$

**47.** $\dfrac{\tan\frac{x}{2} + \cot\frac{x}{2}}{\cot\frac{x}{2} - \tan\frac{x}{2}}$ appears to be equivalent to

sec $x$.

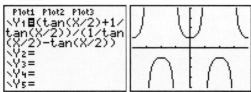

$$\frac{\tan\frac{x}{2} + \cot\frac{x}{2}}{\cot\frac{x}{2} - \tan\frac{x}{2}} = \frac{\dfrac{\sin\frac{x}{2}}{\cos\frac{x}{2}} + \dfrac{\cos\frac{x}{2}}{\sin\frac{x}{2}}}{\dfrac{\cos\frac{x}{2}}{\sin\frac{x}{2}} - \dfrac{\sin\frac{x}{2}}{\cos\frac{x}{2}}}$$

$$= \frac{\dfrac{\sin\frac{x}{2}}{\cos\frac{x}{2}} + \dfrac{\cos\frac{x}{2}}{\sin\frac{x}{2}}}{\dfrac{\cos\frac{x}{2}}{\sin\frac{x}{2}} - \dfrac{\sin\frac{x}{2}}{\cos\frac{x}{2}}} \cdot \frac{\sin\frac{x}{2}\cos\frac{x}{2}}{\sin\frac{x}{2}\cos\frac{x}{2}}$$

$$= \frac{\sin^2\frac{x}{2} + \cos^2\frac{x}{2}}{\cos^2\frac{x}{2} - \sin^2\frac{x}{2}} = \frac{1}{\cos 2\left(\frac{x}{2}\right)}$$

$$= \frac{1}{\cos x} = \sec x$$

**49.** Verify $\sec^2\dfrac{x}{2} = \dfrac{2}{1+\cos x}$ is an identity.

$$\sec^2\frac{x}{2} = \frac{1}{\cos^2\frac{x}{2}} = \frac{1}{\left(\pm\sqrt{\dfrac{1+\cos x}{2}}\right)^2}$$

$$= \frac{1}{\dfrac{1+\cos x}{2}} = \frac{2}{1+\cos x}$$

**51.** Verify $\sin^2\dfrac{x}{2} = \dfrac{\tan x - \sin x}{2\tan x}$ is an identity.

Work with the left side.

$$\sin^2\frac{x}{2} = \left(\pm\sqrt{\frac{1-\cos x}{2}}\right)^2 = \frac{1-\cos x}{2}$$

Work with the right side.

$$\frac{\tan x - \sin x}{2\tan x} = \frac{\dfrac{\sin x}{\cos x} - \sin x}{2 \cdot \dfrac{\sin x}{\cos x}}$$

$$= \frac{\dfrac{\sin x}{\cos x} - \sin x}{2 \cdot \dfrac{\sin x}{\cos x}} \cdot \frac{\cos x}{\cos x}$$

$$= \frac{\sin x - \cos x \sin x}{2\sin x}$$

$$= \frac{\sin x (1 - \cos x)}{2\sin x} = \frac{1-\cos x}{2}$$

Since $\sin^2\dfrac{x}{2} = \dfrac{1-\cos x}{2} = \dfrac{\tan x - \sin x}{2\tan x}$, the

statement has been verified.

**53.** Verify $\dfrac{2}{1+\cos x} - \tan^2\dfrac{x}{2} = 1$ is an identity.

$$\frac{2}{1+\cos x} - \tan^2\frac{x}{2} = \frac{2}{1+\cos x} - \left(\pm\sqrt{\frac{1-\cos x}{1+\cos x}}\right)^2$$

$$= \frac{2}{1+\cos x} - \frac{1-\cos x}{1+\cos x}$$

$$= \frac{2-1+\cos x}{1+\cos x}$$

$$= \frac{1+\cos x}{1+\cos x} = 1$$

**55.** Verify $1 - \tan^2\dfrac{\theta}{2} = \dfrac{2\cos\theta}{1+\cos\theta}$ is an identity.

$$1 - \tan^2\frac{\theta}{2} = 1 - \left(\frac{\sin\theta}{1+\cos\theta}\right)^2 = 1 - \frac{\sin^2\theta}{(1+\cos\theta)^2}$$

$$= \frac{(1+\cos\theta)^2 - \sin^2\theta}{(1+\cos\theta)^2}$$

$$= \frac{1 + 2\cos\theta + \cos^2\theta - \sin^2\theta}{(1+\cos\theta)^2}$$

$$= \frac{1 + 2\cos\theta + \cos^2\theta - \left(1 - \cos^2\theta\right)}{(1+\cos\theta)^2}$$

$$= \frac{1 + 2\cos\theta + 2\cos^2\theta - 1}{(1+\cos\theta)^2}$$

$$= \frac{2\cos^2\theta + 2\cos\theta}{(1+\cos\theta)^2}$$

$$= \frac{2\cos\theta(1+\cos\theta)}{(1+\cos\theta)^2} = \frac{2\cos\theta}{1+\cos\theta}$$

**57.** $\tan \dfrac{A}{2} = \dfrac{\sin A}{1+\cos A} = \dfrac{\sin A}{1+\cos A} \cdot \dfrac{1-\cos A}{1-\cos A}$

$\qquad = \dfrac{\sin A(1-\cos A)}{1-\cos^2 A} = \dfrac{\sin A(1-\cos A)}{\sin^2 A}$

$\qquad = \dfrac{1-\cos A}{\sin A}$

**59.** $\sin \dfrac{\theta}{2} = \dfrac{1}{m}, \ m = \dfrac{5}{4}$

Since $\sin \dfrac{\theta}{2} = \dfrac{1}{\frac{5}{4}} = \dfrac{4}{5}$ and $\sin^2 \dfrac{\theta}{2} = \dfrac{1-\cos \theta}{2}$,

we have

$\left(\dfrac{4}{5}\right)^2 = \dfrac{1-\cos \theta}{2} \Rightarrow \dfrac{16}{25} = \dfrac{1-\cos \theta}{2} \Rightarrow$

$\dfrac{32}{25} = 1-\cos \theta \Rightarrow \dfrac{7}{25} = -\cos \theta \Rightarrow$

$\cos \theta = -\dfrac{7}{25}$

Thus, we have $\theta = \cos^{-1}\left(-\dfrac{7}{25}\right) \approx 106°$.

**61.** $\sin \dfrac{\theta}{2} = \dfrac{1}{m}, \ \theta = 60°$

$\sin \dfrac{60°}{2} = \dfrac{1}{m} \Rightarrow \sin 30° = \dfrac{1}{m}$

$\dfrac{1}{2} = \dfrac{1}{m} \Rightarrow m = 2$

**63.** $AB = BD$ because they are both radii of the circle.

**65.** The sum of the measures of angles $DAB$ and $ADB$ is $180° - 150° = 30°$.
$m\angle DAB = m\angle ADB$, so the measure of each is $15°$.

**67.** $AD^2 = AC^2 + CD^2 \Rightarrow$

$AD^2 = 1^2 + \left(2+\sqrt{3}\right)^2 = 1+4+4\sqrt{3}+3$

$\qquad = 8 + 4\sqrt{3} = \left(\sqrt{6}+\sqrt{2}\right)^2 \Rightarrow$

$AD = \sqrt{6} + \sqrt{2}$

**69.** $m\angle EAD = 90°$ (an angle inscribed in a semicircle is a right angle), so
$m\angle EAC + m\angle CAD = 90° \Rightarrow m\angle EAC = 15°$.

$\cos 15° = \dfrac{AC}{AE} = \dfrac{1}{AE} \Rightarrow AE = \dfrac{1}{\cos 15°}$

$AE = \dfrac{1}{\cos 15°} = \dfrac{1}{\frac{\sqrt{6}+\sqrt{2}}{4}} = \dfrac{4}{\sqrt{6}+\sqrt{2}}$

$\qquad = \dfrac{4}{\sqrt{6}+\sqrt{2}} \cdot \dfrac{\sqrt{6}-\sqrt{2}}{\sqrt{6}-\sqrt{2}} = \dfrac{4\left(\sqrt{6}-\sqrt{2}\right)}{6-2}$

$\qquad = \dfrac{4\left(\sqrt{6}-\sqrt{2}\right)}{4} = \sqrt{6}-\sqrt{2}$

$ED = 2BD = 4$. In $\triangle AED$,

$\sin 15° = \dfrac{AE}{ED} = \dfrac{\sqrt{6}-\sqrt{2}}{4}$

## Summary Exercises on Verifying Trigonometric Identities

For the following exercises, other solutions are possible.

**1.** Verify $\tan \theta + \cot \theta = \sec \theta \csc \theta$ is an identity.

$\tan \theta + \cot \theta = \dfrac{\sin \theta}{\cos \theta} + \dfrac{\cos \theta}{\sin \theta}$

$\qquad = \dfrac{\sin^2 \theta}{\cos \theta \sin \theta} + \dfrac{\cos^2 \theta}{\cos \theta \sin \theta}$

$\qquad = \dfrac{\sin^2 \theta + \cos^2 \theta}{\cos \theta \sin \theta}$

$\qquad = \dfrac{1}{\cos \theta \sin \theta}$

$\qquad = \dfrac{1}{\cos \theta} \cdot \dfrac{1}{\sin \theta}$

$\qquad = \sec \theta \csc \theta$

**3.** Verify $\tan \dfrac{x}{2} = \csc x - \cot x$ is an identity.

Starting on the right side, we have

$\csc x - \cot x = \dfrac{1}{\sin x} - \dfrac{\cos x}{\sin x} = \dfrac{1-\cos x}{\sin x} = \tan \dfrac{x}{2}$

**5.** Verify $\dfrac{\sin t}{1+\cos t} = \dfrac{1-\cos t}{\sin t}$ is an identity.

$\dfrac{\sin t}{1+\cos t} = \dfrac{\sin t}{1+\cos t} \cdot \dfrac{1-\cos t}{1-\cos t}$

$\qquad = \dfrac{\sin t(1-\cos t)}{1-\cos^2 t} = \dfrac{\sin t(1-\cos t)}{\sin^2 t}$

$\qquad = \dfrac{1-\cos t}{\sin t}$

**7.** Verify $\sin 2\theta = \dfrac{2\tan\theta}{1+\tan^2\theta}$ is an identity.

Starting on the right side, we have

$$\frac{2\tan\theta}{1+\tan^2\theta} = \frac{2\tan\theta}{\sec^2\theta} = \frac{2\cdot\dfrac{\sin\theta}{\cos\theta}}{\dfrac{1}{\cos^2\theta}}$$

$$= 2\cdot\frac{\sin\theta}{\cos\theta}\cdot\frac{\cos^2\theta}{1}$$

$$= 2\sin\theta\cos\theta = \sin 2\theta$$

**9.** Verify $\cot\theta - \tan\theta = \dfrac{2\cos^2\theta - 1}{\sin\theta\cos\theta}$ is an identity.

$$\cot\theta - \tan\theta = \frac{\cos\theta}{\sin\theta} - \frac{\sin\theta}{\cos\theta}$$

$$= \frac{\cos^2\theta}{\sin\theta\cos\theta} - \frac{\sin^2\theta}{\sin\theta\cos\theta}$$

$$= \frac{\cos^2\theta - \sin^2\theta}{\sin\theta\cos\theta}$$

$$= \frac{\cos^2\theta - \left(1 - \cos^2\theta\right)}{\sin\theta\cos\theta}$$

$$= \frac{\cos^2\theta - 1 + \cos^2\theta}{\sin\theta\cos\theta}$$

$$= \frac{2\cos^2\theta - 1}{\sin\theta\cos\theta}$$

**11.** Verify $\dfrac{\sin(x+y)}{\cos(x-y)} = \dfrac{\cot x + \cot y}{1 + \cot x\cot y}$ is an identity.

$$\frac{\sin(x+y)}{\cos(x-y)} = \frac{\sin x\cos y + \cos x\sin y}{\cos x\cos y + \sin x\sin y}$$

$$= \frac{\sin x\cos y + \cos x\sin y}{\cos x\cos y + \sin x\sin y}\cdot\frac{\dfrac{1}{\cos x\cos y}}{\dfrac{1}{\cos x\cos y}}$$

$$= \frac{\dfrac{\sin x\cos y}{\cos x\cos y} + \dfrac{\cos x\sin y}{\cos x\cos y}}{\dfrac{\cos x\cos y}{\cos x\cos y} + \dfrac{\sin x\sin y}{\cos x\cos y}}$$

$$= \frac{\dfrac{\sin x}{\cos x} + \dfrac{\sin y}{\cos y}}{1 + \dfrac{\sin x}{\cos x}\cdot\dfrac{\sin y}{\cos y}} = \frac{\cot x + \cot y}{1 + \cot x\cot y}$$

**13.** Verify $\dfrac{\sin\theta + \tan\theta}{1 + \cos\theta} = \tan\theta$ is an identity.

$$\frac{\sin\theta + \tan\theta}{1 + \cos\theta} = \frac{\sin\theta + \dfrac{\sin\theta}{\cos\theta}}{1 + \cos\theta}$$

$$= \frac{\sin\theta + \dfrac{\sin\theta}{\cos\theta}}{1 + \cos\theta}\cdot\frac{\cos\theta}{\cos\theta}$$

$$= \frac{\sin\theta\cos\theta + \sin\theta}{\cos\theta(1 + \cos\theta)}$$

$$= \frac{\sin\theta(\cos\theta + 1)}{\cos\theta(1 + \cos\theta)} = \frac{\sin\theta}{\cos\theta} = \tan\theta$$

**15.** Verify $\cos x = \dfrac{1 - \tan^2\dfrac{x}{2}}{1 + \tan^2\dfrac{x}{2}}$ is an identity.

$$\frac{1 - \tan^2\dfrac{x}{2}}{1 + \tan^2\dfrac{x}{2}} = \frac{1 - \left(\dfrac{1-\cos x}{\sin x}\right)^2}{1 + \left(\dfrac{1-\cos x}{\sin x}\right)^2} = \frac{1 - \dfrac{(1-\cos x)^2}{\sin^2 x}}{1 + \dfrac{(1-\cos x)^2}{\sin^2 x}} = \frac{1 - \dfrac{(1-\cos x)^2}{\sin^2 x}}{1 + \dfrac{(1-\cos x)^2}{\sin^2 x}}\cdot\frac{\sin^2 x}{\sin^2 x} = \frac{\sin^2 x - (1-\cos x)^2}{\sin^2 x + (1-\cos x)^2}$$

$$= \frac{\sin^2 x - \left(1 - 2\cos x + \cos^2 x\right)}{\sin^2 x + \left(1 - 2\cos x + \cos^2 x\right)} = \frac{\sin^2 x - 1 + 2\cos x - \cos^2 x}{\sin^2 x + 1 - 2\cos x + \cos^2 x} = \frac{\left(1 - \cos^2 x\right) - 1 + 2\cos x - \cos^2 x}{\left(\sin^2 x + \cos^2 x\right) + 1 - 2\cos x}$$

$$= \frac{1 - \cos^2 x - 1 + 2\cos x - \cos^2 x}{1 + 1 - 2\cos x} = \frac{2\cos x - 2\cos^2 x}{2 - 2\cos x} = \frac{2\cos x(1 - \cos x)}{2(1 - \cos x)} = \cos x$$

**17.** Verify $\dfrac{\tan^2 t + 1}{\tan t \csc^2 t} = \tan t$ is an identity.

$$\frac{\tan^2 t + 1}{\tan t \csc^2 t} = \frac{\dfrac{\sin^2 t}{\cos^2 t} + 1}{\dfrac{\sin t}{\cos t} \cdot \dfrac{1}{\sin^2 t}} = \frac{\dfrac{\sin^2 t}{\cos^2 t} + 1}{\dfrac{1}{\cos t \sin t}} = \frac{\dfrac{\sin^2 t}{\cos^2 t} + 1}{\dfrac{1}{\cos t \sin t}} \cdot \frac{\cos^2 t \sin t}{\cos^2 t \sin t} = \frac{\sin^3 t + \cos^2 t \sin t}{\cos t}$$

$$= \frac{\sin t \left(\sin^2 t + \cos^2 t\right)}{\cos t} = \frac{\sin t \left(1\right)}{\cos t} = \frac{\sin t}{\cos t} = \tan t$$

**19.** Verify $\tan 4\theta = \dfrac{2 \tan 2\theta}{2 - \sec^2 2\theta}$ is an identity.

$$\frac{2 \tan 2\theta}{2 - \sec^2 2\theta} = \frac{2 \cdot \dfrac{\sin 2\theta}{\cos 2\theta}}{2 - \dfrac{1}{\cos^2 2\theta}} = \frac{2 \cdot \dfrac{\sin 2\theta}{\cos 2\theta}}{2 - \dfrac{1}{\cos^2 2\theta}} \cdot \frac{\cos^2 2\theta}{\cos^2 2\theta} = \frac{2 \sin 2\theta \cos 2\theta}{2 \cos^2 2\theta - 1} = \frac{\sin\left[2(2\theta)\right]}{\cos\left[2(2\theta)\right]} = \frac{\sin 4\theta}{\cos 4\theta} = \tan 4\theta$$

**21.** Verify $\dfrac{\cot s - \tan s}{\cos s + \sin s} = \dfrac{\cos s - \sin s}{\sin s \cos s}$ is an identity.

$$\frac{\cot s - \tan s}{\cos s + \sin s} = \frac{\dfrac{\cos s}{\sin s} - \dfrac{\sin s}{\cos s}}{\cos s + \sin s} = \frac{\dfrac{\cos s}{\sin s} - \dfrac{\sin s}{\cos s}}{\cos s + \sin s} \cdot \frac{\sin s \cos s}{\sin s \cos s} = \frac{\cos^2 s - \sin^2 s}{\left(\cos s + \sin s\right)\sin s \cos s}$$

$$= \frac{\left(\cos s + \sin s\right)\left(\cos s - \sin s\right)}{\left(\cos s + \sin s\right)\sin s \cos s} = \frac{\cos s - \sin s}{\sin s \cos s}$$

**23.** Verify $\dfrac{\tan(x+y) - \tan y}{1 + \tan(x+y)\tan y} = \tan x$ is an identity.

$$\frac{\tan(x+y) - \tan y}{1 + \tan(x+y)\tan y} = \frac{\dfrac{\tan x + \tan y}{1 - \tan x \tan y} - \tan y}{1 + \dfrac{\tan x + \tan y}{1 - \tan x \tan y} \cdot \tan y} = \frac{\dfrac{\tan x + \tan y}{1 - \tan x \tan y} - \tan y}{1 + \dfrac{\tan x + \tan y}{1 - \tan x \tan y} \cdot \tan y} \cdot \frac{1 - \tan x \tan y}{1 - \tan x \tan y}$$

$$= \frac{\tan x + \tan y - \tan y \left(1 - \tan x \tan y\right)}{1 - \tan x \tan y + \left(\tan x + \tan y\right)\tan y} = \frac{\tan x + \tan x \tan^2 y}{1 - \tan x \tan y + \tan x \tan y + \tan^2 y}$$

$$= \frac{\tan x \left(1 + \tan^2 y\right)}{1 + \tan^2 y} = \tan x$$

**25.** Verify $\dfrac{\cos^4 x - \sin^4 x}{\cos^2 x} = 1 - \tan^2 x$ is an identity.

$$\frac{\cos^4 x - \sin^4 x}{\cos^2 x} = \frac{\left(\cos^2 x + \sin^2 x\right)\left(\cos^2 x - \sin^2 x\right)}{\cos^2 x} = \frac{(1)\left(\cos^2 x - \sin^2 x\right)}{\cos^2 x} = \frac{\cos^2 x - \sin^2 x}{\cos^2 x}$$

$$= \frac{\cos^2 x}{\cos^2 x} - \frac{\sin^2 x}{\cos^2 x} = 1 - \tan^2 x$$

**27.** Verify $\dfrac{2\left(\sin x - \sin^3 x\right)}{\cos x} = \sin 2x$ is an identity.

$$\frac{2\left(\sin x - \sin^3 x\right)}{\cos x} = \frac{2 \sin x \left(1 - \sin^2 x\right)}{\cos x} = \frac{2 \sin x \cos^2 x}{\cos x} = 2 \sin x \cos x = \sin 2x$$

**29.** Verify $\sin(60° + x) + \sin(60° - x) = \sqrt{3}\cos x$ is an identity.

$$\sin(60° + x) + \sin(60° - x) = (\sin 60° \cos x + \cos 60° \sin x) + (\sin 60° \cos x - \cos 60° \sin x)$$

$$= 2\sin 60° \cos x = 2\left(\frac{\sqrt{3}}{2}\right)\cos x = \sqrt{3}\cos x$$

**31.** Verify $\dfrac{\cos(x+y) + \cos(y-x)}{\sin(x+y) - \sin(y-x)} = \cot x$ is an identity.

$$\frac{\cos(x+y) + \cos(y-x)}{\sin(x+y) - \sin(y-x)} = \frac{(\cos x \cos y - \sin x \sin y) + (\cos y \cos x + \sin y \sin x)}{(\sin x \cos y + \cos x \sin y) - (\sin y \cos x - \cos y \sin x)} = \frac{2\cos x \cos y}{2\cos y \sin x} = \frac{\cos x}{\sin x} = \cot x$$

**33.** Verify $\sin^3\theta + \cos^3\theta + \sin\theta\cos^2\theta + \sin^2\cos\theta = \sin\theta + \cos\theta$ is an identity.

$$\sin^3\theta + \cos^3\theta + \sin\theta\cos^2\theta + \sin^2\cos\theta = \left(\sin^3\theta + \sin\theta\cos^2\theta\right) + \left(\cos^3\theta + \sin^2\cos\theta\right)$$

$$= \sin\theta\left(\sin^2\theta + \cos^2\theta\right) + \cos\theta\left(\cos^2\theta + \sin^2\theta\right)$$

$$= \left(\sin\theta + \cos\theta\right)\left(\sin^2\theta + \cos^2\theta\right) = \sin\theta + \cos\theta$$

## Chapter 5: Review Exercises

**1.** Since $\sec x = \dfrac{1}{\cos x}$, the correct choice is B.

**3.** Since $\tan x = \dfrac{\sin x}{\cos x}$, the correct choice is C.

**5.** Since $\tan^2 x = \dfrac{1}{\cot^2 x}$, the correct choice is D.

**7.** $\sec^2\theta - \tan^2\theta = \dfrac{1}{\cos^2\theta} - \dfrac{\sin^2\theta}{\cos^2\theta}$

$$= \frac{1 - \sin^2\theta}{\cos^2\theta} = \frac{\cos^2\theta}{\cos^2\theta} = 1$$

**9.** $\tan^2\theta\left(1 + \cot^2\theta\right) = \dfrac{\sin^2\theta}{\cos^2\theta}\left(1 + \dfrac{\cos^2\theta}{\sin^2\theta}\right)$

$$= \frac{\sin^2\theta}{\cos^2\theta}\left(\frac{\sin^2\theta + \cos^2\theta}{\sin^2\theta}\right)$$

$$= \frac{\sin^2\theta}{\cos^2\theta}\left(\frac{1}{\sin^2\theta}\right) = \frac{1}{\cos^2\theta}$$

**11.** $\tan\theta - \sec\theta\csc\theta = \dfrac{\sin\theta}{\cos\theta} - \dfrac{1}{\cos\theta}\cdot\dfrac{1}{\sin\theta}$

$$= \frac{\sin\theta}{\cos\theta} - \frac{1}{\sin\theta\cos\theta}$$

$$= \frac{\sin^2\theta}{\sin\theta\cos\theta} - \frac{1}{\sin\theta\cos\theta}$$

$$= \frac{\sin^2\theta - 1}{\sin\theta\cos\theta} = \frac{\left(1 - \cos^2\theta\right) - 1}{\sin\theta\cos\theta}$$

$$= \frac{-\cos^2\theta}{\sin\theta\cos\theta} = -\frac{\cos\theta}{\sin\theta}$$

**13.** $\cos x = \dfrac{3}{5}$, $x$ is in quadrant IV.

$$\sin^2 x = 1 - \cos^2 x = 1 - \left(\frac{3}{5}\right)^2 = 1 - \frac{9}{25} = \frac{16}{25}$$

Since $x$ is in quadrant IV, $\sin x < 0$.

$$\sin x = -\sqrt{\frac{16}{25}} = -\frac{4}{5}$$

$$\tan x = \frac{\sin x}{\cos x} = \frac{-\frac{4}{5}}{\frac{3}{5}} = -\frac{4}{3}$$

$$\cot(-x) = -\cot x = \frac{1}{-\tan x} = \frac{1}{-\left(-\frac{4}{3}\right)} = \frac{3}{4}$$

**15.** Use the fact that $165° = 180° - 15°$.

$$\sin 165° = \sin 180° \cos 15° - \cos 180° \sin 15°$$

$$= 0 \cdot \cos 15° - 1 \cdot \sin 15° = \sin 15°$$

$$= \sin(45° - 30°)$$

$$= \sin 45° \cos 30° - \cos 45° \sin 30°$$

$$= \frac{\sqrt{2}}{2}\cdot\frac{\sqrt{3}}{2} - \frac{\sqrt{2}}{2}\cdot\frac{1}{2} = \frac{\sqrt{6} - \sqrt{2}}{4}$$

$$\cos 165° = \cos 180° \cos 15° + \sin 180° \sin 15°$$
$$= -1 \cdot \cos 15° + 0 \cdot \sin 15° = -\cos 15°$$
$$= -\cos (45° - 30°)$$
$$= -(\cos 45° \cos 30° + \sin 45° \sin 30°)$$
$$= -\left( \frac{\sqrt{2}}{2} \cdot \frac{\sqrt{3}}{2} + \frac{\sqrt{2}}{2} \cdot \frac{1}{2} \right)$$
$$= -\left( \frac{\sqrt{6} + \sqrt{2}}{4} \right) = \frac{-\sqrt{6} - \sqrt{2}}{4}$$

$$\tan 165° = \frac{\tan 180° - \tan 15°}{1 + \tan 180° \tan 15°} = \frac{0 - \tan 15°}{1 + 0 \cdot \tan 15°}$$
$$= -\tan 15° = -\frac{\tan 45° - \tan 30°}{1 + \tan 45° \tan 30°}$$
$$= -\frac{1 - \frac{\sqrt{3}}{3}}{1 + \frac{\sqrt{3}}{3}} = -\frac{1 - \frac{\sqrt{3}}{3}}{1 + \frac{\sqrt{3}}{3}} \cdot \frac{3}{3}$$
$$= -\frac{3 - \sqrt{3}}{3 + \sqrt{3}} = -\frac{3 - \sqrt{3}}{3 + \sqrt{3}} \cdot \frac{3 - \sqrt{3}}{3 - \sqrt{3}}$$
$$= -\frac{9 - 3\sqrt{3} - 3\sqrt{3} + 3}{9 - 3} = -\frac{12 - 6\sqrt{3}}{6}$$
$$= -(2 - \sqrt{3}) = -2 + \sqrt{3}$$

$$\cot 165° = \frac{1}{\tan 165°} = \frac{1}{-2 + \sqrt{3}}$$
$$= \frac{1}{-2 + \sqrt{3}} \cdot \frac{-2 - \sqrt{3}}{-2 - \sqrt{3}} = \frac{-2 - \sqrt{3}}{4 - 3}$$
$$= -2 - \sqrt{3}$$

$$\sec 165° = \frac{1}{\cos 165°} = \frac{1}{\frac{-\sqrt{6} - \sqrt{2}}{4}} = \frac{4}{-\sqrt{6} - \sqrt{2}}$$
$$= \frac{4}{-\sqrt{6} - \sqrt{2}} \cdot \frac{-\sqrt{6} + \sqrt{2}}{-\sqrt{6} + \sqrt{2}}$$
$$= \frac{4\left(-\sqrt{6} + \sqrt{2}\right)}{6 - 2} = -\sqrt{6} + \sqrt{2}$$

$$\csc 165° = \frac{1}{\sin 165°} = \frac{1}{\frac{\sqrt{6} - \sqrt{2}}{4}} = \frac{4}{\sqrt{6} - \sqrt{2}}$$
$$= \frac{4}{\sqrt{6} - \sqrt{2}} \cdot \frac{\sqrt{6} + \sqrt{2}}{\sqrt{6} + \sqrt{2}} = \frac{4\left(\sqrt{6} + \sqrt{2}\right)}{6 - 2}$$
$$= \sqrt{6} + \sqrt{2}$$

**17.** E. $\cos 210° = \cos(150° + 60°)$
$$= \cos 150° \cos 60° - \sin 150° \sin 60°$$

**19.** J. $\tan(-35°) = \cot\left[90° - (-35°)\right] = \cot 125°$

**21.** I. $\cos 35° = \cos(-35°)$

**23.** H. $\sin 75° = \sin(15° + 60°)$
$$= \sin 15° \cos 60° + \cos 15° \sin 60°$$

**25.** G. $\cos 300° = \cos 2(150°)$
$$= \cos^2 150° - \sin^2 150°$$

**27.** Find $\sin(x + y)$, $\cos(x - y)$, and $\tan(x + y)$, given $\sin x = -\dfrac{3}{5}$, $\cos y = -\dfrac{7}{25}$, $x$ and $y$ are in quadrant III. Since $x$ and $y$ are in quadrant III, $\cos x$ and $\sin y$ are negative.

$$\cos x = -\sqrt{1 - \sin^2 x} = -\sqrt{1 - \left(-\frac{3}{5}\right)^2}$$
$$= -\sqrt{1 - \frac{9}{25}} = -\sqrt{\frac{16}{25}} = -\frac{4}{5}$$

$$\sin y = -\sqrt{1 - \cos^2 y} = -\sqrt{1 - \left(-\frac{7}{25}\right)^2}$$
$$= -\sqrt{1 - \frac{49}{625}} = -\sqrt{\frac{596}{625}} = -\frac{24}{25}$$

$$\sin(x + y) = \sin x \cos y + \cos x \sin y$$
$$= \left(-\frac{3}{5}\right)\left(-\frac{7}{25}\right) + \left(-\frac{4}{5}\right)\left(-\frac{24}{25}\right)$$
$$= \frac{21}{125} + \frac{96}{125} = \frac{117}{125}$$

$$\cos(x - y) = \cos x \cos y + \sin x \sin y$$
$$= \left(-\frac{4}{5}\right)\left(-\frac{7}{25}\right) + \left(-\frac{3}{5}\right)\left(-\frac{24}{25}\right)$$
$$= \frac{28}{125} + \frac{72}{125} = \frac{100}{125} = \frac{4}{5}$$

To find $\tan(x + y)$, first find $\cos(x + y)$.

$$\cos(x + y) = \cos x \cos y - \sin x \sin y$$
$$= \left(-\frac{4}{5}\right)\left(-\frac{7}{25}\right) - \left(-\frac{3}{5}\right)\left(-\frac{24}{25}\right)$$
$$= \frac{28}{125} - \frac{72}{125} = -\frac{44}{125}$$

$$\tan(x + y) = \frac{\sin(x + y)}{\cos(x + y)} = \frac{\frac{117}{125}}{-\frac{44}{5}} \cdot \frac{125}{125} = -\frac{117}{44}$$

Note that using the formula
$$\tan(x + y) = \frac{\tan x + \tan y}{1 - \tan x \tan y}, \text{ we have}$$
$$\tan(x + y) = \frac{\frac{3}{4} + \frac{24}{7}}{1 - \left(\frac{3}{4}\right)\left(\frac{24}{7}\right)} = \frac{21 + 96}{28 - 72} = -\frac{117}{44}$$

To find the quadrant of $x + y$, notice that $\sin(x + y) > 0$, which implies $x + y$ is in quadrant I or II. Also $\tan(x + y) < 0$, which implies that $x + y$ is in quadrant II or IV. Therefore, $x + y$ is in quadrant II.

**29.** Find $\sin(x+y)$, $\cos(x-y)$, and $\tan(x+y)$, given $\sin x = -\dfrac{1}{2}$, $\cos y = -\dfrac{2}{5}$, $x$ and $y$ are in quadrant III.

Since $x$ and $y$ are in quadrant III, $\cos x$ and $\sin y$ are negative.

$$\cos x = -\sqrt{1-\sin^2 x} = -\sqrt{1-\left(-\frac{1}{2}\right)^2}$$

$$= -\sqrt{1-\frac{1}{4}} = -\sqrt{\frac{3}{4}} = -\frac{\sqrt{3}}{2}$$

$$\sin y = -\sqrt{1-\cos^2 y} = -\sqrt{1-\left(-\frac{2}{5}\right)^2}$$

$$= -\sqrt{1-\frac{4}{25}} = -\sqrt{\frac{21}{25}} = -\frac{\sqrt{21}}{5}$$

$$\sin(x+y) = \sin x \cos y + \cos x \sin y$$

$$= \left(-\frac{1}{2}\right)\left(-\frac{2}{5}\right) + \left(-\frac{\sqrt{3}}{2}\right)\left(-\frac{\sqrt{21}}{5}\right)$$

$$= \frac{2}{10} + \frac{\sqrt{63}}{10} = \frac{2+3\sqrt{7}}{10}$$

$$\cos(x-y) = \cos x \cos y + \sin x \sin y$$

$$= \left(-\frac{\sqrt{3}}{2}\right)\left(-\frac{2}{5}\right) + \left(-\frac{1}{2}\right)\left(-\frac{\sqrt{21}}{5}\right)$$

$$= \frac{2\sqrt{3}}{10} + \frac{\sqrt{21}}{10} = \frac{2\sqrt{3}+\sqrt{21}}{10}$$

To find $\tan(x+y)$, first find $\cos(x+y)$.

$$\cos(x+y) = \cos x \cos y - \sin x \sin y$$

$$= \left(-\frac{2}{5}\right)\left(-\frac{\sqrt{3}}{2}\right) - \left(-\frac{1}{2}\right)\left(-\frac{\sqrt{21}}{5}\right)$$

$$= \frac{2\sqrt{3}}{10} - \frac{\sqrt{21}}{10} = \frac{2\sqrt{3}-\sqrt{21}}{10}$$

$$\tan(x+y) = \frac{\sin(x+y)}{\cos(x+y)} = \frac{\frac{2+3\sqrt{7}}{10}}{\frac{2\sqrt{3}-\sqrt{21}}{10}}$$

$$= \frac{2+3\sqrt{7}}{2\sqrt{3}-\sqrt{21}}$$

Using the formula $\tan(x+y) = \dfrac{\tan x + \tan y}{1-\tan x \tan y}$, we have

$$\tan(x+y) = \frac{\frac{\sqrt{3}}{3}+\frac{\sqrt{21}}{2}}{1-\left(\frac{\sqrt{3}}{3}\right)\left(\frac{\sqrt{21}}{2}\right)} = \frac{2\sqrt{3}+3\sqrt{21}}{6-3\sqrt{7}}$$

$$= -\frac{75\sqrt{3}+24\sqrt{21}}{27}$$

The two forms of $\tan(x+y)$ are equal.

To find the quadrant of $x+y$, notice that $\sin(x+y) > 0$, which implies $x+y$ is in quadrant I or II. Also $\tan(x+y) < 0$, which implies that $x+y$ is in quadrant II or IV. Therefore, $x+y$ is in quadrant II.

**31.** Find $\sin(x+y)$, $\cos(x-y)$, and $\tan(x+y)$, given $\sin x = \dfrac{1}{10}$, $\cos y = \dfrac{4}{5}$, $x$ is in quadrant I and $y$ is in quadrant IV.

Since $x$ is in quadrant I, $\cos x$ is positive.

$$\cos x = \sqrt{1-\sin^2 x} = \sqrt{1-\left(\frac{1}{10}\right)^2}$$

$$= \sqrt{1-\frac{1}{100}} = \sqrt{\frac{99}{100}} = \frac{3\sqrt{11}}{10}$$

Since $y$ is in quadrant IV, $\sin y$ is negative.

$$\sin y = -\sqrt{1-\cos^2 y} = -\sqrt{1-\left(\frac{4}{5}\right)^2}$$

$$= -\sqrt{1-\frac{16}{25}} = -\sqrt{\frac{9}{25}} = -\frac{3}{5}$$

$$\sin(x+y) = \sin x \cos y + \cos x \sin y$$

$$= \left(\frac{1}{10}\right)\left(\frac{4}{5}\right) + \left(\frac{3\sqrt{11}}{10}\right)\left(-\frac{3}{5}\right)$$

$$= \frac{4}{50} - \frac{9\sqrt{11}}{50} = \frac{4-9\sqrt{11}}{50}$$

$$\cos(x-y) = \cos x \cos y + \sin x \sin y$$

$$= \left(\frac{3\sqrt{11}}{10}\right)\left(\frac{4}{5}\right) + \left(\frac{1}{10}\right)\left(-\frac{3}{5}\right)$$

$$= \frac{12\sqrt{11}}{50} - \frac{3}{50} = \frac{12\sqrt{11}-3}{50}$$

To find $\tan(x+y)$, first find $\cos(x+y)$.

$$\cos(x+y) = \cos x \cos y - \sin x \sin y$$

$$= \left(\frac{3\sqrt{11}}{10}\right)\left(\frac{4}{5}\right) - \left(\frac{1}{10}\right)\left(-\frac{3}{5}\right)$$

$$= \frac{12\sqrt{11}}{50} + \frac{3}{50} = \frac{12\sqrt{11}+3}{50}$$

$$\tan(x+y) = \frac{\sin(x+y)}{\cos(x+y)} = \frac{\frac{4-9\sqrt{11}}{50}}{\frac{12\sqrt{11}+3}{50}}$$

$$= \frac{4-9\sqrt{11}}{12\sqrt{11}+3}$$

To find $\tan(x+y)$ using the formula

$\tan(x+y) = \dfrac{\tan x + \tan y}{1 - \tan x \tan y}$ , we have

$\tan x = \dfrac{\sin x}{\cos x} = \dfrac{\frac{1}{10}}{\frac{3\sqrt{11}}{10}} = \dfrac{1}{3\sqrt{11}} \cdot \dfrac{\sqrt{11}}{\sqrt{11}} = \dfrac{\sqrt{11}}{33}$

$\tan y = \dfrac{\sin y}{\cos y} = \dfrac{-\frac{3}{5}}{\frac{4}{5}} = -\dfrac{3}{4}$

$\tan(x+y) = \dfrac{\tan x + \tan y}{1 - \tan x \tan y} = \dfrac{\frac{\sqrt{11}}{33} + \left(-\frac{3}{4}\right)}{1 - \left(\frac{\sqrt{11}}{33}\right)\left(-\frac{3}{4}\right)}$

$= \dfrac{4\sqrt{11} - 99}{132 + 3\sqrt{11}}$

The two forms of $\tan(x+y)$ are equal.

To find the quadrant of $x+y$, notice that $\sin(x+y) < 0$, which implies $x + y$ is in quadrant III or IV. Also $\tan(x+y) < 0$, which implies that $x + y$ is in quadrant II or IV. Therefore, $x + y$ is in quadrant IV.

**33.** Find $\sin\theta$ and $\cos\theta$, given $\cos 2\theta = -\dfrac{3}{4}$,

$90° < 2\theta < 180°$.
$90° < 2\theta < 180° \Rightarrow 45° < \theta < 90° \Rightarrow \theta$ is in quadrant I, so $\sin\theta$ and $\cos\theta$ are both positive.

$\cos 2\theta = 1 - 2\sin^2\theta \Rightarrow -\dfrac{3}{4} = 1 - 2\sin^2\theta \Rightarrow$

$-\dfrac{7}{4} = -2\sin^2\theta \Rightarrow \dfrac{7}{8} = \sin^2\theta \Rightarrow$

$\sin\theta = \sqrt{\dfrac{7}{8}} = \dfrac{\sqrt{7}}{2\sqrt{2}} = \dfrac{\sqrt{14}}{4}$

$\cos\theta = \sqrt{1 - \sin^2\theta} = \sqrt{1 - \dfrac{7}{8}} = \sqrt{\dfrac{1}{8}} = \dfrac{1}{\sqrt{8}}$

$= \dfrac{1}{2\sqrt{2}} = \dfrac{\sqrt{2}}{4}$

**35.** Find $\sin 2x$ and $\cos 2x$, given $\tan x = 3$, $\sin x < 0$.
Since $\tan x > 0$ and $\sin x < 0$, $x$ is in quadrant III, and $2x$ is in quadrant I or II.

$\tan 2x = \dfrac{2\tan x}{1 - \tan^2 x} = \dfrac{2(3)}{1 - 3^2} = -\dfrac{6}{8} = -\dfrac{3}{4}$

Since $\tan 2x < 0$, $2x$ is in quadrant II. Thus, $\sin 2x > 0$ and $\cos 2x < 0$.

$\sec 2x = -\sqrt{1 + \tan^2 x} = -\sqrt{1 + \left(-\dfrac{3}{4}\right)^2}$

$= -\sqrt{1 + \dfrac{9}{16}} = -\sqrt{\dfrac{25}{16}} = -\dfrac{5}{4} \Rightarrow$

$\cos 2x = \dfrac{1}{-\frac{5}{4}} = -\dfrac{4}{5}$

$\sin 2x = \sqrt{1 - \cos^2(2x)} = \sqrt{1 - \left(-\dfrac{4}{5}\right)^2}$

$= \sqrt{1 - \dfrac{16}{25}} = \sqrt{\dfrac{9}{25}} = \dfrac{3}{5}$

**37.** Find $\cos\dfrac{\theta}{2}$, given $\cos\theta = -\dfrac{1}{2}$, $90° < \theta < 180°$.

Since $90° < \theta < 180° \Rightarrow 45° < \dfrac{\theta}{2} < 90°$, $\dfrac{\theta}{2}$ is

in quadrant I and $\cos\dfrac{\theta}{2} > 0$.

$\cos\dfrac{\theta}{2} = \sqrt{\dfrac{1 + \left(-\frac{1}{2}\right)}{2}} = \sqrt{\dfrac{2-1}{4}} = \sqrt{\dfrac{1}{4}} = \dfrac{1}{2}$

**39.** Find $\tan x$, given $\tan 2x = 2$, with $\pi < x < \dfrac{3\pi}{2}$.

$\tan 2x = \dfrac{2\tan x}{1 - \tan^2 x} \Rightarrow 2 = \dfrac{2\tan x}{1 - \tan^2 x} \Rightarrow$
$2\tan x = 2\left(1 - \tan^2 x\right)$, if $\tan x \neq \pm 1$

Thus, $2\left(\tan^2 x + 2\tan x - 1\right) = 0 \Rightarrow$

$\tan^2 x + 2\tan x - 1 = 0$, so we can use the quadratic formula to solve for $\tan x$.

$x = \dfrac{-b \pm \sqrt{b^2 - 4ac}}{2a} \Rightarrow$

$\tan x = \dfrac{-1 \pm \sqrt{1^2 - 4(1)(-1)}}{2} = \dfrac{-1 \pm \sqrt{5}}{2}$

Since $x$ is in quadrant III, $\tan x > 0$, so

$\tan x = \dfrac{-1 + \sqrt{5}}{2}$

**41.** Find $\tan\dfrac{x}{2}$, given $\sin x = .8$, with $0 < x < \dfrac{\pi}{2}$.

$\cos x = \pm\sqrt{1 - \sin^2 x} = \pm\sqrt{1 - .8^2} = \sqrt{1 - .64}$
$= \pm\sqrt{.36} = \pm.6$

Since $x$ is in quadrant I, $\cos x > 0$, so $\cos x = .6$

$\tan\dfrac{x}{2} = \dfrac{1 - \cos x}{\sin x} = \dfrac{1 - .6}{.8} = \dfrac{.4}{.8} = .5$

Exercises 43–48 are graphed in the following window:

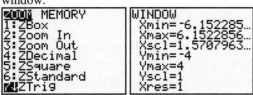

**43.** $-\dfrac{\sin 2x + \sin x}{\cos 2x - \cos x}$ appears to be equivalent to $\cot \dfrac{x}{2}$.

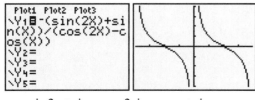

$$-\frac{\sin 2x + \sin x}{\cos 2x - \cos x} = -\frac{2\sin x \cos x + \sin x}{2\cos^2 x - 1 - \cos x}$$
$$= -\frac{\sin x(2\cos x + 1)}{2\cos^2 x - \cos x - 1}$$
$$= -\frac{\sin x(2\cos x + 1)}{(2\cos x + 1)(\cos x - 1)}$$
$$= -\frac{\sin x}{\cos x - 1} = \frac{\sin x}{1 - \cos x}$$
$$= \frac{1}{\dfrac{1 - \cos x}{\sin x}} = \frac{1}{\tan \dfrac{x}{2}} = \cot \frac{x}{2}$$

**45.** $\dfrac{\sin x}{1 - \cos x}$ appears to be equivalent to $\cot \dfrac{x}{2}$.

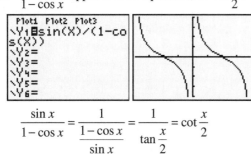

$$\frac{\sin x}{1 - \cos x} = \frac{1}{\dfrac{1 - \cos x}{\sin x}} = \frac{1}{\tan \dfrac{x}{2}} = \cot \frac{x}{2}$$

**47.** $\dfrac{2\left(\sin x - \sin^3 x\right)}{\cos x}$ appears to be equivalent to $\sin 2x$.

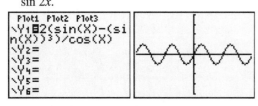

$$\frac{2\left(\sin x - \sin^3 x\right)}{\cos x} = \frac{2\sin x\left(1 - \sin^2 x\right)}{\cos x}$$
$$= \frac{2\sin x \cos^2 x}{\cos x}$$
$$= 2\sin x \cos x = \sin 2x$$

**49.** Verify $\sin^2 x - \sin^2 y = \cos^2 y - \cos^2 x$ is an identity.
$$\sin^2 x - \sin^2 y = \left(1 - \cos^2 x\right) - \left(1 - \cos^2 y\right)$$
$$= 1 - \cos^2 x - 1 + \cos^2 y$$
$$= \cos^2 y - \cos^2 x$$

**51.** Verify $\dfrac{\sin^2 x}{2 - 2\cos x} = \cos^2 \dfrac{x}{2}$ is an identity.
$$\frac{\sin^2 x}{2 - 2\cos x} = \frac{1 - \cos^2 x}{2(1 - \cos x)}$$
$$= \frac{(1 - \cos x)(1 + \cos x)}{2(1 - \cos x)}$$
$$= \frac{1 + \cos x}{2} = \cos^2 \frac{x}{2}$$

**53.** Verify $2\cos A - \sec A = \cos A - \dfrac{\tan A}{\csc A}$ is an identity.
Work with the right side.
$$\cos A - \frac{\tan A}{\csc A} = \cos A - \frac{\dfrac{\sin A}{\cos A}}{\dfrac{1}{\sin A}} = \cos A - \frac{\sin^2 A}{\cos A}$$
$$= \frac{\cos^2 A}{\cos A} - \frac{\sin^2 A}{\cos A}$$
$$= \frac{\cos^2 A - \sin^2 A}{\cos A}$$
$$= \frac{\cos^2 A - \left(1 - \cos^2 A\right)}{\cos A}$$
$$= \frac{2\cos^2 A - 1}{\cos A} = 2\cos A - \frac{1}{\cos A}$$
$$= 2\cos A - \sec A$$

**55.** Verify $1 + \tan^2 \alpha = 2\tan \alpha \csc 2\alpha$ is an identity.
Work with the right side.
$$2\tan \alpha \csc 2\alpha = \frac{2\tan \alpha}{\sin 2\alpha} = \frac{2 \cdot \dfrac{\sin \alpha}{\cos \alpha}}{2\sin \alpha \cos \alpha}$$
$$= \frac{2\sin \alpha}{2\sin \alpha \cos^2 \alpha} = \frac{1}{\cos^2 \alpha}$$
$$= \sec^2 \alpha = 1 + \tan^2 \alpha$$

**57.** Verify $\tan\theta\sin 2\theta = 2-2\cos^2\theta$ is an identity.

$$\tan\theta\sin 2\theta = \tan\theta\left(2\sin\theta\cos\theta\right)$$
$$= \frac{\sin\theta}{\cos\theta}\left(2\sin\theta\cos\theta\right) = 2\sin^2\theta$$
$$= 2\left(1-\cos^2\theta\right) = 2-2\cos^2\theta$$

**59.** Verify $2\tan x\csc 2x - \tan^2 x = 1$ is an identity.

$$2\tan x\csc 2x - \tan^2 x$$
$$= 2\tan x\frac{1}{\sin 2x} - \tan^2 x$$
$$= 2\cdot\frac{\sin x}{\cos x}\cdot\frac{1}{2\sin x\cos x} - \frac{\sin^2 x}{\cos^2 x}$$
$$= \frac{1}{\cos^2 x} - \frac{\sin^2 x}{\cos^2 x} = \frac{1-\sin^2 x}{\cos^2 x} = \frac{\cos^2 x}{\cos^2 x} = 1$$

**61.** Verify $\tan\theta\cos^2\theta = \dfrac{2\tan\theta\cos^2\theta - \tan\theta}{1-\tan^2\theta}$ is an identity.

Work with the right side.

$$\frac{2\tan\theta\cos^2\theta - \tan\theta}{1-\tan^2\theta}$$
$$= \frac{\tan\theta\left(2\cos^2\theta - 1\right)}{1-\tan^2\theta}$$
$$= \frac{\tan\theta\left(2\cos^2\theta - 1\right)}{1-\dfrac{\sin^2\theta}{\cos^2\theta}}\cdot\frac{\cos^2\theta}{\cos^2\theta}$$
$$= \frac{\tan\theta\cos^2\theta\left(2\cos^2\theta - 1\right)}{\cos^2\theta - \sin^2\theta}$$
$$= \frac{\tan\theta\cos^2\theta\left(2\cos^2\theta - 1\right)}{2\cos^2\theta - 1} = \tan\theta\cos^2\theta$$

**63.** Verify $\dfrac{\sin^2 x - \cos^2 x}{\csc x} = 2\sin^3 x - \sin x$ is an identity.

$$\frac{\sin^2 x - \cos^2 x}{\csc x} = \frac{\sin^2 x - \left(1-\sin^2 x\right)}{\dfrac{1}{\sin x}}$$
$$= \frac{2\sin^2 x - 1}{\dfrac{1}{\sin x}}\cdot\frac{\sin x}{\sin x}$$
$$= \left(2\sin^2 x - 1\right)\sin x$$
$$= 2\sin^3 x - \sin x$$

**65.** Verify $\tan 4\theta = \dfrac{2\tan 2\theta}{2-\sec^2 2\theta}$ is an identity.

$$\tan 4\theta = \tan\left[2\left(2\theta\right)\right] = \frac{2\tan 2\theta}{1-\tan^2 2\theta}$$
$$= \frac{2\tan 2\theta}{1-\left(\sec^2 2\theta - 1\right)} = \frac{2\tan 2\theta}{2-\sec^2 2\theta}$$

**67.** Verify $\tan\left(\dfrac{x}{2}+\dfrac{\pi}{4}\right) = \sec x + \tan x$ is an identity. Working with the left side, we have

$$\tan\left(\frac{x}{2}+\frac{\pi}{4}\right) = \frac{\tan\frac{x}{2}+\tan\frac{\pi}{4}}{1-\tan\frac{x}{2}\tan\frac{\pi}{4}} = \frac{\tan\frac{x}{2}+1}{1-\tan\frac{x}{2}}$$

Working with the right side, we have

$$\sec x + \tan x = \frac{1}{\cos x} + \frac{\sin x}{\cos x} = \frac{1+\sin x}{\cos x}$$
$$= \frac{\left(\cos^2\frac{x}{2}+\sin^2\frac{x}{2}\right)+\sin\left[2\left(\frac{x}{2}\right)\right]}{\cos\left[2\left(\frac{x}{2}\right)\right]}$$
$$= \frac{\cos^2\frac{x}{2}+\sin^2\frac{x}{2}+2\sin\frac{x}{2}\cos\frac{x}{2}}{\cos^2 x - \sin^2 x}$$
$$= \frac{\left(\cos\frac{x}{2}+\sin\frac{x}{2}\right)^2}{\left(\cos\frac{x}{2}+\sin\frac{x}{2}\right)\left(\cos\frac{x}{2}-\sin\frac{x}{2}\right)}$$
$$= \frac{\cos\frac{x}{2}+\sin\frac{x}{2}}{\cos\frac{x}{2}-\sin\frac{x}{2}}\cdot\frac{\dfrac{1}{\cos\frac{x}{2}}}{\dfrac{1}{\cos\frac{x}{2}}}$$
$$= \frac{\dfrac{\cos\frac{x}{2}}{\cos\frac{x}{2}}+\dfrac{\sin\frac{x}{2}}{\cos\frac{x}{2}}}{\dfrac{\cos\frac{x}{2}}{\cos\frac{x}{2}}-\dfrac{\sin\frac{x}{2}}{\cos\frac{x}{2}}} = \frac{1+\tan\frac{x}{2}}{1-\tan\frac{x}{2}}$$

Since $\tan\left(\dfrac{x}{2}+\dfrac{\pi}{4}\right) = \dfrac{\tan\frac{x}{2}+1}{1-\tan\frac{x}{2}} = \sec x + \tan x$,

the statement is verified.

**69.** Verify $-\cot\dfrac{x}{2} = \dfrac{\sin 2x + \sin x}{\cos 2x - \cos x}$ is an identity.

Work with the right side.

$$\dfrac{\sin 2x + \sin x}{\cos 2x - \cos x} = \dfrac{2\sin x\cos x + \sin x}{\left(2\cos^2 x - 1\right) - \cos x}$$

$$= \dfrac{\sin x\left(2\cos x + 1\right)}{2\cos^2 x - \cos x - 1}$$

$$= \dfrac{\sin x\left(2\cos x + 1\right)}{\left(2\cos x + 1\right)\left(\cos x - 1\right)}$$

$$= \dfrac{\sin x}{1 - \cos x} = -\dfrac{\sin x}{\cos x - 1}$$

$$= -\dfrac{1}{\frac{\sin x}{\cos x - 1}} = -\dfrac{1}{\tan\frac{x}{2}} = -\cot\dfrac{x}{2}$$

**71. (a)** When $h = 0$,

$$D = \dfrac{v^2\sin\theta\cos\theta + v\cos\theta\sqrt{(v\sin\theta)^2 + 64\cdot 0}}{32}$$

$$= \dfrac{v^2\sin\theta\cos\theta + v\cos\theta\sqrt{(v\sin\theta)^2}}{32}$$

$$= \dfrac{v^2\sin\theta\cos\theta + v^2\sin\theta\cos\theta}{32}$$

$$= \dfrac{2v^2\sin\theta\cos\theta}{32} = \dfrac{v^2\sin 2\theta}{32}$$

This is dependent on both the velocity and the angle at which the object is thrown.

**(b)** Using the result from part (a), we have

$$D = \dfrac{v^2\sin 2\theta}{32} = \dfrac{36^2\sin(2\cdot 30°)}{32}$$

$$= \dfrac{1296\sin(60°)}{32} = \dfrac{1296\cdot\frac{\sqrt{3}}{2}}{32}$$

$$= \dfrac{81\sqrt{3}}{4} \approx 35\text{ ft}$$

## Chapter 5 Test

**1.** $\cos x = \dfrac{24}{25}$, $x$ is in quadrant IV.

$$\sin^2 x = 1 - \cos^2 x = 1 - \left(\dfrac{24}{25}\right)^2 = 1 - \dfrac{576}{625} = \dfrac{49}{625}$$

Since $x$ is in quadrant IV, $\sin x < 0$.

$$\sin x = -\sqrt{\dfrac{49}{625}} = -\dfrac{7}{25}$$

$$\tan x = \dfrac{\sin x}{\cos x} = \dfrac{-\frac{7}{25}}{\frac{24}{25}} = -\dfrac{7}{24}$$

$$\cot x = \dfrac{1}{\tan x} = \dfrac{1}{-\frac{7}{24}} = -\dfrac{24}{7}$$

$$\sec x = \dfrac{1}{\cos x} = \dfrac{1}{\frac{24}{25}} = \dfrac{25}{24}$$

$$\csc x = \dfrac{1}{\sin x} = \dfrac{1}{-\frac{7}{25}} = -\dfrac{25}{7}$$

**2.** $\sec\theta - \sin\theta\tan\theta = \dfrac{1}{\cos\theta} - \sin\theta\cdot\dfrac{\sin\theta}{\cos\theta}$

$$= \dfrac{1 - \sin^2\theta}{\cos\theta} = \dfrac{\cos^2\theta}{\cos\theta} = \cos\theta$$

**3.** $\tan^2 x - \sec^2 x = \dfrac{\sin^2 x}{\cos^2 x} - \dfrac{1}{\cos^2 x}$

$$= \dfrac{\sin^2 x - 1}{\cos^2 x} = -\dfrac{1 - \sin^2 x}{\cos^2 x}$$

$$= -\dfrac{\cos^2 x}{\cos^2 x} = -1$$

**4.** $\cos\dfrac{5\pi}{12} = \cos\left(\dfrac{\pi}{6} + \dfrac{\pi}{4}\right)$

$$= \cos\dfrac{\pi}{6}\cos\dfrac{\pi}{4} - \sin\dfrac{\pi}{6}\sin\dfrac{\pi}{4}$$

$$= \dfrac{\sqrt{3}}{2}\left(\dfrac{\sqrt{2}}{2}\right) - \dfrac{1}{2}\left(\dfrac{\sqrt{2}}{2}\right) = \dfrac{\sqrt{6} - \sqrt{2}}{4}$$

**5. (a)** $\cos(270° - x)$

$$= \cos 270°\cos x + \sin 270°\sin x$$

$$= 0\cdot\cos x + (-1)\sin x = 0 - \sin x$$

$$= -\sin x$$

**(b)** $\tan(\pi + x) = \dfrac{\tan\pi + \tan x}{1 - \tan\pi\tan x} = \tan x$

**6.** $\sin(-22.5°) = \pm\sqrt{\dfrac{1 - \cos(-45°)}{2}} = \pm\sqrt{\dfrac{1 - \frac{\sqrt{2}}{2}}{2}}$

$$= \pm\sqrt{\dfrac{2 - \sqrt{2}}{4}} = \pm\dfrac{\sqrt{2 - \sqrt{2}}}{2}$$

Since $-22.5°$ is in quadrant IV, $\sin(-22.5°)$ is

negative. Thus, $\sin(-22.5°) = -\dfrac{\sqrt{2 - \sqrt{2}}}{2}$.

**7.** $\cot\dfrac{x}{2} - \cot x$ appears to be equivalent to $\csc x$.

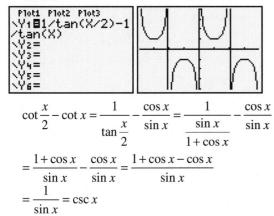

$$\cot\dfrac{x}{2} - \cot x = \dfrac{1}{\tan\dfrac{x}{2}} - \dfrac{\cos x}{\sin x} = \dfrac{1}{\dfrac{\sin x}{1+\cos x}} - \dfrac{\cos x}{\sin x}$$

$$= \dfrac{1+\cos x}{\sin x} - \dfrac{\cos x}{\sin x} = \dfrac{1+\cos x - \cos x}{\sin x}$$

$$= \dfrac{1}{\sin x} = \csc x$$

**8.** Find $\sin(A+B)$, $\cos(A+B)$, and $\tan(A-B)$, given $\sin A = \dfrac{5}{13}$, $\cos B = -\dfrac{3}{5}$, $A$ is in quadrant I and $B$ is in quadrant II. Thus, $\cos A > 0$ and $\sin B > 0$.

$$\cos A = \sqrt{1-\sin^2 A} = \sqrt{1-\left(\dfrac{5}{13}\right)^2}$$

$$= \sqrt{1-\dfrac{25}{169}} = \sqrt{\dfrac{144}{169}} = \dfrac{12}{13}$$

$$\sin B = \sqrt{1-\cos^2 B} = \sqrt{1-\left(-\dfrac{3}{5}\right)^2}$$

$$= \sqrt{1-\dfrac{9}{25}} = \sqrt{\dfrac{16}{25}} = \dfrac{4}{5}$$

**(a)** $\sin(A+B) = \sin A\cos B + \cos A\sin B$

$$= \left(\dfrac{5}{13}\right)\left(-\dfrac{3}{5}\right) + \left(\dfrac{12}{13}\right)\left(\dfrac{4}{5}\right)$$

$$= -\dfrac{15}{65} + \dfrac{48}{65} = \dfrac{33}{65}$$

**(b)** $\cos(A+B) = \cos A\cos B - \sin A\sin B$

$$= \left(\dfrac{12}{13}\right)\left(-\dfrac{3}{5}\right) - \left(\dfrac{5}{13}\right)\left(\dfrac{4}{5}\right)$$

$$= -\dfrac{36}{65} - \dfrac{20}{65} = -\dfrac{56}{65}$$

**(c)** To use the formula

$$\tan(A+B) = \dfrac{\tan A + \tan B}{1-\tan A\tan B},\text{ first find}$$

$\tan A$ and $\tan B$:

$$\tan A = \dfrac{\sin A}{\cos A} = \dfrac{\dfrac{5}{13}}{\dfrac{12}{13}} = \dfrac{5}{12}$$

$$\tan B = \dfrac{\sin B}{\cos B} = \dfrac{\dfrac{4}{5}}{-\dfrac{3}{5}} = -\dfrac{4}{3}$$

$$\tan(A-B) = \dfrac{\dfrac{5}{12} - \left(-\dfrac{4}{3}\right)}{1+\left(\dfrac{5}{12}\right)\left(-\dfrac{4}{3}\right)}$$

$$= \dfrac{15+48}{36-20} = \dfrac{63}{16}$$

**(d)** To find the quadrant of $A+B$, notice that $\sin(A+B) > 0$, which implies $x+y$ is in quadrant I or II. Also $\cos(A+B) < 0$, which implies that $A+B$ is in quadrant II or III. Therefore, $A+B$ is in quadrant II.

**9.** Given $\cos\theta = -\dfrac{3}{5}, \dfrac{\pi}{2} < \theta < \pi$

Since $\theta$ is in quadrant II, $\sin\theta > 0$, $2\theta$ is in quadrant III or quadrant IV, and

$$\dfrac{\pi}{4} < \dfrac{\theta}{2} < \dfrac{\pi}{2} \Rightarrow \dfrac{\theta}{2}\text{ is in quadrant I. Also}$$

$$\sin\theta = \sqrt{1-\cos^2\theta} = \sqrt{1-\left(-\dfrac{3}{5}\right)^2} = \sqrt{\dfrac{16}{25}} = \dfrac{4}{5}$$

$$\tan\theta = \dfrac{\sin\theta}{\cos\theta} = \dfrac{\dfrac{4}{5}}{-\dfrac{3}{5}} = -\dfrac{4}{3}$$

**(a)** $\cos 2\theta = 2\cos^2\theta - 1 = 2\left(-\dfrac{3}{5}\right)^2 - 1 = -\dfrac{7}{25}$

Note that $2\theta$ is in quadrant III because $\cos 2\theta < 0$.

**(b)** $\sin 2\theta = 2\sin\theta\cos\theta = 2\left(\dfrac{4}{5}\right)\left(-\dfrac{3}{5}\right) = -\dfrac{24}{25}$

**(c)** $\tan 2\theta = \dfrac{2\tan\theta}{1-\tan^2\theta} = \dfrac{2\left(-\dfrac{4}{3}\right)}{1-\left(\dfrac{4}{3}\right)^2} = \dfrac{-\dfrac{8}{3}}{1-\dfrac{16}{9}}$

$$= \dfrac{-24}{9-16} = \dfrac{24}{7}$$

**(d)** $\cos\dfrac{1}{2}\theta = \sqrt{\dfrac{1+\cos\theta}{2}} = \sqrt{\dfrac{1+\left(-\dfrac{3}{5}\right)}{2}}$

$$= \sqrt{\dfrac{5-3}{10}} = \sqrt{\dfrac{2}{10}} = \dfrac{1}{\sqrt{5}} = \dfrac{\sqrt{5}}{5}$$

(e)  $\tan\frac{1}{2}\theta = \dfrac{\sin\theta}{1+\cos\theta} = \dfrac{\frac{4}{5}}{1-\frac{3}{5}} = \dfrac{4}{5-3} = 2$

**10.** Verify $\sec^2 B = \dfrac{1}{1-\sin^2 B}$ is an identity.
Work with the right side.
$\dfrac{1}{1-\sin^2 B} = \dfrac{1}{\cos^2 B} = \sec^2 B$

**11.** Verify $\tan^2 x - \sin^2 x = (\tan x \sin x)^2$ is an identity.

$\tan^2 x - \sin^2 x = \dfrac{\sin^2 x}{\cos^2 x} - \sin^2 x$

$= \dfrac{\sin^2 x - \sin^2 x \cos^2 x}{\cos^2 x}$

$= \dfrac{\sin^2 x (1 - \cos^2 x)}{\cos^2 x} = \dfrac{\sin^2 x \sin^2 x}{\cos^2 x}$

$= \tan^2 x \sin^2 x = (\tan x \sin x)^2$

**12.** Verify $\dfrac{\tan x - \cot x}{\tan x + \cot x} = 2\sin^2 x - 1$ is an identity.

$\dfrac{\tan x - \cot x}{\tan x + \cot x} = \dfrac{\dfrac{\sin x}{\cos x} - \dfrac{\cos x}{\sin x}}{\dfrac{\sin x}{\cos x} + \dfrac{\cos x}{\sin x}}$

$= \dfrac{\dfrac{\sin x}{\cos x} - \dfrac{\cos x}{\sin x}}{\dfrac{\sin x}{\cos x} + \dfrac{\cos x}{\sin x}} \cdot \dfrac{\cos x \sin x}{\cos x \sin x}$

$= \dfrac{\sin^2 x - \cos^2 x}{\sin^2 + \cos^2 x}$

$= \sin^2 x - \cos^2 x$

$= \sin^2 x - (1 - \sin^2 x)$

$= 2\sin^2 x - 1$

**13.** Verify $\cos 2A = \dfrac{\cot A - \tan A}{\csc A \sec A}$ is an identity.
Work with the right side.

$\dfrac{\cot A - \tan A}{\csc A \sec A} = \dfrac{\dfrac{\cos A}{\sin A} - \dfrac{\sin A}{\cos A}}{\left(\dfrac{1}{\sin A}\right)\left(\dfrac{1}{\cos A}\right)} \cdot \dfrac{\sin A \cos A}{\sin A \cos A}$

$= \cos^2 A - \sin^2 A = \cos 2A$

**14.** Verify $\dfrac{\sin 2x}{\cos 2x + 1} = \tan x$ is an identity.

$\dfrac{\sin 2x}{\cos 2x + 1} = \dfrac{2\sin x \cos x}{(2\cos^2 x - 1) + 1}$

$= \dfrac{2\sin x \cos x}{2\cos^2 x} = \dfrac{\sin x}{\cos x} = \tan x$

**15. (a)** $V = 163\sin\omega t$. $\sin x = \cos\left(\dfrac{\pi}{2} - x\right) \Rightarrow$

$V = 163\cos\left(\dfrac{\pi}{2} - \omega t\right).$

**(b)** $V = 163\sin\omega t = 163\sin 120\pi t$

$= 163\cos\left(\dfrac{\pi}{2} - 120\pi t\right) \Rightarrow$ the

maximum voltage occurs when

$\cos\left(\dfrac{\pi}{2} - 120\pi t\right) = 1$. Thus, the

maximum voltage is $V = 163$ volts.

$\cos\left(\dfrac{\pi}{2} - 120\pi t\right) = 1$ when

$\dfrac{\pi}{2} - 120\pi t = 2k\pi$, where $k$ is any

integer. The first maximum occurs when

$\dfrac{\pi}{2} - 120\pi t = 0 \Rightarrow$

$\dfrac{\pi}{2} = 120\pi t \Rightarrow \dfrac{1}{120\pi} \cdot \dfrac{\pi}{2} = t \Rightarrow t = \dfrac{1}{240}$

The maximum voltage will first occur at

$\dfrac{1}{240}$ sec .

# Chapter 6

## Inverse Circular Functions and Trigonometric Equations

### Section 6.1: Inverse Circular Functions

1. For a function to have an inverse, it must be <u>one-to-one</u>.

3. The range of $y = \cos^{-1} x$ equals the <u>domain</u> of $y = \cos x$.

5. If a function $f$ has an inverse and $f(\pi) = -1$, then $f^{1}(-1) = \underline{\pi}$.

7. (a) $[-1, 1]$

   (b) $\left[ -\dfrac{\pi}{2}, \dfrac{\pi}{2} \right]$

   (c) increasing

   (d) $-2$ is not in the domain.

   (d) $-\dfrac{4\pi}{3}$ is not in the range.

9. (a) $(-\infty, \infty)$

   (b) $\left( -\dfrac{\pi}{2}, \dfrac{\pi}{2} \right)$

   (c) increasing

   (d) no

11. $\cos^{-1} \dfrac{1}{a}$

13. $y = \sin^{-1} 0$

    $\sin y = 0, -\dfrac{\pi}{2} \le y \le \dfrac{\pi}{2}$

    Since $\sin 0 = 0, \ y = 0$.

15. $y = \cos^{-1}(-1)$

    $\cos y = -1, 0 \le y \le \pi$

    Since $\cos \pi = -1, y = \pi$.

17. $y = \sin^{-1}(-1)$

    $\sin y = -1, -\dfrac{\pi}{2} \le y \le \dfrac{\pi}{2}$

    Since $\sin \dfrac{\pi}{2} = -1, y = -\dfrac{\pi}{2}$.

19. $y = \arctan 0$

    $\tan y = 0, -\dfrac{\pi}{2} < y < \dfrac{\pi}{2}$

    Since $\tan 0 = 0, y = 0$.

21. $y = \arccos 0$

    $\cos y = 0, 0 \le y \le \pi$

    Since $\cos \dfrac{\pi}{2} = 0, \ y = \dfrac{\pi}{2}$.

23. $y = \sin^{-1} \dfrac{\sqrt{2}}{2}$

    $\sin y = \dfrac{\sqrt{2}}{2}, -\dfrac{\pi}{2} \le y \le \dfrac{\pi}{2}$

    Since $\sin \dfrac{\pi}{4} = \dfrac{\sqrt{2}}{2}, \ y = \dfrac{\pi}{4}$.

25. $y = \arccos\left( -\dfrac{\sqrt{3}}{2} \right)$

    $\cos y = -\dfrac{\sqrt{3}}{2}, 0 \le y \le \pi$

    Since $\cos \dfrac{5\pi}{6} = -\dfrac{\sqrt{3}}{2}, y = \dfrac{5\pi}{6}$.

27. $y = \cot^{-1}(-1)$

    $\cot y = -1, 0 < y < \pi$

    $y$ is in quadrant II. The reference angle is $\dfrac{\pi}{4}$.

    Since $\cot \dfrac{3\pi}{4} = 1, \ y = \dfrac{3\pi}{4}$.

29. $y = \csc^{-1}(-2)$

    $\csc y = -2, -\dfrac{\pi}{2} \le y \le \dfrac{\pi}{2}, y \ne 0$

    $y$ is in quadrant IV. The reference angle is $\dfrac{\pi}{6}$.

    Since $\csc\left( -\dfrac{\pi}{6} \right) = -2, y = -\dfrac{\pi}{6}$.

**31.** $y = \text{arc sec}\left(\dfrac{2\sqrt{3}}{3}\right)$

$\sec y = \dfrac{2\sqrt{3}}{3},\ 0 \le y \le \pi,\ y \ne \dfrac{\pi}{2}$

Since $\sec \dfrac{\pi}{6} = \dfrac{2\sqrt{3}}{3},\ y = \dfrac{\pi}{6}.$

**33.** $y = \sec^{-1} 1$

$\sec\ y = 1,\ 0 \le y \le \pi,\ y \ne \dfrac{\pi}{2}$

Since $\sec 0 = 1,\ y = 0.$

**35.** $\theta = \arctan(-1)$

$\tan \theta = -1,\ -90° < \theta < 90°$

$\theta$ is in quadrant IV. The reference angle is $45°$. Thus, $\theta = -45°$.

**37.** $\theta = \arcsin\left(-\dfrac{\sqrt{3}}{2}\right)$

$\sin \theta = -\dfrac{\sqrt{3}}{2},\ -90° \le \theta \le 90°$

$\theta$ is in quadrant IV. The reference angle is $60°$. $\theta = -60°$.

**39.** $\theta = \cot^{-1}\left(-\dfrac{\sqrt{3}}{3}\right)$

$\cot \theta = -\dfrac{\sqrt{3}}{3},\ 0° < \theta < 180°$

$\theta$ is in quadrant II. The reference angle is $60°$. $\theta = 180° - 60° = 120°$

**41.** $\theta = \sec^{-1}(-2)$

$\sec \theta = -2,\ 0° \le \theta \le 180°,\ \theta \ne 90°$

$\theta$ is in quadrant II. The reference angle is $60°$. $\theta = 180° - 60° = 120°$.

**43.** $\theta = \tan^{-1} \sqrt{3}$

$\tan \theta = \sqrt{3},\ -90° < \theta < 90°$

Since $\tan 60° = \sqrt{3}, \theta = 60°.$

**45.** $\theta = \sin^{-1} 2$

$\sin \theta = 2, 0° \le \theta \le 180°$

There is no angle $\theta$ such that $\sin \theta = 2$.

For Exercises 47–55, be sure that your calculator is in degree mode. Keystroke sequences may vary based on the type and/or model of calculator being used.

**47.** $\theta = \sin^{-1}(-.13349122)$

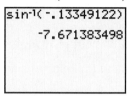

$\sin^{-1}(-.13349122) = -7.6713835°$

**49.** $\theta = \arccos(-.39876459)$

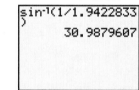

$\arccos(-.39876459) \approx 113.500970°$

**51.** $\theta = \csc^{-1} 1.9422833$

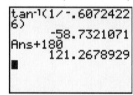

$\csc^{-1} 1.9422833 \approx 30.987961°$

**53.** $\theta = \cot^{-1}(-.60724226)$

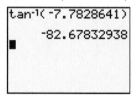

Note that $\theta$ is in quadrant II because $\cot^{-1}\theta$ is defined for $0° < \theta < 180°$.

$\cot^{-1}(-.60724226) \approx 121.267893°$

**55.** $\theta = \tan^{-1}(-7.7828641)$

Note that $\tan^{-1} y$ is defined for $-90° < y < 90°$.

$\tan^{-1}(-7.7828641) \approx -82.678329°$

For Exercises 57–65, be sure that your calculator is in radian mode. Keystroke sequences may vary based on the type and/or model of calculator being used.

**57.**  $y = \arctan 1.1111111$

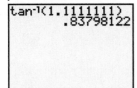

arctan $1.1111111 \approx .83798122$

**59.**  $y = \cot^{-1}\left(-.92170128\right)$

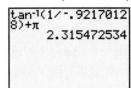

$\cot^{-1}\left(-.91270128\right) \approx 2.3154725$

**61.**  $y = \arcsin .92837781$

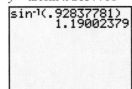

arcsin $.92837781 \approx 1.1900238$

**63.**  $y = \cos^{-1}\left(-.32647891\right)$

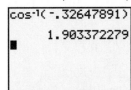

$\cos^{-1}\left(-.32647891\right) \approx 1.9033723$

**65.**  $y = \cot^{-1}\left(-36.874610\right)$

Note that $\theta$ is in quadrant II because $\cot^{-1}\theta$ is defined for $0 < \theta < \pi$.

$\cot^{-1}\left(-36.874610\right) \approx 3.1144804$

**67.**

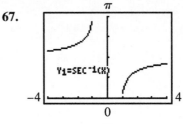

**69.**

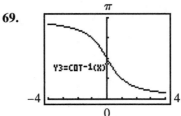

**71.**

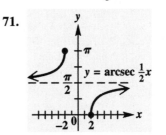

Domain: $(-\infty, -2] \cup [2, \infty)$

Range: $\left[0, \dfrac{\pi}{2}\right) \cup \left(\dfrac{\pi}{2}, \pi\right]$

**73.**  1.003 is not in the domain of $y = \sin^{-1} x$. (Alternatively, you could state that 1.003 is not in the range of $y = \sin x$.)

The graphs for Exercise 75 is in the standard window.

**75.**  It is the graph of $y = x$.

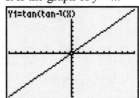

**77.**  $\tan\left(\arccos\dfrac{3}{4}\right)$

Let $\omega = \arccos\dfrac{3}{4}$, so that $\cos\omega = \dfrac{3}{4}$. Since arccos is defined only in quadrants I and II, and $\dfrac{3}{4}$ is positive, $\omega$ is in quadrant I. Sketch $\omega$ and label a triangle with the side opposite $\omega$ equal to $\sqrt{4^2 - 3^2} = \sqrt{16 - 9} = \sqrt{7}$.

*(continued on next page)*

*(continued from page 147)*

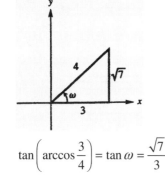

$$\tan\left(\arccos\frac{3}{4}\right) = \tan\omega = \frac{\sqrt{7}}{3}$$

**79.** $\cos(\tan^{-1}(-2))$

Let $\omega = \tan^{-1}(-2)$, so that $\tan\omega = -2$. Since

$\tan^{-1}$ is defined only in quadrants I and IV, and $-2$ is negative, $\omega$ is in quadrant IV. Sketch $\omega$ and label a triangle with the hypotenuse equal to

$$\sqrt{(-2)^2 + 1} = \sqrt{4+1} = \sqrt{5}.$$

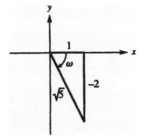

$$\cos(\tan^{-1}(-2)) = \cos\omega = \frac{\sqrt{5}}{5}$$

**81.** $\sin\left(2\tan^{-1}\frac{12}{5}\right)$

Let $\omega = \tan^{-1}\frac{12}{5}$, so that $\tan\omega = \frac{12}{5}$. Since

$\tan^{-1}\omega$ is defined only in quadrants I and IV,

and $\frac{12}{5}$ is positive, $\omega$ is in quadrant I.

Sketch $\omega$ and label a right triangle with the hypotenuse equal to

$$\sqrt{12^2 + 5^2} = \sqrt{144+25} = \sqrt{169} = 13.$$

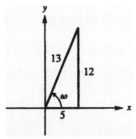

$$\sin\omega = \frac{12}{13}; \cos\omega = \frac{5}{13}$$

$$\sin\left(2\tan^{-1}\frac{12}{5}\right) = \sin(2\omega) = 2\sin\omega\cos\omega$$

$$= 2\left(\frac{12}{13}\right)\left(\frac{5}{13}\right) = \frac{120}{169}$$

**83.** $\cos\left(2\arctan\frac{4}{3}\right)$

Let $\omega = \arctan\frac{4}{3}$, so that $\tan\omega = \frac{4}{3}$. Since

arctan is defined only in quadrants I and IV,

and $\frac{4}{3}$ is positive, $\omega$ is in quadrant I. Sketch

$\omega$ and label a triangle with the hypotenuse

equal to $\sqrt{4^2 + 3^2} = \sqrt{16+9} = \sqrt{25} = 5$.

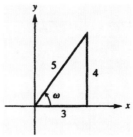

$$\cos\omega = \frac{3}{5}; \sin\omega = \frac{4}{5}$$

$$\cos\left(2\arctan\frac{4}{3}\right) = \cos(2\omega) = \cos^2\omega - \sin^2\omega$$

$$= \left(\frac{3}{5}\right)^2 - \left(\frac{4}{5}\right)^2$$

$$= \frac{9}{25} - \frac{16}{25} = -\frac{7}{25}$$

**85.** $\sin\left(2\cos^{-1}\frac{1}{5}\right)$

Let $\theta = \cos^{-1}\frac{1}{5}$, so that $\cos\theta = \frac{1}{5}$. The

inverse cosine function yields values only in

quadrants I and II, and since $\frac{1}{5}$ is positive, $\theta$

is in quadrant I. Sketch $\theta$ and label the sides

of the right triangle. By the Pythagorean

theorem, the length opposite to $\theta$ will be

$$\sqrt{5^2 - 1^2} = \sqrt{24} = 2\sqrt{6}.$$

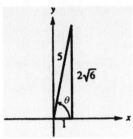

From the figure,

$\sin\theta = \dfrac{2\sqrt{6}}{5}$. Then,

$$\sin\left(2\cos^{-1}\dfrac{1}{5}\right) = \sin 2\theta = 2\sin\theta\cos\theta$$

$$= 2\left(\dfrac{2\sqrt{6}}{5}\right)\left(\dfrac{1}{5}\right) = \dfrac{4\sqrt{6}}{25}$$

**87.** $\sec(\sec^{-1} 2)$

Since secant and inverse secant are inverse functions, $\sec\left(\sec^{-1} 2\right) = 2$.

**89.** $\cos\left(\tan^{-1}\dfrac{5}{12} - \cot^{-1}\dfrac{3}{4}\right)$

Let $\alpha = \tan^{-1}\dfrac{5}{12}$ and $\beta = \tan^{-1}\dfrac{4}{3}$. Then

$\tan\alpha = \dfrac{5}{12}$ and $\tan\beta = \dfrac{3}{4}$. Sketch angles $\alpha$ and $\beta$, both in quadrant I.

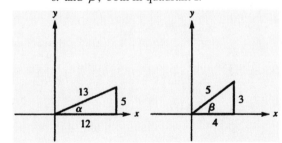

We have $\sin\alpha = \dfrac{5}{13}$, $\cos\alpha = \dfrac{12}{13}$, $\sin\beta = \dfrac{3}{5}$,

and $\cos\beta = \dfrac{4}{5}$.

$$\cos\left(\tan^{-1}\dfrac{5}{12} - \tan^{-1}\dfrac{3}{4}\right)$$
$$= \cos(\alpha - \beta) = \cos\alpha\cos\beta + \sin\alpha\sin\beta$$
$$= \left(\dfrac{12}{13}\right)\left(\dfrac{4}{5}\right) + \left(\dfrac{5}{13}\right)\left(\dfrac{3}{5}\right) = \dfrac{48}{65} + \dfrac{15}{65} = \dfrac{63}{65}$$

**91.** $\sin\left(\sin^{-1}\dfrac{1}{2} + \tan^{-1}(-3)\right)$

Let $\sin^{-1}\dfrac{1}{2} = A$ and $\tan^{-1}(-3) = B$.

Then $\sin A = \dfrac{1}{2}$ and $\tan B = -3$. Sketch angle $A$ in quadrant I and angle $B$ in quadrant IV.

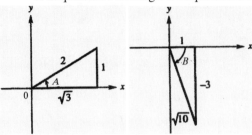

We have $\cos A = \dfrac{\sqrt{3}}{2}$, $\sin A = \dfrac{1}{2}$,

$\cos B = \dfrac{1}{\sqrt{10}} = \dfrac{\sqrt{10}}{10}$, and

$\sin B = \dfrac{-3}{\sqrt{10}} = -\dfrac{3\sqrt{10}}{10}$.

$$\sin\left(\sin^{-1}\dfrac{1}{2} + \tan^{-1}(-3)\right)$$
$$= \sin(A + B) = \sin A\cos B + \cos A\sin B$$
$$= \dfrac{1}{2}\cdot\dfrac{1}{\sqrt{10}} + \dfrac{\sqrt{3}}{2}\cdot\dfrac{-3}{\sqrt{10}}$$
$$= \dfrac{1 - 3\sqrt{3}}{2\sqrt{10}} = \dfrac{\sqrt{10} - 3\sqrt{30}}{20}$$

For Exercises 93–95, your calculator could be in either degree or radian mode. Keystroke sequences may vary based on the type and/or model of calculator being used.

**93.** $\cos\left(\tan^{-1}.5\right)$

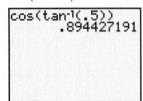

$\cos(\tan^{-1}.5) \approx .894427191$

**95.** tan (arcsin .12251014)

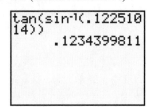

tan (arcsin .12251014) $\approx$ .1234399811

**97.** $\sin\left(\arccos u\right)$

Let $\theta = \arccos u$, so $\cos\theta = u = \dfrac{u}{1}$. Since

$u > 0, \ 0 < \theta < \dfrac{\pi}{2}$.

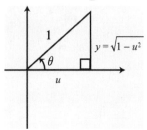

Since $y > 0$, from the Pythagorean theorem,

$y = \sqrt{1^2 - u^2} = \sqrt{1 - u^2}$.

Therefore, $\sin\theta = \dfrac{\sqrt{1 - u^2}}{1} = \sqrt{1 - u^2}$. Thus,

$\sin\left(\arccos u\right) = \sqrt{1 - u^2}$.

**99.** $\cos\left(\arcsin u\right)$

Let $\theta = \arcsin u$, so $\sin\theta = u = \dfrac{u}{1}$. Since

$u > 0, \ 0 < \theta < \dfrac{\pi}{2}$.

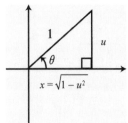

Since $x > 0$, from the Pythagorean theorem,

$x = \sqrt{1^2 - u^2} = \sqrt{1 - u^2}$.

Therefore, $\cos\theta = \dfrac{\sqrt{1 - u^2}}{1} = \sqrt{1 - u^2}$. Thus,

$\cos\left(\arcsin u\right) = \sqrt{1 - u^2}$

**101.** $\sin\left(2\sec^{-1}\dfrac{u}{2}\right)$

Let $\theta = \sec^{-1}\dfrac{u}{2}$, so $\sec\theta = \dfrac{u}{2}$. Since $u > 0$,

$0 < \theta < \dfrac{\pi}{2}$.

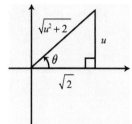

Since $y > 0$, from the Pythagorean theorem,

$y = \sqrt{u^2 - 2^2} = \sqrt{u^2 - 4}$.

Now find $\sin 2\theta$.

$\sin\theta = \dfrac{\sqrt{u^2 - 4}}{u}$ and $\cos\theta = \dfrac{2}{u}$ Thus,

$\sin\left(2\sec^{-1}\dfrac{u}{2}\right) = \sin 2\theta = 2\sin\theta\cos\theta$

$= 2\left(\dfrac{\sqrt{u^2 - 4}}{u}\right)\left(\dfrac{2}{u}\right)$

$= \dfrac{4\sqrt{u^2 - 4}}{u^2}$

**103.** $\tan\left(\sin^{-1}\dfrac{u}{\sqrt{u^2 + 2}}\right)$

Let $\theta = \sin^{-1}\dfrac{u}{\sqrt{u^2 + 2}}$, so $\sin\theta = \dfrac{u}{\sqrt{u^2 + 2}}$.

Since $u > 0, \ 0 < \theta < \dfrac{\pi}{2}$.

Since $x > 0$, from the Pythagorean theorem,

$$x = \sqrt{\left(\sqrt{u^2+2}\right)^2 - u^2} = \sqrt{u^2+2-u^2} = \sqrt{2}.$$

Therefore, $\tan \theta = \dfrac{u}{\sqrt{2}} = \dfrac{u\sqrt{2}}{2}$. Thus,

$$\tan\left(\sin^{-1}\frac{u}{\sqrt{u^2+2}}\right) = \frac{u\sqrt{2}}{2}.$$

**105.** $\sec\left(\operatorname{arc\,cot}\dfrac{\sqrt{4-u^2}}{u}\right)$

Let $\theta = \operatorname{arc\,cot}\dfrac{\sqrt{4-u^2}}{u}$, so $\cot\theta = \dfrac{\sqrt{4-u^2}}{u}$.

Since $u > 0$, $0 < \theta < \dfrac{\pi}{2}$.

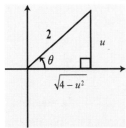

From the Pythagorean theorem,

$$r = \sqrt{\left(\sqrt{4-u^2}\right)^2 + u^2}$$
$$= \sqrt{4 - u^2 + u^2} = \sqrt{4} = 2.$$

Therefore, $\sec\theta = \dfrac{2}{\sqrt{4-u^2}} = \dfrac{2\sqrt{4-u^2}}{4-u^2}$.

Thus, $\sec\left(\operatorname{arc\,cot}\dfrac{\sqrt{4-u^2}}{u}\right) = \dfrac{2\sqrt{4-u^2}}{4-u^2}$.

**107. (a)** $\theta = \arcsin\sqrt{\dfrac{42^2}{2\left(42^2\right)+64(0)}}$

$$= \arcsin\sqrt{\frac{42^2}{2\left(42^2\right)+0}} = \arcsin\sqrt{\frac{42^2}{2\left(42^2\right)}}$$

$$= \arcsin\sqrt{\frac{1}{2}} = \arcsin\frac{1}{\sqrt{2}}$$

$$= \arcsin\frac{\sqrt{2}}{2} = 45°$$

**(b)** $\theta = \arcsin\sqrt{\dfrac{v^2}{2v^2+64(6)}}$

$$= \arcsin\sqrt{\frac{v^2}{2v^2+384}}$$

As $v$ gets larger and larger,

$$\sqrt{\frac{v^2}{2v^2+384}} \approx \sqrt{\frac{1}{2}} = \frac{\sqrt{2}}{2}. \text{ Thus,}$$

$$\theta \approx \arcsin\frac{\sqrt{2}}{2} = 45°.$$

The equation of the asymptote is $\theta = 45°$.

**109.** $\theta = \tan^{-1}\left(\dfrac{x}{x^2+2}\right)$

**(a)** $x = 1$,

$$\theta = \tan^{-1}\left(\frac{1}{1^2+2}\right) = \tan^{-1}\left(\frac{1}{3}\right) \approx 18°$$

**(b)** $x = 2$,

$$\theta = \tan^{-1}\left(\frac{2}{2^2+2}\right) = \tan^{-1}\frac{2}{6}$$

$$= \tan^{-1}\frac{1}{3} \approx 18°$$

**(c)** $x = 3$,

$$\theta = \tan^{-1}\left(\frac{3}{3^2+2}\right) = \tan^{-1}\frac{3}{11} \approx 15°$$

**(d)** $\tan(\theta+\alpha) = \dfrac{1+1}{x} = \dfrac{2}{x}$ and $\tan\alpha = \dfrac{1}{x}$

$$\tan(\theta+\alpha) = \frac{\tan\theta+\tan\alpha}{1-\tan\theta\tan\alpha}$$

$$\frac{2}{x} = \frac{\tan\theta+\dfrac{1}{x}}{1-\tan\theta\left(\dfrac{1}{x}\right)}$$

$$\frac{2}{x} = \frac{x\tan\theta+1}{x-\tan\theta}$$

$$2(x-\tan\theta) = x(x\tan\theta+1)$$

$$2x-2\tan\theta = x^2\tan\theta+x$$

$$2x-x = x^2\tan\theta+2\tan\theta$$

$$x = \tan\theta\left(x^2+2\right)$$

$$\tan\theta = \frac{x}{x^2+2}$$

$$\theta = \tan^{-1}\left(\frac{x}{x^2+2}\right)$$

**(e)** If we graph $y_1 = \tan^{-1}\left(\dfrac{x}{x^2+2}\right)$ using a graphing calculator, the maximum value of the function occurs when $x$ is 1.4142151 m. (Note: Due to the computational routine, there may be a discrepancy in the last few decimal places.)

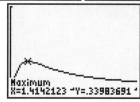

**(f)** $x = \sqrt{(1)(2)} = \sqrt{2}$

## Section 6.2: Trigonometric Equations I

**1.** Solve the linear equation for $\cot x$.

**3.** Solve the quadratic equation for $\sec x$ by factoring.

**5.** Solve the quadratic equation for $\sin x$ using the quadratic formula.

**7.** Use the identity to rewrite as an equation with one trigonometric function.

**9.** $-30°$ is not in the interval $[0°, 360°)$.

**11.** $2\cot x + 1 = -1 \Rightarrow 2\cot x = -2 \Rightarrow \cot x = -1$

Over the interval $[0, 2\pi)$, the equation $\cot x = -1$ has two solutions, the angles in quadrants II and IV that have a reference angle of $\dfrac{\pi}{4}$. These are $\dfrac{3\pi}{4}$ and $\dfrac{7\pi}{4}$.

Solution set: $\left\{\dfrac{3\pi}{4}, \dfrac{7\pi}{4}\right\}$

**13.** $2\sin x + 3 = 4 \Rightarrow 2\sin x = 1 \Rightarrow \sin x = \dfrac{1}{2}$

Over the interval $[0, 2\pi)$, the equation $\sin x = \dfrac{1}{2}$ has two solutions, the angles in quadrants I and II that have a reference angle of $\dfrac{\pi}{6}$. These are $\dfrac{\pi}{6}$ and $\dfrac{5\pi}{6}$.

Solution set: $\left\{\dfrac{\pi}{6}, \dfrac{5\pi}{6}\right\}$

**15.** $\tan^2 x + 3 = 0 \Rightarrow \tan^2 x = -3$

The square of a real number cannot be negative, so this equation has no solution.

Solution set: $\varnothing$

**17.** $\left(\cot x - 1\right)\left(\sqrt{3}\cot x + 1\right) = 0$

$\cot x - 1 = 0 \Rightarrow \cot x = 1$ or

$\sqrt{3}\cot x + 1 = 0 \Rightarrow \sqrt{3}\cot x = -1 \Rightarrow$

$\cot x = -\dfrac{1}{\sqrt{3}} \Rightarrow \cot x = -\dfrac{\sqrt{3}}{3}$

Over the interval $[0, 2\pi)$, the equation $\cot x = 1$ has two solutions, the angles in quadrants I and III that have a reference angle of $\dfrac{\pi}{4}$. These are $\dfrac{\pi}{4}$ and $\dfrac{5\pi}{4}$. In the same interval, $\cot x = -\dfrac{\sqrt{3}}{3}$ also has two solutions. The angles in quadrants II and IV that have a reference angle of $\dfrac{\pi}{3}$ are $\dfrac{2\pi}{3}$ and $\dfrac{5\pi}{3}$.

Solution set: $\left\{\dfrac{\pi}{4}, \dfrac{2\pi}{3}, \dfrac{5\pi}{4}, \dfrac{5\pi}{3}\right\}$

**19.** $\cos^2 x + 2\cos x + 1 = 0$

$\cos^2 x + 2\cos x + 1 = 0 \Rightarrow \left(\cos x + 1\right)^2 = 0 \Rightarrow$

$\cos x + 1 = 0 \Rightarrow \cos x = -1$

Over the interval $[0, 2\pi)$, the equation $\cos x = -1$ has one solution. This solution is $\pi$. Solution set: $\{\pi\}$

**21.** $-2\sin^2 x = 3\sin x + 1$

$$-2\sin^2 x = 3\sin x + 1$$
$$2\sin^2 x + 3\sin x + 1 = 0$$
$$\left(2\sin x + 1\right)\left(\sin x + 1\right) = 0$$

$2\sin x + 1 = 0 \Rightarrow \sin x = -\dfrac{1}{2}$ or

$\sin x + 1 = 0 \Rightarrow \sin x = -1$

Over the interval $[0, 2\pi)$, the equation

$\sin x = -\dfrac{1}{2}$ has two solutions. The angles in

quadrants III and IV that have a reference

angle of $\dfrac{\pi}{6}$ are $\dfrac{7\pi}{6}$ and $\dfrac{11\pi}{6}$.

In the same interval, $\sin x = -1$ when the angle

is $\dfrac{3\pi}{2}$. Solution set: $\left\{ \dfrac{7\pi}{6}, \dfrac{3\pi}{2}, \dfrac{11\pi}{6} \right\}$

**23.** $\left( \cot\theta - \sqrt{3} \right)\left( 2\sin\theta + \sqrt{3} \right) = 0$

$\cot\theta - \sqrt{3} = 0 \Rightarrow \cot\theta = \sqrt{3}$ or
$2\sin\theta + \sqrt{3} = 0 \Rightarrow 2\sin\theta = -\sqrt{3} \Rightarrow$

$\sin\theta = -\dfrac{\sqrt{3}}{2}$

Over the interval $[0°, 360°)$, the equation

$\cot\theta = \sqrt{3}$ has two solutions, the angles in
quadrants I and III that have a reference angle
of $30°$. These are $30°$ and $210°$. In the same

interval, the equation $\sin\theta = -\dfrac{\sqrt{3}}{2}$ has two

solutions, the angles in quadrants III and IV
that have a reference angle of $60°$. These are
$240°$ and $300°$.
Solution set: $\left\{ 30°,\ 210°,\ 240°,\ 300° \right\}$

**25.** $2\sin\theta - 1 = \csc\theta$

$2\sin\theta - 1 = \csc\theta$

$2\sin\theta - 1 = \dfrac{1}{\sin\theta}$

$2\sin^2\theta - \sin\theta = 1$

$2\sin^2\theta - \sin\theta - 1 = 0$

$\left( 2\sin\theta + 1 \right)\left( \sin\theta - 1 \right) = 0$

$2\sin\theta + 1 = 0 \Rightarrow \sin\theta = -\dfrac{1}{2}$ or

$\sin\theta - 1 = 0 \Rightarrow \sin\theta = 1$
Over the interval $[0°, 360°)$, the equation

$\sin\theta = -\dfrac{1}{2}$ has two solutions, the angles in

quadrants III and IV that have a reference
angle of $30°$. These are $210°$ and $330°$. In
the same interval, the only angle $\theta$ for which
$\sin\theta = 1$ is $90°$.
Solution set: $\left\{ 90°,\ 210°,\ 330° \right\}$

**27.** $\tan\theta - \cot\theta = 0$

$\tan\theta - \cot\theta = 0 \Rightarrow \tan\theta - \dfrac{1}{\tan\theta} = 0 \Rightarrow$

$\tan^2\theta - 1 = 0 \Rightarrow \tan^2\theta = 1 \Rightarrow \tan\theta = \pm 1$
Over the interval $[0°, 360°)$, the equation

$\tan\theta = 1$ has two solutions, the angles in
quadrants I and III that have a reference angle
of $45°$. These are $45°$ and $225°$.
In the same interval, the equation $\tan\theta = -1$
has two solutions, the angles in quadrants II
and IV that have a reference angle of $45°$.
These are $135°$ and $315°$.
Solution set: $\left\{ 45°, 135°, 225°, 315° \right\}$

**29.** $\csc^2\theta - 2\cot\theta = 0$

$\csc^2\theta - 2\cot\theta = 0$

$\left( 1 + \cot^2\theta \right) - 2\cot\theta = 0$

$\cot^2\theta - 2\cot\theta + 1 = 0$

$\left( \cot\theta - 1 \right)^2 = 0$

$\cot\theta - 1 = 0 \Rightarrow \cot\theta = 1$
Over the interval $[0°, 360°)$, the equation

$\cot\theta = 1$ has two solutions, the angles in
quadrants I and III that have a reference angle
of $45°$. These are $45°$ and $225°$.
Solution set: $\left\{ 45°,\ 225° \right\}$

**31.** $2\tan^2\theta \sin\theta - \tan^2\theta = 0$

$2\tan^2\theta \sin\theta - \tan^2\theta = 0$

$\tan^2\theta \left( 2\sin\theta - 1 \right) = 0$

$\tan^2\theta = 0$

$\tan\theta = 0$ or $2\sin\theta - 1 = 0 \Rightarrow$

$2\sin\theta = 1 \Rightarrow \sin\theta = \dfrac{1}{2}$

Over the interval $[0°, 360°)$, the equation

$\tan\theta = 0$ has two solutions. These are
$0°$ and $180°$. In the same interval, the

equation $\sin\theta = \dfrac{1}{2}$ has two solutions, the

angles in quadrants I and II that have a
reference angle of $30°$. These are
$30°$ and $150°$.
Solution set: $\left\{ 0°, 30°, 150°, 180° \right\}$

**33.** $\sec^2 \theta \tan \theta = 2 \tan \theta$

$$\sec^2 \theta \tan \theta = 2 \tan \theta$$
$$\sec^2 \theta \tan \theta - 2 \tan \theta = 0$$
$$\tan \theta \left( \sec^2 \theta - 2 \right) = 0$$
$$\tan \theta = 0 \text{ or } \sec^2 \theta - 2 = 0 \Rightarrow$$
$$\sec^2 \theta = 2 \Rightarrow \sec \theta = \pm \sqrt{2}$$

Over the interval $[0°, 360°)$, the equation $\tan \theta = 0$ has two solutions. These are $0°$ and $180°$. In the same interval, the equation $\sec \theta = \sqrt{2}$ has two solutions, the angles in quadrants I and IV that have a reference angle of $45°$. These are $45°$ and $315°$. Finally, the equation $\sec \theta = -\sqrt{2}$ has two solutions, the angles in quadrants II and III that have a reference angle of $45°$. These are $135°$ and $225°$.

Solution set:
$$\{0°, 45°, 135°, 180°, 225°, 315°\}$$

For Exercises 35–41, make sure your calculator is in degree mode.

**35.** $9 \sin^2 \theta - 6 \sin \theta = 1$

$$9 \sin^2 \theta - 6 \sin \theta = 1 \Rightarrow 9 \sin^2 \theta - 6 \sin \theta - 1 = 0$$

We use the quadratic formula with $a = 9$, $b = -6$, and $c = -1$.

$$\sin \theta = \frac{6 \pm \sqrt{36 - 4(9)(-1)}}{2(9)} = \frac{6 \pm \sqrt{36 + 36}}{18}$$
$$= \frac{6 \pm \sqrt{72}}{18} = \frac{6 \pm 6\sqrt{2}}{18} = \frac{1 \pm \sqrt{2}}{3}$$

Since $\sin \theta = \dfrac{1 + \sqrt{2}}{3} > 0$ (and less than 1), we will obtain two angles. One angle will be in quadrant I and the other will be in quadrant II. Using a calculator, if

$$\sin \theta = \frac{1 + \sqrt{2}}{3} \approx .80473787, \text{ the quadrant I}$$

angle will be approximately $53.6°$. The quadrant II angle will be approximately $180° - 53.6° = 126.4°$. Since

$$\sin \theta = \frac{1 - \sqrt{2}}{3} < 0 \text{ (and greater than } -1\text{), we}$$

will obtain two angles. One angle will be in quadrant III and the other will be in quadrant IV. Using a calculator, if

$$\sin \theta = \frac{1 - \sqrt{2}}{3} \approx -.13807119, \text{ then}$$

$$\theta \approx -7.9°.$$

Since this solution is not in the interval $[0°, 360°)$, we must use it as a reference angle to find angles in the interval.

Our reference angle will be $7.9°$. The angle in quadrant III will be approximately $180° + 7.9° = 187.9°$. The angle in quadrant IV will be approximately $360° - 7.9° = 352.1°$.

Solution set: $\{53.6°, 126.4°, 187.9°, 352.1°\}$

**37.** $\tan^2 \theta + 4 \tan \theta + 2 = 0$

We use the quadratic formula with $a = 1$, $b = 4$, and $c = 2$.

$$\tan \theta = \frac{-4 \pm \sqrt{16 - 4(1)(2)}}{2(1)} = \frac{-4 \pm \sqrt{16 - 8}}{2}$$
$$= \frac{-4 \pm \sqrt{8}}{2} = \frac{-4 \pm 2\sqrt{2}}{2} = -2 \pm \sqrt{2}$$

Since $\tan \theta = -2 + \sqrt{2} < 0$, we will obtain two angles. One angle will be in quadrant II and the other will be in quadrant IV. Using a calculator, if $\tan \theta = -2 + \sqrt{2} = -.5857864$, then $\theta \approx -30.4°$. Since this solution is not in the interval $[0°, 360°)$, we must use it as a reference angle to find angles in the interval. Our reference angle will be $30.4°$. The angle in quadrant II will be approximately $180° - 30.4° = 149.6°$. The angle in quadrant IV will be approximately $360° - 30.4° = 329.6°$.

Since $\tan \theta = -2 - \sqrt{2} < 0$, we will obtain two angles. One angle will be in quadrant II and the other will be in quadrant IV. Using a calculator, if $\tan \theta = -2 - \sqrt{2} = -3.4142136$, then $\theta \approx -73.7°$. Since this solution is not in the interval $[0°, 360°)$, we must use it as a reference angle to find angles in the interval. Our reference angle will be $73.7°$. The angle in quadrant II will be approximately $180° - 73.7° = 106.3°$. The angle in quadrant IV will be approximately $360° - 73.7° = 286.3°$.

Solution set:
$$\{106.3°, 149.6°, 286.3°, 329.6°\}$$

**39.** $\sin^2\theta - 2\sin\theta + 3 = 0$

We use the quadratic formula with $a = 1$, $b = -2$, and $c = 3$.

$$\sin\theta = \frac{2 \pm \sqrt{4 - (4)(1)(3)}}{2(1)} = \frac{2 \pm \sqrt{4-12}}{2}$$

$$= \frac{2 \pm \sqrt{-8}}{2} = \frac{2 \pm 2i\sqrt{2}}{2} = 1 \pm i\sqrt{2}$$

Since $1 \pm i\sqrt{2}$ is not a real number, the equation has no real solutions.

Solution set: $\varnothing$

**41.** $\cot\theta + 2\csc\theta = 3$

$$\cot\theta + 2\csc\theta = 3 \Rightarrow \frac{\cos\theta}{\sin\theta} + \frac{2}{\sin\theta} = 3$$

$$\cos\theta + 2 = 3\sin\theta$$

$$(\cos\theta + 2)^2 = (3\sin\theta)^2$$

$$\cos^2\theta + 4\cos\theta + 4 = 9\sin^2\theta$$

$$\cos^2\theta + 4\cos\theta + 4 = 9(1 - \cos^2\theta)$$

$$\cos^2\theta + 4\cos\theta + 4 = 9 - 9\cos^2\theta$$

$$10\cos^2\theta + 4\cos\theta - 5 = 0$$

We use the quadratic formula with $a = 10$, $b = 4$, and $c = -5$.

$$\cos\theta = \frac{-4 \pm \sqrt{4^2 - 4(10)(-5)}}{2(10)}$$

$$= \frac{-4 \pm \sqrt{16 + 200}}{20} = \frac{-4 \pm \sqrt{216}}{20}$$

$$= \frac{-4 \pm 6\sqrt{6}}{20} = \frac{-2 \pm 3\sqrt{6}}{10}$$

Since $\cos\theta = \dfrac{-2 + 3\sqrt{6}}{10} > 0$ (and less than 1),

we will obtain two angles. One angle will be in quadrant I and the other will be in quadrant IV. Using a calculator, if

$\cos\theta = \dfrac{-2 + 3\sqrt{6}}{10} \approx .53484692$, the quadrant I

angle will be approximately 57.7°. The quadrant IV angle will be approximately $360° - 57.7° = 302.3°$.

Since $\cos\theta = \dfrac{-2 - 3\sqrt{6}}{10} < 0$ (and greater than

$-1$), we will obtain two angles. One angle will be in quadrant II and the other will be in quadrant III. Using a calculator, if

$\cos\theta = \dfrac{-2 - 3\sqrt{6}}{10} \approx -.93484692$, the quadrant

II angle will be approximately 159.2°.

The reference angle is $180° - 159.2° = 20.8°$. Thus, the quadrant III angle will be approximately $180° + 20.8° = 200.8°$. Since the solution was found by squaring both sides of an equation, we must check that each proposed solution is a solution of the original equation. 302.3° and 200.8° do not satisfy our original equation. Thus, they are not elements of the solution set.

Solution set: $\{57.7°,\ 159.2°\}$

In Exercises 43–45, if you are using a calculator, make sure it is in radian mode.

**43.** $3\sin^2 x - \sin x - 1 = 0$

We use the quadratic formula with $a = 3$, $b = -1$, and $c = -1$.

$$\sin x = \frac{-(-1) \pm \sqrt{(-1)^2 - 4(3)(-1)}}{2(3)}$$

$$= \frac{1 \pm \sqrt{1 + 12}}{6} = \frac{1 \pm \sqrt{13}}{6}$$

Since $\sin x = \dfrac{1 + \sqrt{13}}{6} > 0$ (and less than 1), we

will obtain two angles. One angle will be in quadrant I and the other will be in quadrant II. Using a calculator, if

$\sin x = \dfrac{1 + \sqrt{13}}{6} \approx .76759188$, the quadrant I

angle will be approximately .8751. The quadrant II angle will be approximately

$\pi - .88 \approx 2.2665$. Since $\sin x = \dfrac{1 - \sqrt{13}}{6} < 0$

(and greater than $-1$), we will obtain two angles. One angle will be in quadrant III and the other will be in quadrant IV. Using a calculator,

if $\sin x = \dfrac{1 - \sqrt{13}}{6} \approx -.43425855$, then

$x \approx -.4492$. Since this solution is not in the interval $[0, 2\pi)$, we must use it as a reference angle to find angles in the interval. Our reference angle will be .4492. The angle in quadrant III will be approximately $\pi + .4492 \approx 3.5908$. The angle in quadrant IV will be approximately $2\pi - .4492 \approx 5.8340$. Thus, the solution set is

$\{.8751 + 2n\pi,\ 2.2665 + 2n\pi,\ 3.5908 + 2n\pi,$

and $5.8340 + 2n\pi$, where $n$ is any integer$\}$.

**45.** $4\cos^2 x - 1 = 0$

$4\cos^2 x - 1 = 0 \Rightarrow \cos^2 x = \dfrac{1}{4} \Rightarrow \cos x = \pm\dfrac{1}{2}$

Over the interval $[0, 2\pi)$, the equation

$\cos x = \dfrac{1}{2}$ has two solutions. The angles in

quadrants I and IV that have a reference angle

of $\dfrac{\pi}{3}$ are $\dfrac{\pi}{3}$ and $\dfrac{5\pi}{3}$. In the same interval,

$\cos x = -\dfrac{1}{2}$ has two solutions. The angles in

quadrants II and III that have a reference angle

of $\dfrac{\pi}{3}$ are $\dfrac{2\pi}{3}$ and $\dfrac{4\pi}{3}$. Thus, the solutions

are $\dfrac{\pi}{3} + 2n\pi, \dfrac{2\pi}{3} + 2n\pi, \dfrac{4\pi}{3} + 2n\pi$, and

$\dfrac{5\pi}{3} + 2n\pi$, where $n$ is any integer. The

solution set is $\left\{ \dfrac{\pi}{3} + n\pi \text{ and } \dfrac{2\pi}{3} + n\pi, \text{ where} \right.$

$n$ is any integer$\}$.

In Exercises 47–49, if you are using a calculator, make sure it is in degree mode.

**47.** $5\sec^2 \theta = 6\sec\theta$

$5\sec^2 \theta = 6\sec\theta \Rightarrow 5\sec^2 \theta - 6\sec\theta = 0 \Rightarrow$
$\sec\theta\left(5\sec\theta - 6\right) = 0$

$\sec\theta = 0$ or $5\sec\theta - 6 = 0 \Rightarrow \sec\theta = \dfrac{6}{5}$

$\sec\theta = 0$ is an impossible values since the secant function must be either $\geq 1$ or $\leq -1$.

Since $\sec\theta = \dfrac{6}{5} > 1$, we will obtain two

angles. One angle will be in quadrant I and the other will be in quadrant IV. Using a

calculator, if $\sec\theta = \dfrac{6}{5} = 1.2$, the quadrant I

angle will be approximately 33.6°. The quadrant IV angle will be approximately $360° - 33.6° = 326.4°$. Thus, the solution set is $\{ 33.6° + 360°n \text{ and } 326.4° + 360°n, \text{ where}$ $n$ is any integer$\}$.

**49.** $\dfrac{2\tan\theta}{3 - \tan^2\theta} = 1$

$\dfrac{2\tan\theta}{3 - \tan^2\theta} = 1$

$2\tan\theta = 3 - \tan^2\theta$

$\tan^2\theta + 2\tan\theta - 3 = 0$

$\left(\tan\theta - 1\right)\left(\tan\theta + 3\right) = 0$

$\tan\theta - 1 = 0 \Rightarrow \tan\theta = 1$ or
$\tan\theta + 3 = 0 \Rightarrow \tan\theta = -3$

Over the interval $[0°, 360°)$, the equation

$\tan\theta = 1$ has two solutions 45° and 225°. Over the same interval, the equation $\tan\theta = -3$ has two solutions that are approximately $-71.6° + 180° = 108.4°$ and $-71.6° + 360° = 288.4°$. Thus, the solutions are $45° + 360°n$, $108.4° + 360°n$, $225° + 360°n$ and $288.4° + 360°n$, where $n$ is any integer. Since the period of the tangent function is 180°, the solution set can also be written as $\{ 45° + n \cdot 180°$, $108.4° + n \cdot 180°$, where $n$ is any integer$\}$.

**51.** The $x$-intercept method is shown in the following windows.

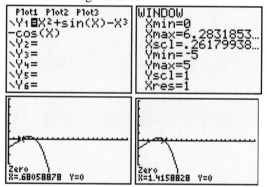

Solution set: $\{.6806, 1.4159\}$

**53.** $P = A\sin\left(2\pi ft + \phi\right)$

**(a)** $0 = .004\sin\left[ 2\pi\left(261.63\right)t + \dfrac{\pi}{7} \right]$

$0 = \sin\left(1643.87t + .45\right)$

Since $1643.87t + .45 = n\pi$, we have

$t = \dfrac{n\pi - .45}{1643.87}$, where $n$ is any integer.

If $n = 0$, then $t \approx .000274$. If $n = 1$, then $t \approx .00164$. If $n = 2$, then $t \approx .00355$. If $n = 3$, then $t \approx .00546$. The only solutions for $t$ in the interval $[0, .005]$ are .00164 and .00355.

**(b)** We must solve the trigonometric equation $P = 0$ to determine when $P \leq 0$. From the graph we can estimate that $P \leq 0$ on the interval [.00164, .00355].

**(c)** $P < 0$ implies that there is a decrease in pressure so an eardrum would be vibrating outward.

**55.** $V = \cos 2\pi t, 0 \leq t \leq \dfrac{1}{2}$

**(a)** $V = 0, \cos 2\pi t = 0 \Rightarrow 2\pi t = \cos^{-1} 0 \Rightarrow$

$2\pi t = \dfrac{\pi}{2} \Rightarrow t = \dfrac{\frac{\pi}{2}}{2\pi} = \dfrac{1}{4} \sec$

**(b)** $V = .5, \cos 2\pi t = .5 \Rightarrow 2\pi t = \cos^{-1}(.5) \Rightarrow$

$2\pi t = \dfrac{\pi}{3} \Rightarrow t = \dfrac{\frac{\pi}{3}}{2\pi} = \dfrac{1}{6} \sec$

**(c)** $V = .25, \cos 2\pi t = .25 \Rightarrow$
$2\pi t = \cos^{-1}(.25) \Rightarrow 2\pi t \approx 1.3181161 \Rightarrow$
$t \approx \dfrac{1.3181161}{2\pi} \approx .21 \sec$

**57.** $s(t) = \sin t + 2\cos t$

**(a)** $s(t) = \dfrac{2 + \sqrt{3}}{2}$

$s(t) = \dfrac{2}{2} + \dfrac{\sqrt{3}}{2} = 2\left(\dfrac{1}{2}\right) + \dfrac{\sqrt{3}}{2}$

$= 2\cos\left(\dfrac{\pi}{3}\right) + \sin\left(\dfrac{\pi}{3}\right)$

One such value is $\dfrac{\pi}{3}$.

**(b)** $s(t) = \dfrac{3\sqrt{2}}{2}$

$s(t) = \dfrac{2\sqrt{2}}{2} + \dfrac{\sqrt{2}}{2} = 2\left(\dfrac{\sqrt{2}}{2}\right) + \dfrac{\sqrt{2}}{2}$

$= 2\cos\left(\dfrac{\pi}{4}\right) + \sin\left(\dfrac{\pi}{4}\right)$

One such value is $\dfrac{\pi}{4}$.

## Section 6.3: Trigonometric Equations II

**1.** Since $2x = \dfrac{2\pi}{3}, 2\pi, \dfrac{8\pi}{3} \Rightarrow$

$x = \dfrac{2\pi}{6}, \dfrac{2\pi}{2}, \dfrac{8\pi}{6} \Rightarrow x = \dfrac{\pi}{3}, \pi, \dfrac{4\pi}{3}$, the

solution set is $\left\{\dfrac{\pi}{3}, \pi, \dfrac{4\pi}{3}\right\}$.

**3.** Since $3\theta = 180°, 630°, 720°, 930° \Rightarrow$
$\theta = 60°, 210°, 240°, 310°$, the solution set is $\{60°, 210°, 240°, 310°\}$.

**5.** $\dfrac{\tan 2\theta}{2} \neq \tan \theta$ for all values of $\theta$.

**7.** $\cos 2x = \dfrac{\sqrt{3}}{2}$

Since $0 \leq x < 2\pi$, $0 \leq 2x < 4\pi$. Thus,

$2x = \dfrac{\pi}{6}, \dfrac{11\pi}{6}, \dfrac{13\pi}{6}, \dfrac{23\pi}{6} \Rightarrow$

$x = \dfrac{\pi}{12}, \dfrac{11\pi}{12}, \dfrac{13\pi}{12}, \dfrac{23\pi}{12}$.

Solution set: $\dfrac{\pi}{12}, \dfrac{11\pi}{12}, \dfrac{13\pi}{12}, \dfrac{23\pi}{12}$

**9.** $\sin 3x = -1$
Since $0 \leq x < 2\pi$, $0 \leq 3x < 6\pi$. Thus,

$3x = \dfrac{3\pi}{2}, \dfrac{7\pi}{2}, \dfrac{11\pi}{2} \Rightarrow x = \dfrac{\pi}{2}, \dfrac{7\pi}{6}, \dfrac{11\pi}{6}$.

Solution set: $\left\{\dfrac{\pi}{2}, \dfrac{7\pi}{6}, \dfrac{11\pi}{6}\right\}$

**11.** $3\tan 3x = \sqrt{3} \Rightarrow \tan 3x = \dfrac{\sqrt{3}}{3}$

Since $0 \leq x < 2\pi$, $0 \leq 3x < 6\pi$.

Thus, $3x = \dfrac{\pi}{6}, \dfrac{7\pi}{6}, \dfrac{13\pi}{6}, \dfrac{19\pi}{6}, \dfrac{25\pi}{6}, \dfrac{31\pi}{6}$

implies $x = \dfrac{\pi}{18}, \dfrac{7\pi}{18}, \dfrac{13\pi}{18}, \dfrac{19\pi}{18}, \dfrac{25\pi}{18}, \dfrac{31\pi}{18}$.

Solution set:

$\left\{\dfrac{\pi}{18}, \dfrac{7\pi}{18}, \dfrac{13\pi}{18}, \dfrac{19\pi}{18}, \dfrac{25\pi}{18}, \dfrac{31\pi}{18}\right\}$

**13.** $\sqrt{2}\cos 2x = -1 \Rightarrow \cos 2x = \dfrac{-1}{\sqrt{2}} = -\dfrac{\sqrt{2}}{2}$

Since $0 \le x < 2\pi$, $0 \le 2x < 4\pi$. Thus,

$2x = \dfrac{3\pi}{4}, \dfrac{5\pi}{4}, \dfrac{11\pi}{4}, \dfrac{13\pi}{4} \Rightarrow$

$x = \dfrac{3\pi}{8}, \dfrac{5\pi}{8}, \dfrac{11\pi}{8}, \dfrac{13\pi}{8}$

Solution set: $\left\{\dfrac{3\pi}{8}, \dfrac{5\pi}{8}, \dfrac{11\pi}{8}, \dfrac{13\pi}{8}\right\}$

**15.** $\sin\dfrac{x}{2} = \sqrt{2} - \sin\dfrac{x}{2}$

$\sin\dfrac{x}{2} = \sqrt{2} - \sin\dfrac{x}{2} \Rightarrow \sin\dfrac{x}{2} + \sin\dfrac{x}{2} = \sqrt{2} \Rightarrow$

$2\sin\dfrac{x}{2} = \sqrt{2} \Rightarrow \sin\dfrac{x}{2} = \dfrac{\sqrt{2}}{2}$

Since $0 \le x < 2\pi$, $0 \le \dfrac{x}{2} < \pi$. Thus,

$\dfrac{x}{2} = \dfrac{\pi}{4}, \dfrac{3\pi}{4} \Rightarrow x = \dfrac{\pi}{2}, \dfrac{3\pi}{2}$.

Solution set: $\left\{\dfrac{\pi}{2}, \dfrac{3\pi}{2}\right\}$

**17.** $\sin x = \sin 2x$

$\sin x = \sin 2x \Rightarrow \sin x = 2\sin x\cos x \Rightarrow$

$\sin x - 2\sin x\cos x = 0 \Rightarrow \sin x(1 - 2\cos x) = 0$

Over the interval $[0, 2\pi)$, we have

$1 - 2\cos x = 0 \Rightarrow -2\cos x = -1 \Rightarrow$

$\cos x = \dfrac{1}{2} \Rightarrow x = \dfrac{\pi}{3}$ or $\dfrac{5\pi}{3}$

$\sin x = 0 \Rightarrow x = 0$ or $\pi$

Solution set: $\left\{0, \dfrac{\pi}{3}, \pi, \dfrac{5\pi}{3}\right\}$

**19.** $8\sec^2\dfrac{x}{2} = 4 \Rightarrow \sec^2\dfrac{x}{2} = \dfrac{1}{2} \Rightarrow \sec\dfrac{x}{2} = \pm\dfrac{\sqrt{2}}{2}$

Since $-\dfrac{\sqrt{2}}{2}$ is not in the interval $(-\infty, -1]$

and $\dfrac{\sqrt{2}}{2}$ is not in the interval $[1, \infty)$, this

equation has no solution. Solution set: $\varnothing$

**21.** $\sin\dfrac{x}{2} = \cos\dfrac{x}{2}$

$\sin\dfrac{x}{2} = \cos\dfrac{x}{2} \Rightarrow \sin^2\dfrac{x}{2} = \cos^2\dfrac{x}{2} \Rightarrow$

$\sin^2\dfrac{x}{2} = 1 - \sin^2\dfrac{x}{2} \Rightarrow 2\sin^2\dfrac{x}{2} = 1$

$\sin^2\dfrac{x}{2} = \dfrac{1}{2} \Rightarrow \sin\dfrac{x}{2} = \pm\sqrt{\dfrac{1}{2}} \Rightarrow \sin\dfrac{x}{2} = \pm\dfrac{\sqrt{2}}{2}$

Since $0 \le x < 2\pi$, $0 \le \dfrac{x}{2} < \pi$.

If $\sin\dfrac{x}{2} = \dfrac{\sqrt{2}}{2}$, $\dfrac{x}{2} = \dfrac{\pi}{4}, \dfrac{3\pi}{4} \Rightarrow x = \dfrac{\pi}{2}, \dfrac{3\pi}{2}$.

If $\sin\dfrac{x}{2} = -\dfrac{\sqrt{2}}{2}$, there are no solutions in the

interval $[0, \pi)$. Because the solution was

found by squaring an equation, the proposed

solutions must be checked.

Check $x = \dfrac{\pi}{2}$      Check $x = \dfrac{3\pi}{2}$

$\sin\dfrac{x}{2} = \cos\dfrac{x}{2}$      $\sin\dfrac{x}{2} = \cos\dfrac{x}{2}$

$\sin\dfrac{\frac{\pi}{2}}{2} = \cos\dfrac{\frac{\pi}{2}}{2}$ ?      $\sin\dfrac{\frac{3\pi}{2}}{2} = \cos\dfrac{\frac{3\pi}{2}}{2}$ ?

$\sin\dfrac{\pi}{4} = \cos\dfrac{\pi}{4}$ ?      $\sin\dfrac{3\pi}{4} = \cos\dfrac{3\pi}{4}$ ?

$\dfrac{\sqrt{2}}{2} = \dfrac{\sqrt{2}}{2}$   True      $\dfrac{\sqrt{2}}{2} = -\dfrac{\sqrt{2}}{2}$   False

$\dfrac{\pi}{2}$ is a solution.      $\dfrac{3\pi}{2}$ is not a solution.

Solution set: $\left\{\dfrac{\pi}{2}\right\}$

**23.** $\cos 2x + \cos x = 0$

We choose an identity for $\cos 2x$ that involves

only the cosine function.

$\cos 2x + \cos x = 0$

$(2\cos^2 x - 1) + \cos x = 0$

$2\cos^2 x + \cos x - 1 = 0$

$(2\cos x - 1)(\cos x + 1) = 0 \Rightarrow$

$\cos x = \dfrac{1}{2}$ or $\cos x = -1$

$\cos x = \dfrac{1}{2} \Rightarrow x = \dfrac{\pi}{3}$ or $\dfrac{5\pi}{3}$

$\cos x = -1 \Rightarrow x = \pi$

Solution set: $\left\{\dfrac{\pi}{3}, \pi, \dfrac{5\pi}{3}\right\}$

**25.** $\sqrt{2}\sin 3\theta - 1 = 0$

$\sqrt{2}\sin 3\theta - 1 = 0 \Rightarrow \sqrt{2}\sin 3\theta = 1 \Rightarrow$

$\sin 3\theta = \dfrac{1}{\sqrt{2}} \Rightarrow \sin 3\theta = \dfrac{\sqrt{2}}{2}$

Since $0 \le \theta < 360°, \quad 0° \le 3\theta < 1080°.$

In quadrant I and II, sine is positive. Thus,
$3\theta = 45°, 135°, 405°, 495°, 765°, 855° \Rightarrow$
$\theta = 15°, 45°, 135°, 165°, 255°, 285°$

Solution set: $\{15°, 45°, 135°, 165°, 255°, 285°\}$

**27.** $\cos\dfrac{\theta}{2} = 1$

Since $0 \le \theta < 360°, \quad 0° \le \dfrac{\theta}{2} < 180°.$ Thus,

$\dfrac{\theta}{2} = 0° \Rightarrow \theta = 0°.$ Solution set: $\{0°\}$

**29.** $2\sqrt{3}\sin\dfrac{\theta}{2} = 3$

$2\sqrt{3}\sin\dfrac{\theta}{2} = 3 \Rightarrow \sin\dfrac{\theta}{2} = \dfrac{3}{2\sqrt{3}} \Rightarrow$

$\sin\dfrac{\theta}{2} = \dfrac{3\sqrt{3}}{6} \Rightarrow \sin\dfrac{\theta}{2} = \dfrac{\sqrt{3}}{2}$

Since $0 \le \theta < 360°, \quad 0° \le \dfrac{\theta}{2} < 180°.$ Thus,

$\dfrac{\theta}{2} = 60°, 120° \Rightarrow \theta = 120°, 240°.$

Solution set: $\{120°, 240°\}$

**31.** $2\sin\theta = 2\cos 2\theta$

$2\sin\theta = 2\cos 2\theta \Rightarrow \sin\theta = \cos 2\theta \Rightarrow$
$\sin\theta = 1 - 2\sin^2\theta \Rightarrow 2\sin^2\theta + \sin\theta - 1 = 0 \Rightarrow$
$(2\sin\theta - 1)(\sin\theta + 1) = 0 \Rightarrow$
$2\sin\theta - 1 = 0$ or $\sin\theta + 1 = 0$

Over the interval $[0°, 360°),$ we have

$2\sin\theta - 1 = 0 \Rightarrow 2\sin\theta = 1 \Rightarrow \sin\theta = \dfrac{1}{2} \Rightarrow$

$\theta = 30°$ or $150°$
$\sin\theta + 1 = 0 \Rightarrow \sin\theta = -1 \Rightarrow \theta = 270°$
Solution set: $\{30°, 150°, 270°\}$

In Exercises 33–39, we are to find all solutions.

**33.** $1 - \sin\theta = \cos 2\theta$

$1 - \sin\theta = \cos 2\theta \Rightarrow 1 - \sin\theta = 1 - 2\sin^2\theta \Rightarrow$
$2\sin^2\theta - \sin\theta = 0 \Rightarrow \sin\theta(2\sin\theta - 1) = 0$

Over the interval $[0°, 360°),$ we have

$\sin\theta = 0 \Rightarrow \theta = 0°$ or $180°.$

$2\sin\theta - 1 = 0 \Rightarrow \sin\theta = \dfrac{1}{2} \Rightarrow \theta = 30°$ or $150°$

Solution set: $\{0° + 360°n, \ 30° + 360°n,$
$150° + 360°n, \ 180° + 360°n,$

where $n$ is any integer$\}$ or

$\{180°n, \ 30° + 360°n, 150° + 360°n,$

where $n$ is any integer$\}$

**35.** $\csc^2\dfrac{\theta}{2} = 2\sec\theta$

$\csc^2\dfrac{\theta}{2} = 2\sec\theta \Rightarrow \dfrac{1}{\sin^2\dfrac{\theta}{2}} = \dfrac{2}{\cos\theta} \Rightarrow$

$2\sin^2\dfrac{\theta}{2} = \cos\theta$

$2\left(\dfrac{1 - \cos\theta}{2}\right) = \cos\theta \Rightarrow 1 - \cos\theta = \cos\theta \Rightarrow$

$1 = 2\cos\theta \Rightarrow \cos\theta = \dfrac{1}{2}$

Over the interval $[0°, 360°),$ we have

$\cos\theta = \dfrac{1}{2} \Rightarrow \theta = 60°$ or $300°$

Solution set:
$\{60° + 360°n, \ 300° + 360°n,$

where $n$ is any integer$\}$

**37.** $2 - \sin 2\theta = 4\sin 2\theta$

$2 - \sin 2\theta = 4\sin 2\theta \Rightarrow 2 = 5\sin 2\theta \Rightarrow$

$\sin 2\theta = \dfrac{2}{5} \Rightarrow \sin 2\theta = .4$

Since $0 \le \theta < 360°, \quad 0° \le 2\theta < 720°.$ In
quadrant I and II, sine is positive.
$\sin 2\theta = .4 \Rightarrow 2\theta = 23.6°, 156.4°, 383.6°, 516.4°$
Thus, $\theta = 11.8°, 78.2°, 191.8°, 258.2°.$
Solution set:
$\{11.8° + 360°n, \ 78.2° + 360°n, \ 191.8° + 360°n,$

$258.2° + 360°n, \text{where } n \text{ is any integer}\}$ or

$\{11.8° + 180°n, \ 78.2° + 180°n,$

where $n$ is any integer$\}$

**39.** $2\cos^2 2\theta = 1 - \cos 2\theta$

$2\cos^2 2\theta = 1 - \cos 2\theta$

$2\cos^2 2\theta + \cos 2\theta - 1 = 0$
$(2\cos 2\theta - 1)(\cos 2\theta + 1) = 0$

(*continued on next page*)

(*continued from page 159*)

Since $0 \le \theta < 360° \Rightarrow 0° \le 2\theta < 720°$, we have $2\cos 2\theta - 1 = 0 \Rightarrow 2\cos 2\theta = 1 \Rightarrow$

$\cos 2\theta = \dfrac{1}{2}$. Thus,

$2\theta = 60°,\ 300°,\ 420°,\ 660° \Rightarrow$

$\theta = 30°,\ 150°,\ 210°,\ 330°$ or

$\cos 2\theta + 1 = 0 \Rightarrow \cos 2\theta = -1$

$2\theta = 180°,\ 540° \Rightarrow \theta = 90°,\ 270°$

Solution set:

$\{30° + 360°n,\ 90° + 360°n,$

$150° + 360°n,\ 210° + 360°n,\ 270° + 360°n,$

$330° + 360°n$, where $n$ is any integer$\}$ or

$\{30° + 180°n,\ 90° + 180°n,\ 150° + 180°n,$

where $n$ is any integer$\}$

**41.** The *x*-intercept method is shown in the following windows.

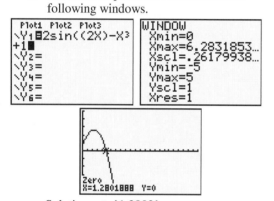

Solution set: $\{1.2802\}$

**43. (a)**

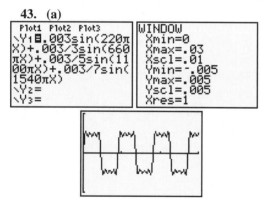

**(b)** The graph is periodic, and the wave has "jagged square" tops and bottoms.

**(c)** The eardrum is moving outward when $P < 0$.

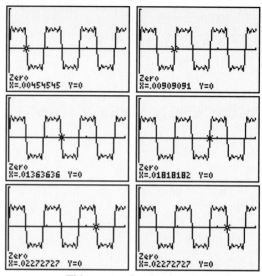

This occurs for the time intervals $(.0045, .0091), (.0136, .0182),$

$(.0227, .0273).$

**45. (a)**

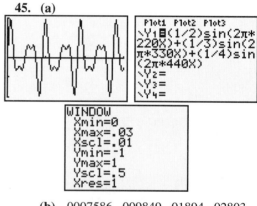

**(b)** .0007586, .009849, .01894, .02803

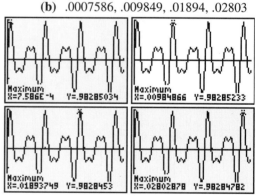

**(c)** 110 Hz

**(d)**

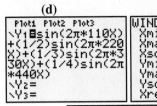

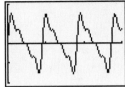

**47.** $i = I_{max} \sin 2\pi ft$

Let $i = 40$, $I_{max} = 100$, $f = 60$.

$40 = 100 \sin\left[2\pi(60)t\right]$

$40 = 100 \sin 120\pi t \Rightarrow .4 = \sin 120\pi t$

Using calculator,

$120\pi t \approx .4115168 \Rightarrow t \approx \dfrac{.4115168}{120\pi} \Rightarrow$

$t \approx .0010916 \Rightarrow t \approx .001 \sec$

**49.** $i = I_{max} \sin 2\pi ft$

Let $i = I_{max}$, $f = 60$.

$I_{max} = I_{max} \sin\left[2\pi(60)t\right] \Rightarrow 1 = \sin 120\pi t \Rightarrow$

$120\pi t = \dfrac{\pi}{2} \Rightarrow 120t = \dfrac{1}{2} \Rightarrow t = \dfrac{1}{240} \approx .004 \sec$

# Chapter 6 Quiz
**(Sections 6.1–6.3)**

**1.** Domain: $[-1, 1]$; range: $\left[0, \pi\right]$

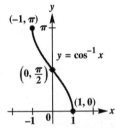

**3. (a)**

$\theta = \arccos .92341853 \approx 22.568922°$

**(b)**

$\theta = \cot^{-1}(-1.08886767) \approx 137.431085°$

**5.** $2\sin\theta - \sqrt{3} = 0 \Rightarrow 2\sin\theta = \sqrt{3} \Rightarrow \sin\theta = \dfrac{\sqrt{3}}{2}$

Over the interval $\left[0°, 360°\right)$, the equation $\sin\theta = \dfrac{\sqrt{3}}{2}$ has two solutions, the angles in quadrants I and II that have a reference angle of 60°. These are 60° and 120°.

Solution set: $\left\{60°,\ 120°\right\}$

**7.** $\tan^2 x - 5\tan x + 3 = 0$

We use the quadratic formula with $a = 1$, $b = -5$, and $c = 3$.

$\tan x = \dfrac{-(-5) \pm \sqrt{(-5)^2 - 4(1)(3)}}{2(1)}$

$= \dfrac{5 \pm \sqrt{25 - 12}}{2} = \dfrac{5 \pm \sqrt{13}}{2}$

Since $\tan x = \dfrac{5 + \sqrt{13}}{2} > 0$, we will obtain two angles, one in quadrant I and the other in quadrant III. Using a calculator, we find $x \approx 1.3424$ and $x \approx 4.4840$.

Since $\tan x = \dfrac{5 - \sqrt{13}}{2} > 0$, we will obtain two angles, one in quadrant I and the other in quadrant III. Using a calculator, we find $x \approx .6089$ and $x \approx 3.7505$.

Solution set: $\left\{.6089, 1.3424, 3.7505, 4.4840\right\}$

**9.** $\cos\dfrac{x}{2} + \sqrt{3} = -\cos\dfrac{x}{2} \Rightarrow 2\cos\dfrac{x}{2} = -\sqrt{3} \Rightarrow$

$\cos\dfrac{x}{2} = -\dfrac{\sqrt{3}}{2}$

If $0 \le x < 2\pi$, then $0 \le \dfrac{x}{2} < \pi$. Over the interval $\left[0, \pi\right)$, $\dfrac{x}{2} = \dfrac{5\pi}{6}, \dfrac{7\pi}{6} \Rightarrow x = \dfrac{5\pi}{3}, \dfrac{7\pi}{3}$.

Since this is a cosine function with period $4\pi$, the solution set is $\left\{\dfrac{5\pi}{3} + 4n\pi, \dfrac{7\pi}{3} + 4n\pi\right\}$.

## Section 6.4: Equations Involving Inverse Trigonometric Functions

1. Since $\arcsin 0 = 0$, the correct choice is C.

3. Since $\arccos\left(-\dfrac{\sqrt{2}}{2}\right) = \dfrac{3\pi}{4}$, the correct choice is C.

5. $y = 5\cos x \Rightarrow \dfrac{y}{5} = \cos x \Rightarrow x = \arccos\dfrac{y}{5}$

7. $2y = \cot 3x \Rightarrow 3x = \text{arccot } 2y \Rightarrow$
   $x = \dfrac{1}{3}\text{arccot } 2y$

9. $y = 3\tan 2x \Rightarrow \dfrac{y}{3} = \tan 2x \Rightarrow 2x = \arctan\dfrac{y}{3} \Rightarrow$
   $x = \dfrac{1}{2}\arctan\dfrac{y}{3}$

11. $y = 6\cos\dfrac{x}{4} \Rightarrow \dfrac{y}{6} = \cos\dfrac{x}{4} \Rightarrow \dfrac{x}{4} = \arccos\dfrac{y}{6} \Rightarrow$
    $x = 4\arccos\dfrac{y}{6}$

13. $y = -2\cos 5x \Rightarrow -\dfrac{y}{2} = \cos 5x \Rightarrow$
    $5x = \arccos\left(-\dfrac{y}{2}\right) \Rightarrow x = \dfrac{1}{5}\arccos\left(-\dfrac{y}{2}\right)$

15. $y = \cos(x+3) \Rightarrow x+3 = \arccos y \Rightarrow$
    $x = -3 + \arccos y$

17. $y = \sin x - 2 \Rightarrow y + 2 = \sin x \Rightarrow$
    $x = \arcsin(y+2)$

19. $y = 2\sin x - 4 \Rightarrow y + 4 = 2\sin x \Rightarrow$
    $\dfrac{y+4}{2} = \sin x \Rightarrow x = \arcsin\left(\dfrac{y+4}{2}\right)$

21. $y = \sqrt{2} + 3\sec 2x \Rightarrow y - \sqrt{2} = 3\sec 2x \Rightarrow$
    $\dfrac{y-\sqrt{2}}{3} = \sec 2x \Rightarrow 2x = \sec^{-1}\left(\dfrac{y-\sqrt{2}}{3}\right) \Rightarrow$
    $x = \dfrac{1}{2}\sec^{-1}\left(\dfrac{y-\sqrt{2}}{3}\right)$

23. First, $\sin x - 2 \neq \sin(x-2)$. If you think of the graph of $y = \sin x - 2$, this represents the graph of $f(x) = \sin x$, shifted 2 units down. If you think of the graph of $y = \sin(x-2)$, this represents the graph of $f(x) = \sin x$, shifted 2 units right.

25. $-4\arcsin x = \pi \Rightarrow \arcsin x = -\dfrac{\pi}{4} \Rightarrow$
    $x = \sin\left(-\dfrac{\pi}{4}\right) = -\dfrac{\sqrt{2}}{2}$
    Solution set: $\left\{-\dfrac{\sqrt{2}}{2}\right\}$

27. $\dfrac{4}{3}\cos^{-1}\dfrac{y}{4} = \pi \Rightarrow \cos^{-1}\dfrac{y}{4} = \dfrac{3\pi}{4} \Rightarrow$
    $\dfrac{y}{4} = \cos\dfrac{3\pi}{4} \Rightarrow \dfrac{y}{4} = -\dfrac{\sqrt{2}}{2} \Rightarrow y = -2\sqrt{2}$
    Solution set: $\left\{-2\sqrt{2}\right\}$

29. $2\arccos\left(\dfrac{y-\pi}{3}\right) = 2\pi$
    $2\arccos\left(\dfrac{y-\pi}{3}\right) = 2\pi \Rightarrow$
    $\arccos\left(\dfrac{y-\pi}{3}\right) = \pi \Rightarrow \dfrac{y-\pi}{3} = \cos\pi$
    $\dfrac{y-\pi}{3} = -1 \Rightarrow y - \pi = -3 \Rightarrow y = \pi - 3$
    Solution set: $\left\{\pi - 3\right\}$

31. $\arcsin x = \arctan\dfrac{3}{4}$

    Let $\arctan\dfrac{3}{4} = u$, so $\tan u = \dfrac{3}{4}$, $u$ is in quadrant I. Sketch a triangle and label it. The hypotenuse is $\sqrt{3^2 + 4^2} = \sqrt{9+16} = \sqrt{25} = 5$.

    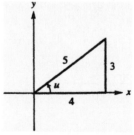

Therefore, $\sin u = \dfrac{3}{r} = \dfrac{3}{5}$. This equation

becomes $\arcsin x = u$, or $x = \sin u$. Thus,

$x = \dfrac{3}{5}$.

Solution set: $\left\{ \dfrac{3}{5} \right\}$

**33.** $\cos^{-1} x = \sin^{-1} \dfrac{3}{5}$

Let $\sin^{-1} \dfrac{3}{5} = u$, so $\sin u = \dfrac{3}{5}$. Sketch a

triangle and label it. The hypotenuse is

$\sqrt{3^2 + 4^2} = \sqrt{9 + 16} = \sqrt{25} = 5$.

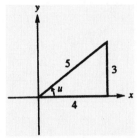

Therefore, $\cos u = \dfrac{4}{5}$. The equation becomes

$\cos^{-1} x = u$, or $x = \cos u$. Thus, $x = \dfrac{4}{5}$.

Solution set: $\left\{ \dfrac{4}{5} \right\}$

**35.** $\sin^{-1} x - \tan^{-1} 1 = -\dfrac{\pi}{4}$

$\sin^{-1} x - \tan^{-1} 1 = -\dfrac{\pi}{4} \Rightarrow$

$\sin^{-1} x = \tan^{-1} 1 - \dfrac{\pi}{4} \Rightarrow \sin^{-1} x = \dfrac{\pi}{4} - \dfrac{\pi}{4} \Rightarrow$

$\sin^{-1} x = 0 \Rightarrow \sin 0 = x \Rightarrow x = 0$

Solution set: $\{0\}$

**37.** $\arccos x + 2 \arcsin \dfrac{\sqrt{3}}{2} = \pi$

$\arccos x + 2 \arcsin \dfrac{\sqrt{3}}{2} = \pi \Rightarrow$

$\arccos x = \pi - 2 \arcsin \dfrac{\sqrt{3}}{2} \Rightarrow$

$\arccos x = \pi - 2 \left( \dfrac{\pi}{3} \right)$

$\arccos x = \pi - \dfrac{2\pi}{3} \Rightarrow \arccos x = \dfrac{\pi}{3} \Rightarrow$

$x = \cos \dfrac{\pi}{3} \Rightarrow x = \dfrac{1}{2}$

Solution set: $\left\{ \dfrac{1}{2} \right\}$

**39.** $\arcsin 2x + \arccos x = \dfrac{\pi}{6}$

$\arcsin 2x + \arccos x = \dfrac{\pi}{6}$

$\arcsin 2x = \dfrac{\pi}{6} - \arccos x$

$2x = \sin \left( \dfrac{\pi}{6} - \arccos x \right)$

Use the identity
$\sin(A - B) = \sin A \cos B - \cos A \sin B$.

$2x = \sin \dfrac{\pi}{6} \cos(\arccos x) - \cos \dfrac{\pi}{6} \sin(\arccos x)$

Let $u = \arccos x$. Thus, $\cos u = x = \dfrac{x}{1}$.

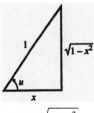

$\sin u = \sqrt{1 - x^2}$

$2x = \sin \dfrac{\pi}{6} \cdot \cos u - \cos \dfrac{\pi}{6} \sin u \Rightarrow$

$2x = \dfrac{1}{2} x - \dfrac{\sqrt{3}}{2} \left( \sqrt{1 - x^2} \right) \Rightarrow$

$4x = x - \sqrt{3} \cdot \sqrt{1 - x^2}$

$3x = -\sqrt{3} \cdot \sqrt{1 - x^2}$

$(3x)^2 = \left( -\sqrt{3} \cdot \sqrt{1 - x^2} \right)^2 \Rightarrow 9x^2 = 3\left( 1 - x^2 \right)$

$9x^2 = 3 - 3x^2 \Rightarrow 12x^2 = 3$

$x^2 = \dfrac{3}{12} = \dfrac{1}{4} \Rightarrow x = \pm \dfrac{1}{2}$

Check these proposed solutions since they
were found by squaring both side of an
equation.

*(continued on next page)*

(*continued from page 163*)

Check $x = \dfrac{1}{2}$.

$$\arcsin 2x + \arccos x = \frac{\pi}{6}$$

$$\arcsin\left(2 \cdot \frac{1}{2}\right) + \arccos\left(\frac{1}{2}\right) = \frac{\pi}{6} \ ?$$

$$\frac{\pi}{2} + \frac{\pi}{3} = \frac{\pi}{6} \ ?$$

$$\frac{5\pi}{6} = \frac{\pi}{6} \ \text{False}$$

$\dfrac{1}{2}$ is not a solution.

Check $x = -\dfrac{1}{2}$.

$$\arcsin 2x + \arccos x = \frac{\pi}{6}$$

$$\arcsin\left(2 \cdot -\frac{1}{2}\right) + \arccos\left(-\frac{1}{2}\right) = \frac{\pi}{6} \ ?$$

$$-\frac{\pi}{2} + \frac{2\pi}{3} = \frac{\pi}{6} \ ?$$

$$\frac{\pi}{6} = \frac{\pi}{6} \ \text{True}$$

$-\dfrac{1}{2}$ is a solution.

Solution set: $\left\{-\dfrac{1}{2}\right\}$

**41.**  $\cos^{-1} x + \tan^{-1} x = \dfrac{\pi}{2}$

$$\cos^{-1} x + \tan^{-1} x = \frac{\pi}{2}$$

$$\cos^{-1} x = \frac{\pi}{2} - \tan^{-1} x$$

$$x = \cos\left(\frac{\pi}{2} - \tan^{-1} x\right)$$

Use the identity
$\cos(A - B) = \cos A \cos B + \sin A \sin B.$

$$x = \cos\frac{\pi}{2}\cos\left(\tan^{-1} x\right) + \sin\frac{\pi}{2}\sin\left(\tan^{-1} x\right)$$

$$x = 0 \cdot \cos\left(\tan^{-1} x\right) + 1 \cdot \sin\left(\tan^{-1} x\right)$$

$$x = \sin\left(\tan^{-1} x\right)$$

Let $u = \tan^{-1} x$. So, $\tan u = x$.

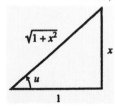

From the triangle, we find $\sin u = \dfrac{x}{\sqrt{1+x^2}}$, so

the equation $x = \sin\left(\tan^{-1} x\right)$ becomes

$x = \dfrac{x}{\sqrt{1+x^2}}$. Solve this equation.

$$x = \frac{x}{\sqrt{1+x^2}} \Rightarrow x\sqrt{1+x^2} = x$$

$$x\sqrt{1+x^2} - x = 0 \Rightarrow x\left(\sqrt{1+x^2} - 1\right) = 0$$

$$x = 0 \text{ or } \sqrt{1+x^2} - 1 = 0 \Rightarrow \sqrt{1+x^2} = 1 \Rightarrow$$

$$1 + x^2 = 1 \Rightarrow x^2 = 0 \Rightarrow x = 0$$

Solution set: $\{0\}$

**43.**

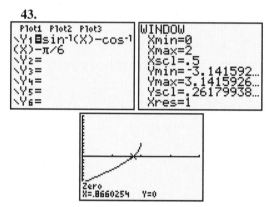

**45.**  The $x$-intercept method is shown in the following windows.

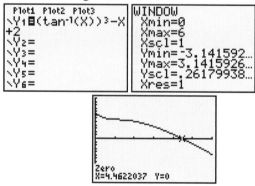

Solution set: $\{4.4622\}$

**47.**  $A = \sqrt{\begin{array}{c}\left(A_1\cos\phi_1 + A_2\cos\phi_2\right)^2 \\ + \left(A_1\sin\phi_1 + A_2\sin\phi_2\right)^2\end{array}}$  and

$$\phi = \arctan\left(\frac{A_1\sin\phi_1 + A_2\sin\phi_2}{A_1\cos\phi + A_2\cos\phi_2}\right)$$

Make sure your calculator is in radian mode.

**(a)** Let $A_1 = .0012, \phi_1 = .052, A_2 = .004,$ and $\phi_2 = .61.$

$$A = \sqrt{\begin{array}{c}(.0012\cos.052 + .004\cos.61)^2 \\ + (.0012\sin.052 + .004\sin.61)^2\end{array}}$$
$$\approx .00506$$

$$\phi = \arctan\left(\frac{.0012\sin.052 + .004\sin.61}{.0012\cos.052 + .004\cos.61}\right)$$
$$\approx .484$$

If $f = 220,$ then $P = A\sin(2\pi ft + \phi)$

becomes $P = .00506\sin(440\pi t + .484).$

**(b)**

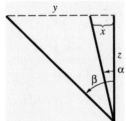

The two graphs are the same.

**49. (a)** $\tan\alpha = \dfrac{x}{z}$ and $\tan\beta = \dfrac{x+y}{z}$

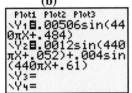

**(b)** Since

$$\tan\alpha = \frac{x}{z} \Rightarrow z\tan\alpha = x \Rightarrow z = \frac{x}{\tan\alpha} \text{ and}$$

$$\tan\beta = \frac{x+y}{z} \Rightarrow z\tan\beta = x+y \Rightarrow$$

$$z = \frac{x+y}{\tan\beta}, \text{ we have } \frac{x}{\tan\alpha} = \frac{x+y}{\tan\beta}$$

**(c)** $(x+y)\tan\alpha = x\tan\beta \Rightarrow$

$$\tan\alpha = \frac{x\tan\beta}{x+y} \Rightarrow \alpha = \arctan\left(\frac{x\tan\beta}{x+y}\right)$$

**(d)** $x\tan\beta = (x+y)\tan\alpha$

$$\tan\beta = \frac{(x+y)\tan\alpha}{x}$$

$$\beta = \arctan\left(\frac{(x+y)\tan\alpha}{x}\right)$$

**51. (a)** $E = E_{\max}\sin 2\pi ft \Rightarrow \dfrac{E}{E_{\max}} = \sin 2\pi ft \Rightarrow$

$$2\pi ft = \arcsin\frac{E}{E_{\max}} \Rightarrow$$

$$t = \frac{1}{2\pi f}\arcsin\frac{E}{E_{\max}}$$

**(b)** Let $E_{\max} = 12, E = 5,$ and $f = 100.$

$$t = \frac{1}{2\pi(100)}\arcsin\frac{5}{12}$$

$$= \frac{1}{200\pi}\arcsin\frac{5}{12} \approx .00068\sec$$

**53.** $y = \dfrac{1}{3}\sin\dfrac{4\pi t}{3}$

**(a)** $3y = \sin\dfrac{4\pi t}{3} \Rightarrow \dfrac{4\pi t}{3} = \arcsin 3y \Rightarrow$

$$4\pi t = 3\arcsin 3y \Rightarrow t = \frac{3}{4\pi}\arcsin 3y$$

**(b)** If $y = .3$ radian,

$$t = \frac{3}{4\pi}\arcsin.9 \Rightarrow t \approx .27\sec.$$

## Chapter 6: Review Exercises

**1.**

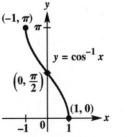

Domain: $[-1,1]$; range: $\left[-\dfrac{\pi}{2}, \dfrac{\pi}{2}\right]$

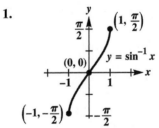

Domain: $[-1,1]$; range: $[0,\pi]$

*(continued on next page)*

(*continued from page 165*)

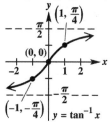

Domain: $[-\infty, \infty]$; range: $\left[-\dfrac{\pi}{2}, \dfrac{\pi}{2}\right]$

**3.** False. $\arcsin\left(-\dfrac{1}{2}\right) = -\dfrac{\pi}{6}$, not $\dfrac{11\pi}{6}$.

**5.** $y = \sin^{-1}\dfrac{\sqrt{2}}{2} \Rightarrow \sin y = \dfrac{\sqrt{2}}{2}$

Since $-\dfrac{\pi}{2} \le y \le \dfrac{\pi}{2}$, $y = \dfrac{\pi}{4}$.

**7.** $y = \tan^{-1}\left(-\sqrt{3}\right) \Rightarrow \tan y = -\sqrt{3}$

Since $-\dfrac{\pi}{2} < y < \dfrac{\pi}{2}$, $y = -\dfrac{\pi}{3}$.

**9.** $y = \cos^{-1}\left(-\dfrac{\sqrt{2}}{2}\right) \Rightarrow \cos y = -\dfrac{\sqrt{2}}{2}$

Since $0 \le y \le \pi$, $y = \dfrac{3\pi}{4}$.

**11.** $y = \sec^{-1}(-2) \Rightarrow \sec y = -2$

Since $0 \le y \le \pi$, $y \ne \dfrac{\pi}{2} \Rightarrow y = \dfrac{2\pi}{3}$.

**13.** $y = \operatorname{arccot}(-1) \Rightarrow \cot y = -1$

Since $0 < y < \pi$, $y = \dfrac{3\pi}{4}$.

**15.** $\theta = \arcsin\left(-\dfrac{\sqrt{3}}{2}\right) \Rightarrow \sin\theta = -\dfrac{\sqrt{3}}{2}$

Since $-90° \le \theta \le 90°$, $\theta = -60°$.

For Exercises 17–21, be sure that your calculator is in degree mode. Keystroke sequences may vary based on the type and/or model of calculator being used.

**17.** $\theta = \arctan 1.7804675$

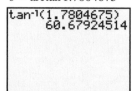

$\theta = 60.67924514°$

**19.** $\theta = \cos^{-1} .80396577$

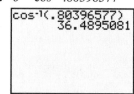

$\theta \approx 36.4895081°$

**21.** $\theta = \operatorname{arc\,sec} 3.4723155$

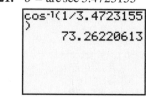

$\theta \approx 73.26220613°$

**23.** $\cos\left(\arccos(-1)\right) = \cos\pi = -1$ or

$\cos\left(\arccos(-1)\right) = \cos 180° = -1$

**25.** $\arccos\left(\cos\dfrac{3\pi}{4}\right) = \arccos\left(-\dfrac{\sqrt{2}}{2}\right) = \dfrac{3\pi}{4}$

**27.** $\tan^{-1}\left(\tan\dfrac{\pi}{4}\right) = \tan^{-1}\dfrac{\sqrt{2}}{2} = \dfrac{\pi}{4}$

**29.** $\sin\left(\arccos\dfrac{3}{4}\right)$

Let $\omega = \arccos\dfrac{3}{4}$, so that $\cos\omega = \dfrac{3}{4}$. Since arccos is defined only in quadrants I and II, and $\dfrac{3}{4}$ is positive, $\omega$ is in quadrant I. Sketch $\omega$ and label a triangle with the side opposite $\omega$ equal to $\sqrt{4^2 - 3^2} = \sqrt{16-9} = \sqrt{7}$.

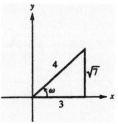

$\sin\left(\arccos\dfrac{3}{4}\right) = \sin\omega = \dfrac{\sqrt{7}}{4}$

**31.** $\cos\left(\csc^{-1}(-2)\right)$

Let $\omega = \csc^{-1}(-2)$, so that $\csc\omega = -2$. Since

$-\dfrac{\pi}{2} \le \omega \le \dfrac{\pi}{2}$ and $\omega \ne 0$, and $\csc\omega = -2$

(negative), $\omega$ is in quadrant IV. Sketch $\omega$
and label a triangle with side adjacent to $\omega$
equal to $\sqrt{2^2 - (-1)^2} = \sqrt{4-1} = \sqrt{3}$.

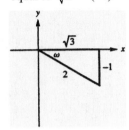

$\cos\left(\csc^{-1}(-2)\right) = \cos\omega = \dfrac{\sqrt{3}}{2}$

**33.** $\tan\left(\arcsin\dfrac{3}{5} + \arccos\dfrac{5}{7}\right)$

Let $\omega_1 = \arcsin\dfrac{3}{5}$, $\omega_2 = \arccos\dfrac{5}{7}$. Sketch

angles $\omega_1$ and $\omega_2$ in quadrant I.

The side adjacent to $\omega_1$ is

$\sqrt{5^2 - 3^2} = \sqrt{25-9} = \sqrt{16} = 4$. The side

opposite $\omega_2$ is

$\sqrt{7^2 - 5^2} = \sqrt{49 - 25} = \sqrt{24} = 2\sqrt{6}.$

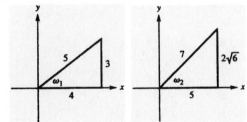

We have $\tan\omega_1 = \dfrac{3}{4}$ and $\tan\omega_2 = \dfrac{2\sqrt{6}}{5}$.

$\tan\left(\arcsin\dfrac{3}{5} + \arccos\dfrac{5}{7}\right)$

$= \tan\left(\omega_1 + \omega_2\right) = \dfrac{\tan\omega_1 + \tan\omega_2}{1 - \tan\omega_1\tan\omega_2}$

$= \dfrac{\dfrac{3}{4} + \dfrac{2\sqrt{6}}{5}}{1 - \left(\dfrac{3}{4}\right)\left(\dfrac{2\sqrt{6}}{5}\right)} = \dfrac{\dfrac{15 + 8\sqrt{6}}{20}}{\dfrac{20 - 6\sqrt{6}}{20}}$

$= \dfrac{15 + 8\sqrt{6}}{20 - 6\sqrt{6}} = \dfrac{15 + 8\sqrt{6}}{20 - 6\sqrt{6}} \cdot \dfrac{20 + 6\sqrt{6}}{20 + 6\sqrt{6}}$

$= \dfrac{588 + 250\sqrt{6}}{184} = \dfrac{294 + 125\sqrt{6}}{92}$

**35.** $\tan\left(\text{arcsec}\,\dfrac{\sqrt{u^2 + 1}}{u}\right)$

Let $\theta = \text{arcsec}\,\dfrac{\sqrt{u^2 + 1}}{u}$, so $\sec\theta = \dfrac{\sqrt{u^2 + 1}}{u}$.

If $u > 0$, $0 < \theta < \dfrac{\pi}{2}$.

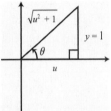

From the Pythagorean theorem,

$y = \sqrt{\left(\sqrt{u^2 + 1}\right)^2 - u^2} = \sqrt{u^2 + 1 - u^2} = \sqrt{1} = 1.$

Therefore $\tan\theta = \dfrac{1}{u}$. Thus,

$\tan\left(\text{arcsec}\,\dfrac{\sqrt{u^2 + 1}}{u}\right) = \dfrac{1}{u}.$

**37.** $2\tan x - 1 = 0$

$2\tan x - 1 = 0 \Rightarrow 2\tan x = 1 \Rightarrow \tan x = \dfrac{1}{2}$

Over the interval $[0, 2\pi)$, the equation

$\tan x = \dfrac{1}{2}$ has two solutions. One solution is in

quadrant I and the other is in quadrant III.
Using a calculator, the quadrant I solution is
approximately .4636. The quadrant III solution
would be approximately $.4636 + \pi \approx 3.6052$.

Solution set: $\{.4636, 3.6052\}$

**39.** $\tan x = \cot x$

Use the identity $\cot x = \dfrac{1}{\tan x}$, $\tan x \neq 0$.

$\tan x = \cot x \Rightarrow \tan x = \dfrac{1}{\tan x} \Rightarrow$

$\tan^2 x = 1 \Rightarrow \tan x = \pm 1$

Over the interval $[0, 2\pi)$, the equation

$\tan x = 1$ has two solutions. One solution is in quadrant I and the other is in quadrant III.

These solutions are $\dfrac{\pi}{4}$ and $\dfrac{5\pi}{4}$. In the same

interval, the equation $\tan x = -1$ has two solutions. One solution is in quadrant II and the other is in quadrant IV. These solutions are

$\dfrac{3\pi}{4}$ and $\dfrac{7\pi}{4}$. Solution set: $\left\{\dfrac{\pi}{4}, \dfrac{3\pi}{4}, \dfrac{5\pi}{4}, \dfrac{7\pi}{4}\right\}$

**41.** $\tan^2 2x - 1 = 0$

$\tan^2 2x - 1 = 0 \Rightarrow \tan^2 2x = 1 \Rightarrow \tan 2x = \pm 1$
Since $0 \leq x < 2\pi, 0 \leq 2x < 4\pi$. Thus,

$2x = \dfrac{\pi}{4}, \dfrac{3\pi}{4}, \dfrac{5\pi}{4}, \dfrac{7\pi}{4}, \dfrac{9\pi}{4}, \dfrac{11\pi}{4}, \dfrac{13\pi}{4}, \dfrac{15\pi}{4}$

implies

$x = \dfrac{\pi}{8}, \dfrac{3\pi}{8}, \dfrac{5\pi}{8}, \dfrac{7\pi}{8}, \dfrac{9\pi}{8}, \dfrac{11\pi}{8}, \dfrac{13\pi}{8}, \dfrac{15\pi}{8}$.

Solution set:

$\left\{\dfrac{\pi}{8}, \dfrac{3\pi}{8}, \dfrac{5\pi}{8}, \dfrac{7\pi}{8}, \dfrac{9\pi}{8}, \dfrac{11\pi}{8}, \dfrac{13\pi}{8}, \dfrac{15\pi}{8}\right\}$

**43.** $\cos 2x + \cos x = 0$

$\cos 2x + \cos x = 0 \Rightarrow 2\cos^2 x - 1 + \cos x = 0 \Rightarrow$
$2\cos^2 x + \cos x - 1 = 0 \Rightarrow$
$(2\cos x - 1)(\cos x + 1) = 0$

$2\cos x - 1 = 0 \Rightarrow 2\cos x = 1 \Rightarrow \cos x = \dfrac{1}{2}$ or

$\cos x + 1 = 0 \Rightarrow \cos x = -1$

Over the interval $[0, 2\pi)$, the equation

$\cos x = \dfrac{1}{2}$ has two solutions. The angles in

quadrants I and IV that have a reference angle

of $\dfrac{\pi}{3}$ are $\dfrac{\pi}{3}$ and $\dfrac{5\pi}{3}$. In the same interval,

$\cos x = -1$ when the angle is $\pi$.
Solution set:

$\left\{\dfrac{\pi}{3} + 2n\pi, \ \pi + 2n\pi, \ \dfrac{5\pi}{3} + 2n\pi, \right.$

where $n$ is any integer$\Big\}$

**45.** $\sin^2 \theta + 3\sin\theta + 2 = 0$

$\sin^2 \theta + 3\sin\theta + 2 = 0$
$(\sin\theta + 2)(\sin\theta + 1) = 0$

In the interval $[0°, 360°)$, we have

$\sin\theta + 1 = 0 \Rightarrow \sin\theta = -1 \Rightarrow \theta = 270°$ and
$\sin\theta + 2 = 0 \Rightarrow \sin\theta = -2 < -1 \Rightarrow$ no solution

Solution set: $\{270°\}$

**47.** $\sin 2\theta = \cos 2\theta + 1$

$\sin 2\theta = \cos 2\theta + 1$
$(\sin 2\theta)^2 = (\cos 2\theta + 1)^2$
$\sin^2 2\theta = \cos^2 2\theta + 2\cos 2\theta + 1$
$1 - \cos^2 2\theta = \cos^2 2\theta + 2\cos 2\theta + 1$
$2\cos^2 2\theta + 2\cos 2\theta = 0$
$\cos^2 2\theta + \cos 2\theta = 0$
$\cos 2\theta (\cos 2\theta + 1) = 0$

Since $0° \leq \theta < 360°$, $0° \leq 2\theta < 720°$.

$\cos 2\theta = 0 \Rightarrow 2\theta = 90°, 270°, 450°, 630° \Rightarrow$
$\theta = 45°, 135°, 225°, 315°$

$\cos 2\theta + 1 = 0 \Rightarrow \cos 2\theta = -1$
$2\theta = 180°, 540° \Rightarrow \theta = 90°, 270°$

Possible values for $\theta$ are
$\theta = 45°, 90°, 135°, 225°, 270°, 315°$.

All proposed solutions must be checked since the solutions were found by squaring an equation. A value for $\theta$ will be a solution if
$\sin 2\theta - \cos 2\theta = 1$.
$\theta = 45°, 2\theta = 90° \Rightarrow$
$\sin 90° - \cos 90° = 1 - 0 = 1$
$\theta = 90°, 2\theta = 180° \Rightarrow$
$\sin 180° - \cos 180° = 0 - (-1) = 1$
$\theta = 135°, 2\theta = 270° \Rightarrow$
$\sin 270° - \cos 270° = -1 - 0 \neq 1$
$\theta = 225°, 2\theta = 450° \Rightarrow$
$\sin 450° - \cos 450° = 1 - 0 = 1$
$\theta = 270°, 2\theta = 540° \Rightarrow$
$\sin 540° - \cos 540° = 0 - (-1) = 1$
$\theta = 315°, 2\theta = 630° \Rightarrow$
$\sin 630° - \cos 630° = -1 - 0 \neq 1$
Thus, $\theta = 45°, 90°, 225°, 270°$.
Solution set: $\{45°, 90°, 225°, 270°\}$

**49.** $3\cos^2 \theta + 2\cos\theta - 1 = 0$
$(3\cos\theta - 1)(\cos\theta + 1) = 0$

In the interval $[0°, 360°)$, we have

$3\cos\theta - 1 = 0 \Rightarrow \cos\theta = \dfrac{1}{3} \Rightarrow$

$\theta \approx 70.5°$ and $289.5°$ (using a calculator)

$\cos\theta + 1 = 0 \Rightarrow \cos\theta = -1 \Rightarrow \theta = 180°$

Solution set: $\{70.5°, 180°, 289.5°\}$

**51.** $4y = 2\sin x \Rightarrow 2y = \sin x \Rightarrow x = \arcsin 2y$

**53.** $2y = \tan(3x + 2) \Rightarrow 3x + 2 = \arctan 2y \Rightarrow$

$3x = \arctan 2y - 2 \Rightarrow x = \left(\dfrac{1}{3}\arctan 2y\right) - \dfrac{2}{3}$

**55.** $\dfrac{4}{3}\arctan\dfrac{x}{2} = \pi \Rightarrow \arctan\dfrac{x}{2} = \dfrac{3\pi}{4}$

But, by definition, the range of arctan is

$\left(-\dfrac{\pi}{2}, \dfrac{\pi}{2}\right)$. So, this equation has no solution.

Solution set: $\varnothing$

**57.** $\arccos x + \arctan 1 = \dfrac{11\pi}{12}$

$\arccos x + \arctan 1 = \dfrac{11\pi}{12}$

$\arccos x = \dfrac{11\pi}{12} - \arctan 1$

$\arccos x = \dfrac{11\pi}{12} - \dfrac{\pi}{4} = \dfrac{8\pi}{12} = \dfrac{2\pi}{3}$

$\cos\dfrac{2\pi}{3} = x \Rightarrow x = -\dfrac{1}{2}$

Solution set: $\left\{-\dfrac{1}{2}\right\}$

**59. (a)** Let $\alpha$ be the angle to the left of $\theta$.

Thus, we have

$\tan(\alpha + \theta) = \dfrac{5 + 10}{x}$

$\alpha + \theta = \arctan\left(\dfrac{15}{x}\right)$

$\theta = \arctan\left(\dfrac{15}{x}\right) - \alpha$

$\theta = \arctan\left(\dfrac{15}{x}\right) - \arctan\left(\dfrac{5}{x}\right)$

**(b)** The maximum occurs at approximately 8.66026 ft. There may be a discrepancy in the final digits.

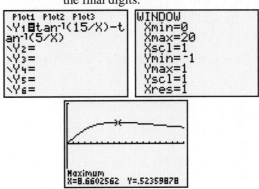

**61.** If $\theta_1 > 48.8°$, then $\theta_2 > 90°$ and the light beam is completely underwater.

**63.**

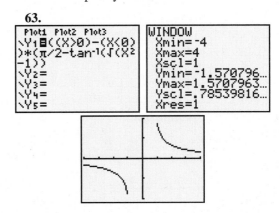

## Chapter 6 Test

**1.**

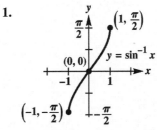

Domain: $[-1, 1]$; range: $\left[-\dfrac{\pi}{2}, \dfrac{\pi}{2}\right]$

**2. (a)** $y = \arccos\left(-\dfrac{1}{2}\right) \Rightarrow \cos y = -\dfrac{1}{2}$

Since $0 \le y \le \pi$, $y = \dfrac{2\pi}{3}$.

**(b)** $y = \sin^{-1}\left(-\dfrac{\sqrt{3}}{2}\right) \Rightarrow \sin y = -\dfrac{\sqrt{3}}{2}$

Since $-\dfrac{\pi}{2} \le y \le \dfrac{\pi}{2}$, $y = -\dfrac{\pi}{3}$.

**(c)** $y = \tan^{-1} 0 \Rightarrow \tan y = 0$

Since $-\dfrac{\pi}{2} < y < \dfrac{\pi}{2}$, $y = 0$.

**(d)** $y = \text{arcsec}(-2) \Rightarrow \sec y = -2$

Since $0 \le y \le \pi$ and $y \ne \dfrac{\pi}{2}$, $y = \dfrac{2\pi}{3}$.

**3. (a)** $\theta = \arccos \dfrac{\sqrt{3}}{2} \Rightarrow \cos \theta = \dfrac{\sqrt{3}}{2}$

Since $0 \le y \le 180°$, $y = 30°$.

**(b)** $\theta = \tan^{-1}(-1) \Rightarrow \tan \theta = -1$

Since $-90° < y < 90°$, $y = -45°$.

**(c)** $\theta = \cot^{-1}(-1) \Rightarrow \cot \theta = -1$

Since $0° < y < 180°$, $y = 135°$.

**(d)** $\theta = \csc^{-1}\left(-\dfrac{2\sqrt{3}}{3}\right) \Rightarrow \csc \theta = -\dfrac{2\sqrt{3}}{3}$

Since $-90 \le y \le 90°$, $\theta = -60°$.

**4. (a)**

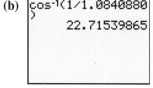

$\sin^{-1} .67610476 \approx 42.54°$

**(b)**

```
cos-1(1/1.0840880
)
       22.71539865
```

$\sec^{-1} 1.0840880 \approx 22.72°$

**(c)**

```
tan-1(1/(-.712558
6))
        -54.52790091
Ans+180
        125.4720991
```

$\cot^{-1}(-.7125586) \approx 125.47°$

**5. (a)** $\cos\left(\arcsin \dfrac{2}{3}\right)$

Let $\arcsin \dfrac{2}{3} = u$, so that $\sin u = \dfrac{2}{3}$. Since arcsine is defined only in quadrants I and IV, and $\dfrac{2}{3}$ is positive, $u$ is in quadrant I. Sketch $u$ and label a triangle with the side adjacent $u$ equal to

$$\sqrt{3^2 - 2^2} = \sqrt{9-4} = \sqrt{5}.$$

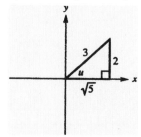

$$\cos\left(\arcsin \dfrac{2}{3}\right) = \cos u = \dfrac{\sqrt{5}}{3}$$

**(b)** $\sin\left(2\cos^{-1} \dfrac{1}{3}\right)$

Let $\theta = \cos^{-1} \dfrac{1}{3}$, so that $\cos \theta = \dfrac{1}{3}$. Since arccosine is defined only in quadrants I and II, and $\dfrac{1}{3}$ is positive, $\theta$ is in quadrant I. Sketch $\theta$ and label a triangle with the side opposite to $\theta$ equal to

$$\theta = \sqrt{3^2 - (-1)^2} = \sqrt{9-1} = \sqrt{8} = 2\sqrt{2}.$$

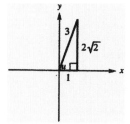

Thus, $\sin \theta = \dfrac{2\sqrt{2}}{3}$ and

$$\sin\left(2\cos^{-1} \dfrac{1}{3}\right) = \sin 2\theta.$$

$$\sin\left(2\cos^{-1} \dfrac{1}{3}\right) = \sin 2\theta = 2\sin\theta\cos\theta$$

$$= 2\left(\dfrac{2\sqrt{2}}{3}\right)\left(\dfrac{1}{3}\right) = \dfrac{4\sqrt{2}}{9}$$

**6.** Since $-1 \le \sin\theta \le 1$, there is no value of $\theta$ for which $\sin\theta = 3$. Thus, $\sin^{-1}3$ is not defined.

**7.** $\arcsin\left(\sin\dfrac{5\pi}{6}\right) = \arcsin\left(\dfrac{1}{2}\right) = \dfrac{\pi}{6} \ne \dfrac{5\pi}{6}$

**8.** $\tan(\arcsin u)$

Let $\theta = \arcsin u$, so $\sin\theta = u = \dfrac{u}{1}$. If $u > 0$,

$0 < \theta < \dfrac{\pi}{2}$.

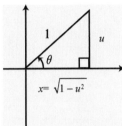

$x = \sqrt{1 - u^2}$

From the Pythagorean theorem,

$x = \sqrt{1^2 - u^2} = \sqrt{1 - u^2}$ . Therefore,

$\tan\theta = \dfrac{u}{\sqrt{1 - u^2}} = \dfrac{u\sqrt{1 - u^2}}{1 - u^2}$ . Thus,

$\tan(\arcsin u) = \dfrac{u\sqrt{1 - u^2}}{1 - u^2}$ .

**9.** $-3\sec\theta + 2\sqrt{3} = 0 \Rightarrow \sec\theta = \dfrac{2\sqrt{3}}{3}$

Over the interval $[0, 360°)$, the equation

$\sec x = \dfrac{2\sqrt{3}}{3}$ has two solutions. One solution is in quadrant I and the other is in quadrant IV. These solutions are 30° and 330°.

Solution set: $\{30°, 330°\}$

**10.** $\sin^2\theta = \cos^2\theta + 1$
$\sin^2\theta = 1 - \sin^2\theta + 1$
$2\sin^2\theta = 2 \Rightarrow \sin^2\theta = 1 \Rightarrow \sin\theta = \pm 1$
Over the interval $[0, 360°)$, the equation
$\sin\theta = 1$ has one solution, 90°. Over the interval $[0, 2\pi)$, the equation $\sin\theta = -1$ has one solution, 270°. Solution set: $\{90°, 270°\}$

**11.** $\csc^2\theta - 2\cot\theta = 4$
$1 + \cot^2\theta - 2\cot\theta = 4$
$\cot^2\theta - 2\cot\theta - 3 = 0$
$(\cot\theta - 3)(\cot\theta + 1) = 0 \Rightarrow$
$\cot\theta = 3$ or $\cot\theta = -1$
Over the interval $[0, 360°)$, the equation $\cot\theta = 3$ has two solutions, 18.4° and 198.4° (found using a calculator.) Over the interval $[0, 2\pi)$, the equation $\cot\theta = -1$ has two solutions, 135° and 315°.
Solution set: $\{18.4°, 135°, 198.4°, 315°\}$

**12.**
$$\cos x = \cos 2x$$
$$\cos x = 2\cos^2 x - 1$$
$$2\cos^2 x - \cos x - 1 = 0$$
$$(2\cos x + 1)(\cos x - 1) = 0 \Rightarrow$$
$$\cos x = -\dfrac{1}{2} \text{ or } \cos x = 1$$
Over the interval $[0, 2\pi)$, the equation
$\cos x = -\dfrac{1}{2}$ has two solutions, $\dfrac{2\pi}{3}$ and $\dfrac{4\pi}{3}$ .

Over the interval $[0, 2\pi)$, the equation
$\cos x = 1$ has one solution, 0.
Solution set: $\left\{0, \dfrac{2\pi}{3}, \dfrac{4\pi}{3}\right\}$

**13.** $\sqrt{2}\cos 3x - 1 = 0 \Rightarrow \cos 3x = \dfrac{1}{\sqrt{2}} = \dfrac{\sqrt{2}}{2}$
Since $0 \le x < 2\pi, 0 \le 3x < 6\pi$. Thus
$3x = \dfrac{\pi}{4}, \dfrac{7\pi}{4}, \dfrac{9\pi}{4}, \dfrac{15\pi}{4}, \dfrac{17\pi}{4}, \dfrac{23\pi}{4} \Rightarrow$
$x = \dfrac{\pi}{12}, \dfrac{7\pi}{12}, \dfrac{3\pi}{4}, \dfrac{5\pi}{4}, \dfrac{17\pi}{12}, \dfrac{23\pi}{12}$
Solution set: $\left\{\dfrac{\pi}{12}, \dfrac{7\pi}{12}, \dfrac{3\pi}{4}, \dfrac{5\pi}{4}, \dfrac{17\pi}{12}, \dfrac{23\pi}{12}\right\}$

**14.** $\sin x \cos x = \dfrac{1}{3} \Rightarrow 2\sin x \cos x = \dfrac{2}{3} \Rightarrow$
$\sin 2x = \dfrac{2}{3}$
Since $0 \le x < 2\pi, 0 \le 2x < 4\pi$. Use a calculator to find $2x$:
$2x \approx .72672, 2.4118, 7.0129, 8.69505 \Rightarrow$
$x \approx .3649, 1.2059, 3.5605, 4.3475$
Solution set: $\{.3649, 1.2059, 3.5605, 4.3475\}$

**15.** $\sin^2 \theta = -\cos 2\theta$

$\sin^2 \theta = -\left(1 - 2\sin^2 \theta\right)$

$\sin^2 \theta = 1 \Rightarrow \sin \theta = \pm 1$

Over the interval $[0, 360°)$, the equation $\sin \theta = 1$ has one solution, 90°. Over the interval $[0, 360°)$, the equation $\sin \theta = -1$ has one solution, 270°. Since $270° = 90° + 180°$, the solution is $90° + 180°n$, where $n$ is any integer.

**16.** $2\sqrt{3} \sin \dfrac{x}{2} = 3 \Rightarrow \sin \dfrac{x}{2} = \dfrac{3}{2\sqrt{3}} \Rightarrow$

$\dfrac{x}{2} = \sin^{-1} \dfrac{3}{2\sqrt{3}} \Rightarrow \dfrac{x}{2} = \sin^{-1} \dfrac{\sqrt{3}}{2}$

Since $0 \le x < 2\pi \Rightarrow 0° \le \dfrac{x}{2} < \pi$, we have

$\dfrac{x}{2} = \dfrac{\pi}{3}, \dfrac{2\pi}{3} \Rightarrow x = \dfrac{2\pi}{3}, \dfrac{4\pi}{3}$

Solution set:

$\left\{ \dfrac{2\pi}{3} + 2\pi n, \dfrac{4\pi}{3} + 2\pi n, \right.$

$\left. \text{where } n \text{ is any integer} \right\}$

**17.** **(a)** $y = \cos 3x \Rightarrow 3x = \arccos y \Rightarrow$

$x = \dfrac{1}{3} \arccos y$

**(b)** $\arcsin x = \arctan \dfrac{4}{3}$

Let $\omega = \arctan \dfrac{4}{3}$. Then $\tan \omega = \dfrac{4}{3}$.

Sketch $\omega$ in quadrant I and label a triangle with the hypotenuse equal to

$\sqrt{4^2 + 3^2} = \sqrt{16 + 9} = \sqrt{25} = 5.$

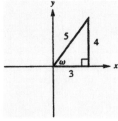

Thus, we have

$\arcsin x = \arctan \dfrac{4}{3} \Rightarrow \arcsin x = \omega \Rightarrow$

$x = \sin \omega = \dfrac{4}{5}$

**18.** $y = \dfrac{\pi}{8} \cos \left[ \pi \left( t - \dfrac{1}{3} \right) \right]$

Solve $0 = \dfrac{\pi}{8} \cos \left[ \pi \left( t - \dfrac{1}{3} \right) \right]$.

$0 = \dfrac{\pi}{8} \cos \left[ \pi \left( t - \dfrac{1}{3} \right) \right] \Rightarrow 0 = \cos \left[ \pi \left( t - \dfrac{1}{3} \right) \right]$

$\pi \left( t - \dfrac{1}{3} \right) = \arccos 0 \Rightarrow \pi \left( t - \dfrac{1}{3} \right) = \dfrac{\pi}{2} + n\pi$

$t - \dfrac{1}{3} = \dfrac{1}{2} + n \Rightarrow t = \dfrac{1}{2} + \dfrac{1}{3} + n \Rightarrow t = \dfrac{5}{6} + n,$

where $n$ is any integer

In the interval $[0, \pi)$, $n = 0, 1$, and 2 provide valid values for $t$. Thus, we have

$t = \dfrac{5}{6}, \dfrac{5}{6} + 1, \dfrac{5}{6} + 2 \Rightarrow t = \dfrac{5}{6} \text{sec}, \dfrac{11}{6} \text{sec}, \dfrac{17}{6} \text{sec}$

# Chapter 7

## Applications of Trigonometry and Vectors

### Section 7.1: Oblique Triangles and the Law of Sines

#### Connections (page 307)

$$X = \frac{(a-h)x}{f\sec\theta - y\sin\theta}, Y = \frac{(a-h)y\cos\theta}{f\sec\theta - y\sin\theta}$$

1. House: $(x_H, y_H) = (.9, 3.5)$; elevation 150 ft

   Forest fire: $(x_F, y_F) = (2.1, -2.4)$; elevation 690 ft

   $a = 7400$ ft; $f = 6$ in.; $\theta = 4.1°$

   Coordinates of house:

   $$X = \frac{(a-h)x}{f\sec\theta - y\sin\theta}$$

   $$X = \frac{(7400-150)\cdot.9}{6\sec 4.1° - 3.5\sin 4.1°}$$

   $$= \frac{6525}{6\sec 4.1° - 3.5\sin 4.1°}$$

   $$\approx 1131.8 \text{ ft}$$

   $$Y = \frac{(a-h)y\cos\theta}{f\sec\theta - y\sin\theta}$$

   $$Y = \frac{(7400-150)\cdot 3.5\cos 4.1°}{6\sec 4.1° - 3.5\sin 4.1°}$$

   $$= \frac{25,375\cos 4.1°}{6\sec 4.1° - 3.5\sin 4.1°}$$

   $$\approx 4390.2 \text{ ft}$$

   Coordinates of forest fire:

   $$X = \frac{(a-h)x}{f\sec\theta - y\sin\theta}$$

   $$X = \frac{(7400-690)\cdot 2.1}{6\sec 4.1° - (-2.4)\sin 4.1°}$$

   $$= \frac{14,091}{6\sec 4.1° + 2.4\sin 4.1°}$$

   $$\approx 2277.5 \text{ ft}$$

   $$Y = \frac{(a-h)y\cos\theta}{f\sec\theta - y\sin\theta}$$

   $$Y = \frac{(7400-690)\cdot(-2.4)\cos 4.1°}{6\sec 4.1° - (-2.4)\sin 4.1°}$$

   $$= \frac{-16,104\cos 4.1°}{6\sec 4.1° + 2.4\sin 4.1°}$$

   $$\approx -2596.2 \text{ ft}$$

2. The points needed to find the distance are $(1131.8, 4390.2)$ and $(2277.5, -2596.2)$. Using the distance formula

   $$d = \sqrt{(x_2 - x_1)^2 + (y_2 - y_1)^2}\text{ , we have}$$

   $$d = \sqrt{(2277.5 - 1131.8)^2 + (-2596.2 - 4390.2)^2}$$

   $$= \sqrt{1145.7^2 + (-6986.4)^2}$$

   $$= \sqrt{50,122,413.45} \approx 7079.7 \text{ ft}$$

#### Exercises

1. A: $\dfrac{a}{b} = \dfrac{\sin A}{\sin B}$ can be rewritten as

   $\dfrac{a}{\sin A} = \dfrac{b}{\sin B}$. This is a valid proportion.

   B: $\dfrac{a}{\sin A} = \dfrac{b}{\sin B}$ is a valid proportion.

   C: $\dfrac{\sin A}{a} = \dfrac{b}{\sin B}$ cannot be rewritten as

   $\dfrac{a}{\sin A} = \dfrac{b}{\sin B}$. $\dfrac{\sin A}{a} = \dfrac{b}{\sin B}$ is not a valid proportion.

   D: $\dfrac{\sin A}{a} = \dfrac{\sin B}{b}$ is a valid proportion.

3. The measure of angle $C$ is
   $180° - (60° + 75°) = 180° - 135° = 45°$.

   $$\frac{a}{\sin A} = \frac{c}{\sin C} \Rightarrow \frac{a}{\sin 60°} = \frac{\sqrt{2}}{\sin 45°} \Rightarrow$$

   $$a = \frac{\sqrt{2}\sin 60°}{\sin 45°} = \frac{\sqrt{2}\cdot\frac{\sqrt{3}}{2}}{\frac{\sqrt{2}}{2}}$$

   $$= \sqrt{2}\cdot\frac{\sqrt{3}}{2}\cdot\frac{2}{\sqrt{2}} = \sqrt{3}$$

5. $A = 37°$, $B = 48°$, $c = 18$ m

   $C = 180° - A - B \Rightarrow$
   $C = 180° - 37° - 48° = 95°$

   *(continued on next page)*

173

*(continued from page 173)*

$$\frac{b}{\sin B} = \frac{c}{\sin C} \Rightarrow \frac{b}{\sin 48°} = \frac{18}{\sin 95°} \Rightarrow$$
$$b = \frac{18\sin 48°}{\sin 95°} \approx 13 \text{ m}$$
$$\frac{a}{\sin A} = \frac{c}{\sin C} \Rightarrow \frac{a}{\sin 37°} = \frac{18}{\sin 95°} \Rightarrow$$
$$a = \frac{18\sin 37°}{\sin 95°} \approx 11 \text{ m}$$

**7.** $A = 27.2°$, $C = 115.5°$, $c = 76.0$ ft

$$B = 180° - A - C \Rightarrow$$
$$B = 180° - 27.2° - 115.5° = 37.3°$$
$$\frac{a}{\sin A} = \frac{c}{\sin C} \Rightarrow \frac{a}{\sin 27.2°} = \frac{76.0}{\sin 115.5°} \Rightarrow$$
$$a = \frac{76.0\sin 27.2°}{\sin 115.5°} \approx 38.5 \text{ ft}$$
$$\frac{b}{\sin B} = \frac{c}{\sin C} \Rightarrow \frac{b}{\sin 37.3°} = \frac{76.0}{\sin 115.5°} \Rightarrow$$
$$b = \frac{76.0\sin 37.3°}{\sin 115.5°} \approx 51.0 \text{ ft}$$

**9.** $A = 68.41°$, $B = 54.23°$, $a = 12.75$ ft

$$C = 180° - A - B - C$$
$$= 180° - 68.41° - 54.23° = 57.36°$$
$$\frac{a}{\sin A} = \frac{b}{\sin B} \Rightarrow \frac{12.75}{\sin 68.41°} = \frac{b}{\sin 54.23°} \Rightarrow$$
$$b = \frac{12.75\sin 54.23°}{\sin 68.41°} \approx 11.13 \text{ ft}$$
$$\frac{a}{\sin A} = \frac{c}{\sin C} \Rightarrow \frac{12.75}{\sin 68.41°} = \frac{c}{\sin 57.36°} \Rightarrow$$
$$c = \frac{12.75\sin 57.36°}{\sin 68.41°} \approx 11.55 \text{ ft}$$

**11.** $A = 87.2°, b = 75.9$ yd, $C = 74.3°$

$$B = 180° - A - C \Rightarrow$$
$$B = 180° - 87.2° - 74.3° = 18.5°$$
$$\frac{a}{\sin A} = \frac{b}{\sin B} \Rightarrow \frac{a}{\sin 87.2°} = \frac{75.9}{\sin 18.5°} \Rightarrow$$
$$a = \frac{75.9\sin 87.2°}{\sin 18.5°} \approx 239 \text{ yd}$$
$$\frac{b}{\sin B} = \frac{c}{\sin C} \Rightarrow \frac{75.9}{\sin 18.5°} = \frac{c}{\sin 74.3°} \Rightarrow$$
$$c = \frac{75.9\sin 74.3°}{\sin 18.5°} \approx 230 \text{ yd}$$

**13.** $B = 20°50'$, $AC = 132$ ft, $C = 103°10'$

$$A = 180° - B - C$$
$$A = 180° - 20°50' - 103°10' \Rightarrow A = 56°00'$$
$$\frac{AC}{\sin B} = \frac{AB}{\sin C} \Rightarrow \frac{132}{\sin 20°50'} = \frac{AB}{\sin 103°10'} \Rightarrow$$
$$AB = \frac{132\sin 103°10'}{\sin 20°50'} \approx 361 \text{ ft}$$
$$\frac{BC}{\sin A} = \frac{AC}{\sin B} \Rightarrow \frac{BC}{\sin 56°00'} = \frac{132}{\sin 20°50'} \Rightarrow$$
$$BC = \frac{132\sin 56°00'}{\sin 20°50'} \approx 308 \text{ ft}$$

**15.** $A = 39.70°, C = 30.35°$, $b = 39.74$ m

$$B = 180° - A - C \Rightarrow$$
$$B = 180° - 39.70° - 30.35° \Rightarrow$$
$$B = 109.95° \approx 110.0° \text{ (rounded)}$$
$$\frac{a}{\sin A} = \frac{b}{\sin B} \Rightarrow \frac{a}{\sin 39.70°} = \frac{39.74}{\sin 109.95°} \Rightarrow$$
$$a = \frac{39.74\sin 39.70°}{\sin 110.0°} \approx 27.01 \text{ m}$$
$$\frac{b}{\sin B} = \frac{c}{\sin C} \Rightarrow \frac{39.74}{\sin 109.95°} = \frac{c}{\sin 30.35°} \Rightarrow$$
$$c = \frac{39.74\sin 30.35°}{\sin 110.0°} \approx 21.37 \text{ m}$$

**17.** $B = 42.88°$, $C = 102.40°$, $b = 3974$ ft

$$A = 180° - B - C \Rightarrow$$
$$A = 180° - 42.88° - 102.40° = 34.72°$$
$$\frac{a}{\sin A} = \frac{b}{\sin B} \Rightarrow \frac{a}{\sin 34.72°} = \frac{3974}{\sin 42.88°} \Rightarrow$$
$$a = \frac{3974\sin 34.72°}{\sin 42.88°} \approx 3326 \text{ ft}$$
$$\frac{b}{\sin B} = \frac{c}{\sin C} \Rightarrow \frac{3974}{\sin 42.88°} = \frac{c}{\sin 102.40°} \Rightarrow$$
$$c = \frac{3974\sin 102.40°}{\sin 42.88°} \approx 5704 \text{ ft}$$

**19.** $A = 39°54', a = 268.7$ m, $B = 42°32'$

$$C = 180° - A - B \Rightarrow$$
$$C = 180° - 39°54' - 42°32' = 97°34'$$
$$\frac{a}{\sin A} = \frac{b}{\sin B} \Rightarrow \frac{268.7}{\sin 39°54'} = \frac{b}{\sin 42°32'} \Rightarrow$$
$$b = \frac{268.7\sin 42°32'}{\sin 39°54'} \approx 283.2 \text{ m}$$
$$\frac{a}{\sin A} = \frac{c}{\sin C} \Rightarrow \frac{268.7}{\sin 39°54'} = \frac{c}{\sin 97°54'} \Rightarrow$$
$$c = \frac{268.7\sin 97°54'}{\sin 39°54'} \approx 415.2 \text{ m}$$

**21.** With three sides we have,

$$\frac{a}{\sin A}=\frac{b}{\sin B}=\frac{c}{\sin C}.$$ This does not provide enough information to solve the triangle. Whenever you choose two out of the three ratios to create a proportion, you are missing two pieces of information.

**23.** This is not a valid statement. It is true that if you have the measures of two angles, the third can be found. However, if you do not have at least one side, the triangle cannot be uniquely determined. If we consider the congruence axiom involving two angles, ASA, an included side must be considered.

**25.** $B=112°10';\ C=15°20';\ BC=354$ m

$A=180°-B-C$

$A=180°-112°10'-15°20'$

$=179°60'-127°30'=52°30'$

$$\frac{BC}{\sin A}=\frac{AB}{\sin C}$$

$$\frac{354}{\sin 52°30'}=\frac{AB}{\sin 15°20'}$$

$$AB=\frac{354\sin 15°20'}{\sin 52°30'}\approx 118\text{ m}$$

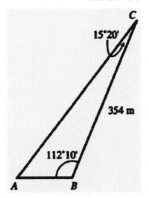

**27.** Let $d$ = the distance the ship traveled between the two observations; $L$ = the location of the lighthouse.

$L=180°-38.8°-44.2°=97.0°$

$$\frac{d}{\sin 97°}=\frac{12.5}{\sin 44.2°}$$

$$d=\frac{12.5\sin 97°}{\sin 44.2°}\approx 17.8\text{ km}$$

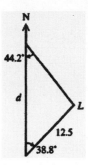

**29.** $$\frac{x}{\sin 54.8°}=\frac{12.0}{\sin 70.4°}$$

$$x=\frac{12.0\sin 54.8°}{\sin 70.4°}\approx 10.4\text{ in.}$$

**31.** We cannot find $\theta$ directly because the length of the side opposite angle $\theta$ is not given. Redraw the triangle shown in the figure and label the third angle as $\alpha$.

$$\frac{\sin\alpha}{1.6+2.7}=\frac{\sin 38°}{1.6+3.6}$$

$$\frac{\sin\alpha}{4.3}=\frac{\sin 38°}{5.2}$$

$$\sin\alpha=\frac{4.3\sin 38°}{5.2}\approx .50910468$$

$$\alpha\approx\sin^{-1}(.50910468)\approx 31°$$

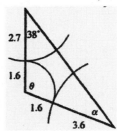

Thus,

$\theta=180°-38°-\alpha\approx 180°-38°-31°=111°$

**33.** Let $x$ = the distance to the lighthouse at bearing N 37° E; $y$ = the distance to the lighthouse at bearing N 25° E.

$\theta=180°-37°=143°$

$\alpha=180°-\theta-25°$

$=180°-143°-25°=12°$

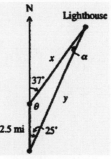

*(continued on next page)*

*(continued from page 175)*

$$\frac{2.5}{\sin \alpha} = \frac{x}{\sin 25°} \Rightarrow x = \frac{2.5 \sin 25°}{\sin 12°} \approx 5.1 \text{ mi}$$

$$\frac{2.5}{\sin \alpha} = \frac{y}{\sin \theta} \Rightarrow \frac{2.5}{\sin 12°} = \frac{y}{\sin 143°} \Rightarrow$$

$$y = \frac{2.5 \sin 143°}{\sin 12°} \approx 7.2 \text{ mi}$$

**35.** Angle $C$ is equal to the difference between the angles of elevation.
$C = B - A = 52.7430° - 52.6997° = .0433°$
The distance $BC$ to the moon can be determined using the law of sines.

$$\frac{BC}{\sin A} = \frac{AB}{\sin C}$$

$$\frac{BC}{\sin 52.6997°} = \frac{398}{\sin .0433°}$$

$$BC = \frac{398 \sin 52.6997°}{\sin .0433°}$$

$$BC \approx 418,930 \text{ km}$$

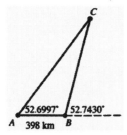

If one finds distance $AC$, then we have

$$\frac{AC}{\sin B} = \frac{AB}{\sin C}$$

$$\frac{AC}{\sin(180° - 52.7430°)} = \frac{398}{\sin .0433°}$$

$$\frac{AC}{\sin 127.2570°} = \frac{398}{\sin .0433°}$$

$$AC = \frac{398 \sin 127.2570°}{\sin .0433°}$$

$$\approx 419,171 \text{ km}$$

In either case the distance is approximately 419,000 km compared to the actual value of 406,000 km.

**37.**

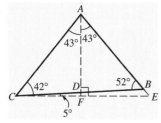

We need to determine the length of $CB$, which is the sum of $CD$ and $DB$. Triangle $ACE$ is isosceles, so

$$m\angle ACE = m\angle AEC = \frac{180° - 86°}{2} = 47°. \text{ It}$$

follows that $m\angle ACD = 47° - 5° = 42°$. $m\angle CAD = m\angle DAB = 43°$ because the photograph was taken with no tilt and $AD$ is the bisector of $\angle CAB$. The length of $AD$ is 3500 ft, and
$m\angle ABC = 180° - (86° + 42°) = 52°$. Using the law of sines, we have

$$CD = \frac{3500 \sin 43°}{\sin 42°} \approx 3567.3 \text{ ft and}$$

$$DB = \frac{3500 \sin 43°}{\sin 52°} \approx 3029.1 \text{ ft}. \text{ Thus,}$$

$$CB = CD + DB \approx 3567.3 + 3029.1$$

$$\approx 6596.4 \approx 6600 \text{ ft}$$

**39.** To find the area of the triangle, use $A = \frac{1}{2}bh$,

with $b = 1$ and $h = \sqrt{3}$. $A = \frac{1}{2}(1)(\sqrt{3}) = \frac{\sqrt{3}}{2}$

Now use $A = \frac{1}{2}ab \sin C$, with $a = \sqrt{3}$, $b = 1$,

and $C = 90°$.

$$A = \frac{1}{2}(\sqrt{3})(1)\sin 90° = \frac{1}{2}(\sqrt{3})(1)(1) = \frac{\sqrt{3}}{2}$$

**41.** To find the area of the triangle, use $A = \frac{1}{2}bh$,

with $b = 1$ and $h = \sqrt{2}$. $A = \frac{1}{2}(1)(\sqrt{2}) = \frac{\sqrt{2}}{2}$

Now use $A = \frac{1}{2}ab \sin C$, with $a = 2$, $b = 1$,

and $C = 45°$.

$$A = \frac{1}{2}(2)(1)\sin 45° = \frac{1}{2}(2)(1)\left(\frac{\sqrt{2}}{2}\right) = \frac{\sqrt{2}}{2}$$

**43.** $A = 42.5°$, $b = 13.6$ m, $c = 10.1$ m
Angle $A$ is included between sides $b$ and $c$. Thus, we have

$$A = \frac{1}{2}bc \sin A = \frac{1}{2}(13.6)(10.1)\sin 42.5°$$

$$\approx 46.4 \text{ m}^2$$

**45.** $B = 124.5°$, $a = 30.4$ cm, $c = 28.4$ cm

Angle $B$ is included between sides $a$ and $c$. Thus, we have

$$A = \frac{1}{2}ac \sin B = \frac{1}{2}(30.4)(28.4)\sin 124.5°$$

$$\approx 356 \text{ cm}^2$$

**47.** $A = 56.80°$, $b = 32.67$ in., $c = 52.89$ in.

Angle $A$ is included between sides $b$ and $c$. Thus, we have

$$A = \frac{1}{2}bc \sin A = \frac{1}{2}(32.67)(52.89)\sin 56.80°$$

$$\approx 722.9 \text{ in.}^2$$

**49.** $A = 30.50°$, $b = 13.00$ cm, $C = 112.60°$

In order to use the area formula, we need to find either $a$ or $c$.

$B = 180° - A - C \Rightarrow$

$B = 180° - 30.50° - 112.60° = 36.90°$

Finding $a$:

$$\frac{a}{\sin A} = \frac{b}{\sin B} \Rightarrow \frac{a}{\sin 30.5°} = \frac{13.00}{\sin 36.90°} \Rightarrow$$

$$a = \frac{13.00 \sin 30.5°}{\sin 36.90°} \approx 10.9890 \text{ cm}$$

$$A = \frac{1}{2}ab \sin C$$

$$= \frac{1}{2}(10.9890)(13.00)\sin 112.60°$$

$$\approx 65.94 \text{ cm}^2$$

Finding $c$:

$$\frac{b}{\sin B} = \frac{c}{\sin C} \Rightarrow \frac{13.00}{\sin 36.9°} = \frac{c}{\sin 112.6°} \Rightarrow$$

$$c = \frac{13.00 \sin 112.6°}{\sin 36.9°} \approx 19.9889 \text{ cm}$$

$$A = \frac{1}{2}bc \sin A = \frac{1}{2}(19.9889)(13.00)\sin 30.5°$$

$$\approx 65.94 \text{ cm}^2$$

**51.** $A = \frac{1}{2}ab \sin C$

$$= \frac{1}{2}(16.1)(15.2)\sin 125° \approx 100 \text{ m}^2$$

**53.** Since $\dfrac{a}{\sin A} = \dfrac{b}{\sin B} = \dfrac{c}{\sin C} = 2r$ and $r = \dfrac{1}{2}$ (since the diameter is 1), we have

$$\frac{a}{\sin A} = \frac{b}{\sin B} = \frac{c}{\sin C} = 2\left(\frac{1}{2}\right) = 1$$

Then, $a = \sin A$, $b = \sin B$, and $c = \sin C$.

**55.** Since triangles $ACD$ and $BCD$ are right triangles, we have $\tan \alpha = \dfrac{x}{d + BC}$ and

$\tan \beta = \dfrac{x}{BC}$. Since $\tan \beta = \dfrac{x}{BC} \Rightarrow$

$BC = \dfrac{x}{\tan \beta}$, we can substitute into

$\tan \alpha = \dfrac{x}{d + BC}$ and solve for $x$.

$$\tan \alpha = \frac{x}{d + BC} \Rightarrow \tan \alpha = \frac{x}{d + \dfrac{x}{\tan \beta}} \Rightarrow$$

$$\frac{\sin \alpha}{\cos \alpha} = \frac{x}{d + \dfrac{x \cos \beta}{\sin \beta}} \cdot \frac{\sin \beta}{\sin \beta}$$

$$\frac{\sin \alpha}{\cos \alpha} = \frac{x \sin \beta}{d \sin \beta + x \cos \beta}$$

$$\sin \alpha \left(d \sin \beta + x \cos \beta\right) = x \sin \beta \left(\cos \alpha\right)$$

$$d \sin \alpha \sin \beta + x \sin \alpha \cos \beta = x \cos \alpha \sin \beta$$

$$d \sin \alpha \sin \beta = x \cos \alpha \sin \beta - x \sin \alpha \cos \beta$$

$$= -x \left(\sin \alpha \cos \beta - \cos \alpha \sin \beta\right)$$

$$= -x \sin \left(a - \beta\right)$$

$$= x \sin \left[-(\alpha - \beta)\right]$$

$$d \sin \alpha \sin \beta = x \sin \left(\beta - \alpha\right)$$

$$\frac{d \sin \alpha \sin \beta}{\sin \left(\beta - \alpha\right)} = x$$

## Section 7.2: The Ambiguous Case of the Law of Sines

**1.** Having three angles will not yield a unique triangle. The correct choice is A.

**3.** The vertical distance from the point $(3, 4)$ to the $x$-axis is 4.

   **(a)** If $h$ is more than 4, two triangles can be drawn. But $h$ must be less than 5 for both triangles to be on the positive $x$-axis. So, $4 < h < 5$.

   **(b)** If $h = 4$, then exactly one triangle is possible. If $h > 5$, then only one triangle is possible on the positive $x$-axis.

   **(c)** If $h < 4$, then no triangle is possible, since the side of length $h$ would not reach the $x$-axis.

**5.** $a = 50$, $b = 26$, $A = 95°$

$$\frac{\sin A}{a} = \frac{\sin B}{b} \Rightarrow \frac{\sin 95°}{50} = \frac{\sin B}{26} \Rightarrow$$

$$\sin B = \frac{50 \sin 95°}{26} \approx .51802124 \Rightarrow B \approx 31.2°$$

Another possible value for $B$ is $180° - 21.2° = 148.8°$, but this is too large to be in a triangle that also has a $95°$ angle. Therefore, only one triangle is possible.

**7.** $a = 31$, $b = 26$, $B = 48°$

$$\frac{\sin A}{a} = \frac{\sin B}{b} \Rightarrow \frac{\sin A}{31} = \frac{\sin 48°}{26} \Rightarrow$$

$$\sin A = \frac{31 \sin 48°}{26} \approx .88605729 \Rightarrow A \approx 62.4°$$

Another possible value for $A$ is $180° - 62.4° = 117.6°$. Therefore, two triangles are possible.

**9.** $a = 50$, $b = 61$, $A = 58°$

$$\frac{\sin A}{a} = \frac{\sin B}{b} \Rightarrow \frac{\sin 58°}{50} = \frac{\sin B}{61} \Rightarrow$$

$$\sin B = \frac{61 \sin 58°}{50} \approx 1.03461868$$

Since $\sin B > 1$ is impossible, no triangle is possible for the given parts.

**11.** $a = \sqrt{6}$, $b = 2$, $A = 60°$

$$\frac{\sin B}{b} = \frac{\sin A}{a} \Rightarrow \frac{\sin B}{2} = \frac{\sin 60°}{\sqrt{6}} \Rightarrow$$

$$\sin B = \frac{2 \sin 60°}{\sqrt{6}} = \frac{2 \cdot \frac{\sqrt{3}}{2}}{\sqrt{6}} = \frac{\sqrt{3}}{\sqrt{6}} = \sqrt{\frac{3}{6}}$$

$$= \sqrt{\frac{1}{2}} = \frac{1}{\sqrt{2}} \cdot \frac{\sqrt{2}}{2}$$

$$\sin B = \frac{\sqrt{2}}{2} \Rightarrow B = 45°$$

There is another angle between $0°$ and $180°$ whose sine is $\frac{\sqrt{2}}{2}$: $180° - 45° = 135°$.

However, this is too large because $A = 60°$ and $60° + 135° = 195° > 180°$, so there is only one solution, $B = 45°$.

**13.** $A = 29.7°$, $b = 41.5$ ft, $a = 27.2$ ft

$$\frac{\sin B}{b} = \frac{\sin A}{a} \Rightarrow \frac{\sin B}{41.5} = \frac{\sin 29.7°}{27.2} \Rightarrow$$

$$\sin B = \frac{41.5 \sin 29.7°}{27.2} \approx .75593878$$

There are two angles $B$ between $0°$ and $180°$ that satisfy the condition. Since $\sin B \approx .75593878$, to the nearest tenth value of $B$ is $B_1 = 49.1°$.

Supplementary angles have the same sine value, so another possible value of $B$ is $B_2 = 180° - 49.1° = 130.9°$. This is a valid angle measure for this triangle since $A + B_2 = 29.7° + 130.9° = 160.6° < 180°$.

Solving separately for angles $C_1$ and $C_2$ we have the following.

$$C_1 = 180° - A - B_1$$
$$= 180° - 29.7° - 49.1° = 101.2°$$
$$C_2 = 180° - A - B_2$$
$$= 180° - 29.7° - 130.9° = 19.4°$$

**15.** $C = 41°20'$, $b = 25.9$ m, $c = 38.4$ m

$$\frac{\sin B}{b} = \frac{\sin C}{c} \Rightarrow \frac{\sin B}{25.9} = \frac{\sin 41°20'}{38.4} \Rightarrow$$

$$\sin B = \frac{25.9 \sin 41°20'}{38.4} \approx .44545209$$

There are two angles $B$ between $0°$ and $180°$ that satisfy the condition. Since $\sin B \approx .44545209$, $B_1 \approx 26.5° = 26°30'$. Supplementary angles have the same sine value, so another possible value of $B$ is $B_2 = 180° - 26°30' = 153°30'$. This is not a valid angle measure for this triangle because $C + B_2 = 41°20' + 153°30' = 194°30' > 180°$.

Thus, $A = 180° - 26°30' - 41°20' = 112°10'$.

**17.** $B = 74.3°$, $a = 859$ m, $b = 783$ m

$$\frac{\sin A}{a} = \frac{\sin B}{b} \Rightarrow \frac{\sin A}{859} = \frac{\sin 74.3°}{783} \Rightarrow$$

$$\sin A = \frac{859 \sin 74.3°}{783} \approx 1.0561331$$

Since $\sin A > 1$ is impossible, no such triangle exists.

**19.** $A = 142.13°$, $b = 5.432$ ft, $a = 7.297$ ft

$$\frac{\sin B}{b} = \frac{\sin A}{a} \Rightarrow \sin B = \frac{b \sin A}{a} \Rightarrow$$

$$\sin B = \frac{5.432 \sin 142.13°}{7.297} \approx .45697580 \Rightarrow$$

$$B \approx 27.19°$$

Because angle $A$ is obtuse, angle $B$ must be acute, so this is the only possible value for $B$ and there is one triangle with the given measurements.

$$C = 180° - A - B$$
$$= 180° - 142.13° - 27.19° = 10.68°$$

Thus, $B \approx 27.19°$ and $C \approx 10.68°$.

**21.** $A = 42.5°$, $a = 15.6$ ft, $b = 8.14$ ft

$$\frac{\sin B}{b} = \frac{\sin A}{a} \Rightarrow \frac{\sin B}{8.14} = \frac{\sin 42.5°}{15.6} \Rightarrow$$

$$\sin B = \frac{8.14 \sin 42.5°}{15.6} \approx .35251951$$

There are two angles $B$ between $0°$ and $180°$ that satisfy the condition. Since $\sin B \approx .35251951$, to the nearest tenth value of $B$ is $B_1 = 20.6°$. Supplementary angles have the same sine value, so another possible value of $B$ is $B_2 = 180° - 20.6° = 159.4°$. This is not a valid angle measure for this triangle since $A + B_2 = 42.5° + 159.4° = 201.9° > 180°$. Thus, $C = 180° - 42.5° - 20.6° = 116.9°$. Solving for $c$, we have the following.

$$\frac{c}{\sin C} = \frac{a}{\sin A} \Rightarrow \frac{c}{\sin 116.9°} = \frac{15.6}{\sin 42.5°} \Rightarrow$$

$$c = \frac{15.6 \sin 116.9°}{\sin 42.5°} \approx 20.6 \text{ ft}$$

**23.** $B = 72.2°$, $b = 78.3$ m, $c = 145$ m

$$\frac{\sin C}{c} = \frac{\sin B}{b} \Rightarrow \frac{\sin C}{145} = \frac{\sin 72.2°}{78.3} \Rightarrow$$

$$\sin C = \frac{145 \sin 72.2°}{78.3} \approx 1.7632026$$

Since $\sin C > 1$ is impossible, no such triangle exists.

**25.** $A = 38°40'$, $a = 9.72$ km, $b = 11.8$ km

$$\frac{\sin B}{b} = \frac{\sin A}{a} \Rightarrow \frac{\sin B}{11.8} = \frac{\sin 38°40'}{9.72} \Rightarrow$$

$$\sin B = \frac{11.8 \sin 38°40'}{9.72} \approx .75848811$$

There are two angles $B$ between $0°$ and $180°$ that satisfy the condition. Since $\sin B \approx .75848811$, to the nearest tenth value of $B$ is $B_1 = 49.3° \approx 49°20'$. Supplementary angles have the same sine value, so another possible value of $B$ is

$B_2 = 180° - 49°20'$
$\quad = 179°60' - 49°20' = 130°40'$

This is a valid angle measure for this triangle since

$A + B_2 = 38°40' + 130°40' = 169°20' < 180°$.

Solving separately for triangles $AB_1C_1$ and $AB_2C_2$ we have the following.

$AB_1C_1:$

$C_1 = 180° - A - B_1 = 180° - 38°40' - 49°20'$
$\quad = 180° - 88°00' = 92°00'$

$$\frac{c_1}{\sin C_1} = \frac{a}{\sin A} \Rightarrow \frac{c_1}{\sin 92°00'} = \frac{9.72}{\sin 38°40'} \Rightarrow$$

$$c_1 = \frac{9.72 \sin 92°00'}{\sin 38°40'} \approx 15.5 \text{ km}$$

$AB_2C_2:$

$C_2 = 180° - A - B_2$
$\quad = 180° - 38°40' - 130°40' = 10°40'$

$$\frac{c_2}{\sin C_2} = \frac{a}{\sin A} \Rightarrow \frac{c_2}{\sin 10°40'} = \frac{9.72}{\sin 38°40'} \Rightarrow$$

$$c_2 = \frac{9.72 \sin 10°40'}{\sin 38°40'} \approx 2.88 \text{ km}$$

**27.** $A = 96.80°$, $b = 3.589$ ft, $a = 5.818$ ft

$$\frac{\sin B}{b} = \frac{\sin A}{a} \Rightarrow \frac{\sin B}{3.589} = \frac{\sin 96.80°}{5.818} \Rightarrow$$

$$\sin B = \frac{3.589 \sin 96.80°}{5.818} \approx .61253922$$

There are two angles $B$ between $0°$ and $180°$ that satisfy the condition. Since $\sin B \approx .61253922$, $B_1 \approx 37.77°$. Supplementary angles have the same sine value, so another possible value of $B$ is $B_2 = 180° - 37.77° = 142.23°$. This is not a valid angle measure for this triangle since $A + B_2 = 96.80° + 142.23° = 239.03° > 180°$. Thus $C = 180° - 96.80° - 37.77° = 45.43°$. Solving for $c$, we have the following.

$$\frac{c}{\sin C} = \frac{a}{\sin A} \Rightarrow \frac{c}{\sin 45.43°} = \frac{5.8518}{\sin 96.80°} \Rightarrow$$

$$b = \frac{5.8518 \sin 45.43°}{\sin 96.80°} \approx 4.174 \text{ ft}$$

**29.** $B = 39.68°$, $a = 29.81$ m, $b = 23.76$ m

$$\frac{\sin A}{a} = \frac{\sin B}{b} \Rightarrow \frac{\sin A}{29.81} = \frac{\sin 39.68°}{23.76} \Rightarrow$$

$$\sin A = \frac{29.81 \sin 39.68°}{23.76} \approx .80108002$$

There are two angles $A$ between $0°$ and $180°$ that satisfy the condition. Since $\sin A \approx .80108002$, to the nearest hundredth value of $A$ is $A_1 = 53.23°$. Supplementary angles have the same sine value, so another possible value of $A$ is $A_2 = 180° - 53.23° = 126.77°$. This is a valid angle measure for this triangle since $B + A_2 = 39.68° + 126.77° = 166.45° < 180°$.

*(continued on next page)*

*(continued from page 179)*

Solving separately for triangles $A_1BC_1$ and $A_2BC_2$ we have the following.

$A_1BC_1$ :

$$C_1 = 180° - A_1 - B$$
$$= 180° - 53.23° - 39.68° = 87.09°$$

$$\frac{c_1}{\sin C_1} = \frac{b}{\sin B} \Rightarrow \frac{c_1}{\sin 87.09°} = \frac{23.76}{\sin 39.68°} \Rightarrow$$

$$c_1 = \frac{23.76 \sin 87.09°}{\sin 39.68°} \approx 37.16 \text{ m}$$

$A_2BC_2$:

$$C_2 = 180° - A_2 - B$$
$$= 180° - 126.77° - 39.68° = 13.55°$$

$$\frac{c_2}{\sin C_2} = \frac{b}{\sin B} \Rightarrow \frac{c_2}{\sin 13.55°} = \frac{23.76}{\sin 39.68°} \Rightarrow$$

$$c_1 = \frac{23.76 \sin 13.55°}{\sin 39.68°} \approx 8.719 \text{ m}$$

**31.** $a = \sqrt{5}, c = 2\sqrt{5}, A = 30°$

$$\frac{\sin C}{c} = \frac{\sin A}{a} \Rightarrow \frac{\sin C}{2\sqrt{5}} = \frac{\sin 30°}{\sqrt{5}} \Rightarrow$$

$$\sin C = \frac{2\sqrt{5} \sin 30°}{\sqrt{5}} = \frac{2\sqrt{5} \cdot \frac{1}{2}}{\sqrt{5}} = \frac{\sqrt{5}}{\sqrt{5}} = 1 \Rightarrow$$

$$C = 90°$$

This is a right triangle.

**33.** Answers will vary.

**35.** $Y = 43°30', Z = 95°30', XY = 960$ m

We are looking for $XZ$.

$$\frac{XZ}{\sin Y} = \frac{XY}{\sin Z} \Rightarrow \frac{XZ}{\sin 43°30'} = \frac{960}{\sin 95°30'} \Rightarrow$$

$$XZ = \frac{960 \sin 43°30'}{\sin 95°30'} \approx 664 \text{ m}$$

**37.** The height of the building is $CD$.

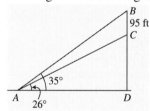

In right triangle $ABD$, we have $m\angle B = 90° - 35° = 55°$. In triangle $ABC$, we have $m\angle CAB = 35° - 26° = 9°$. This gives

$$\frac{BC}{\sin \angle CAB} = \frac{AC}{\sin \angle B} \Rightarrow \frac{95}{\sin 9°} = \frac{AC}{\sin 55°} \Rightarrow$$

$$AC = \frac{95 \sin 55°}{\sin 9°} \approx 497.5$$

In triangle $ACD$,

$$\sin \angle CAD = \frac{CD}{AC} \Rightarrow \sin 26° = \frac{CD}{497.5} \Rightarrow$$
$$CD = 497.5 \sin 26° \approx 218 \text{ ft}$$

**39.** Prove that $\dfrac{a+b}{b} = \dfrac{\sin A + \sin B}{\sin B}$ .

Start with the law of sines.

$$\frac{a}{\sin A} = \frac{b}{\sin B} \Rightarrow a = \frac{b \sin A}{\sin B}$$

Substitute for $a$ in the expression $\dfrac{a+b}{b}$ .

$$\frac{a+b}{b} = \frac{\frac{b \sin A}{\sin B} + b}{b} = \frac{\frac{b \sin A}{\sin B} + b}{b} \cdot \frac{\sin B}{\sin B}$$

$$= \frac{b \sin A + b \sin B}{b \sin B} = \frac{b(\sin A + \sin B)}{\sin B}$$

$$= \frac{\sin A + \sin B}{\sin B}$$

## Section 7.3: The Law of Cosines

### Connections (page 324)

1. Answers will vary.

2. Answers will vary.

### Exercises

**1.** $a$, $b$, and $C$

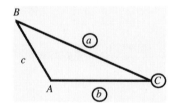

(a) SAS

(b) law of cosines

**3.** $a$, $b$, and $A$

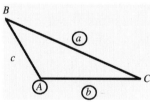

(a) SSA

(b) law of sines

**5.** $A$, $B$, and $c$

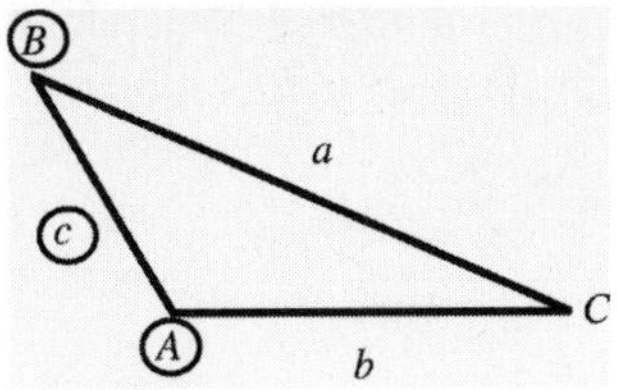


**(a)** ASA

**(b)** law of sines

**7.** $a$, $B$, and $C$

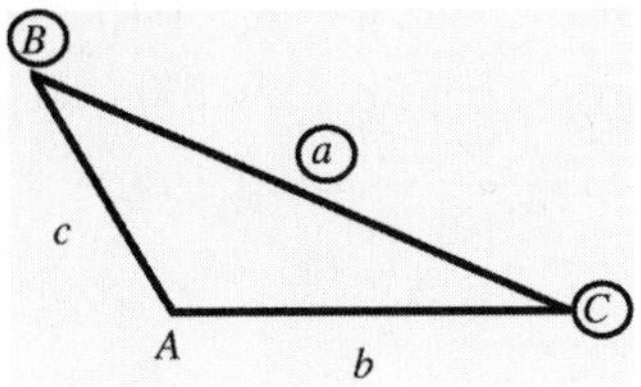


**(a)** ASA

**(b)** law of sines

**9.** $a^2 = 1^2 + \left(4\sqrt{2}\right)^2 - 2(1)\left(4\sqrt{2}\right)\cos 45°$

$= 1 + 32 - 8\sqrt{2}\left(\dfrac{\sqrt{2}}{2}\right) = 33 - 8 = 25$

$a = \sqrt{25} = 5$

**11.** $\cos\theta = \dfrac{b^2 + c^2 - a^2}{2bc} = \dfrac{3^2 + 5^2 - 7^2}{2(3)(5)}$

$= \dfrac{9 + 25 - 49}{30} = -\dfrac{1}{2} \Rightarrow \theta = 120°$

**13.** $A = 121°$, $b = 5$, $c = 3$

Start by finding $a$ with the law of cosines.

$a^2 = b^2 + c^2 - 2bc\cos A \Rightarrow$

$a^2 = 5^2 + 3^2 - 2(5)(3)\cos 121° \approx 49.5 \Rightarrow$

$a \approx 7.04 \approx 7.0$

Of the remaining angles $B$ and $C$, $C$ must be smaller since it is opposite the shorter of the two sides $b$ and $c$. Therefore, $C$ cannot be obtuse.

$\dfrac{\sin A}{a} = \dfrac{\sin C}{C} \Rightarrow \dfrac{\sin 121°}{7.04} = \dfrac{\sin C}{3} \Rightarrow$

$\sin C = \dfrac{3\sin 121°}{7.04} \approx .36527016 \Rightarrow C \approx 21.4°$

Thus, $B = 180° - 121° - 21.4° = 37.6°$.

**15.** $a = 12$, $b = 10$, $c = 10$

We can use the law of cosines to solve for any angle of the triangle. Since $b$ and $c$ have the same measure, so do $B$ and $C$ since this would be an isosceles triangle.

If we solve for $B$, we obtain

$b^2 = a^2 + c^2 - 2ac\cos B$

$\cos B = \dfrac{a^2 + c^2 - b^2}{2ac}$

$\cos B = \dfrac{12^2 + 10^2 - 10^2}{2(12)(10)} = \dfrac{144}{240} = \dfrac{3}{5} \Rightarrow$

$B \approx 53.1°$

Therefore, $C = B \approx 53.1°$ and $A = 180° - 53.1° - 53.1° = 73.8°$. If we solve for $A$ directly, however, we obtain

$a^2 = b^2 + c^2 - 2bc\cos A$

$\cos A = \dfrac{b^2 + c^2 - a^2}{2bc}$

$\cos A = \dfrac{10^2 + 10^2 - 12^2}{2(10)(10)} = \dfrac{56}{200} = \dfrac{7}{5}$

$A \approx 73.7°$

The angles may not sum to 180° due to rounding.

**17.** $B = 55°$, $a = 90$, $c = 100$

Start by finding $b$ with the law of cosines.

$b^2 = a^2 + c^2 - 2ac\cos B$

$b^2 = 90^2 + 100^2 - 2(90)(100)\cos 55° \approx 7775.6$

$b \approx 88.18$

(will be rounded as 88.2) Of the remaining angles $A$ and $C$, $A$ must be smaller since it is opposite the shorter of the two sides $a$ and $c$. Therefore, $A$ cannot be obtuse.

$\dfrac{\sin A}{a} = \dfrac{\sin B}{b} \Rightarrow \dfrac{\sin A}{90} = \dfrac{\sin 55°}{88.18} \Rightarrow$

$\sin A = \dfrac{90\sin 55°}{88.18} \approx .83605902 \Rightarrow A \approx 56.7°$

Thus, $C = 180° - 55° - 56.7° = 68.3°$.

**19.** $A = 41.4°$, $b = 2.78$ yd, $c = 3.92$ yd

First find $a$.

$a^2 = b^2 + c^2 - 2bc\cos A$

$a^2 = 2.78^2 + 3.92^2 - 2(2.78)(3.92)\cos 41.4°$

$\approx 6.7460 \Rightarrow a \approx 2.60$ yd

Find $B$ next, since angle $B$ is smaller than angle $C$ (because $b < c$), and thus angle $B$ must be acute.

$\dfrac{\sin B}{b} = \dfrac{\sin A}{a} \Rightarrow \dfrac{\sin B}{2.78} = \dfrac{\sin 41.4°}{2.597}$

$\sin B = \dfrac{2.78\sin 41.4°}{2.597} \approx .707091182 \Rightarrow$

$B \approx 45.1°$

Finally, $C = 180° - 41.4° - 45.1° = 93.5°$.

**21.** $C = 45.6°$, $b = 8.94$ m, $a = 7.23$ m

First find $c$.

$c^2 = a^2 + b^2 - 2ab \cos C \Rightarrow$

$c^2 = 7.23^2 + 8.94^2 - 2(7.23)(8.94) \cos 45.6°$

$\approx 41.7493 \Rightarrow c \approx 6.46$ m

Find $A$ next, since angle $A$ is smaller than angle $B$ (because $a < b$), and thus angle $A$ must be acute.

$\dfrac{\sin A}{a} = \dfrac{\sin C}{c} \Rightarrow \dfrac{\sin A}{7.23} = \dfrac{\sin 45.6°}{6.461} \Rightarrow$

$\sin A = \dfrac{7.23 \sin 45.6°}{6.461} \approx .79951052 \Rightarrow$

$A \approx 53.1°$

Finally, $B = 180° - 53.1° - 45.6° = 81.3°$.

**23.** $a = 9.3$ cm, $b = 5.7$ cm, $c = 8.2$ cm

We can use the law of cosines to solve for any of angle of the triangle. We solve for $A$, the largest angle. We will know that $A$ is obtuse if $\cos A < 0$.

$a^2 = b^2 + c^2 - 2bc \cos A \Rightarrow$

$\cos A = \dfrac{5.7^2 + 8.2^2 - 9.3^2}{2(5.7)(8.2)} \approx .14163457 \Rightarrow$

$A \approx 82°$

Find $B$ next, since angle $B$ is smaller than angle $C$ (because $b < c$), and thus angle $B$ must be acute.

$\dfrac{\sin B}{b} = \dfrac{\sin A}{a} \Rightarrow \dfrac{\sin B}{5.7} = \dfrac{\sin 82°}{9.3} \Rightarrow$

$\sin B = \dfrac{5.7 \sin 82°}{9.3} \approx .60693849 \Rightarrow B \approx 37°$

Thus, $C = 180° - 82° - 37° = 61°$.

**25.** $a = 42.9$ m, $b = 37.6$ m, $c = 62.7$ m

Angle $C$ is the largest, so find it first.

$c^2 = a^2 + b^2 - 2ab \cos C \Rightarrow$

$\cos C = \dfrac{42.9^2 + 37.6^2 - 62.7^2}{2(42.9)(37.6)} \approx -.20988940 \Rightarrow$

$C \approx 102.1° \approx 102°10'$

Find $B$ next, since angle $B$ is smaller than angle $A$ (because $b < a$), and thus angle $B$ must be acute.

$\dfrac{\sin B}{b} = \dfrac{\sin C}{c} \Rightarrow \dfrac{\sin B}{37.6} = \dfrac{\sin 102.1°}{62.7} \Rightarrow$

$\sin B = \dfrac{37.6 \sin 102.1°}{62.7} \approx .58635805 \Rightarrow$

$B \approx 35.9° \approx 35°50'$

Thus,

$A = 180° - 35°50' - 102°10'$

$= 180° - 138° = 42°00'$

**27.** $AB = 1240$ ft, $AC = 876$ ft, $BC = 965$ ft

Let $AB = c$, $AC = b$, $BC = a$

Angle $C$ is the largest, so find it first.

$c^2 = a^2 + b^2 - 2ab \cos C \Rightarrow$

$\cos C = \dfrac{965^2 + 876^2 - 1240^2}{2(965)(876)} \approx .09522855 \Rightarrow$

$C \approx 84.5°$ or $84°30'$

Find $B$ next, since angle $B$ is smaller than angle $A$ (because $b < a$), and thus angle $B$ must be acute.

$\dfrac{\sin B}{b} = \dfrac{\sin C}{c} \Rightarrow \dfrac{\sin B}{876} = \dfrac{\sin 84°30'}{1240} \Rightarrow$

$\sin B = \dfrac{876 \sin 84°30'}{1240} \approx .70319925 \Rightarrow$

$B \approx 44.7°$ or $44°40'$

Thus,

$A = 180° - 44°40' - 84°30'$

$= 179°60' - 129°10' = 50°50'$

**29.** $A = 80°40'$ $b = 143$ cm, $c = 89.6$ cm

First find $a$.

$a^2 = b^2 + c^2 - 2bc \cos A \Rightarrow$

$a^2 = 143^2 + 89.6^2 - 2(143)(89.6) \cos 80°40'$

$\approx 24,321.25 \Rightarrow a \approx 156$ cm

Find $C$ next, since angle $C$ is smaller than angle $B$ (because $c < b$), and thus angle $C$ must be acute.

$\dfrac{\sin C}{c} = \dfrac{\sin A}{a} \Rightarrow \dfrac{\sin C}{89.6} = \dfrac{\sin 80°40'}{156.0} \Rightarrow$

$\sin C = \dfrac{89.6 \sin 80°40'}{156.0} \approx .56675534 \Rightarrow$

$C \approx 34.5° = 34°30'$

Finally, $B = 180° - 80°40' - 34°30'$

$= 179°60' - 115°10' = 64°50'$

**31.** $B = 74.80°$, $a = 8.919$ in., $c = 6.427$ in.

First find $b$.

$b^2 = a^2 + c^2 - 2ac \cos B \Rightarrow$

$b^2 = 8.919^2 + 6.427^2$

$\qquad - 2(8.919)(6.427) \cos 74.80°$

$\approx 90.7963 \Rightarrow b \approx 9.529$ in.

Find $C$ next, since angle $C$ is smaller than angle $A$ (because $c < a$), and thus angle $C$ must be acute.

$\dfrac{\sin C}{c} = \dfrac{\sin B}{b} \Rightarrow \dfrac{\sin C}{6.427} = \dfrac{\sin 74.80°}{9.5287} \Rightarrow$

$\sin C = \dfrac{6.427 \sin 74.80°}{9.5287} \approx .65089267 \Rightarrow$

$C \approx 40.61°$

Thus, $A = 180° - 74.80° - 40.61° = 64.59°$.

**33.** $A = 112.8°$, $b = 6.28$ m, $c = 12.2$ m
First find $a$.
$$a^2 = b^2 + c^2 - 2bc \cos A \Rightarrow$$
$$a^2 = 6.28^2 + 12.2^2 - 2(6.28)(12.2)\cos 112.8°$$
$$\approx 247.658 \Rightarrow a \approx 15.7 \text{ m}$$
Find $B$ next, since angle $B$ is smaller than angle $C$ (because $b < c$ ), and thus angle $B$ must be acute.
$$\frac{\sin B}{b} = \frac{\sin A}{a} \Rightarrow \frac{\sin B}{6.28} = \frac{\sin 112.8°}{15.74} \Rightarrow$$
$$\sin B = \frac{6.28 \sin 112.8°}{15.74} \approx .36780817 \Rightarrow$$
$$B \approx 21.6°$$
Finally, $C = 180° - 112.8° - 21.6° = 45.6°$.

**35.** $a = 3.0$ ft, $b = 5.0$ ft, $c = 6.0$ ft
Angle $C$ is the largest, so find it first.
$$c^2 = a^2 + b^2 - 2ab \cos C \Rightarrow$$
$$\cos C = \frac{3.0^2 + 5.0^2 - 6.0^2}{2(3.0)(5.0)} = -\frac{2}{30} = -\frac{1}{15}$$
$$\approx -.06666667 \Rightarrow C \approx 94°$$
Find $A$ next, since angle $A$ is smaller than angle $B$ (because $a < b$ ), and thus angle $A$ must be acute.
$$\frac{\sin A}{a} = \frac{\sin C}{c} \Rightarrow \frac{\sin A}{3} = \frac{\sin 94°}{6} \Rightarrow$$
$$\sin A = \frac{3 \sin 94°}{6} \approx .49878203 \Rightarrow A \approx 30°$$
Thus, $B = 180° - 30° - 94° = 56°$.

**37.** There are three ways to apply the law of cosines when $a = 3, b = 4$, and $c = 10$.
Solving for $A$:
$$a^2 = b^2 + c^2 - 2bc \cos A \Rightarrow$$
$$\cos A = \frac{4^2 + 10^2 - 3^2}{2(4)(10)} = \frac{107}{80} = 1.3375$$
Solving for $B$:
$$b^2 = a^2 + c^2 - 2ac \cos B \Rightarrow$$
$$\cos B = \frac{3^2 + 10^2 - 4^2}{2(3)(10)} = \frac{93}{60} = \frac{31}{20} = 1.55$$
Solving for $C$:
$$c^2 = a^2 + b^2 - 2ab \cos C \Rightarrow$$
$$\cos C = \frac{3^2 + 4^2 - 10^2}{2(3)(4)} = \frac{-75}{24} = -\frac{25}{8} = -3.125$$
Since the cosine of any angle of a triangle must be between –1 and 1, a triangle cannot have sides 3, 4, and 10.

**39.** Find $AB$, or $c$, in the following triangle.

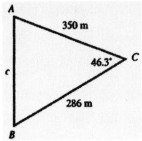

$$c^2 = a^2 + b^2 - 2ab \cos C$$
$$c^2 = 286^2 + 350^2 - 2(286)(350)\cos 46.3°$$
$$c^2 = 65,981.3 \Rightarrow c \approx 257$$
The length of $AB$ is 257 m.

**41.** Find $AC$, or $b$, in the following triangle.

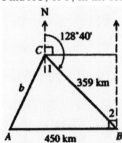

$$m\angle 1 = 180° - 128°40' = 51°20'$$
Angles 1 and 2 are alternate interior angles formed when parallel lines (the north lines) are cut by a transversal, line $BC$, so
$$m\angle 2 = m\angle 1 = 51°20'.$$
$$m\angle ABC = 90° - m\angle 2 = 90° - 51°20' = 38°40'$$
$$b^2 = a^2 + c^2 - 2ac \cos B \Rightarrow$$
$$b^2 = 359^2 + 450^2 - 2(359)(450)\cos 38°40'$$
$$\approx 79,106 \Rightarrow b \approx 281 \text{ km}$$
$C$ is about 281 km from $A$.

**43.** Sketch a triangle showing the situation as follows.

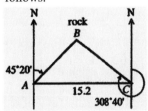

$$m\angle A = 90° - 45°20' = 44°40'$$
$$m\angle C = 308°40' - 270° = 38°40'$$
$$m\angle B = 180° - A - C$$
$$= 180° - 44°40' - 38°40' = 96°40'$$

(*continued on next page*)

*(continued from page 183)*

Since we have only one side of a triangle, use the law of sines to find $BC = a$.

$$\frac{a}{\sin A} = \frac{b}{\sin B} \Rightarrow \frac{a}{\sin 44° \ 40'} = \frac{15.2}{\sin 96° \ 40'} \Rightarrow$$

$$a = \frac{15.2 \sin 44° \ 40'}{\sin 96° \ 40'} \approx 10.8$$

The distance between the ship and the rock is about 10.8 miles.

**45.** Use the law of cosines to find the angle, $\theta$.

$$\cos \theta = \frac{20^2 + 16^2 - 13^2}{2(20)(16)} = \frac{487}{640} \approx .76093750 \Rightarrow$$

$$\theta \approx 40°$$

**47.** Let $A$ = the angle between the beam and the 45-ft cable.

$$\cos A = \frac{45^2 + 90^2 - 60^2}{2(45)(90)} = \frac{6525}{8100} = \frac{29}{36}$$

$$\approx .80555556 \Rightarrow A \approx 36°$$

Let $B$ = the angle between the beam and the 60-ft cable.

$$\cos B = \frac{90^2 + 60^2 - 45^2}{2(90)(60)} = \frac{9675}{10,800} = \frac{43}{48}$$

$$\approx .89583333 \Rightarrow B \approx 26°$$

**49.** Let $A$ = home plate; $B$ = first base; $C$ = second base; $D$ = third base; $P$ = pitcher's rubber. Draw $AC$ through $P$, draw $PB$ and $PD$.

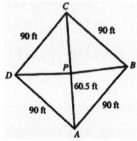

In triangle $ABC$, angle $m\angle B = 90°$, and $m\angle A = m\angle C = 45°$.

$$AC = \sqrt{90^2 + 90^2} = \sqrt{2 \cdot 90^2} = 90\sqrt{2} \text{ and}$$

$$PC = 90\sqrt{2} - 60.5 \approx 66.8 \text{ ft}$$

In triangle $APB$, $m\angle A = 45°$.

$$PB^2 = AP^2 + AB^2 - 2(AP)(AB)\cos A$$

$$PB^2 = 60.5^2 + 90^2 - 2(60.5)(90)\cos 45°$$

$$PB^2 \approx 4059.86 \Rightarrow PB \approx 63.7 \text{ ft}$$

Since triangles $APB$ and $APD$ are congruent, $PB = PD = 63.7$ ft.

The distance to second base is 66.8 ft and the distance to both first and third base is 63.7 ft.

**51.** Find the distance of the ship from point $A$.

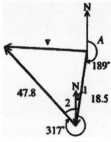

$$m\angle 1 = 189° - 180° = 9°$$

$$m\angle 2 = 360° - 317° = 43°$$

$$m\angle 1 + m\angle 2 = 9° + 43° = 52°$$

Use the law of cosines to find $v$.

$$v^2 = 47.8^2 + 18.5^2 - 2(47.8)(18.5)\cos 52°$$

$$\approx 1538.23 \Rightarrow v \approx 39.2 \text{ km}$$

**53.** $\cos A = \dfrac{17^2 + 21^2 - 9^2}{2(17)(21)} = \dfrac{649}{714} \approx .90896359 \Rightarrow$

$A \approx 25°$

Thus, the bearing of $B$ from is $325° + 25° = 350°$.

**55.** Let $c$ = the length of the property line that cannot be directly measured.

Using the law of cosines, we have

$$c^2 = 14.0^2 + 13.0^2 - 2(14.0)(13.0)\cos 70°$$

$$\approx 240.5 \Rightarrow c \approx 15.5 \text{ ft}$$

(rounded to three significant digits)

The length of the property line is approximately $18.0 + 15.5 + 14.0 = 47.5$ feet

**57.** Using the law of cosines we can solve for the measure of angle $A$.

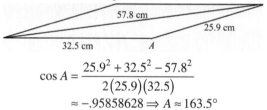

$$\cos A = \frac{25.9^2 + 32.5^2 - 57.8^2}{2(25.9)(32.5)}$$

$$\approx -.95858628 \Rightarrow A \approx 163.5°$$

**59.** Find $x$ using the law of cosines.

$$x^2 = 25^2 + 25^2 - 2(25)(25)\cos 52° \approx 480 \Rightarrow$$

$$x \approx 22 \text{ ft}$$

**61.** Let $a$ be the length of the segment from $(0, 0)$ to $(6, 8)$. Use the distance formula.

$$a = \sqrt{(6-0)^2 + (8-0)^2} = \sqrt{6^2 + 8^2}$$
$$= \sqrt{36 + 64} = \sqrt{100} = 10$$

Let $b$ be the length of the segment from $(0, 0)$ to $(4, 3)$.

$$b = \sqrt{(4-0)^2 + (3-0)^2} = \sqrt{4^2 + 3^2}$$
$$= \sqrt{16 + 9} = \sqrt{25} = 5$$

Let $c$ be the length of the segment from $(4, 3)$ to $(6, 8)$.

$$c = \sqrt{(6-4)^2 + (8-3)^2} = \sqrt{2^2 + 5^2}$$
$$= \sqrt{4 + 25} = \sqrt{29}$$

$$\cos\theta = \frac{a^2 + b^2 - c^2}{2ab} \Rightarrow$$

$$\cos\theta = \frac{10^2 + 5^2 - \left(\sqrt{29}\right)^2}{2(10)(5)}$$

$$= \frac{100 + 25 - 29}{100} = .96 \Rightarrow \theta \approx 16.26°$$

**63.** Using $A = \dfrac{1}{2}bh \Rightarrow$

$$A = \frac{1}{2}(16)\left(3\sqrt{3}\right) = 24\sqrt{3} \approx 41.57.$$

To use Heron's Formula, first find the semiperimeter,

$$s = \frac{1}{2}(a+b+c) = \frac{1}{2}(6+14+16) = \frac{1}{2} \cdot 36 = 18.$$

Now find the area of the triangle.

$$A = \sqrt{s(s-a)(s-b)(s-c)}$$
$$= \sqrt{18(18-6)(18-14)(18-16)}$$
$$= \sqrt{18(12)(4)(2)} = \sqrt{1728}$$
$$= 24\sqrt{3} \approx 41.57$$

Both formulas give the same area.

**65.** $a = 12$ m, $b = 16$ m, $c = 25$ m

$$s = \frac{1}{2}(a+b+c) = \frac{1}{2}(12+16+25)$$
$$= \frac{1}{2} \cdot 53 = 26.5$$

$$A = \sqrt{s(s-a)(s-b)(s-c)}$$
$$= \sqrt{26.5(26.5-12)(26.5-16)(26.5-25)}$$
$$= \sqrt{26.5(14.5)(10.5)(1.5)} \approx 78 \text{ m}^2$$

(rounded to two significant digits)

**67.** $a = 154$ cm, $b = 179$ cm, $c = 183$ cm

$$s = \frac{1}{2}(a+b+c) = \frac{1}{2}(154+179+183)$$
$$= \frac{1}{2} \cdot 516 = 258$$

$$A = \sqrt{s(s-a)(s-b)(s-c)}$$
$$= \sqrt{258(258-154)(258-179)(258-183)}$$
$$= \sqrt{258(104)(79)(75)} \approx 12,600 \text{ cm}^2$$

(rounded to three significant digits)

**69.** $a = 76.3$ ft, $b = 109$ ft, $c = 98.8$ ft

$$s = \frac{1}{2}(a+b+c) = \frac{1}{2}(76.3+109+98.8)$$
$$= \frac{1}{2} \cdot 284.1 = 142.05$$

$$A = \sqrt{s(s-a)(s-b)(s-c)}$$
$$= \sqrt{\begin{array}{c}142.05(142.05-76.3)(142.05-109) \cdot \\ (142.05-98.8)\end{array}}$$
$$= \sqrt{142.05(65.75)(33.05)(43.25)} \approx 3650 \text{ ft}^2$$

(rounded to three significant digits)

**71.** $AB = 22.47928$ mi, $AC = 28.14276$ mi, $A = 58.56989°$

This is SAS, so use the law of cosines.

$$BC^2 = AC^2 + AB^2 - 2(AC)(AB)\cos A$$
$$BC^2 = 28.14276^2 + 22.47928^2$$
$$\quad - 2(28.14276)(22.47928)\cos 58.56989°$$
$$BC^2 \approx 637.55393$$
$$BC \approx 25.24983$$

$BC$ is approximately 25.24983 mi.
(rounded to seven significant digits)

**73.** Perimeter: $9 + 10 + 17 = 36$ feet, so the semiperimeter is $\dfrac{1}{2} \cdot 36 = 18$ feet.

Use Heron's Formula to find the area.

$$A = \sqrt{s(s-a)(s-b)(s-c)}$$
$$= \sqrt{18(18-9)(18-10)(18-17)}$$
$$= \sqrt{18(9)(8)(1)} = \sqrt{1296} = 36 \text{ ft}$$

Since the perimeter and area both equal 36 feet, the triangle is a *perfect triangle*.

**75.** Find the area of the Bermuda Triangle using Heron's Formula.

$$s = \frac{1}{2}(a+b+c) = \frac{1}{2}(850+925+1300)$$

$$= \frac{1}{2} \cdot 3075 = 1537.5$$

$$A = \sqrt{s(s-a)(s-b)(s-c)}$$

$$= \sqrt{\begin{array}{c}1537.5(1537.5-850) \cdot \\ (1537.5-925)(1537.5-1300)\end{array}}$$

$$= \sqrt{1537.5(687.5)(612.5)(237.5)}$$

$$\approx 392,128.82$$

The area of the Bermuda Triangle is about $390,000 \ \text{mi}^2$.

**77.** **(a)** Using the law of sines, we have

$$\frac{\sin C}{c} = \frac{\sin A}{a} \Rightarrow \frac{\sin C}{15} = \frac{\sin 60°}{13} \Rightarrow$$

$$\sin C = \frac{15 \sin 60°}{13} = \frac{15}{13} \cdot \frac{\sqrt{3}}{2} \approx .99926008$$

There are two angles $C$ between $0°$ and $180°$ that satisfy the condition. Since $\sin C \approx .99926008$, to the nearest tenth value of $C$ is $C_1 = 87.8°$.

Supplementary angles have the same sine value, so another possible value of $C$ is $B_2 = 180° - 87.8° = 92.2°$.

**(b)** By the law of cosines, we have

$$\cos C = \frac{a^2 + b^2 - c^2}{2ab}$$

$$\cos C = \frac{13^2 + 7^2 - 15^2}{2(13)(7)} = \frac{-7}{182}$$

$$= -\frac{1}{26} \approx -.03846154 \Rightarrow C \approx 92.2°$$

**(c)** With the law of cosines, we are required to find the inverse cosine of a negative number; therefore; we know angle $C$ is greater than $90°$.

**79.** Given point $D$ is on side $\overline{AB}$ of triangle $ABC$ such that $\overline{CD}$ bisects $\angle C$,

$m\angle ACD = m\angle DCB$. Show that $\dfrac{AD}{DB} = \dfrac{b}{a}$.

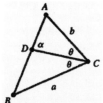

Let $\theta = m\angle ACD$, and $\alpha = m\angle ADC$
Then $\theta = m\angle DCB$ and $= m\angle BDC = 180 - \alpha$.

By the law of sines, we have

$$\frac{\sin \theta}{AD} = \frac{\sin \alpha}{b} \Rightarrow \sin \theta = \frac{AD \sin \alpha}{b} \text{ and}$$

$$\frac{\sin \theta}{DB} = \frac{\sin(180° - \alpha)}{a} \Rightarrow$$

$$\sin \theta = \frac{DB \sin(180° - \alpha)}{a}$$

By substitution, we have

$$\frac{AD \sin \alpha}{b} = \frac{DB \sin(180° - \alpha)}{a}$$

Since $\sin \alpha = \sin(180° - \alpha)$, $\dfrac{AD}{b} = \dfrac{DB}{a}$.

Multiplying both sides by $\dfrac{b}{DB}$, we have

$$\frac{AD}{b} \cdot \frac{b}{DB} = \frac{DB}{a} \cdot \frac{b}{DB} \Rightarrow \frac{AD}{DB} = \frac{b}{a}.$$

**81.**

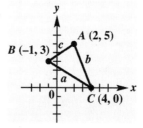

**83.** Use the law of cosines to find the measure of $\angle A$.

$$a^2 = b^2 + c^2 - 2bc \cos A$$

$$\left(\sqrt{34}\right)^2 = \left(\sqrt{29}\right)^2 + \left(\sqrt{13}\right)^2$$

$$- 2\left(\sqrt{29}\right)\left(\sqrt{13}\right)\cos A$$

$$34 = 42 - 2\sqrt{377} \cos A$$

$$-8 = -2\sqrt{377} \cos A$$

$$\frac{4}{\sqrt{377}} = \cos A$$

$$A = \frac{1}{2}ab \sin C = \frac{1}{2}\left(\sqrt{29}\right)\left(\sqrt{13}\right) \sin C$$

$$= \frac{\sqrt{377}}{2} \sin\left(\cos^{-1} \frac{4}{\sqrt{377}}\right) = 9.5 \text{ sq units}$$

**85.** $(x_1, y_1) = (2,5); \ (x_2, y_2) = (-1,3); \ (x_3, y_3) = (4,0)$

$$A = \frac{1}{2}\left|x_1 y_2 - y_1 x_2 + x_2 y_3 - y_2 x_3 + x_3 y_1 - y_3 x_1\right|$$

$$= \frac{1}{2}\left|2(3) - 5(-1) + (-1)(0) - 3(4) + 4(5) - 0(2)\right|$$

$$= 9.5 \text{ sq units}$$

# Chapter 7: Quiz
**(Sections 7.1–7.3)**

1. Using the law of sines, we have

$$\frac{\sin B}{b} = \frac{\sin C}{c} \Rightarrow \frac{\sin 30.6°}{7.42} = \frac{\sin C}{4.54} \Rightarrow$$

$$\sin C = \frac{4.54 \sin 30.6°}{7.42} \approx .311462 \Rightarrow$$

$$C \approx 18.1°$$

$$A = 180° - B - C$$

$$= 180° - 30.6° - 18.1° = 131.3°$$

3. Using the law of cosines, we have

$$c^2 = a^2 + b^2 - 2ab \cos C$$

$$21.2^2 = 28.4^2 + 16.9^2 - 2 \cdot 28.4 \cdot 16.9 \cos C$$

$$-642.73 = -959.92 \cos C$$

$$\frac{642.73}{959.92} = \cos C \Rightarrow$$

$$C \approx 48.0° \text{ (rounded to three}$$
$$\text{significant digits)}$$

5. First find the semiperimeter:

$$s = \frac{1}{2}(19.5 + 21.0 + 22.5) = 31.5$$

Using Heron's formula, we have

$$A = \sqrt{s(s-a)(s-b)(s-c)}$$

$$= \sqrt{31.5(31.5 - 19.5)(31.5 - 21.0)(31.5 - 22.5)}$$

$$= \sqrt{31.5(12)(10.5)(9)}$$

$$= \sqrt{35,721} = 189 \text{ km}^2$$

7. $\angle C = 180° - 111° - 41° = 28°$

Using the law of sines, we have

$$\frac{a}{\sin A} = \frac{c}{\sin C} \Rightarrow \frac{a}{\sin 111°} = \frac{326}{\sin 28°} \Rightarrow$$

$$a = \frac{326 \sin 111°}{\sin 28°} \approx 648$$

$$\frac{b}{\sin B} = \frac{c}{\sin C} \Rightarrow \frac{b}{\sin 41°} = \frac{326}{\sin 28°} \Rightarrow$$

$$b = \frac{326 \sin 41°}{\sin 28°} \approx 456$$

Note that both $a$ and $b$ have been rounded to three significant digits.

# Section 7.4: Vectors, Operations, and the Dot Product

1. Equal vectors have the same magnitude and direction. Equal vectors are **m** and **p**; **n** and **r**.

3. One vector is a positive scalar multiple of another if the two vectors point in the same direction; they may have different magnitudes.
   $\mathbf{m} = 1\mathbf{p}; \ \mathbf{m} = 2\mathbf{t}; \ \mathbf{n} = 1\mathbf{r}; \ \mathbf{p} = 2\mathbf{t}$ or

   $$\mathbf{p} = 1\mathbf{m}; \ \mathbf{t} = \frac{1}{2}\mathbf{m}; \ \mathbf{r} = 1\mathbf{n}; \ \mathbf{t} = \frac{1}{2}\mathbf{p}$$

5.

7.

9.

11.

13.

15.

**17.** $\mathbf{a} + (\mathbf{b} + \mathbf{c}) = (\mathbf{a} + \mathbf{b}) + \mathbf{c}$

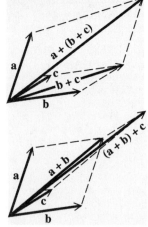

Yes, vector addition is associative.

**19.** Use the figure to find the components of $\mathbf{a}$ and $\mathbf{b}$: $\mathbf{a} = \langle -8, 8 \rangle$ and $\mathbf{b} = \langle 4, 8 \rangle$.

   **(a)** $\mathbf{a} + \mathbf{b} = \langle -8, 8 \rangle + \langle 4, 8 \rangle$
$$= \langle -8 + 4, 8 + 8 \rangle = \langle -4, 16 \rangle$$

   **(b)** $\mathbf{a} - \mathbf{b} = \langle -8, 8 \rangle - \langle 4, 8 \rangle$
$$= \langle -8 - 4, 8 - 8 \rangle = \langle -12, 0 \rangle$$

   **(c)** $-\mathbf{a} = -\langle -8, 8 \rangle = \langle 8, -8 \rangle$

**21.** Use the figure to find the components of $\mathbf{a}$ and $\mathbf{b}$: $\mathbf{a} = \langle 4, 8 \rangle$ and $\mathbf{b} = \langle 4, -8 \rangle$.

   **(a)** $\mathbf{a} + \mathbf{b} = \langle 4, 8 \rangle + \langle 4, -8 \rangle$
$$= \langle 4 + 4, 8 - 8 \rangle = \langle 8, 0 \rangle$$

   **(b)** $\mathbf{a} - \mathbf{b} = \langle 4, 8 \rangle - \langle 4, -8 \rangle$
$$= \langle 4 - 4, 8 - (-8) \rangle = \langle 0, 16 \rangle$$

   **(c)** $-\mathbf{a} = -\langle 4, 8 \rangle = \langle -4, -8 \rangle$

**23.** Use the figure to find the components of $\mathbf{a}$ and $\mathbf{b}$: $\mathbf{a} = \langle -8, 4 \rangle$ and $\mathbf{b} = \langle 8, 8 \rangle$.

   **(a)** $\mathbf{a} + \mathbf{b} = \langle -8, 4 \rangle + \langle 8, 8 \rangle$
$$= \langle -8 + 8, 4 + 8 \rangle = \langle 0, 12 \rangle$$

   **(b)** $\mathbf{a} - \mathbf{b} = \langle -8, 4 \rangle - \langle 8, 8 \rangle$
$$= \langle -8 - 8, 4 - 8 \rangle = \langle -16, -4 \rangle$$

   **(c)** $-\mathbf{a} = -\langle -8, 4 \rangle = \langle 8, -4 \rangle$

**25. (a)** $2\mathbf{a} = 2(2\mathbf{i}) = 4\mathbf{i}$

   **(b)** $2\mathbf{a} + 3\mathbf{b} = 2(2\mathbf{i}) + 3(\mathbf{i} + \mathbf{j})$
$$= 4\mathbf{i} + 3\mathbf{i} + 3\mathbf{j} = 7\mathbf{i} + 3\mathbf{j}$$

   **(c)** $\mathbf{b} - 3\mathbf{a} = \mathbf{i} + \mathbf{j} - 3(2\mathbf{i}) = \mathbf{i} + \mathbf{j} - 6\mathbf{i} = -5\mathbf{i} + \mathbf{j}$

**27. (a)** $2\mathbf{a} = 2\langle -1, 2 \rangle = \langle -2, 4 \rangle$

   **(b)** $2\mathbf{a} + 3\mathbf{b} = 2\langle -1, 2 \rangle + 3\langle 3, 0 \rangle$
$$= \langle -2, 4 \rangle + \langle 9, 0 \rangle$$
$$= \langle -2 + 9, 4 + 0 \rangle = \langle 7, 4 \rangle$$

   **(c)** $\mathbf{b} - 3\mathbf{a} = \langle 3, 0 \rangle - 3\langle -1, 2 \rangle = \langle 3, 0 \rangle - \langle -3, 6 \rangle$
$$= \langle 3 - (-3), 0 - 6 \rangle = \langle 6, -6 \rangle$$

**29.** $|\mathbf{u}| = 12, |\mathbf{w}| = 20, \theta = 27°$

**31.** $|\mathbf{u}| = 20, |\mathbf{w}| = 30, \theta = 30°$

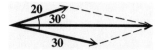

**33.** Magnitude:
$$\sqrt{15^2 + (-8)^2} = \sqrt{225 + 64} = \sqrt{289} = 17$$

Angle: $\tan \theta' = \dfrac{b}{a} \Rightarrow \tan \theta' = \dfrac{-8}{15} \Rightarrow$
$$\theta' = \tan^{-1}\left(-\dfrac{8}{15}\right) \approx -28.1° \Rightarrow$$
$$\theta = -28.1° + 360° = 331.9°$$
($\theta$ lies in quadrant IV)

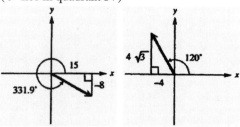

Exercise 33        Exercise 35

**35.** Magnitude:
$$\sqrt{(-4)^2 + (4\sqrt{3})^2} = \sqrt{16 + 48} = \sqrt{64} = 8$$

Angle: $\tan \theta' = \dfrac{b}{a} \Rightarrow \tan \theta' = \dfrac{4\sqrt{3}}{-4} \Rightarrow$
$$\theta' = \tan^{-1}\left(-\sqrt{3}\right) = -60° \Rightarrow$$
$$\theta = -60° + 180° = 120°$$
($\theta$ lies in quadrant II)

In Exercises 37–41, **x** is the horizontal component of **v**, and **y** is the vertical component of **v**. Thus, $|\mathbf{x}|$ is the magnitude of **x** and $|\mathbf{y}|$ is the magnitude of **y**.

**37.** $\alpha = 20°$, $|\mathbf{v}| = 50$

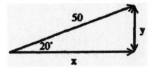

$\mathbf{x} = 50\cos 20° \approx 47 \Rightarrow |\mathbf{x}| \approx 47$
$\mathbf{y} = 50\sin 20° \approx 17 \Rightarrow |\mathbf{y}| \approx 17$

**39.** $\alpha = 35°50'$, $|\mathbf{v}| = 47.8$

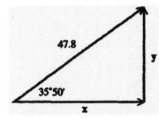

$\mathbf{x} = 47.8\cos 35°50' \approx 38.8 \Rightarrow |\mathbf{x}| \approx 38.8$
$\mathbf{y} = 47.8\sin 35°50' \approx 28.0 \Rightarrow |\mathbf{y}| \approx 28.0$

**41.** $\alpha = 128.5°$, $|\mathbf{v}| = 198$

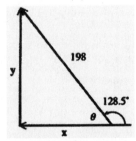

$\mathbf{x} = 198\cos 128.5° \approx -123 \Rightarrow |\mathbf{x}| \approx 123$
$\mathbf{y} = 198\sin 128.5° \approx 155 \Rightarrow |\mathbf{y}| \approx 155$

**43.** $\mathbf{u} = \langle a,b \rangle = \langle 5\cos(30°), 5\sin(30°) \rangle$
$= \left\langle \dfrac{5\sqrt{3}}{2}, \dfrac{5}{2} \right\rangle$

**45.** $\mathbf{v} = \langle a,b \rangle = \langle 4\cos(40°), 4\sin(40°) \rangle$
$\approx \langle 3.0642, 2.5712 \rangle$

**47.** $\mathbf{v} = \langle a,b \rangle = \langle 5\cos(-35°), 5\sin(-35°) \rangle$
$\approx \langle 4.0958, -2.8679 \rangle$

**49.** Forces of 250 newtons and 450 newtons, forming an angle of 85°
$\alpha = 180° - 85° = 95°$
$|\mathbf{v}|^2 = 250^2 + 450^2 - 2(250)(450)\cos 95°$
$|\mathbf{v}|^2 \approx 284,610.04$
$|\mathbf{v}| \approx 533.5$

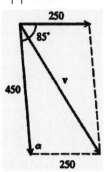

The magnitude of the resulting force is 530 newtons. (rounded to two significant digits)

**51.** Forces of 116 lb and 139 lb, forming an angle of 140° 50′
$\alpha = 180° - 140°50'$
$\quad = 179°60' - 140°50' = 39°10'$
$|\mathbf{v}|^2 = 139^2 + 116^2 - 2(139)(116)\cos 39°10'$
$|\mathbf{v}|^2 \approx 7774.7359$
$|\mathbf{v}| \approx 88.174$

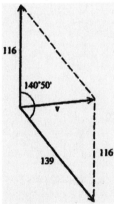

The magnitude of the resulting force is 88.2 lb. (rounded to three significant digits)

**53.** $\alpha = 180° - 40° = 140°$
$|\mathbf{v}|^2 = 40^2 + 60^2 - 2(40)(60)\cos 140°$
$|\mathbf{v}|^2 \approx 8877.0133$
$|\mathbf{v}| \approx 94.2$ lb

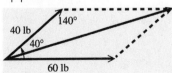

**55.** $\alpha = 180° - 110° = 70°$

$|\mathbf{v}|^2 = 15^2 + 25^2 - 2(15)(25)\cos 70°$

$|\mathbf{v}|^2 \approx 593.48489$

$|\mathbf{v}| \approx 24.4$ lb

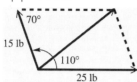

**57.** If $\mathbf{u} = \langle a, b \rangle$ and $\mathbf{v} = \langle c, d \rangle$, then

$\mathbf{u} + \mathbf{v} = \langle a+c, b+d \rangle$.

For Exercises 59–65, $\mathbf{u} = \langle -2, 5 \rangle$ and $\mathbf{v} = \langle 4, 3 \rangle$.

**59.** $\mathbf{u} + \mathbf{v} = \langle -2, 5 \rangle + \langle 4, 3 \rangle$
$= \langle -2+4, 5+3 \rangle = \langle 2, 8 \rangle$

**61.** $-4\mathbf{u} = -4\langle -2, 5 \rangle = \langle -4(-2), -4(5) \rangle = \langle 8, -20 \rangle$

**63.** $3\mathbf{u} - 6\mathbf{v} = 3\langle -2, 5 \rangle - 6\langle 4, 3 \rangle$
$= \langle -6, 15 \rangle - \langle 24, 18 \rangle$
$= \langle -6-24, 15-18 \rangle = \langle -30, -3 \rangle$

**65.** $\mathbf{u} + \mathbf{v} - 3\mathbf{u} = \langle -2, 5 \rangle + \langle 4, 3 \rangle - 3\langle -2, 5 \rangle$
$= \langle -2, 5 \rangle + \langle 4, 3 \rangle - \langle 3(-2), 3(5) \rangle$
$= \langle -2, 5 \rangle + \langle 4, 3 \rangle - \langle -6, 15 \rangle$
$= \langle -2+4, 5+3 \rangle - \langle -6, 15 \rangle$
$= \langle 2, 8 \rangle - \langle -6, 15 \rangle$
$= \langle 2-(-6), 8-15 \rangle = \langle 8, -7 \rangle$

**67.** $\langle -5, 8 \rangle = -5\mathbf{i} + 8\mathbf{j}$

**69.** $\langle 2, 0 \rangle = 2\mathbf{i} + 0\mathbf{j} = 2\mathbf{i}$

**71.** $\langle 6, -1 \rangle \cdot \langle 2, 5 \rangle = 6(2) + (-1)(5) = 12 - 5 = 7$

**73.** $\langle 2, -3 \rangle \cdot \langle 6, 5 \rangle = 2(6) + (-3)(5) = 12 - 15 = -3$

**75.** $4\mathbf{i} = \langle 4, 0 \rangle; 5\mathbf{i} - 9\mathbf{j} = \langle 5, -9 \rangle$

$\langle 4, 0 \rangle \cdot \langle 5, -9 \rangle = 4(5) + 0(-9) = 20 - 0 = 20$

**77.** $\langle 2, 1 \rangle \cdot \langle -3, 1 \rangle$

$\cos\theta = \dfrac{\langle 2, 1 \rangle \cdot \langle -3, 1 \rangle}{\sqrt{2^2 + 1^2} \cdot \sqrt{(-3)^2 + 1^2}} = \dfrac{-6+1}{\sqrt{5} \cdot \sqrt{10}}$

$= \dfrac{-5}{5\sqrt{2}} = \dfrac{-1}{\sqrt{2}} = -\dfrac{\sqrt{2}}{2} \Rightarrow \theta = 135°$

**79.** $\langle 1, 2 \rangle \cdot \langle -6, 3 \rangle$

$\cos\theta = \dfrac{\langle 1, 2 \rangle \cdot \langle -6, 3 \rangle}{\sqrt{1^2 + 2^2} \cdot \sqrt{(-6)^2 + 3^2}} = \dfrac{-6+6}{\sqrt{5}\sqrt{45}}$

$= \dfrac{0}{15} = 0 \Rightarrow \theta = 90°$

**81.** First write the given vectors in component form: $3\mathbf{i} + 4\mathbf{j} = \langle 3, 4 \rangle$ and $\mathbf{j} = \langle 0, 1 \rangle$

$\cos\theta = \dfrac{\langle 3, 4 \rangle \cdot \langle 0, 1 \rangle}{|\langle 3, 4 \rangle||\langle 0, 1 \rangle|} = \dfrac{\langle 3, 4 \rangle \cdot \langle 0, 1 \rangle}{\sqrt{3^2 + 4^2} \cdot \sqrt{0^2 + 1^2}}$

$= \dfrac{0+4}{\sqrt{25} \cdot \sqrt{1}} = \dfrac{4}{5 \cdot 1} = \dfrac{4}{5} = .8 \Rightarrow$

$\theta = \cos^{-1} .8 \approx 36.87°$

For Exercises 83–85, $\mathbf{u} = \langle -2, 1 \rangle$, $\mathbf{v} = \langle 3, 4 \rangle$, and $\mathbf{w} = \langle -5, 12 \rangle$.

**83.** $(3\mathbf{u}) \cdot \mathbf{v} = (3\langle -2, 1 \rangle) \cdot \langle 3, 4 \rangle$
$= \langle -6, 3 \rangle \cdot \langle 3, 4 \rangle = -18 + 12 = -6$

**85.** $\mathbf{u} \cdot \mathbf{v} - \mathbf{u} \cdot \mathbf{w} = \langle -2, 1 \rangle \cdot \langle 3, 4 \rangle - \langle -2, 1 \rangle \cdot \langle -5, 12 \rangle$
$= (-6+4) - (10+12)$
$= -2 - 22 = -24$

**87.** Since $\langle 1, 2 \rangle \cdot \langle -6, 3 \rangle = -6 + 6 = 0$, the vectors are orthogonal.

**89.** Since $\langle 1, 0 \rangle \cdot \langle \sqrt{2}, 0 \rangle = \sqrt{2} + 0 = \sqrt{2} \neq 0$, the vectors are not orthogonal

**91.** $\sqrt{5}\mathbf{i} - 2\mathbf{j} = \langle \sqrt{5}, -2 \rangle; -5\mathbf{i} + 2\sqrt{5}\mathbf{j} = \langle -5, 2\sqrt{5} \rangle$

Since $\langle \sqrt{5}, -2 \rangle \cdot \langle -5, 2\sqrt{5} \rangle = -5\sqrt{5} - 4\sqrt{5}$

$= -9\sqrt{5} \neq 0$, the vectors are not orthogonal.

**93.** Draw a line parallel to the $x$-axis and the vector $\mathbf{u} + \mathbf{v}$ (shown as a dashed line)
Since $\theta_1 = 110°$, its supplementary angle is $70°$. Further, since $\theta_2 = 260°$, the angle $\alpha$ is $260° - 180° = 80°$. Then the angle $CBA$ becomes $180 - (80 + 70) = 180 - 150 = 30°$.

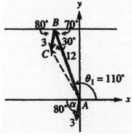

Using the law of cosines, the magnitude of **u** + **v** is found as follows:

$$|\mathbf{u}+\mathbf{v}|^2 = a^2 + c^2 - 2ac\cos B$$

$$|\mathbf{u}+\mathbf{v}|^2 = 3^2 + 12^2 - 2(3)(12)\cos 30°$$

$$= 9 + 144 - 72 \cdot \frac{\sqrt{3}}{2} = 153 - 36\sqrt{3}$$

$$\approx 90.646171$$

Thus, $|\mathbf{u}+\mathbf{v}| \approx 9.5208$.

Using the law of sines, we have

$$\frac{\sin A}{a} = \frac{\sin B}{b} \Rightarrow \frac{\sin A}{3} = \frac{\sin 30°}{9.5208} \Rightarrow$$

$$\sin A = \frac{3\sin 30°}{9.5208} = \frac{3 \cdot \frac{1}{2}}{9.5208} \approx .15754979 \Rightarrow$$

$$A \approx 9.0647°$$

The direction angle of **u** + **v** is
$110° + 9.0647° = 119.0647°$.

**95.** Since $c = 3\cos 260° \approx -.52094453$ and
$d = 3\sin 260° \approx -2.95442326$,
$\langle c, d \rangle \approx \langle -.5209, -2.9544 \rangle$.

**97.** Magnitude:

$$\sqrt{(-4.62518625)^2 + 8.32188819^2} \approx 9.5208$$

Angle:

$$\tan\theta' = \frac{8.32188819}{-4.625186258} \Rightarrow \theta' \approx -60.9353° \Rightarrow$$

$$\theta = -60.9353° + 180° = 119.0647°$$

($\theta$ lies in quadrant II)

## Section 7.5: Applications of Vectors

**1.** Find the direction and magnitude of the equilibrant.
Since $A = 180° - 28.2° = 151.8°$, we can use the law of cosines to find the magnitude of the resultant, **v**.

$$|\mathbf{v}|^2 = 1240^2 + 1480^2 - 2(1240)(1480)\cos 151.8°$$

$$\approx 6962736.2 \Rightarrow |\mathbf{v}| \approx 2639 \text{ lb}$$

(will be rounded as 2640)
Use the law of sines to find $\alpha$.

$$\frac{\sin\alpha}{1240} = \frac{\sin 151.8°}{2639}$$

$$\sin\alpha = \frac{1240\sin 151.8°}{2639} \approx .22203977$$

$$\alpha \approx 12.8°$$

Thus, we have 2640 lb at an angle of
$\theta = 180° - 12.8° = 167.2°$ with the 1480-lb force.

**3.** Let $\alpha$ = the angle between the forces.
To find $\alpha$, use the law of cosines to find $\theta$.

$$786^2 = 692^2 + 423^2 - 2(692)(423)\cos\theta$$

$$\cos\theta = \frac{692^2 + 423^2 - 786^2}{2(692)(423)} \approx .06832049$$

$$\theta \approx 86.1°$$

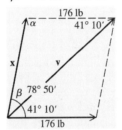

Thus, $\alpha = 180° - 86.1° = 93.9°$.

**5.** Use the parallelogram rule. In the figure, **x** represents the second force and **v** is the resultant.

$$\alpha = 180° - 78°50'$$
$$= 179°60' - 78°50' = 101°10'$$
and
$$\beta = 78°50' - 41°10' = 37°40'$$

Using the law of sines, we have

$$\frac{|\mathbf{x}|}{\sin 41°10'} = \frac{176}{\sin 37°40'} \Rightarrow$$

$$|\mathbf{x}| = \frac{176\sin 41°10'}{\sin 37°40'} \approx 190$$

$$\frac{|\mathbf{v}|}{\sin\alpha} = \frac{176}{\sin 37°40'} \Rightarrow$$

$$|\mathbf{v}| = \frac{176\sin 101°10'}{\sin 37°40'} \approx 283$$

Thus, the magnitude of the second force is about 190 lb and the magnitude of the resultant is about 283 lb.

**7.** Let $\theta$ = the angle that the hill makes with the horizontal.
The 80-lb downward force has a 25-lb component parallel to the hill. The two right triangles are similar and have congruent angles.

$$\sin\theta = \frac{25}{80} = \frac{5}{16} = .3125 \Rightarrow \theta \approx 18°$$

(*continued on next page*)

(*continued from page 191*)

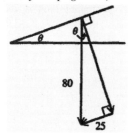

9.  Find the force needed to pull a 60-ton monolith along the causeway.
    The force needed to pull 60 tons is equal to the magnitude of **x**, the component parallel to the causeway.

    $$\sin 2.3° = \frac{|\mathbf{x}|}{60} \Rightarrow |\mathbf{x}| = 60\sin 2.3° \approx 2.4 \text{ tons}$$

    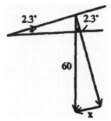

    The force needed is 2.4 tons.

11. Like Example 3 on page 345 of the text, angle *B* equals angle $\theta$, and here the magnitude of vector **BA** represents the weight of the stump grinder. The vector **AC** equals vector **BE**, which represents the force required to hold the stump grinder on the incline. Thus, we have

    $$\sin B = \frac{18}{60} = \frac{3}{10} = .3 \Rightarrow B \approx 17.5°$$

13. Find the weight of the crate and the tension on the horizontal rope.

    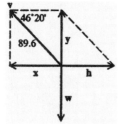

    **v** has horizontal component **x** and vertical component **y**. The resultant **v** + **h** also has vertical component **y**. The resultant balances the weight of the crate, so its vertical component is the equilibrant of the crate's weight: $|\mathbf{w}| = |\mathbf{y}| = 89.6\sin 46°20' \approx 64.8$ lb

Since the crate is not moving side-to-side, **h**, the horizontal tension on the rope, is the opposite of **x**.

$$|\mathbf{h}| = |\mathbf{x}| = 89.6\cos 46°20' \approx 61.9 \text{ lb}$$

The weight of the crate is 64.8 lb; the tension is 61.9 lb.

15. Refer to the diagram below. In order for the ship to turn due east, the ship must turn the measure of angle *CAB*, which is $90° - 34° = 56°$. Angle *DAB* is therefore $180° - 56° = 124°$.

    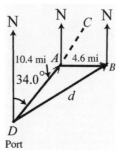

    Port

    Using the law of cosines, we can solve for the distance the ship is from port as follows.

    $$d^2 = 10.4^2 + 4.6^2 - 2(10.4)(4.6)\cos 124°$$
    $$\approx 182.824 \Rightarrow d \approx 13.52$$

    hus, the distance the ship is from port is 13.5 mi. (rounded to three significant digits) To find the bearing, we first seek the measure of angle *ADB*, which we will refer to as angle *D*. Using the law of cosines we have

    $$\cos D = \frac{13.52^2 + 10.4^2 - 4.6^2}{2(13.52)(10.4)} \approx .95937073 \Rightarrow$$
    $$D \approx 16.4°$$

    Thus, the bearing is
    $34.0° + D = 34.0° + 16.4° = 50.4°$.

17. Find the distance of the ship from point *A*.
    Angle 1 = $189° - 180° = 9°$
    Angle 2 = $360° - 317° = 43°$
    Angle 1 + Angle 2 = $9° + 43° = 52°$

    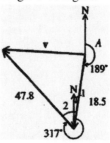

    Use the law of cosines to find $|\mathbf{v}|$.

    $$|\mathbf{v}|^2 = 47.8^2 + 18.5^2 - 2(47.8)(18.5)\cos 52°$$
    $$\approx 1538.23 \Rightarrow |\mathbf{v}| \approx 39.2 \text{ km}$$

**19.** Let $x$ = be the actual speed of the motorboat; $y$ = the speed of the current.

$$\sin 10° = \frac{y}{20.0} \Rightarrow y = 20.0 \sin 10° \approx 3.5$$

$$\cos 10° = \frac{x}{20.0} \Rightarrow x = 20.0 \cos 10° \approx 19.7$$

The speed of the current is 3.5 mph and the actual speed of the motorboat is 19.7 mph.

**21.** Let $\mathbf{v}$ = the ground speed vector.
Find the bearing and ground speed of the plane.
Angle $A = 233° - 114° = 119°$
Use the law of cosines to find $|\mathbf{v}|$.

$$|\mathbf{v}|^2 = 39^2 + 450^2 - 2(39)(450)\cos 119°$$
$$|\mathbf{v}|^2 \approx 221,037.82$$
$$|\mathbf{v}| \approx 470.1$$

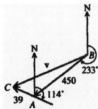

The ground speed is 470 mph. (rounded to two significant digits)
Use the law of sines to find angle $B$.

$$\frac{\sin B}{39} = \frac{\sin 119°}{470.1} \Rightarrow$$

$$\sin B = \frac{39 \sin 119°}{470.1} \approx .07255939 \Rightarrow B \approx 4°$$

Thus, the bearing is
$B + 233° = 4° + 233° = 237°.$

**23.** Let $|\mathbf{x}|$ = the airspeed and $|\mathbf{d}|$ = the ground speed.

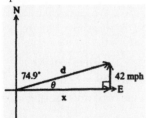

$\theta = 90° - 74.9° = 15.1°$

$$\frac{|\mathbf{x}|}{42} = \cot 15.1° \Rightarrow$$

$$|\mathbf{x}| = 42 \cot 15.1° = \frac{42}{\tan 15.1°} \approx 156 \text{ mph}$$

$$\frac{|\mathbf{d}|}{42} = \csc 15.1° \Rightarrow$$

$$|\mathbf{d}| = 42 \csc 15.1° = \frac{42}{\sin 15.1°} \approx 161 \text{ mph}$$

**25.** Let $\mathbf{c}$ = the ground speed vector.

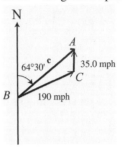

By alternate interior angles, angle $A = 64°30'$.
Use the law of sines to find $B$.

$$\frac{\sin B}{35.0} = \frac{\sin A}{190} \Rightarrow$$

$$\sin B = \frac{35.0 \sin 64°30'}{190} \approx .16626571 \Rightarrow$$

$$B \approx 9.57° \approx 9°30'$$

Thus, the bearing is
$64°30' + B = 64°30' + 9°30' = 74°00'.$
Since $C = 180° - A - B$
$= 180° - 64.50° - 9.57° = 105.93°,$ we use the law of sines to find the ground speed.

$$\frac{|\mathbf{c}|}{\sin C} = \frac{35.0}{\sin B} \Rightarrow$$

$$|\mathbf{c}| = \frac{35.0 \sin 105.93°}{\sin 9.57°} \approx 202 \text{ mph}$$

The bearing is $74°00'$; the ground speed is 202 mph.

**27.** Let $\mathbf{v}$ = the airspeed vector

The ground speed is $\frac{400 \text{ mi}}{2.5 \text{ hr}} = 160 \text{ mph}.$

angle $BAC = 328° - 180° = 148°$
Using the law of cosines to find $|\mathbf{v}|$, we have

$$|\mathbf{v}|^2 = 11^2 + 160^2 - 2(11)(160)\cos 148°$$
$$|\mathbf{v}|^2 \approx 28,706.1$$
$$|\mathbf{v}| \approx 169.4$$

The airspeed must be 170 mph. (rounded to two significant digits)

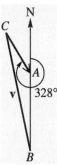

Use the law of sines to find $B$.

$$\frac{\sin B}{11} = \frac{\sin 148°}{169.4} \Rightarrow \sin B = \frac{11\sin 148°}{169.4} \Rightarrow$$
$$\sin B \approx .03441034 \Rightarrow B \approx 2.0°$$

The bearing must be approximately
$360° - 2.0° = 358°$.

**29.** Find the ground speed and resulting bearing.
Angle $A = 245° - 174° = 71°$

Use the law of cosines to find $|\mathbf{v}|$.

$$|\mathbf{v}|^2 = 30^2 + 240^2 - 2(30)(240)\cos 71°$$
$$|\mathbf{v}|^2 \approx 53,811.8$$
$$|\mathbf{v}| \approx 232.1$$

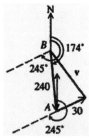

The ground speed is 230 km per hr. (rounded
to two significant digits)
Use the law of sines to find angle $B$.

$$\frac{\sin B}{30} = \frac{\sin 71°}{230} \Rightarrow$$
$$\sin B = \frac{30\sin 71°}{230} \approx .12332851 \Rightarrow$$
$$B \approx 7°$$

Thus, the bearing is
$174° - B = 174° - 7° = 167°$.

**31.** $R = i - 2j$ and $A = .5i + j$

**(a)** Write the given vector in component
form. $R = i - 2j = \langle 1, -2 \rangle$ and
$$A = .5i + j = \langle .5, 1 \rangle$$

$$|R| = \sqrt{1^2 + (-2)^2} = \sqrt{1+4} = \sqrt{5} \approx 2.2$$
and
$$|A| = \sqrt{.5^2 + 1^2} = \sqrt{.25 + 1} = \sqrt{1.25} \approx 1.1$$

About 2.2 in. of rain fell. The area of the
opening of the rain gauge is about 1.1
in.$^2$.

**(b)** $V = |R \cdot A| = |\langle 1, -2 \rangle \cdot \langle .5, 1 \rangle|$
$$= |.5 + (-2)| = |-1.5| = 1.5$$

The volume of rain was 1.5 in.$^3$.

**(c)** **R** and **A** should be parallel and point in
opposite directions.

## Summary Exercises on Applications of Trigonometry and Vectors

**1.** Consider the diagram below.

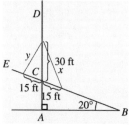

If we extend the flagpole, a right triangle $CAB$
is formed. Thus, the measure of angle $BCA$ is
$90° - 20° = 70°$. Since angle $DCB$ and $BCA$
are supplementary, the measure of angle $DCB$
is $180° - 70° = 110°$. We can now use the law
of cosines to find the measure of the support
wire on the right, $x$.

$$x^2 = 30^2 + 15^2 - 2(30)(15)\cos 110°$$
$$\approx 1432.818 \approx 37.85 \approx 38 \text{ ft}$$

Now, to find the length of the support wire on
the left, we have different ways to find it. One
way would be to use the approximation for $x$
and use the law of cosines. To avoid using the
approximate value, we will find $y$ with the
same method as for $x$.

Since angle $DCB$ and $DCE$ are supplementary,
the measure of angle $DCE$ is
$180° - 110° = 70°$. We can now use the law of
cosines to find the measure of the support wire
on the left, $y$.

$$y^2 = 30^2 + 15^2 - 2(30)(15)\cos 70°$$
$$\approx 817.182 \approx 28.59 \approx 29 \text{ ft}$$

The length of the two wires are about 29 ft and
38 ft.

**3.** Using the law of sines, we can find the measure of $\angle W$. Then find the measure of $\angle P$, and use the law of sines to find $CW$.

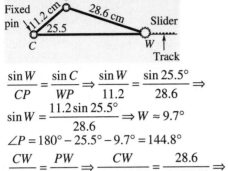

$$\frac{\sin W}{CP} = \frac{\sin C}{WP} \Rightarrow \frac{\sin W}{11.2} = \frac{\sin 25.5°}{28.6} \Rightarrow$$

$$\sin W = \frac{11.2 \sin 25.5°}{28.6} \Rightarrow W \approx 9.7°$$

$$\angle P = 180° - 25.5° - 9.7° = 144.8°$$

$$\frac{CW}{\sin P} = \frac{PW}{\sin C} \Rightarrow \frac{CW}{\sin 144.8°} = \frac{28.6}{\sin 25.5°} \Rightarrow$$

$$CW = \frac{28.6 \sin 144.8°}{\sin 25.5°} \approx 38.3 \text{ cm}$$

**5.** Let $x$ be the new speed of the balloon. $\theta$ is the angle the balloon makes with horizontal.

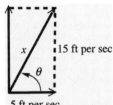

To find $x$, use the Pythagorean theorem.

$$x^2 = 15^2 + 5^2 \Rightarrow x^2 = 225 + 25 \Rightarrow x^2 = 250 \Rightarrow$$

$$x = \sqrt{250} = 5\sqrt{10} \approx 15.8 \text{ ft per sec}$$

To find $\theta$, solve the following

$$\tan \theta = \frac{15}{5} \Rightarrow \tan \theta = 3 \Rightarrow \theta = \tan^{-1} 3 \approx 71.6°$$

**7.** Let $h$ be the height (vertical) of the airplane above the ground;
$x$ is the distance between points $B$ and $D$ as labeled in the diagram.

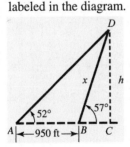

Since angle $ABD$ and $DBC$ form a line, angle $ABD$ is the supplement of angle $DBC$. Thus, the measure of angle $ABD$ is $180° - 57° = 123°$. Since the angles of a triangle must add up to $180°$, the measure of angle $ADB$ is $180° - 123° - 52° = 5°$.

$$\frac{x}{\sin 52°} = \frac{950}{\sin 5°} \Rightarrow x = \frac{950 \sin 52°}{\sin 5°} \approx 8589.34$$

$$\sin 57° = \frac{h}{8589.34} \Rightarrow$$

$$h = 8589.34 \sin 57° \approx 7203.6$$

The airplane is approximately 7200 ft above the ground. (rounded to two significant digits)

**9.**

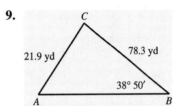

Using the law of sines, we have

$$\frac{\sin A}{BC} = \frac{\sin B}{AC} \Rightarrow \frac{\sin A}{78.3} = \frac{\sin 38°50'}{21.9} \Rightarrow$$

$$\sin A = \frac{78.3 \sin 38°50'}{21.9} \approx 2.24$$

Since $-1 \le \sin A \le 1$, the triangle cannot exist.

## Chapter 7: Review Exercises

**1.** Find $b$, given $C = 74.2°$, $c = 96.3$ m, $B = 39.5°$.
Use the law of sines to find $b$.

$$\frac{b}{\sin B} = \frac{c}{\sin C} \Rightarrow \frac{b}{\sin 39.5°} = \frac{96.3}{\sin 74.2°} \Rightarrow$$

$$b = \frac{96.3 \sin 39.5°}{\sin 74.2°} \approx 63.7 \text{ m}$$

**3.** Find $B$, given $C = 51.3°$, $c = 68.3$ m, $b = 58.2$ m.
Use the law of sines to find $B$.

$$\frac{\sin B}{b} = \frac{\sin C}{c} \Rightarrow \frac{\sin B}{58.2} = \frac{\sin 51.3°}{68.3} \Rightarrow$$

$$\sin B = \frac{58.2 \sin 51.3°}{68.3} \approx .66502269$$

There are two angles $B$ between $0°$ and $180°$ that satisfy the condition. Since $\sin B \approx .66502269$, to the nearest tenth value of $B$ is $B_1 = 41.7°$. Supplementary angles have the same sine value, so another possible value of $B$ is $B_2 = 180° - 41.7° = 138.3°$. This is not a valid angle measure for this triangle since $C + B_2 = 51.3° + 138.3° = 189.6° > 180°$. Thus, $B = 41.7°$.

**5.** Find $A$, given $B = 39°50'$, $b = 268$ m, $a = 340$ m.

Use the law of sines to find $A$.

$$\frac{\sin A}{a} = \frac{\sin B}{b} \Rightarrow \frac{\sin A}{340} = \frac{\sin 39°50'}{268} \Rightarrow$$

$$\sin A = \frac{340 \sin 39°50'}{268} \approx .81264638$$

There are two angles $A$ between $0°$ and $180°$ that satisfy the condition. Since $\sin A \approx .81264638$, to the nearest tenth value of $A$ is $A_1 = 54.4° \approx 54°20'$. Supplementary angles have the same sine value, so another possible value of $A$ is $A_2 = 180° - 54°20'$ $= 179°60' - 54°20' = 125°40'$. This is a valid angle measure for this triangle since $B + A_2 = 39°50' + 125°40' = 165°30' < 180°$.

$A = 54°20'$ or $A = 125°40'$

**7.** No; If you are given two angles of a triangle, then the third angle is known since the sum of the measures of the three angles is $180°$. Since you are also given one side, there will only be one triangle that will satisfy the conditions.

**9.** $a = 10$, $B = 30°$

**(a)** The value of $b$ that forms a right triangle would yield exactly one value for $A$. That is, $b = 10 \sin 30° = 5$. Also, any value of $b$ greater than or equal to 10 would yield a unique value for $A$.

**(b)** Any value of $b$ between 5 and 10, would yield two possible values for $A$.

**(c)** If $b$ is less than 5, then no value for $A$ is possible.

**11.** Find $A$, given $a = 86.14$ in., $b = 253.2$ in., $c = 241.9$ in.

Use the law of cosines to find $A$.

$$a^2 = b^2 + c^2 - 2bc \cos A \Rightarrow$$

$$\cos A = \frac{b^2 + c^2 - a^2}{2bc}$$

$$= \frac{253.2^2 + 241.9^2 - 86.14^2}{2(253.2)(241.9)}$$

$$\approx .94046923$$

Thus, $A \approx 19.87°$ or $19°52'$.

**13.** Find $a$, given $A = 51°20'$, $c = 68.3$ m, $b = 58.2$ m.

Use the law of cosines to find $a$.

$$a^2 = b^2 + c^2 - 2bc \cos A \Rightarrow$$

$$a^2 = 58.2^2 + 68.3^2 - 2(58.2)(68.3)\cos 51°20'$$

$$\approx 3084.99 \Rightarrow a \approx 55.5 \text{ m}$$

**15.** Find $a$, given $A = 60°$, $b = 5$cm, $c = 21$ cm.

Use the law of cosines to find $a$.

$$a^2 = b^2 + c^2 - 2bc \cos A \Rightarrow$$

$$a^2 = 5^2 + 21^2 - 2(5)(21)\cos 60° = 361 \Rightarrow$$

$$a = 19 \text{ cm}$$

**17.** Solve the triangle, given $A = 25.2°$, $a = 6.92$ yd, $b = 4.82$ yd.

$$\frac{\sin B}{b} = \frac{\sin A}{a} \Rightarrow \sin B = \frac{b \sin A}{a} \Rightarrow$$

$$\sin B = \frac{4.82 \sin 25.2°}{6.92} \approx .29656881$$

There are two angles $B$ between $0°$ and $180°$ that satisfy the condition. Since $\sin B \approx .29656881$, to the nearest tenth value of $B$ is $B_1 = 17.3°$. Supplementary angles have the same sine value, so another possible value of $B$ is $B_2 = 180° - 17.3° = 162.7°$. This is not a valid angle measure for this triangle since $A + B_2 = 25.2° + 162.7° = 187.9° > 180°$.

$$C = 180° - A - B \Rightarrow$$

$$C = 180° - 25.2° - 17.3° \Rightarrow C = 137.5°$$

Use the law of sines to find $c$.

$$\frac{c}{\sin C} = \frac{a}{\sin A} \Rightarrow \frac{c}{\sin 137.5°} = \frac{6.92}{\sin 25.2°} \Rightarrow$$

$$c = \frac{6.92 \sin 137.5°}{\sin 25.2°} \approx 11.0 \text{ yd}$$

**19.** Solve the triangle, given $a = 27.6$ cm, $b = 19.8$ cm, $C = 42°30'$.

This is a SAS case, so using the law of cosines.

$$c^2 = a^2 + b^2 - 2ab \cos C \Rightarrow$$

$$c^2 = 27.6^2 + 19.8^2 - 2(27.6)(19.8)\cos 42°30'$$

$$\approx 347.985 \Rightarrow c \approx 18.65 \text{ cm}$$

(will be rounded as 18.7)

Of the remaining angles $A$ and $B$, $B$ must be smaller since it is opposite the shorter of the two sides $a$ and $b$. Therefore, $B$ cannot be obtuse.

$$\frac{\sin B}{b} = \frac{\sin C}{c} \Rightarrow \frac{\sin B}{19.8} = \frac{\sin 42°30'}{18.65} \Rightarrow$$

$$\sin B = \frac{19.8 \sin 42°30'}{18.65} \approx .717124859 \Rightarrow$$

$$B \approx 45.8° \approx 45°50'$$

Thus,

$$A = 180° - B - C = 180° - 45°50' - 42°30'$$

$$= 179°60' - 88°20' = 91°40'$$

**21.** Given $b = 840.6$ m, $c = 715.9$ m, $A = 149.3°$, find the area.
Angle $A$ is included between sides $b$ and $c$. Thus, we have

$$A = \frac{1}{2}bc\sin A$$
$$= \frac{1}{2}(840.6)(715.9)\sin 149.3° \approx 153,600 \text{ m}^2$$

(rounded to four significant digits)

**23.** Given $a = .913$ km, $b = .816$ km, $c = .582$ km, find the area.
Use Heron's formula to find the area.

$$s = \frac{1}{2}(a+b+c) = \frac{1}{2}(.913+.816+.582)$$
$$= \frac{1}{2}\cdot 2.311 = 1.1555$$

$$A = \sqrt{s(s-a)(s-b)(s-c)}$$
$$= \sqrt{\begin{array}{c}1.1555(1.1555-.913)\cdot \\ (1.1555-.816)(1.1555-.582)\end{array}}$$
$$= \sqrt{1.1555(.2425)(.3395)(.5735)}$$
$$\approx .234 \text{ km}^2 \text{ (rounded to three significant digits)}$$

**25.** Since $B = 58.4°$ and $C = 27.9°$,
$A = 180° - B - C = 180° - 58.4° - 27.9° = 93.7°$.
Using the law of sines, we have

$$\frac{AB}{\sin C} = \frac{125}{\sin A} \Rightarrow \frac{AB}{\sin 27.9°} = \frac{125}{\sin 93.7°} \Rightarrow$$
$$AB = \frac{125\sin 27.9°}{\sin 93.7°} \approx 58.61$$

The canyon is 58.6 feet across. (rounded to three significant digits)

**27.** Let $AC$ = the height of the tree.
Angle $A = 90° - 8.0° = 82°$
Angle $C = 180° - B - A = 30°$
Use the law of sines to find $AC = b$.

$$\frac{b}{\sin B} = \frac{c}{\sin C}$$
$$\frac{b}{\sin 68°} = \frac{7.0}{\sin 30°} \Rightarrow b = \frac{7.0\sin 68°}{\sin 30°} \Rightarrow$$
$$b \approx 12.98$$

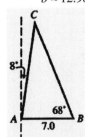

The tree is 13 meters tall. (rounded to two significant digits)

**29.** Let $h$ = the height of tree.
$\theta = 27.2° - 14.3° = 12.9°$
$\alpha = 90° - 27.2° = 62.8°$

$$\frac{h}{\sin\theta} = \frac{212}{\sin\alpha}$$
$$\frac{h}{\sin 12.9°} = \frac{212}{\sin 62.8°}$$
$$h = \frac{212\sin 12.9°}{\sin 62.8°}$$
$$h \approx 53.21$$

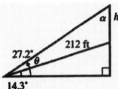

The height of the tree is 53.2 ft. (rounded to three significant digits)

**31.** Let $x$ = the distance between the boats.
In 3 hours the first boat travels $3(36.2) = 108.6$ km and the second travels $3(45.6) = 136.8$ km.
Use the law of cosines to find $x$.

$$x^2 = 108.6^2 + 136.8^2$$
$$\qquad - 2(108.6)(136.8)\cos 54°10'$$
$$\approx 13,113.359 \Rightarrow x \approx 115 \text{ km}$$

They are 115 km apart.

**33.** Use the distance formula to find the distances between the points.
Distance between $(-8, 6)$ and $(0, 0)$:

$$\sqrt{(-8-0)^2+(6-0)^2} = \sqrt{(-8)^2+6^2}$$
$$= \sqrt{64+36} = \sqrt{100} = 10$$

Distance between $(-8, 6)$ and $(3, 4)$:

$$\sqrt{(-8-3)^2+(6-4)^2} = \sqrt{(-11)^2+2^2}$$
$$= \sqrt{121+4} = \sqrt{125}$$
$$= 5\sqrt{5} \approx 11.18$$

Distance between $(3, 4)$ and $(0, 0)$:

$$\sqrt{(3-0)^2+(4-0)^2} = \sqrt{3^2+4^2}$$
$$= \sqrt{9+16} = \sqrt{25} = 5$$

$$s \approx \frac{1}{2}(10+11.18+5) = \frac{1}{2}\cdot 26.18 = 13.09$$

$$A = \sqrt{s(s-a)(s-b)(s-c)}$$
$$= \sqrt{\begin{array}{c}13.09(13.09-10)\cdot \\ (13.09-11.18)(13.09-5)\end{array}}$$
$$= \sqrt{13.09(3.09)(1.91)(8.09)}$$
$$\approx 25 \text{ sq units (rounded to two significant digits)}$$

**35.** $\mathbf{a} - \mathbf{b}$

**37.** (a) true    (b) false

**39.**    $\alpha = 180° - 52° = 128°$

$|\mathbf{v}|^2 = 100^2 + 130^2 - 2(100)(130)\cos 128°$

$|\mathbf{v}|^2 \approx 42907.2$

$|\mathbf{v}| \approx 207 \text{ lb}$

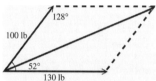

**41.** $|\mathbf{v}| = 964, \ \theta = 154°20'$

horizontal:

$x = |\mathbf{v}|\cos\theta = 964\cos 154° \ 20' \approx 869$

vertical: $y = |\mathbf{v}|\sin\theta = 964\sin 154° \ 20' \approx 418$

**43.** $\mathbf{u} = \langle -9, 12 \rangle$

magnitude:

$|\mathbf{u}| = \sqrt{(-9)^2 + 12^2} = \sqrt{81 + 144} = \sqrt{225} = 15$

Angle: $\tan\theta' = \dfrac{b}{a} \Rightarrow \tan\theta' = \dfrac{12}{-9} \Rightarrow$

$\theta' = \tan^{-1}\left(-\dfrac{4}{3}\right) \approx -53.1° \Rightarrow$

$\theta = -53.1° + 180° = 126.9°$

($\theta$ lies in quadrant II)

**45.** $\mathbf{a} = \langle 3, -2 \rangle, \ \mathbf{b} = \langle -1, 3 \rangle$

$\mathbf{a} \cdot \mathbf{b} = \langle 3, -2 \rangle \cdot \langle -1, 3 \rangle = 3(-1) + (-2) \cdot 3$

$\qquad = -3 - 6 = -9$

$\cos\theta = \dfrac{\mathbf{a} \cdot \mathbf{b}}{|\mathbf{a}||\mathbf{b}|} \Rightarrow \cos\theta = \dfrac{-9}{|\langle 3, -2 \rangle||\langle -1, 3 \rangle|}$

$\qquad = \dfrac{-9}{\sqrt{3^2 + (-2)^2}\sqrt{(-1)^2 + 3^2}}$

$\qquad = -\dfrac{9}{\sqrt{9+4}\sqrt{1+9}} = -\dfrac{9}{\sqrt{13}\sqrt{10}}$

$\qquad = -\dfrac{9}{\sqrt{130}} \approx -.78935222$

Thus, $\theta \approx 142.1°$.

**47.** $|\mathbf{u}| = \sqrt{5^2 + 12^2} = \sqrt{169} = 13$

$\mathbf{v} = \dfrac{\mathbf{u}}{|\mathbf{u}|} = \dfrac{\langle 5, 12 \rangle}{13} = \left\langle \dfrac{5}{13}, \dfrac{12}{13} \right\rangle$

**49.** Let $|\mathbf{x}|$ be the resultant force.

$\theta = 180° - 15° - 10° = 155°$

$|\mathbf{x}|^2 = 12^2 + 18^2 - 2(12)(18)\cos 155°$

$|\mathbf{x}|^2 \approx 859.5 \Rightarrow |\mathbf{x}| \approx 29$

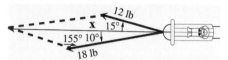

The magnitude of the resultant force on Jessie and the sled is 29 lb.

**51.** Let $\mathbf{v} =$ the ground speed vector.

$\alpha = 212° - 180° = 32°$ and $\beta = 50°$ because they are alternate interior angles. Angle opposite to 520 is $\alpha + \beta = 82°$.

Using the law of sines, we have

$\dfrac{\sin\theta}{37} = \dfrac{\sin 82°}{520} \Rightarrow \sin\theta = \dfrac{37\sin 82°}{520}$

$\sin\theta \approx .07046138 \Rightarrow \theta \approx 4°$

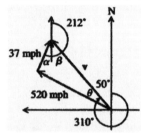

Thus, the bearing is $360° - 50° - \theta = 306°$. The angle opposite $\mathbf{v}$ is $180° - 82° - 4° = 94°$. Using the laws of sines, we have

$\dfrac{|\mathbf{v}|}{\sin 94°} = \dfrac{520}{\sin 82°} \Rightarrow$

$|\mathbf{v}| = \dfrac{520\sin 94°}{\sin 82°} \approx 524 \text{ mph}$

The pilot should fly on a bearing of 306°. Her actual speed is 524 mph.

**53.** Refer to the diagram below. In each of the triangles $ABP$ and $PBC$, we know two angles and one side. Solve each triangle using the law of sines.

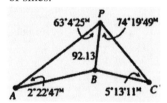

$AB = \dfrac{92.13\sin 63° \ 4' \ 25''}{\sin 2° \ 22' \ 47''} \approx 1978.28 \text{ ft}$ and

$BC = \dfrac{92.13\sin 74° \ 19' \ 49''}{\sin 5° \ 13' \ 11''} \approx 975.05 \text{ ft}$

**55.** $A = 30°, B = 60°, C = 90°, a = 7, b = 7\sqrt{3}, c = 14$

Mollweide's formula: $\dfrac{a-b}{c} = \dfrac{\cos\frac{1}{2}(A-B)}{\cos\frac{1}{2}C}$

$$\frac{7-7\sqrt{3}}{14} \overset{?}{=} \frac{\sin\frac{1}{2}(30°-60°)}{\cos\frac{1}{2}(90°)}$$

$$\frac{1-\sqrt{3}}{2} \overset{?}{=} \frac{\sin(-15°)}{\cos 45°} \Rightarrow \frac{1-\sqrt{3}}{2} \overset{?}{=} -\frac{\sin 15°}{\cos 45°}$$

Using the half-angle formula, we have

$$\sin 15° = \sin\left(\frac{1}{2}\cdot 30°\right) = \sqrt{\frac{1-\cos 30°}{2}}$$

$$= \sqrt{\frac{1-\frac{\sqrt{3}}{2}}{2}} = \sqrt{\frac{2-\sqrt{3}}{4}} = \frac{\sqrt{2-\sqrt{3}}}{2}$$

Continuing, we have

$$\frac{1-\sqrt{3}}{2} \overset{?}{=} -\frac{\sin(15°)}{\cos 45°} \Rightarrow$$

$$\frac{1-\sqrt{3}}{2} \overset{?}{=} \frac{\frac{\sqrt{2-\sqrt{3}}}{2}}{\frac{\sqrt{2}}{2}} \Rightarrow$$

$$\frac{1-\sqrt{3}}{2} \overset{?}{=} \frac{\sqrt{2-\sqrt{3}}}{\sqrt{2}} \Rightarrow$$

$$\frac{\sqrt{(1-\sqrt{3})^2}}{\sqrt{2^2}} \overset{?}{=} \frac{\sqrt{2-\sqrt{3}}}{\sqrt{2}} \Rightarrow$$

$$\sqrt{\frac{(1-\sqrt{3})^2}{2^2}} \overset{?}{=} \sqrt{\frac{2-\sqrt{3}}{2}} \Rightarrow$$

$$\sqrt{\frac{1-2\sqrt{3}+3}{4}} \overset{?}{=} \sqrt{\frac{2-\sqrt{3}}{2}} \Rightarrow$$

$$\sqrt{\frac{4-2\sqrt{3}}{4}} \overset{?}{=} \sqrt{\frac{2-\sqrt{3}}{2}} \Rightarrow$$

$$\sqrt{\frac{2-\sqrt{3}}{2}} = \sqrt{\frac{2-\sqrt{3}}{2}}$$

## Chapter 7 Test

**1.** Find $C$, given $A = 25.2°$, $a = 6.92$ yd, $b = 4.82$ yd.
Use the law of sines to first find the measure of angle $B$.

$$\frac{\sin 25.2°}{6.92} = \frac{\sin B}{4.82} \Rightarrow$$

$$\sin B = \frac{4.82\sin 25.2°}{6.92} \approx .29656881 \Rightarrow$$

$$B \approx 17.3°$$

Use the fact that the angles of a triangle sum to 180° to find the measure of angle $C$.
$C = 180 - A - B = 180° - 25.2° - 17.3° = 137.5°$
Angle $C$ measures 137.5°.

**2.** Find $c$, given $C = 118°$, $b = 131$ km, $a = 75$ km.
Using the law of cosines to find the length of $c$.

$$c^2 = a^2 + b^2 - 2ab\cos C \Rightarrow c^2$$
$$= 75^2 + 131^2 - 2(75)(131)\cos 118°$$
$$\approx 32011.12 \Rightarrow c \approx 178.9 \text{ km}$$

$c$ is approximately 179 km. (rounded to two significant digits)

**3.** Find $B$, given $a = 17.3$ ft, $b = 22.6$ ft, $c = 29.8$ ft.
Using the law of cosines, find the measure of angle $B$.

$$b^2 = a^2 + c^2 - 2ac\cos B \Rightarrow$$
$$\cos B = \frac{a^2+c^2-b^2}{2ac} = \frac{17.3^2+29.8^2-22.6^2}{2(17.3)(29.8)}$$
$$\approx .65617605 \Rightarrow B \approx 49.0°$$

$B$ is approximately 49.0°.

**4.** $a = 14, b = 30, c = 40$
We can use Heron's formula to find the area.

$$s = \frac{1}{2}(a+b+c) = \frac{1}{2}(14+30+40) = 42$$
$$A = \sqrt{s(s-a)(s-b)(s-c)}$$
$$= \sqrt{42(42-14)(42-30)(42-40)}$$
$$= \sqrt{42\cdot 28\cdot 12\cdot 2}$$
$$= \sqrt{28,224} = 168 \text{ sq units}$$

**5.** This is SAS, so we can use the formula

$$A = \frac{1}{2}zy\sin X .$$

$$A = \frac{1}{2}\cdot 6\cdot 12\sin 30° = \frac{1}{2}\cdot 6\cdot 12\cdot \frac{1}{2} = 18 \text{ sq units}$$

**6.** Since $B > 90°$, $b$ must be the longest side of the triangle.

**(a)** $b > 10$

**(b)** none

**(c)** $b \le 10$

7. $A = 60°, b = 30$ m, $c = 45$ m

This is SAS, so use the law of cosines to find

$a$: $a^2 = b^2 + c^2 - 2bc \cos A \Rightarrow$

$a^2 = 30^2 + 45^2 - 2 \cdot 30 \cdot 45 \cos 60° = 1575 \Rightarrow$

$a = 15\sqrt{7} \approx 40$ m

Now use the law of sines to find $B$:

$\dfrac{\sin B}{b} = \dfrac{\sin A}{a} \Rightarrow \dfrac{\sin B}{30} = \dfrac{\sin 60°}{15\sqrt{7}} \Rightarrow$

$\sin B = \dfrac{30 \sin 60°}{15\sqrt{7}} \Rightarrow B \approx 41°$

$C = 180° - A - B = 180° - 60° - 41° = 79°$

8. $b = 1075$ in., $c = 785$ in., $C = 38°30'$

We can use the law of sines.

$\dfrac{\sin B}{b} = \dfrac{\sin C}{c} \Rightarrow \dfrac{\sin B}{1075} = \dfrac{\sin 38°30'}{785} \Rightarrow$

$\sin B = \dfrac{1075 \sin 38°30'}{785} \Rightarrow$

$B_1 \approx 58.5° = 58°30'$ or

$B_2 = 180° - 58°30' = 121°30'$

Solving separately for triangles

$A_1 B_1 C$ and $A_2 B_2 C$ , we have the following.

$A_1 B_1 C$ :

$A_1 = 180° - B_1 - C = 180° - 58°30' - 38°30'$
$\quad = 83°00'$

$\dfrac{a_1}{\sin A_1} = \dfrac{b}{\sin B_1} \Rightarrow \dfrac{a_1}{\sin 83°} = \dfrac{1075}{\sin 58°30'} \Rightarrow$

$a_1 = \dfrac{1075 \sin 83°}{\sin 58°30'} \approx 1251 \approx 1250$ in. (rounded

to three significant digits)

$A_2 B_2 C$ :

$A_2 = 180° - B_2 - C = 180° - 121°30' - 38°30'$
$\quad = 20°00'$

$\dfrac{a_2}{\sin A_2} = \dfrac{b}{\sin B_2} \Rightarrow \dfrac{a_2}{\sin 20°} = \dfrac{1075}{\sin 121°30'} \Rightarrow$

$a_2 = \dfrac{1075 \sin 20°}{\sin 121°30'} \approx 431$ in. (rounded to

three significant digits)

9. magnitude:

$\left| \mathbf{v} \right| = \sqrt{(-6)^2 + 8^2} = \sqrt{36 + 64} = \sqrt{100} = 10$

angle:

$\tan \theta' = \dfrac{y}{x} \Rightarrow$

$\tan \theta' = \dfrac{8}{-6} = -\dfrac{4}{3} \approx -1.33333333 \Rightarrow$

$\theta' \approx -53.1° \Rightarrow \theta = -53.1° + 180° = 126.9°$

($\theta$ lies in quadrant II)

The magnitude $\left| \mathbf{v} \right|$ is 10 and $\theta = 126.9°$.

10.

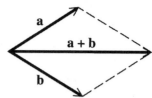

11. $\mathbf{u} = \langle -1, 3 \rangle, \mathbf{v} = \langle 2, -6 \rangle$

(a) $\mathbf{u} + \mathbf{v} = \langle -1, 3 \rangle + \langle 2, -6 \rangle$
$\quad = \langle -1 + 2, 3 + (-6) \rangle = \langle 1, -3 \rangle$

(b) $-3\mathbf{v} = -3\langle 2, -6 \rangle = \langle -3 \cdot 2, -3(-6) \rangle$
$\quad = \langle -6, 18 \rangle$

(c) $\mathbf{u} \cdot \mathbf{v} = \langle -1, 3 \rangle \cdot \langle 2, -6 \rangle = -1(2) + 3(-6)$
$\quad = -2 - 18 = -20$

(d) $\left| \mathbf{u} \right| = \sqrt{(-1)^2 + 3^2} = \sqrt{1 + 9} = \sqrt{10}$

12. Given $A = 24° 50', B = 47° 20'$ and
$AB = 8.4$ mi, first find the measure of angle $C$.
$C = 180° - 47° 20' - 24° 50'$
$\quad = 179°60' - 72° 10' = 107° 50'$

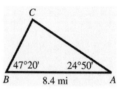

Use this information and the law of sines to
find $AC$.

$\dfrac{AC}{\sin 47° 20'} = \dfrac{8.4}{\sin 107° 50'} \Rightarrow$

$AC = \dfrac{8.4 \sin 47° 20'}{\sin 107° 50'} \approx 6.49$ mi

Drop a perpendicular line from $C$ to segment
$AB$.

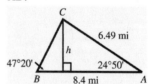

Thus, $\sin 24°50' = \dfrac{h}{6.49} \Rightarrow$

$h \approx 6.49 \sin 24°50' \approx 2.7$ mi.

The balloon is 2.7 miles off the ground.

13. horizontal:

$x = \left| \mathbf{v} \right| \cos \theta = 569 \cos 127.5° \approx -346$ and

vertical: $y = \left| \mathbf{v} \right| \sin \theta = 569 \sin 127.5° \approx 451$

The vector is $\langle -346, 451 \rangle$ .

**14.** Consider the figure below.

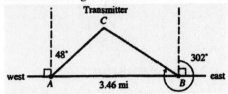

Since the bearing is 48° from $A$, angle $A$ in $ABC$ must be $90° - 48° = 42°$. Since the bearing is 302° from $B$, angle $B$ in $ABC$ must be $302° - 270° = 32°$. The angles of a triangle sum to 180°, so
$C = 180° - A - B = 180° - 42° - 32° = 106°$.
Using the law of sines, we have

$$\frac{b}{\sin B} = \frac{c}{\sin C} \Rightarrow \frac{b}{\sin 32°} = \frac{3.46}{\sin 106°} \Rightarrow$$
$$b = \frac{3.46 \sin 32°}{\sin 106°} \approx 1.91 \text{ mi}$$

The distance from $A$ to the transmitter is 1.91 miles. (rounded to two significant digits)

**15.**

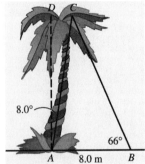

Since $m\angle DAC = 8.0°$,
$m\angle CAD = 90° - 8.0° = 82.0°$. $m\angle B = 66°$, so
$m\angle C = 180° - 82° - 66° = 32°$. Now use the law of sines to find $AC$:

$$\frac{AC}{\sin B} = \frac{AB}{\sin C} \Rightarrow \frac{AC}{\sin 66°} = \frac{8.0}{\sin 32°} \Rightarrow$$
$$AC = \frac{8.0 \sin 66°}{\sin 32°} \approx 13.8 \approx 14 \text{ m}$$

**16.**

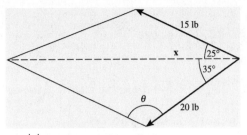

Let $|\mathbf{x}|$ be the equilibrant force.
$\theta = 180° - 35° - 25° = 120°$
Using the law of cosines, we have
$$|\mathbf{x}|^2 = 15^2 + 20^2 - 2 \cdot 15 \cdot 20 \cos 120° \Rightarrow$$
$$|\mathbf{x}|^2 = 925 \Rightarrow \mathbf{x} \approx 30.4 \approx 30 \text{ lb}$$

# Chapter 8

## Complex Numbers, Polar Equations, and Parametric Equations

### Section 8.1: Complex Numbers

1. true

3. true

5. false (Every real number is a complex number.)

7. $-4$ is real and complex.

9. $13i$ is complex, pure imaginary and nonreal complex.

11. $5 + i$ is complex and nonreal complex.

13. $\pi$ is real and complex.

15. $\sqrt{-25} = 5i$ is complex, pure imaginary and nonreal complex.

17. $\sqrt{-25} = i\sqrt{25} = 5i$

19. $\sqrt{-10} = i\sqrt{10}$

21. $\sqrt{-288} = i\sqrt{288} = i\sqrt{144 \cdot 2} = 12i\sqrt{2}$

23. $-\sqrt{-18} = -i\sqrt{18} = -i\sqrt{9 \cdot 2} = -3i\sqrt{2}$

25. $x^2 = -16 \Rightarrow x = \pm\sqrt{-16} = \pm 4i$
    Solution set: $\{\pm 4i\}$

27. $x^2 + 12 = 0 \Rightarrow x^2 = -12 \Rightarrow x = \pm\sqrt{-12} \Rightarrow$
    $x = \pm 2i\sqrt{3}$
    Solution set: $\{\pm 2i\sqrt{3}\}$

29. $3x^2 + 2 = -4x \Rightarrow 3x^2 + 4x + 2 = 0$
    Use the quadratic formula with $a = 3$, $b = 4$, and $c = 2$ to solve for $x$:
    $$x = \frac{-4 \pm \sqrt{4^2 - 4(3)(2)}}{2(3)}$$
    $$= \frac{-4 \pm \sqrt{-8}}{6} = -\frac{2}{3} \pm \frac{\sqrt{2}}{3}i$$
    Solution set: $\left\{-\frac{2}{3} \pm \frac{\sqrt{2}}{3}i\right\}$

31. $x^2 - 6x + 14 = 0$
    Use the quadratic formula with $a = 1$, $b = -6$, and $c = 14$ to solve for $x$:
    $$x = \frac{-(-6) \pm \sqrt{(-6)^2 - 4(1)(14)}}{2(1)}$$
    $$= \frac{6 \pm \sqrt{-20}}{2} = 3 \pm i\sqrt{5}$$
    Solution set: $\{3 \pm i\sqrt{5}\}$

33. $4(x^2 - x) = -7 \Rightarrow 4x^2 - 4x + 7 = 0$
    Use the quadratic formula with $a = 4$, $b = -4$, and $c = 7$ to solve for $x$:
    $$x = \frac{-(-4) \pm \sqrt{(-4)^2 - 4(4)(7)}}{2(4)}$$
    $$= \frac{4 \pm \sqrt{-96}}{8} = \frac{1}{2} \pm \frac{\sqrt{6}}{2}i$$
    Solution set: $\left\{\frac{1}{2} \pm \frac{\sqrt{6}}{2}i\right\}$

35. $x^2 + 1 = -x \Rightarrow x^2 + x + 1 = 0$
    Use the quadratic formula with $a = 1$, $b = 1$, and $c = 1$ to solve for $x$:
    $$x = \frac{-1 \pm \sqrt{1^2 - 4(1)(1)}}{2(1)}$$
    $$= \frac{-1 \pm \sqrt{-3}}{2} = -\frac{1}{2} \pm \frac{\sqrt{3}}{2}i$$
    Solution set: $\left\{-\frac{1}{2} \pm \frac{\sqrt{3}}{2}i\right\}$

37. $\sqrt{-13} \cdot \sqrt{-13} = i\sqrt{13} \cdot i\sqrt{13}$
    $$= i^2\left(\sqrt{13}\right)^2 = -1 \cdot 13 = -13$$

39. $\sqrt{-3} \cdot \sqrt{-8} = i\sqrt{3} \cdot i\sqrt{8} = i^2\sqrt{3 \cdot 8}$
    $$= -1 \cdot \sqrt{24} = -\sqrt{4 \cdot 6} = -2\sqrt{6}$$

41. $\dfrac{\sqrt{-30}}{\sqrt{-10}} = \dfrac{i\sqrt{30}}{i\sqrt{10}} = \sqrt{\dfrac{30}{10}} = \sqrt{3}$

43. $\dfrac{\sqrt{-24}}{\sqrt{8}} = \dfrac{i\sqrt{24}}{\sqrt{8}} = i\sqrt{\dfrac{24}{8}} = i\sqrt{3}$

**45.** $\dfrac{\sqrt{-10}}{\sqrt{-40}} = \dfrac{i\sqrt{10}}{i\sqrt{40}} = \sqrt{\dfrac{10}{40}} = \sqrt{\dfrac{1}{4}} = \dfrac{1}{2}$

**47.** $\dfrac{\sqrt{-6}\cdot\sqrt{-2}}{\sqrt{3}} = \dfrac{i\sqrt{6}\cdot i\sqrt{2}}{\sqrt{3}} = i^2\sqrt{\dfrac{6\cdot 2}{3}}$

$\qquad = -1\cdot\sqrt{\dfrac{12}{3}} = -\sqrt{4} = -2$

**49.** $\dfrac{-6-\sqrt{-24}}{2} = \dfrac{-6-\sqrt{-4\cdot 6}}{2} = \dfrac{-6-2i\sqrt{6}}{2}$

$\qquad = \dfrac{2\left(-3-i\sqrt{6}\right)}{2} = -3 - i\sqrt{6}$

**51.** $\dfrac{10+\sqrt{-200}}{5} = \dfrac{10+\sqrt{-100\cdot 2}}{5}$

$\qquad = \dfrac{10+10i\sqrt{2}}{5} = \dfrac{5\left(2+2i\sqrt{2}\right)}{5}$

$\qquad = 2 + 2i\sqrt{2}$

**53.** $\dfrac{-3+\sqrt{-18}}{24} = \dfrac{-3+\sqrt{-9\cdot 2}}{24} = \dfrac{-3+3i\sqrt{2}}{24}$

$\qquad = \dfrac{3\left(-1+i\sqrt{2}\right)}{24} = \dfrac{-1+i\sqrt{2}}{8}$

$\qquad = -\dfrac{1}{8} + \dfrac{\sqrt{2}}{8}i$

**55.** $(3+2i)+(9-3i) = (3+9)+\left[2+(-3)\right]i$

$\qquad = 12+(-1)i = 12 - i$

**57.** $(-2+4i)-(-4+4i)$

$\qquad = \left[-2-(-4)\right]+(4-4)i$

$\qquad = 2+0i = 2$

**59.** $(2-5i)-(3+4i)-(-1-9i)$

$\qquad = \left[2-3-(-1)\right]+\left[-5-4-(-9)\right]i = 0$

**61.** $-i-2-(6-4i)-(5-2i)$

$\qquad = (-2-6-5)+\left[-1-(-4)-(-2)\right]i$

$\qquad = -13+5i$

**63.** $(2+i)(3-2i)$

$\qquad = 2(3)+2(-2i)+i(3)+i(-2i)$

$\qquad = 6-4i+3i-2i^2 = 6-i-2(-1)$

$\qquad = 6-i+2 = 8-i$

**65.** $(2+4i)(-1+3i)$

$\qquad = 2(-1)+2(3i)+4i(-1)+4i(3i)$

$\qquad = -2+6i-4i+12i^2 = -2+2i+12(-1)$

$\qquad = -2+2i-12 = -14+2i$

**67.** $(3-2i)^2 = 3^2 - 2(3)(2i)+(2i)^2$

$\qquad = 9-12i-4 = 5-12i$

**69.** $(3+i)(3-i) = 3^2 - i^2 = 9-(-1) = 10$

**71.** $(-2-3i)(-2+3i) = (-2)^2-(3i)^2 = 4-9i^2$

$\qquad = 4-9(-1) = 13$

**73.** $\left(\sqrt{6}+i\right)\left(\sqrt{6}-i\right) = \left(\sqrt{6}\right)^2-i^2$

$\qquad = 6-(-1) = 6+1 = 7$

**75.** $i(3-4i)(3+4i) = i\left[(3-4i)(3+4i)\right]$

$\qquad = i\left[3^2-(4i)^2\right]$

$\qquad = i\left[9-16i^2\right]$

$\qquad = i\left[9-16(-1)\right]$

$\qquad = i(9+16) = 25i$

**77.** $3i(2-i)^2 = 3i\left(2^2-2(2i)+i^2\right)$

$\qquad = 3i(4-4i-1) = 3i(3-4i)$

$\qquad = 9i-12i^2 = 9i-12(-1)$

$\qquad = 12+9i$

**79.** $(2+i)(2-i)(4+3i) = \left[(2+i)(2-i)\right](4+3i)$

$\qquad = \left[2^2-i^2\right](4+3i)$

$\qquad = \left[4-(-1)\right](4+3i)$

$\qquad = 5(4+3i) = 20+15i$

**81.** $i^{25} = i^{24}\cdot i = \left(i^4\right)^6\cdot i = 1^6\cdot i = i$

**83.** $i^{22} = i^{20}\cdot i^2 = \left(i^4\right)^5\cdot(-1) = 1^5\cdot(-1) = -1$

**85.** $i^{23} = i^{20}\cdot i^3 = \left(i^4\right)^5\cdot i^3 = 1^5\cdot(-i) = -i$

**87.** $i^{32} = \left(i^4\right)^8 = 1^8 = 1$

**89.** $i^{-13} = i^{-16}\cdot i^3 = \left(i^4\right)^{-4}\cdot i^3 = 1^{-4}\cdot(-i) = -i$

**91.** $\dfrac{1}{i^{-11}} = i^{11} = i^8\cdot i^3 = \left(i^4\right)^2\cdot i^3 = 1^2\cdot(-i) = -i$

**93.** Answers will vary.

**95.** $\dfrac{6+2i}{1+2i} = \dfrac{(6+2i)(1-2i)}{(1+2i)(1-2i)}$

$= \dfrac{6-12i+2i-4i^2}{1^2-(2i)^2} = \dfrac{6-10i-4(-1)}{1-4i^2}$

$= \dfrac{6-10i+4}{1-4(-1)} = \dfrac{10-10i}{1+4} = \dfrac{10-10i}{5}$

$= \dfrac{10}{5} - \dfrac{10}{5}i = 2-2i$

**97.** $\dfrac{2-i}{2+i} = \dfrac{(2-i)(2-i)}{(2+i)(2-i)} = \dfrac{2^2-2(2i)+i^2}{2^2-i^2}$

$= \dfrac{4-4i+(-1)}{4-(-1)} = \dfrac{3-4i}{5} = \dfrac{3}{5} - \dfrac{4}{5}i$

**99.** $\dfrac{1-3i}{1+i} = \dfrac{(1-3i)(1-i)}{(1+i)(1-i)} = \dfrac{1-i-3i+3i^2}{1^2-i^2}$

$= \dfrac{1-4i+3(-1)}{1-(-1)} = \dfrac{1-4i-3}{2}$

$= \dfrac{-2-4i}{2} = \dfrac{-2}{2} - \dfrac{4}{2}i = -1-2i$

**101.** $\dfrac{-5}{i} = \dfrac{-5(-i)}{i(-i)} = \dfrac{5i}{-i^2}$

$= \dfrac{5i}{-(-1)} = \dfrac{5i}{1} = 5i \text{ or } 0+5i$

**103.** $\dfrac{8}{-i} = \dfrac{8\cdot i}{-i\cdot i} = \dfrac{8i}{-i^2}$

$= \dfrac{8i}{-(-1)} = \dfrac{8i}{1} = 8i \text{ or } 0+8i$

**105.** $\dfrac{2}{3i} = \dfrac{2(-3i)}{3i\cdot(-3i)} = \dfrac{-6i}{-9i^2} = \dfrac{-6i}{-9(-1)}$

$= \dfrac{-6i}{9} = -\dfrac{2}{3}i \text{ or } 0-\dfrac{2}{3}i$

Note: In the above solution, we multiplied the numerator and denominator by the complex conjugate of $3i$, namely $-3i$. Since there is a reduction in the end, the same results can be achieved by multiplying the numerator and denominator by $-i$.

**107.** We need to show that $\left(\dfrac{\sqrt{2}}{2} + \dfrac{\sqrt{2}}{2}i\right)^2 = i$.

$\left(\dfrac{\sqrt{2}}{2} + \dfrac{\sqrt{2}}{2}i\right)^2$

$= \left(\dfrac{\sqrt{2}}{2}\right)^2 + 2\cdot\dfrac{\sqrt{2}}{2}\cdot\dfrac{\sqrt{2}}{2}i + \left(\dfrac{\sqrt{2}}{2}i\right)^2$

$= \dfrac{2}{4} + 2\cdot\dfrac{2}{4}i + \dfrac{2}{4}i^2 = \dfrac{1}{2} + i + \dfrac{1}{2}i^2$

$= \dfrac{1}{2} + i + \dfrac{1}{2}(-1) = \dfrac{1}{2} + i - \dfrac{1}{2} = i$

**109.** Let $z = 3-2i$.

$3z - z^2 = 3(3-2i) - (3-2i)^2$

$= 9-6i - \left[3^2 - 2(6i) + (2i)^2\right]$

$= 9-6i - \left(9-12i+4i^2\right)$

$= 9-6i - \left[9-12i+4(-1)\right]$

$= 9-6i - \left[9-12i+(-4)\right]$

$= 9-6i - (5-12i)$

$= 9-6i-5+12i = 4+6i$

**111.** $I = 8+6i, Z = 6+3i$

$E = IZ \Rightarrow$

$E = (8+6i)(6+3i) = 48+24i+36i+18i^2$

$= 48+60i-18 = 30+60i$

**113.** $I = 7+5i, E = 28+54i$

$E = IZ \Rightarrow$

$28+54i = (7+5i)Z$

$Z = \dfrac{28+54i}{7+5i} = \dfrac{28+54i}{7+5i}\cdot\dfrac{7-5i}{7-5i}$

$= \dfrac{196-140i+378i-270i^2}{49-25i^2}$

$= \dfrac{196+238i+270}{49+25} = \dfrac{466+238i}{74}$

$= \dfrac{2(233+119i)}{2(37)} = \dfrac{233}{37} + \dfrac{119i}{37}$

**115.** The resistive part is $50+60=110$, and the reactive part is $15i+17i=32i$, so the total impedance is $Z = 110+32i$.

## Section 8.2: Trigonometric (Polar) Form of Complex Numbers

**1.** The absolute value of a complex number represents the <u>length (or magnitude)</u> of the vector representing it in the complex plane.

**3.**

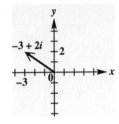

**5.**

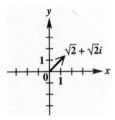

**7.**

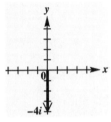

**9.**

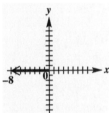

**11.** $1 - 4i$

**13.** $(4 - 3i) + (-1 + 2i) = 3 - i$

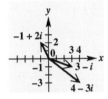

**15.** $(5 - 6i) + (-5 + 3i) = -3i$

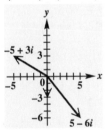

**17.** $-3 + 3i$

**19.** $(2 + 6i) - 2i = 2 + 4i$

**21.** $(7 + 6i) + 3i = 7 + 9i$

**23.** $\left(\dfrac{1}{2} + \dfrac{2}{3}i\right) + \left(\dfrac{2}{3} + \dfrac{1}{2}i\right) = \dfrac{7}{6} + \dfrac{7}{6}i$

**25.** $2(\cos 45° + i \sin 45°) = 2\left(\dfrac{\sqrt{2}}{2} + i\dfrac{\sqrt{2}}{2}\right)$
$$= \sqrt{2} + i\sqrt{2}$$

**27.** $10(\cos 90° + i \sin 90°) = 10(0 + i)$
$$= 0 + 10i = 10i$$

**29.** $4(\cos 240° + i \sin 240°) = 4\left(-\dfrac{1}{2} - i\dfrac{\sqrt{3}}{2}\right)$
$$= -2 - 2i\sqrt{3}$$

**31.** $3 \operatorname{cis} 150° = 3(\cos 150° + i \sin 150°)$
$$= 3\left(-\dfrac{\sqrt{3}}{2} + \dfrac{1}{2}i\right) = -\dfrac{3\sqrt{3}}{2} + \dfrac{3}{2}i$$

**33.** $5 \operatorname{cis} 300° = 5(\cos 300° + i \sin 300°)$
$$= 5\left[\dfrac{1}{2} + \left(-\dfrac{\sqrt{3}}{2}\right)i\right] = \dfrac{5}{2} - \dfrac{5\sqrt{3}}{2}i$$

**35.** $\sqrt{2} \operatorname{cis} 225° = \sqrt{2}(\cos 225° + i \sin 225°)$
$$= \sqrt{2}\left[-\dfrac{\sqrt{2}}{2} + \left(-\dfrac{\sqrt{2}}{2}i\right)\right]$$
$$= -1 - i$$

**37.** $4(\cos(-30°) + i \sin(-30°))$
$$= 4(\cos 30° - i \sin 30°) = 4\left(\dfrac{\sqrt{3}}{2} - \dfrac{1}{2}i\right)$$
$$= 2\sqrt{3} - 2i$$

**39.** $-3 - 3i\sqrt{3}$

Sketch a graph of $-3 - 3i\sqrt{3}$ in the complex plane.

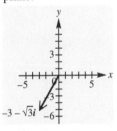

*(continued on next page)*

(*continued from page 205*)

Since $x = -3$ and $y = -3\sqrt{3}$,

$$r = \sqrt{(-3)^2 + \left(-3\sqrt{3}\right)^2} = \sqrt{9 + 27} = \sqrt{36} = 6$$

and $\tan\theta = \dfrac{-3\sqrt{3}}{-3} = \sqrt{3}$. Thus, the reference angle for $\theta$ is 60°. The graph shows that $\theta$ is in quadrant III, so $\theta = 180° + 60° = 240°$. Therefore,

$$-3 - 3i\sqrt{3} = 6\left(\cos 240° + i\sin 240°\right)$$

**41.** $\sqrt{3} - i$

Sketch a graph of $\sqrt{3} - i$ in the complex plane.

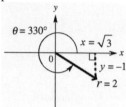

Since $x = \sqrt{3}$ and $y = -1$,

$$r = \sqrt{\left(\sqrt{3}\right)^2 + (-1)^2} = \sqrt{3 + 1} = \sqrt{4} = 2 \text{ and}$$

$\tan\theta = \dfrac{-1}{\sqrt{3}} = -\dfrac{\sqrt{3}}{3}$. Since $\tan\theta = -\dfrac{\sqrt{3}}{3}$, the reference angle for $\theta$ is 30°. The graph shows that $\theta$ is in quadrant IV, so $\theta = 360° - 30° = 330°$. Therefore,

$$\sqrt{3} - i = 2\left(\cos 330° + i\sin 330°\right).$$

**43.** $-5 - 5i$

Sketch a graph of $-5 - 5i$ in the complex plane.

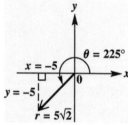

Since $x = -5$ and $y = -5$,

$$r = \sqrt{(-5)^2 + (-5)^2} = \sqrt{25 + 25} = \sqrt{50} = 5\sqrt{2}$$

and $\tan\theta = \dfrac{y}{x} = \dfrac{-5}{-5} = 1$. Since $\tan\theta = 1$, the reference angle for $\theta$ is 45°.

The graph shows that $\theta$ is in quadrant III, so $\theta = 180° + 45° = 225°$. Therefore,

$$-5 - 5i = 5\sqrt{2}\left(\cos 225° + i\sin 225°\right).$$

**45.** $2 + 2i$

Sketch a graph of $2 + 2i$ in the complex plane.

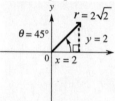

Since $x = 2$ and $y = 2$,

$$r = \sqrt{2^2 + 2^2} = \sqrt{4 + 4} = \sqrt{8} = 2\sqrt{2} \text{ and}$$

$\tan\theta = \dfrac{2}{2} = 1$. Since $\tan\theta = 1$, the reference angle for $\theta$ is 45°. The graph shows that $\theta$ is in quadrant I, so $\theta = 45°$. Therefore,

$$2 + 2i = 2\sqrt{2}\left(\cos 45° + i\sin 45°\right).$$

**47.** $5i = 0 + 5i$

$0 + 5i$ is on the positive $y$-axis, so $\theta = 90°$ and $x = 0$, $y = 5 \Rightarrow$

$$r = \sqrt{0^2 + 5^2} = \sqrt{0 + 25} = 5$$

Thus, $5i = 5\left(\cos 90° + i\sin 90°\right)$.

**49.** $-4 = -4 + 0i$

$-4 + 0i$ is on the negative $x$-axis, so $\theta = 180°$ and $x = -4$, $y = 0 \Rightarrow$

$$r = \sqrt{(-4)^2 + 0^2} = \sqrt{16} = 4$$

Thus, $-4 = 4\left(\cos 180° + i\sin 180°\right).$

**51.** $2 + 3i$

$x = 2$, $y = 3 \Rightarrow r = \sqrt{2^2 + 3^2} = \sqrt{4 + 9} = \sqrt{13}$

$\tan\theta = \dfrac{3}{2}$

$2 + 3i$ is in quadrant I, so

$\theta = 56.31°$.

$$2 + 3i = \sqrt{13}\left(\cos 56.31° + i\sin 56.31°\right)$$

**53.** $3\left(\cos 250° + i\sin 250°\right) = -1.0261 - 2.8191i$

**55.** $12i = 0 + 12i$

$x = 0$, $y = 12 \Rightarrow$

$$r = \sqrt{0^2 + 12^2} = \sqrt{0 + 144} = \sqrt{144} = 12$$

$0 + 12i$ is on the positive $y$-axis, so $0 = 90°$.

$12i = 12(\cos 90° + i\sin 90°)$

**57.** $3+5i$

$x=3, y=5 \Rightarrow r=\sqrt{3^2+5^2}=\sqrt{9+25}=\sqrt{34}$

$\tan\theta=\dfrac{5}{3}$

$3+5i$ is in the quadrant I, so $\theta=59.04°$.

$3+5i=\sqrt{34}\left(\cos 59.04° + i\sin 59.04°\right)$

**59.** Since the modulus represents the magnitude of the vector in the complex plane, $r=1$ would represent a circle of radius one centered at the origin.

**61.** Since the real part of $z=x+yi$ is 1, the graph of $1+yi$ would be the vertical line $x=1$.

**63.** $z=-.2i$

$z^2-1=(-.2i)^2-1=.04i^2-1=.04(-1)-1$
$=-.04-1=-1.04$

The modulus is 1.04.

$(z^2-1)^2-1=(-1.04)^2-1=1.0816-1=.0816$

The modulus is .0816.

$\left[\left(z^2-1\right)^2-1\right]^2-1=(.0816)^2-1$

$=.0665856-1$
$=-.99334144$

The modulus is .99334144 .

The moduli do not exceed 2. Therefore, $z$ is in the Julia set.

**65.** Let $z=r\left(\cos\theta+i\sin\theta\right)$. This represents a vector with magnitude $r$ and angle $\theta$. The conjugate of $z$ is the vector with magnitude $r$ pointing in the $-\theta$ direction. Thus, $r\left[\cos\left(360°-\theta\right)+i\sin\left(360°-\theta\right)\right]$ satisfies these conditions as shown below.

$r\left[\cos\left(360°-\theta\right)+i\sin\left(360°-\theta\right)\right]$

$=r\left[\begin{array}{l}(\cos 360°\cos\theta+\sin 360°\sin\theta)\\ \quad +i(\sin 360°\cos\theta-\cos 360°\sin\theta)\end{array}\right]$

$=r\left[\begin{array}{l}(1\cdot\cos\theta+0\cdot\sin\theta)\\ \quad +i(0\cdot\cos\theta-1\cdot\sin\theta)\end{array}\right]$

$=r\left(\cos\theta-i\sin\theta\right)$

$=r\left[\cos(-\theta)+i\sin(-\theta)\right]$

This represents the conjugate of $z$, which is a reflection over the $x$-axis.

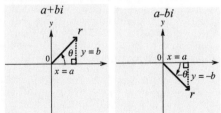

**67.** $(a+bi)-(c+di)=e+0i$, so $b-d=0 \Rightarrow b=d$

Therefore, the terminal points of the vectors corresponding to $a+bi$ and $c+di$ lie on a horizontal line. The answer is B.

**69.** The difference of the vectors is equal to the sum of one of the vectors plus the opposite of the other vector. If this sum is represented by two parallel vectors, the sum of their absolute values will equal the absolute value of their sum. In order for this sum to be represented by two parallel vectors, their difference must be represented by two vectors with opposite directions. The answer is A.

## Section 8.3: The Product and Quotient Theorems

**1.** When multiplying two complex numbers in trigonometric form, we <u>multiply</u> their absolute values and <u>add</u> their arguments.

**2.** When dividing two complex numbers in trigonometric form, we <u>divide</u> their absolute values and <u>subtract</u> their arguments.

**3.** $\left[3\left(\cos 60°+i\sin 60°\right)\right]\left[2\left(\cos 90°+i\sin 90°\right)\right]$

$=3\cdot 2\left[\cos\left(60°+90°\right)+i\sin\left(60°+90°\right)\right]$

$=6\left(\cos 150°+i\sin 150°\right)=6\left(-\dfrac{\sqrt{3}}{2}+\dfrac{1}{2}i\right)$

$=-3\sqrt{3}+3i$

**5.** $\left[4\left(\cos 60°+i\sin 60°\right)\right]\left[6\left(\cos 330°+i\sin 330°\right)\right]$

$=4\cdot 6\left[\left(\cos\left(60°+330°\right)+i\sin\left(60°+330°\right)\right)\right]$

$=24\left(\cos 390°+i\sin 390°\right)$

$=24\left(\cos 30°+i\sin 30°\right)$

$=24\left(\dfrac{\sqrt{3}}{2}+i\dfrac{1}{2}\right)=12\sqrt{3}+12i$

**7.** $\left[2\left(\cos 45°+i\sin 45°\right)\right]\cdot\left[2\left(\cos 225°+i\sin 225°\right)\right]$

$=2\cdot 2\left[\cos\left(45°+225°\right)+i\sin\left(45°+225°\right)\right]$

$=4\left(\cos 270°+i\sin 270°\right)=4\left(0-i\right)$

$=0-4i$ or $-4i$

**9.** $\left[\sqrt{3}\ \text{cis}\ 45°\right]\left[\sqrt{3}\ \text{cis}\ 225°\right]$

$=\sqrt{3}\cdot\sqrt{3}\left[\text{cis}\left(45°+225°\right)\right]$

$=3\ \text{cis}\ 270°=3\left(\cos 270°+i\sin 270°\right)$

$=3\left(0-i\right)=0-3i$ or $-3i$

**11.** $[5 \text{ cis } 90°][3 \text{ cis } 45°] = 5 \cdot 3 \left[ \text{cis} \left( 90° + 45° \right) \right] = 15 \text{ cis } 135°$

$$= 15 \left( \cos 135° + i \sin 135° \right) = 15 \left( -\frac{\sqrt{2}}{2} + \frac{\sqrt{2}}{2} i \right) = -\frac{15\sqrt{2}}{2} + \frac{15\sqrt{2}}{2} i$$

**13.** $\dfrac{4 \left( \cos 120° + i \sin 120° \right)}{2 \left( \cos 150° + i \sin 150° \right)} = \dfrac{4}{2} \left[ \cos \left( 120° - 150° \right) + i \sin \left( 120° - 150° \right) \right]$

$$= 2 \left( \cos \left( -30° \right) + i \sin \left( -30° \right) \right) = 2 \left( \cos 30° - i \sin 30° \right)$$

$$= 2 \left( \frac{\sqrt{3}}{2} - \frac{1}{2} i \right) = \sqrt{3} - i$$

**15.** $\dfrac{10 \left( \cos 230° + i \sin 230° \right)}{5 \left( \cos 50° + i \sin 50° \right)} = \dfrac{10}{5} \left[ \cos \left( 230° - 50° \right) + i \sin \left( 230° - 50° \right) \right]$

$$= 2 \left( \cos 180° + i \sin 180° \right) = 2 \left( -1 + 0 \cdot i \right) = -2 + 0i \text{ or } -2$$

**17.** $\dfrac{3 \text{ cis } 305°}{9 \text{ cis } 65°} = \dfrac{1}{3} \text{ cis } \left( 305° - 65° \right) = \dfrac{1}{3} \left( \text{cis } 240° \right) = \dfrac{1}{3} \left( \cos 240° + i \sin 240° \right) = \dfrac{1}{3} \left( -\dfrac{1}{2} - \dfrac{\sqrt{3}}{2} i \right) = -\dfrac{1}{6} - \dfrac{\sqrt{3}}{6} i$

**19.** $\dfrac{8}{\sqrt{3} + i}$

numerator: $8 = 8 + 0i$ and $r = \sqrt{8^2 + 0^2} = 8$

$\theta = 0°$ since $\cos 0° = 1$ and $\sin 0° = 0$, so

$8 = 8 \text{ cis } 0°$.

denominator: $\sqrt{3} + i$ and

$r = \sqrt{\left( \sqrt{3} \right)^2 + 1^2} = \sqrt{3+1} = \sqrt{4} = 2$

$\tan \theta = \dfrac{1}{\sqrt{3}} = \dfrac{\sqrt{3}}{3}$

Since $x$ and $y$ are both positive, $\theta$ is in quadrant I, so $\theta = 30°$. Thus

$\sqrt{3} + i = 2 \text{ cis } 30°$.

$\dfrac{8}{\sqrt{3} + i} = \dfrac{8 \text{ cis } 0°}{2 \text{ cis } 30°} = \dfrac{8}{2} \text{ cis } \left( 0 - 30° \right)$

$$= 4 \left[ \cos \left( -30° \right) + i \sin \left( -30° \right) \right]$$

$$= 4 \left( \frac{\sqrt{3}}{2} - \frac{1}{2} i \right) = 2\sqrt{3} - 2i$$

**21.** $\dfrac{-i}{1+i}$

numerator: $-i = 0 - i$ and

$r = \sqrt{0^2 + \left( -1 \right)^2} = \sqrt{0+1} = \sqrt{1} = 1$

$\theta = 270°$ since $\cos 270° = 0$ and

$\sin 270° = -1$, so $-i = 1 \text{ cis } 270°$.

denominator: $1 + i$

$r = \sqrt{1^2 + 1^2} = \sqrt{1+1} = \sqrt{2}$ and

$\tan \theta = \dfrac{y}{x} = \dfrac{1}{1} = 1$

Since $x$ and $y$ are both positive, $\theta$ is in quadrant I, so $\theta = 45°$. Thus,

$1 + i = \sqrt{2} \text{ cis } 45°$

$\dfrac{-i}{1+i} = \dfrac{\text{cis } 270°}{\sqrt{2} \text{ cis } 45°} = \dfrac{1}{\sqrt{2}} \text{ cis } \left( 270° - 45° \right)$

$$= \frac{\sqrt{2}}{2} \text{ cis } 225°$$

$$= \frac{\sqrt{2}}{2} \left( \cos 225° + i \sin 225° \right)$$

$$= \frac{\sqrt{2}}{2} \left( -\frac{\sqrt{2}}{2} - i \cdot \frac{\sqrt{2}}{2} \right) = -\frac{1}{2} - \frac{1}{2} i$$

**23.** $\dfrac{2\sqrt{6}-2i\sqrt{2}}{\sqrt{2}-i\sqrt{6}}$

numerator: $2\sqrt{6}-2i\sqrt{2}$ and

$r=\sqrt{\left(2\sqrt{6}\right)^2+\left(-2\sqrt{2}\right)^2}=\sqrt{24+8}$

$=\sqrt{32}=4\sqrt{2}$

$\tan\theta=\dfrac{-2\sqrt{2}}{2\sqrt{6}}=-\dfrac{1}{\sqrt{3}}=-\dfrac{\sqrt{3}}{3}$

Since $x$ is positive and $y$ is negative, $\theta$ is in quadrant IV, so $\theta=-30°$. Thus,

$2\sqrt{6}-2i\sqrt{2}=4\sqrt{2}\,\text{cis}\left(-30°\right).$

denominator: $\sqrt{2}-i\sqrt{6}$ and

$r=\sqrt{\left(\sqrt{2}\right)^2+\left(-\sqrt{6}\right)^2}=\sqrt{2+6}=\sqrt{8}=2\sqrt{2}$

$\tan\theta=\dfrac{-\sqrt{6}}{\sqrt{2}}=-\sqrt{3}$

Since $x$ is positive and $y$ is negative, $\theta$ is in quadrant IV, so $\theta=-60°$. Thus,

$\sqrt{2}-i\sqrt{6}=2\sqrt{2}\,\text{cis}\left(-30°\right)$

$\dfrac{2\sqrt{6}-2i\sqrt{2}}{\sqrt{2}-i\sqrt{6}}=\dfrac{4\sqrt{2}\,\text{cis}\left(-30°\right)}{2\sqrt{2}\,\text{cis}\left(-60°\right)}$

$=\dfrac{4\sqrt{2}}{2\sqrt{2}}\,\text{cis}\left[-30°-\left(-60°\right)\right]$

$=2\,\text{cis}\,30°=2\left(\cos 30°+i\sin 30°\right)$

$=2\left(\dfrac{\sqrt{3}}{2}+i\dfrac{1}{2}\right)=\sqrt{3}+i$

**25.** $\left[2.5\left(\cos 35°+i\sin 35°\right)\right]\left[3.0\left(\cos 50°+i\sin 50°\right)\right]$

$=2.5\cdot 3.0\left[\cos\left(35°+50°\right)+i\sin\left(35°+50°\right)\right]$

$=7.5\left(\cos 85°+i\sin 85°\right)\approx.6537+7.4715i$

**27.** $\left(12\,\text{cis}\,18.5°\right)\left(3\,\text{cis}\,12.5°\right)$

$=12\cdot 3\,\text{cis}\left(18.5°+12.5°\right)$

$=36\,\text{cis}\,31°=36\left(\cos 31°+i\sin 31°\right)$

$\approx 30.8580+18.5414i$

**29.** $\dfrac{45\left(\cos 127°+i\sin 127°\right)}{22.5\left(\cos 43°+i\sin 43°\right)}$

$=\dfrac{45}{22.5}\left[\cos\left(127°-43°\right)+i\sin\left(127°-43°\right)\right]$

$=2\left(\cos 84°+i\sin 84°\right)\approx.2091+1.9890i$

**31.** $\left[2\,\text{cis}\,\dfrac{5\pi}{9}\right]^2=\left[2\,\text{cis}\,\dfrac{5\pi}{9}\right]\left[2\,\text{cis}\,\dfrac{5\pi}{9}\right]$

$=2\cdot 2\,\text{cis}\left(\dfrac{5\pi}{9}+\dfrac{5\pi}{9}\right)$

$=4\left(\cos\dfrac{10\pi}{9}+i\sin\dfrac{10\pi}{9}\right)$

$\approx-3.7588-1.3681i$

In Exercises 33–39, $w=-1+i$ and $z=-1-i$.

**33.** $w\cdot z=\left(-1+i\right)\left(-1-i\right)$

$=-1(-1)+(-1)(-i)+i(-1)+i(-i)$

$=1+i-i-i^2=1-(-1)=2$

**35.** $w\cdot z=\left(\sqrt{2}\,\text{cis}\,135°\right)\left(\sqrt{2}\,\text{cis}\,225°\right)$

$=\sqrt{2}\cdot\sqrt{2}\left[\,\text{cis}\,\left(135°+225°\right)\right]$

$=2\,\text{cis}\,360°=2\,\text{cis}\,0°$

**37.** $\dfrac{w}{z}=\dfrac{-1+i}{-1-i}=\dfrac{-1+i}{-1-i}\cdot\dfrac{-1+i}{-1+i}=\dfrac{1-i-i+i^2}{1-i^2}$

$=\dfrac{1-2i+(-1)}{1-(-1)}=\dfrac{-2i}{2}=-i$

**39.** $\text{cis}\left(-90°\right)=\cos\left(-90°\right)+i\sin\left(-90°\right)$

$=0+i(-1)=0-i=-i$

It is the same.

**41.** The two results will have the same magnitude because in both cases you are finding the product of 2 and 5. We must now determine if the arguments are the same. In the first product, the argument of the product will be $45°+90°=135°$. In the second product, the argument of the product will be $-315°+(-270°)=-585°$. Now, $-585°$ is coterminal with $-585°+2\cdot 360°=135°$. Thus, the two products are the same since they have the same magnitude and argument.

**43.** $E=8\left(\cos 20°+i\sin 20°\right),R=6,X_L=3,$

$I=\dfrac{E}{Z},Z=R+X_L i$

Write $Z=6+3i$ in trigonometric form.

$x=6$, and $y=3\Rightarrow r=\sqrt{6^2+3^2}=\sqrt{36+9}$

$=\sqrt{45}=3\sqrt{5}$.

$\tan\theta=\dfrac{3}{6}=\dfrac{1}{2}$, so $\theta\approx 26.6°$. Thus,

$Z=3\sqrt{5}\,\text{cis}\,26.6°$.

*(continued on next page)*

*(continued from page 209)*

$$I = \frac{8 \text{ cis } 20°}{3\sqrt{5} \text{ cis } 26.6°} = \frac{8}{3\sqrt{5}} \text{ cis } (20° - 26.6°)$$

$$= \frac{8\sqrt{5}}{15} \text{ cis } (-6.6°)$$

$$= \frac{8\sqrt{5}}{15} \left[ \cos(-6.6°) + i \sin(-6.6°) \right]$$

$$\approx 1.18 - .14i$$

**45.** Since $Z_1 = 50 + 25i$ and $Z_2 = 60 + 20i$, we have

$$\frac{1}{Z_1} = \frac{1}{50 + 25i} \cdot \frac{50 - 25i}{50 - 25i} = \frac{50 - 25i}{50^2 - 25^2 i^2}$$

$$= \frac{50 - 25i}{2500 - 625(-1)} = \frac{50 - 25i}{2500 + 625}$$

$$= \frac{50 - 25i}{3125} = \frac{2}{125} - \frac{1}{125} i \quad \text{and}$$

$$\frac{1}{Z_2} = \frac{1}{60 + 20i} \cdot \frac{60 - 20i}{60 - 20i} = \frac{60 - 20i}{60^2 - 20^2 i^2}$$

$$= \frac{60 - 20i}{3600 - 400(-1)} = \frac{60 - 20i}{3600 + 400}$$

$$= \frac{60 - 20i}{4000} = \frac{3}{200} - \frac{1}{200} i$$

$$\frac{1}{Z_1} + \frac{1}{Z_2} = \left( \frac{2}{125} - \frac{1}{125} i \right) + \left( \frac{3}{200} - \frac{1}{200} i \right)$$

$$= \left( \frac{2}{125} + \frac{3}{200} \right) - \left( \frac{1}{125} + \frac{1}{200} \right) i$$

$$= \left( \frac{16}{1000} + \frac{15}{1000} \right) - \left( \frac{8}{1000} + \frac{5}{1000} \right) i$$

$$= \frac{31}{1000} - \frac{13}{1000} i$$

$$Z = \frac{1}{\dfrac{1}{Z_1} + \dfrac{1}{Z_2}} = \frac{1}{\dfrac{31}{1000} - \dfrac{13}{1000} i}$$

$$= \frac{1000}{31 - 13i} \cdot \frac{31 + 13i}{31 + 13i} = \frac{1000(31 + 13i)}{31^2 - 13^2 i^2}$$

$$= \frac{31,000 + 13,000i}{961 - 169(-1)} = \frac{31,000 + 13,000i}{961 + 169}$$

$$= \frac{31,000 + 13,000i}{1130} = \frac{3100}{113} + \frac{1300}{113} i$$

$$\approx 27.43 + 11.5i$$

# Section 8.4: DeMoivre's Theorem; Powers and Roots of Complex Numbers

**1.** $\left[ 3 \left( \cos 30° + i \sin 30° \right) \right]^3$

$$= 3^3 \left[ \cos(3 \cdot 30°) + i \sin(3 \cdot 30°) \right]$$

$$= 27 \left( \cos 90° + i \sin 90° \right)$$

$$= 27 \left( 0 + 1 \cdot i \right) = 0 + 27i \text{ or } 27i$$

**3.** $\left( \cos 45° + i \sin 45° \right)^8$

$$= \left[ \cos(8 \cdot 45°) + i \sin(8 \cdot 45°) \right]$$

$$= \cos 360° + i \sin 360° = 1 + 0 \cdot i \text{ or } 1$$

**5.** $\left[ 3 \text{ cis } 100° \right]^3$

$$= 3^3 \text{ cis } (3 \cdot 100°)$$

$$= 27 \text{ cis } 300° = 27 \left( \cos 300° + i \sin 300° \right)$$

$$= 27 \left( \frac{1}{2} - \frac{\sqrt{3}}{2} i \right) = \frac{27}{2} - \frac{27\sqrt{3}}{2} i$$

**7.** $\left( \sqrt{3} + i \right)^5$

First write $\sqrt{3} + i$ in trigonometric form.

$$r = \sqrt{\left( \sqrt{3} \right)^2 + 1^2} = \sqrt{3 + 1} = \sqrt{4} = 2 \quad \text{and}$$

$$\tan \theta = \frac{1}{\sqrt{3}} = \frac{\sqrt{3}}{3}$$

Because $x$ and $y$ are both positive, $\theta$ is in quadrant I, so $\theta = 30°$.

$$\sqrt{3} + i = 2 \left( \cos 30° + i \sin 30° \right)$$

$$\left( \sqrt{3} + i \right)^5 = \left[ 2 \left( \cos 30° + i \sin 30° \right) \right]^5$$

$$= 2^5 \left[ \cos(5 \cdot 30°) + i \sin(5 \cdot 30°) \right]$$

$$= 32 \left( \cos 150° + i \sin 150° \right)$$

$$= 32 \left( -\frac{\sqrt{3}}{2} + i \frac{1}{2} \right) = -16\sqrt{3} + 16i$$

**9.** $\left( 2\sqrt{2} - 2i\sqrt{2} \right)^6$

First write $2\sqrt{2} - 2i\sqrt{2}$ in trigonometric form.

$$r = \sqrt{\left( 2\sqrt{2} \right)^2 + \left( -2\sqrt{2} \right)^2} = \sqrt{8 + 8} = \sqrt{16} = 4$$

and $\tan \theta = \dfrac{-2\sqrt{2}}{2\sqrt{2}} = -1$

Because $x$ is positive and $y$ is negative, $\theta$ is in quadrant IV, so $\theta = 315°$.

$$2\sqrt{2} - 2i\sqrt{2} = 4 \left( \cos 315° + i \sin 315° \right)^6$$

$$\left(2\sqrt{2} - 2i\sqrt{2}\right)^{6}$$

$$= \left[4\left(\cos 315^{\circ} + i\sin 315^{\circ}\right)\right]^{6}$$

$$= 4^{6}\left[\cos\left(6\cdot 315^{\circ}\right) + i\sin\left(6\cdot 315^{\circ}\right)\right]$$

$$= 4096\left[\cos 1890^{\circ} + i\sin 1890^{\circ}\right]$$

$$= 4096\left(\cos 90^{\circ} + i\sin 90^{\circ}\right)$$

$$= 4096\left(0 + 1\cdot i\right) = 0 + 4096i \text{ or } 4096i$$

**11.** $(-2 - 2i)^{5}$

First write $-2 - 2i$ in trigonometric form.

$$r = \sqrt{(-2)^{2} + (-2)^{2}} = \sqrt{4+4} = \sqrt{8} = 2\sqrt{2} \text{ and}$$

$$\tan\theta = \frac{-2}{-2} = 1$$

Because $x$ and $y$ are both negative, $\theta$ is in quadrant III, so $\theta = 225^{\circ}$.

$$-2 - 2i = 2\sqrt{2}\left(\cos 225^{\circ} + i\sin 225^{\circ}\right)$$

$$(-2 - 2i)^{5}$$

$$= \left[2\sqrt{2}\left(\cos 225^{\circ} + i\sin 225^{\circ}\right)\right]^{5}$$

$$= \left(2\sqrt{2}\right)^{5}\left[\cos\left(5\cdot 225^{\circ}\right) + i\sin\left(5\cdot 225^{\circ}\right)\right]$$

$$= 32\sqrt{32}\left(\cos 1125^{\circ} + i\sin 1125^{\circ}\right)$$

$$= 128\sqrt{2}\left(\cos 45^{\circ} + i\sin 45^{\circ}\right)$$

$$= 128\sqrt{2}\left(\frac{\sqrt{2}}{2} + \frac{\sqrt{2}}{2}i\right) = 128 + 128i$$

**13. (a)** $\cos 0^{\circ} + i\sin 0^{\circ} = 1\left(\cos 0^{\circ} + i\sin 0^{\circ}\right)$

We have $r = 1$ and $\theta = 0^{\circ}$. Since $r^{3}\left(\cos 3\alpha + i\sin 3\alpha\right) = 1\left(\cos 0^{\circ} + i\sin 0^{\circ}\right)$,

then we have $r^{3} = 1 \Rightarrow r = 1$ and

$$3\alpha = 0^{\circ} + 360^{\circ}\cdot k \Rightarrow \alpha = \frac{0^{\circ} + 360^{\circ}\cdot k}{3}$$

$$= 0^{\circ} + 120^{\circ}\cdot k = 120^{\circ}\cdot k, \ k \text{ any integer.}$$

If $k = 0$, then $\alpha = 0^{\circ}$.

If $k = 1$, then $\alpha = 120^{\circ}$.

If $k = 2$, then $\alpha = 240^{\circ}$.

So, the cube roots are

$\cos 0^{\circ} + i\sin 0^{\circ}$, $\cos 120^{\circ} + i\sin 120^{\circ}$,

and $\cos 240^{\circ} + i\sin 240^{\circ}$.

**(b)**

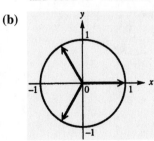

**15. (a)** Find the cube roots of 8 cis 60°

We have $r = 8$ and $\theta = 60^{\circ}$.

Since $r^{3}\left(\cos 3\alpha + i\sin 3\alpha\right)$

$$= 8\left(\cos 60^{\circ} + i\sin 60^{\circ}\right), \text{ we have}$$

$$r^{3} = 8 \Rightarrow r = 2 \text{ and } 3\alpha = 60^{\circ} + 360^{\circ}\cdot k \Rightarrow$$

$$\alpha = \frac{60^{\circ} + 360^{\circ}\cdot k}{3} = 20^{\circ} + 120^{\circ}\cdot k, \ k \text{ any}$$

integer. If $k = 0$, then $\alpha = 20^{\circ} + 0^{\circ} = 20^{\circ}$.

If $k = 1$, then $\alpha = 20^{\circ} + 120^{\circ} = 140^{\circ}$.

If $k = 2$, then $\alpha = 20^{\circ} + 240^{\circ} = 260^{\circ}$.

So, the cube roots are

2 cis 20°, 2 cis 140°, and 2 cis 260°.

**(b)**

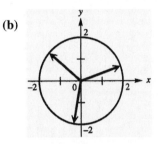

**17. (a)** Find the cube roots of

$-8i = 8\left(\cos 270^{\circ} + i\sin 270^{\circ}\right)$

We have $r = 8$ and $\theta = 270^{\circ}$.

Since $r^{3}\left(\cos 3\alpha + i\sin 3\alpha\right)$

$$= 8\left(\cos 270^{\circ} + i\sin 270^{\circ}\right), \text{ then we have}$$

$$r^{3} = 8 \Rightarrow r = 2 \text{ and}$$

$$3\alpha = 270^{\circ} + 360^{\circ}\cdot k \Rightarrow$$

$$\alpha = \frac{270^{\circ} + 360^{\circ}\cdot k}{3} = 90^{\circ} + 120^{\circ}\cdot k, \ k \text{ any}$$

integer. If $k = 0$, then $\alpha = 90^{\circ} + 0^{\circ} = 90^{\circ}$.

If $k = 1$, then $\alpha = 90^{\circ} + 120^{\circ} = 210^{\circ}$.

If $k = 2$, then $\alpha = 90^{\circ} + 240^{\circ} = 330^{\circ}$.

So, the cube roots are

$2\left(\cos 90^{\circ} + i\sin 90^{\circ}\right)$,

$2\left(\cos 210^{\circ} + i\sin 210^{\circ}\right)$, and

$2\left(\cos 330^{\circ} + i\sin 330^{\circ}\right)$.

**(b)**

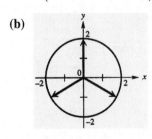

**19. (a)** Find the cube roots of
$-64 = 64(\cos 180° + i \sin 180°)$
We have $r = 64$ and $\theta = 180°$.
Since $r^3\left(\cos 3\alpha + i\sin 3\alpha\right)$
$= 64\left(\cos 180° + i\sin 180°\right)$, then we have

$r^3 = 64 \Rightarrow r = 4$ and
$3\alpha = 180° + 360° \cdot k \Rightarrow$
$\alpha = \dfrac{180° + 360° \cdot k}{3} = 60° + 120° \cdot k,\ k$ any
integer.
If $k = 0$, then $\alpha = 60° + 0° = 60°$.
If $k = 1$, then $\alpha = 60° + 120° = 180°$.
If $k = 2$, then $\alpha = 60° + 240° = 300°$.
So, the cube roots are
$4\left(\cos 60° + i\sin 60°\right)$,
$4\left(\cos 180° + i\sin 180°\right)$, and
$4\left(\cos 300° + i\sin 300°\right)$.

**(b)**

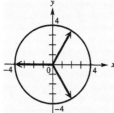

**21. (a)** Find the cube roots of $1 + i\sqrt{3}$.
We have

$r = \sqrt{1^2 + \left(\sqrt{3}\right)^2} = \sqrt{1 + 3} = \sqrt{4} = 2$ and

$\tan\theta = \dfrac{\sqrt{3}}{1} = \sqrt{3}$. Since $\theta$ is in

quadrant I, $\theta = 60°$. Thus,

$1 + i\sqrt{3} = 2\left(\dfrac{1}{2} + i\dfrac{\sqrt{3}}{2}\right)$
$= 2\left(\cos 60° + i\sin 60°\right)$.

Since $r^3\left(\cos 3\alpha + i\sin 3\alpha\right)$
$= 2\left(\cos 60° + i\sin 60°\right)$, then we have
$r^3 = 2 \Rightarrow r = \sqrt[3]{2}$ and
$3\alpha = 60° + 360° \cdot k \Rightarrow$
$\alpha = \dfrac{60° + 360° \cdot k}{3} = 20° + 120° \cdot k,\ k$ any
integer.
If $k = 0$, then $\alpha = 20° + 0° = 20°$.
If $k = 1$, then $\alpha = 20° + 120° = 140°$.
If $k = 2$, then $\alpha = 20° + 240° = 260°$.

So, the cube roots are
$\sqrt[3]{2}\left(\cos 20° + i\sin 20°\right)$,
$\sqrt[3]{2}\left(\cos 140° + i\sin 140°\right)$, and
$\sqrt[3]{2}\left(\cos 260° + i\sin 260°\right)$.

**(b)**

**23. (a)** Find the cube roots of $-2\sqrt{3} + 2i$.

We have $r = \sqrt{\left(-2\sqrt{3}\right)^2 + 2^2} = \sqrt{12 + 4}$

$= \sqrt{16} = 4$ and $\tan\theta = \dfrac{2}{-2\sqrt{3}} = -\dfrac{\sqrt{3}}{3}$.

Since $\theta$ is in quadrant II, $\theta = 150°$. Thus,

$-2\sqrt{3} + 2i = 4\left(-\dfrac{\sqrt{3}}{2} + \dfrac{1}{2}i\right)$
$= 4\left(\cos 150° + i\sin 150°\right)$.

Since $r^3\left(\cos 3\alpha + i\sin 3\alpha\right)$
$= 4\left(\cos 150° + i\sin 150°\right)$, then we have
$r^3 = 4 \Rightarrow r = \sqrt[3]{4}$ and
$3\alpha = 150° + 360° \cdot k \Rightarrow$
$\alpha = \dfrac{150° + 360° \cdot k}{3} = 50° + 120° \cdot k,\ k$ any
integer. If $k = 0$, then $\alpha = 50° + 0° = 50°$.
If $k = 1$, then $\alpha = 50° + 120° = 170°$.
If $k = 2$, then $\alpha = 50° + 240° = 290°$.
So, the cube roots are
$\sqrt[3]{4}\left(\cos 50° + i\sin 50°\right)$,
$\sqrt[3]{4}\left(\cos 170° + i\sin 170°\right)$, and
$\sqrt[3]{4}\left(\cos 290° + i\sin 290°\right)$.

**(b)**

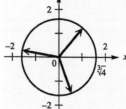

**25.** Find all the second (or square) roots of
$1 = 1(\cos 0° + i\sin 0°)$.

Since $r^2(\cos 2\alpha + i\sin 2\alpha)$

$= 1(\cos 0° + i\sin 0°)$, then we have

$2\alpha = 0° + 360° \cdot k \Rightarrow$

$\alpha = \dfrac{0° + 360° \cdot k}{2} = 0° + 180° \cdot k = 180° \cdot k, \ k$

any integer. If $k = 0$, then $\alpha = 0°$.
If $k = 1$, then $\alpha = 180°$. So, the second roots
of 1 are
$\cos 0° + i\sin 0°$, and $\cos 180° + i\sin 180°$. (or 1
and $-1$)

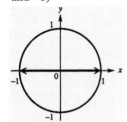

**27.** Find all the sixth roots of
$1 = 1(\cos 0° + i\sin 0°)$.

Since $r^6(\cos 6\alpha + i\sin 6\alpha)$

$= 1(\cos 0° + i\sin 0°)$, then we have

$r^6 = 1 \Rightarrow r = 1$ and $6\alpha = 0° + 360° \cdot k \Rightarrow$

$\alpha = \dfrac{0° + 360° \cdot k}{6} = 0° + 60° \cdot k = 60° \cdot k, \ k$ any

integer. If $k = 0$, then $\alpha = 0°$.
If $k = 1$, then $\alpha = 60°$.
If $k = 2$, then $\alpha = 120°$.
If $k = 3$, then $\alpha = 180°$.
If $k = 4$, then $\alpha = 240°$.
If $k = 5$, then $\alpha = 300°$. So, the sixth roots of
1 are $\cos 0° + i\sin 0°$, $\cos 60° + i\sin 60°$,
$\cos 120° + i\sin 120°$, $\cos 180° + i\sin 180°$,
$\cos 240° + i\sin 240°$, and $\cos 300° + i\sin 300°$.

$\left(\text{or } 1, \ \dfrac{1}{2} + \dfrac{\sqrt{3}}{2}i, \ -\dfrac{1}{2} + \dfrac{\sqrt{3}}{2}i, \ -1,\right.$

$\left.-\dfrac{1}{2} - \dfrac{\sqrt{3}}{2}i, \ \text{and } \dfrac{1}{2} - \dfrac{\sqrt{3}}{2}i\right)$

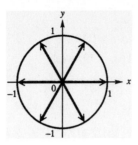

**29.** Find all the third (cube) roots of
$i = 1(\cos 90° + i\sin 90°)$.

Since

$r^3(\cos 3\alpha + i\sin 3\alpha) = 1(\cos 90° + i\sin 90°)$, then

we have $r^3 = 1 \Rightarrow r = 1$ and
$3\alpha = 90° + 360° \cdot k \Rightarrow$

$\alpha = \dfrac{90° + 360° \cdot k}{3} = 30° + 120° \cdot k, \ k$ any integer.

If $k = 0$, then $\alpha = 30° + 0° = 30°$.
If $k = 1$, then $\alpha = 30° + 120° = 150°$.
If $k = 2$, then $\alpha = 30° + 240° = 270°$. So, the

third roots of $i$ are $\cos 30° + i\sin 30°$,
$\cos 150° + i\sin 150°$, and $\cos 270° + i\sin 270°$.

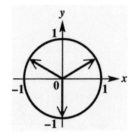

**31.** $x^3 - 1 = 0 \Rightarrow x^3 = 1$
We have $r = 1$ and $\theta = 0°$.

$x^3 = 1 = 1 + 0i = 1(\cos 0° + i\sin 0°)$

Since $r^3(\cos 3\alpha + i\sin 3\alpha)$

$= 1(\cos 0° + i\sin 0°)$, then we have

$r^3 = 1 \Rightarrow r = 1$ and $3\alpha = 0° + 360° \cdot k \Rightarrow$

$\alpha = \dfrac{0° + 360° \cdot k}{3} = 0° + 120° \cdot k = 120° \cdot k, \ k$

any integer. If $k = 0$, then $\alpha = 0°$.
If $k = 1$, then $\alpha = 120°$.
If $k = 2$, then $\alpha = 240°$.
Solution set:
$\{\cos 0° + i\sin 0°, \ \cos 120° + i\sin 120°,$
$\quad \cos 240° + i\sin 240°\}$

or $\left\{1, \ -\dfrac{1}{2} + \dfrac{\sqrt{3}}{2}i, \ -\dfrac{1}{2} - \dfrac{\sqrt{3}}{2}i\right\}$

**33.** $x^3 + i = 0 \Rightarrow x^3 = -i$

We have $r = 1$ and $\theta = 270°$.

$x^3 = -i = 0 - i = 1(\cos 270° + i \sin 270°)$

Since $r^3 (\cos 3\alpha + i \sin 3\alpha)$

$= 1(\cos 270° + i \sin 270°)$, then we have

$r^3 = 1 \Rightarrow r = 1$ and $3\alpha = 270° + 360° \cdot k \Rightarrow$

$\alpha = \dfrac{270° + 360° \cdot k}{3} = 90° + 120° \cdot k$, $k$ any

integer. If $k = 0$, then $\alpha = 90° + 0° = 90°$.

If $k = 1$, then $\alpha = 90° + 120° = 210°$.

If $k = 2$, then $\alpha = 90° + 240° = 330°$.

Solution set:

$\{\cos 90° + i \sin 90°,\ \cos 210° + i \sin 210°,$

$\cos 330° + i \sin 330°\}$ or

$\left\{0,\ -\dfrac{\sqrt{3}}{2} - \dfrac{1}{2}i,\ \dfrac{\sqrt{3}}{2} - \dfrac{1}{2}i\right\}$

**35.** $x^3 - 8 = 0 \Rightarrow x^3 = 8$

We have $r = 8$ and $\theta = 0°$.

$x^3 = 8 = 8 + 0i = 8(\cos 0° + i \sin 0°)$

Since $r^3 (\cos 3\alpha + i \sin 3\alpha)$

$= 8(\cos 0° + i \sin 0°)$, then we have

$r^3 = 8 \Rightarrow r = 2$ and $3\alpha = 0° + 360° \cdot k \Rightarrow$

$\alpha = \dfrac{0° + 360° \cdot k}{3} = 0° + 120° \cdot k = 120° \cdot k$, $k$

any integer. If $k = 0$, then $\alpha = 0°$.

If $k = 1$, then $\alpha = 120°$. If $k = 2$, then $\alpha = 240°$.

Solution set:

$\{2(\cos 0° + i \sin 0°),\ 2(\cos 120° + i \sin 120°),$

$2(\cos 240° + i \sin 240°)\}$ or

$\left\{2,\ -1 + \sqrt{3}i,\ -1 - \sqrt{3}i\right\}$

**37.** $x^4 + 1 = 0 \Rightarrow x^4 = -1$

We have $r = 1$ and $\theta = 180°$.

$x^4 = -1 = -1 + 0i = 1(\cos 180° + i \sin 180°)$

Since $r^4 (\cos 4\alpha + i \sin 4\alpha)$

$= 1(\cos 180° + i \sin 180°)$, then we have

$r^4 = 1 \Rightarrow r = 1$ and $4\alpha = 180° + 360° \cdot k \Rightarrow$

$\alpha = \dfrac{180° + 360° \cdot k}{4} = 45° + 90° \cdot k$, $k$ any

integer.

If $k = 0$, then $\alpha = 45° + 0° = 45°$.

If $k = 1$, then $\alpha = 45° + 90° = 135°$.

If $k = 2$, then $\alpha = 45° + 180° = 225°$.

If $k = 3$, then $\alpha = 45° + 270° = 315°$.

Solution set:

$\{\cos 45° + i \sin 45°,\ \cos 135° + i \sin 135°,$

$\cos 225° + i \sin 225°,\ \cos 315° + i \sin 315°\}$ or

$\left\{\dfrac{\sqrt{2}}{2} + \dfrac{\sqrt{2}}{2}i,\ -\dfrac{\sqrt{2}}{2} + \dfrac{\sqrt{2}}{2}i,\ -\dfrac{\sqrt{2}}{2} - \dfrac{\sqrt{2}}{2}i,\right.$

$\left.\dfrac{\sqrt{2}}{2} - \dfrac{\sqrt{2}}{2}i\right\}$

**39.** $x^4 - i = 0 \Rightarrow x^4 = i$

We have $r = 1$ and $\theta = 90°$.

$x^4 = i = 0 + i = 1(\cos 90° + i \sin 90°)$

Since $r^4 (\cos 4\alpha + i \sin 4\alpha)$

$= 1(\cos 90° + i \sin 90°)$, then we have

$r^4 = 1 \Rightarrow r = 1$ and $4\alpha = 90° + 360° \cdot k \Rightarrow$

$\alpha = \dfrac{90° + 360° \cdot k}{4} = 22.5° + 90° \cdot k$, $k$ any

integer. If $k = 0$, then $\alpha = 22.5° + 0° = 22.5°$.

If $k = 1$, then $\alpha = 22.5° + 90° = 112.5°$.

If $k = 2$, then $\alpha = 22.5° + 180° = 202.5°$.

If $k = 3$, then $\alpha = 22.5° + 270° = 292.5°$.

Solution set:

$\{\cos 22.5° + i \sin 22.5°, \cos 112.5° + i \sin 112.5°,$

$\cos 202.5° + i \sin 202.5°,$

$\cos 292.5° + i \sin 292.5°\}$

**41.** $x^3 - \left(4 + 4i\sqrt{3}\right) = 0 \Rightarrow x^3 = 4 + 4i\sqrt{3}$

We have

$r = \sqrt{4^2 + \left(4\sqrt{3}\right)^2} = \sqrt{16 + 48} = \sqrt{64} = 8$ and

$\tan \theta = \dfrac{4\sqrt{3}}{4} = \sqrt{3}$. Since $\theta$ is in quadrant I,

$\theta = 60°$.

$x^3 = 4 + 4i\sqrt{3} = 8\left(\dfrac{1}{2} + i\dfrac{\sqrt{3}}{2}\right)$

$= 8(\cos 60° + i \sin 60°)$

Since $r^3\left(\cos 3\alpha + i\sin 3\alpha\right)$

$= 8\left(\cos 60° + i\sin 60°\right)$, then we have

$r^3 = 8 \Rightarrow r = 2$ and $3\alpha = 60° + 360° \cdot k \Rightarrow$

$\alpha = \dfrac{60° + 360° \cdot k}{3} = 20° + 120° \cdot k,\ k$ any

integer. If $k = 0$, then $\alpha = 20° + 0° = 20°$.
If $k = 1$, then $\alpha = 20° + 120° = 140°$.
If $k = 2$, then $\alpha = 20° + 240° = 260°$.
Solution set:

$\left\{2\left(\cos 20° + i\sin 20°\right), 2\left(\cos 140° + i\sin 140°\right),\right.$
$\left. 2\left(\cos 260° + i\sin 260°\right)\right\}$

**43.** $x^3 - 1 = 0 \Rightarrow \left(x - 1\right)\left(x^2 + x + 1\right) = 0$

Setting each factor equal to zero, we have
$x - 1 = 0 \Rightarrow x = 1$ and
$x^2 + x + 1 = 0 \Rightarrow$

$x = \dfrac{-1 \pm \sqrt{1^2 - 4\cdot 1\cdot 1}}{2\cdot 1} = \dfrac{-1 \pm \sqrt{-3}}{2}$

$= \dfrac{-1 \pm i\sqrt{3}}{2} = -\dfrac{1}{2} \pm \dfrac{\sqrt{3}}{2}i$

Thus, $x = 1, -\dfrac{1}{2} + \dfrac{\sqrt{3}}{2}i, -\dfrac{1}{2} - \dfrac{\sqrt{3}}{2}i$. We see

that the solutions are the same as Exercise 31.

**45.** De Moivre's theorem states that

$\left(\cos\theta + i\sin\theta\right)^2 = 1^2\left(\cos 2\theta + i\sin 2\theta\right)$
$= \cos 2\theta + i\sin 2\theta$

**47.** Two complex numbers $a + bi$ and $c + di$ are equal only if $a = c$ and $b = d$. Thus, $a = c$ implies $\cos^2\theta - \sin^2\theta = \cos 2\theta$.

**49. (a)** If $z = 0 + 0i$, then $z = 0,\ 0^2 + 0 = 0$,

$0^2 + 0 = 0$, and so on. The calculations repeat as $0, 0, 0, \ldots$, and will never exceed a modulus of 2. The point $\left(0, 0\right)$ is part of the Mandelbrot set. The pixel at the origin should be turned on.

**(b)** If $z = 1 - 1i$, then $\left(1 - i\right)^2 + \left(1 - i\right) = 1 - 3i$.
The modulus of $1 - 3i$ is

$\sqrt{1^2 + \left(-3\right)^2} = \sqrt{1 + 9} = \sqrt{10}$, which is

greater than 2. Therefore, $1 - 1i$ is not part of the Mandelbrot set, and the pixel at $\left(1, -1\right)$ should be left off.

**(c)** If $z = -.5i$, then $\left(-.5i\right)^2 - .5i = -.25 - .5i$;

$\left(-.25 - .5i\right)^2 + \left(-.25 - .5i\right) = -.4375 - .25i$;

$\left(-.4375 - .25i\right)^2 + \left(-.4375 - .25i\right)$
$= -.308593 - .03125i$;

$\left(-.308593 - .03125i\right)^2$
$\qquad + \left(-.308593 - .03125i\right)$
$= -.214339 - .0119629i$;

$\left(-.214339 - .0119629i\right)^2$
$\qquad + \left(-.214339 - .0119629i\right)$
$= -.16854 - .00683466i$

This sequence appears to be approaching the origin, and no number has a modulus greater than 2. Thus, $-.5i$ is part of the Mandelbrot set, and the pixel at $\left(0, -.5\right)$

should be turned on.

**51.** Using the trace function, we find that the other four fifth roots of 1 are:
$.30901699 + .95105652i,$
$-.809017 + .58778525i,$
$-.809017 - .5877853i,$
$.30901699 - .9510565i.$

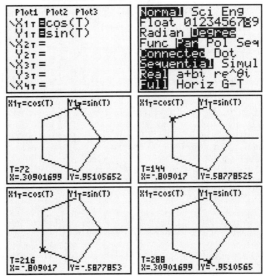

**53.** $2 + 2i\sqrt{3}$ is one cube root of a complex number.

$$r = \sqrt{2^2 + \left(2\sqrt{3}\right)^2} = \sqrt{4 + 12} = 4 \text{ and}$$

$$\tan\theta = \frac{2\sqrt{3}}{2} = \sqrt{3}$$

Because $x$ and $y$ are both positive, $\theta$ is in quadrant I, so $\theta = 60°$.

$$2 + 2i\sqrt{3} = 4\left(\cos 60° + i\sin 60°\right)$$

$$\left(2 + 2i\sqrt{3}\right)^3 = \left[4\left(\cos 60° + i\sin 60°\right)\right]^3$$
$$= 4^3\left[\cos\left(3 \cdot 60°\right) + i\sin\left(3 \cdot 60°\right)\right]$$
$$= 64\left(\cos 180° + i\sin 180°\right)$$
$$= 64\left(-1 + 0 \cdot i\right) = -64 + 0 \cdot i$$

Since the graphs of the other roots of $-64$ must be equally spaced around a circle and the graphs of these roots are all on a circle that has center at the origin and radius 4, the other roots are

$$4\left[\cos\left(60° + 120°\right) + i\sin\left(60° + 120°\right)\right]$$
$$= 4\left(\cos 180° + i\sin 180°\right) = 4\left(-1 + i \cdot 0\right) = -4$$

$$4\left[\cos\left(60° + 240°\right) + i\sin\left(60° + 240°\right)\right]$$

$$= 4\left(\cos 300° + i\sin 300°\right) = 4\left(\frac{1}{2} - \frac{\sqrt{3}}{2}i\right)$$

$$= 2 - 2i\sqrt{3}$$

**55.** $x^2 - 3 + 2i = 0 \Rightarrow x^2 = 3 - 2i$

$$r = \sqrt{3^2 + \left(-2\right)^2} = \sqrt{9 + 4} = \sqrt{13} \Rightarrow$$

$$r^{1/n} = r^{1/2} = \left(\sqrt{13}\right)^{1/2} \approx 1.89883 \text{ and}$$

$\tan\theta = -\frac{2}{3}$. Since $\theta$ is in quadrant IV,

$\theta \approx 326.3099°$ and

$$\alpha = \frac{326.3099° + 360° \cdot k}{2} \approx 163.155° + 180° \cdot k,$$

where $k$ is an integer.

$$x \approx 1.89833\left(\cos 163.155° + i\sin 163.155°\right),$$
$$1.89833\left(\cos 343.155° + i\sin 343.155°\right)$$

Solution set:
$$\left\{-1.8174 + .5503i, 1.8174 - .5503i\right\}$$

**57.** $x^5 + 2 + 3i = 0 \Rightarrow x^5 = -2 - 3i$

$$r = \sqrt{\left(-2\right)^2 + \left(-3\right)^2} = \sqrt{4 + 9} = \sqrt{13} \Rightarrow$$

$$r^{1/n} = r^{1/5} = \left(\sqrt{13}\right)^{1/5} = 13^{1/10} \approx 1.2924$$

and $\tan\theta = \frac{-3}{-2} = 1.5$

Since $\theta$ is in quadrant III, $\theta \approx 236.310°$ and

$$\alpha = \frac{236.310° + 360° \cdot k}{5} = 47.262° + 72° \cdot k,$$

where $k$ is an integer.

$$x \approx 1.29239\left(\cos 47.262° + i\sin 47.262°\right),$$
$$1.29239\left(\cos 119.262° + i\sin 119.2622°\right),$$
$$1.29239\left(\cos 191.262° + i\sin 191.2622°\right),$$
$$1.29239\left(\cos 263.262° + i\sin 263.2622°\right),$$
$$1.29239\left(\cos 335.262° + i\sin 335.2622°\right)$$

Solution set:
$$\{.87708 + .94922i, -.63173 + 1.1275i,$$
$$-1.2675 - .25240i, -.15164 - 1.28347i,$$
$$1.1738 - .54083i\}$$

**59.** The statement, "Every real number must have two real square roots," is false. Consider, for example, the real number –4. Its two square roots are $2i$ and $-2i$, which are not real.

**61.** If $z$ is an $n$th root of 1, then $z^n = 1$. Since

$$1 = \frac{1}{1} = \frac{1}{z^n} = \left(\frac{1}{z}\right)^n, \text{ then } \frac{1}{z} \text{ is also an } n\text{th root}$$

of 1.

**63.** Answers will vary.

# Chapter 8 Quiz
(Sections 8.1–8.4)

**1. (a)** $\sqrt{-24} \cdot \sqrt{-3} = i\sqrt{24} \cdot i\sqrt{3} = i^2\sqrt{72} = -6\sqrt{2}$

**(b)** $\dfrac{\sqrt{-8}}{\sqrt{72}} = \dfrac{i\sqrt{8}}{\sqrt{72}} = \dfrac{i}{\sqrt{9}} = \dfrac{i}{3} = \dfrac{1}{3}i$

**3. (a)** $\left(1 - i\right)^3 = 1^3 - 3\left(1\right)^2 i + 3\left(1\right)i^2 - i^3$
$$= 1 - 3i + 3\left(-1\right) - \left(-i\right)$$
$$= 1 - 3i - 3 + i = -2 - 2i$$

**(b)** $i^{33} = i^{32} \cdot i = \left(i^4\right)^8 \cdot i = 1^8 \cdot i = i$, or $0 + i$

**5. (a)** Sketch a graph of $-4i$ in the complex plane.

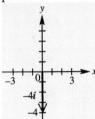

Since $-4i = 0 - 4i$, we have $x = 0$ and $y = -4$, so $r = \sqrt{0^2 + (-4)^2} = 4$. We cannot find $\theta$ by using $\tan \theta = \dfrac{y}{x}$ because $x = 0$. From the graph, we see that $-4i$ is on the negative $y$-axis, so $\theta = 270°$. Thus,

$$-4i = 4(\cos 270° + i \sin 270°)$$

**(b)** $1 - i\sqrt{3}$

Sketch a graph of $1 - i\sqrt{3}$ in the complex plane.

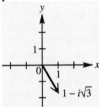

Since $x = 1$ and $y = -\sqrt{3}$,

$$r = \sqrt{1^2 + \left(-\sqrt{3}\right)^2} = \sqrt{1+3} = \sqrt{4} = 2 \text{ and}$$

$\tan \theta = \dfrac{-\sqrt{3}}{1} = -\sqrt{3}$. The graph shows that $\theta$ is in quadrant IV, so $\theta = 300°$. Therefore,

$$1 - i\sqrt{3} = 2(\cos 300° + i \sin 300°)$$

**(c)** $-3 - i$

Sketch a graph of $-3 - i$ in the complex plane.

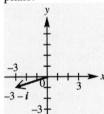

Using a calculator, we find that the reference angle is 18.4°. The graph shows that $\theta$ is in quadrant III, so $\theta = 180° + 18.4° = 198.4°$. Therefore,

$$-3 - i = \sqrt{10}\left(\cos 198.4° + i \sin 198.4°\right).$$

**7.** $w = 12(\cos 80° + i \sin 80°)$,
  $z = 3(\cos 50° + i \sin 50°)$

**(a)** $wz = 12(\cos 80° + i \sin 80°)$
$$\qquad\qquad \cdot 3(\cos 50° + i \sin 50°)$$
$$= 12 \cdot 3\left[\begin{matrix}\cos(80° + 50°) \\ + i\sin(80° + 50°)\end{matrix}\right]$$
$$= 36(\cos 130° + i \sin 130°)$$

**(b)** $\dfrac{w}{z} = \dfrac{12(\cos 80° + i \sin 80°)}{3(\cos 50° + i \sin 50°)}$
$$= 4\left[\cos(80° - 50°) + i\sin(80° - 50°)\right]$$
$$= 4(\cos 30° + i \sin 30°)$$
$$= 4\left(\frac{\sqrt{3}}{2} + \frac{1}{2}i\right) = 2\sqrt{3} + 2i$$

**(c)** $z^3 = \left[3(\cos 50° + i \sin 50°)\right]^3$
$$= 3^3\left[\cos(3 \cdot 50°) + i\sin(3 \cdot 50°)\right]$$
$$= 27(\cos 150° + i \sin 150°)$$
$$= 27\left(-\frac{\sqrt{3}}{2} + \frac{1}{2}i\right) = -\frac{27\sqrt{3}}{2} + \frac{27}{2}i$$

## Section 8.5: Polar Equations and Graphs

**1. (a)** II (since $r > 0$ and $90° < \theta < 180°$)

**(b)** I (since $r > 0$ and $0° < \theta < 90°$)

**(c)** IV (since $r > 0$ and $-90° < \theta < 0°$)

**(d)** III (since $r > 0$ and $180° < \theta < 270°$)

For Exercises 3(b)–12(b), answers may vary.

**3. (a)**

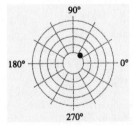

**(b)** Two other pairs of polar coordinates for (1, 45°) are (1, 405°) and (−1, 225°).

(c) Since $x = r\cos\theta \Rightarrow x = 1\cdot\cos 45° = \dfrac{\sqrt{2}}{2}$

and $y = r\sin\theta \Rightarrow y = 1\cdot\sin 45° = \dfrac{\sqrt{2}}{2}$,

the point is $\left(\dfrac{\sqrt{2}}{2}, \dfrac{\sqrt{2}}{2}\right)$.

**5. (a)**

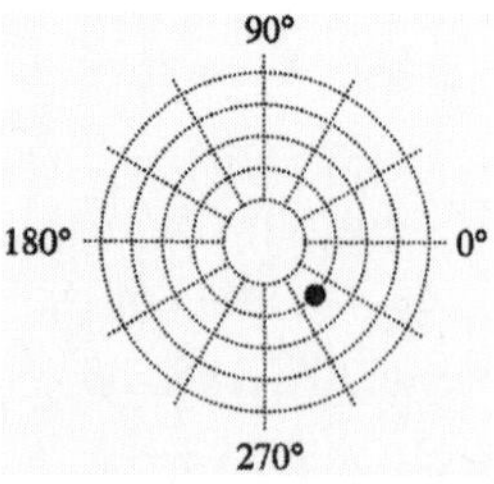

**(b)** Two other pairs of polar coordinates for $(-2, 135°)$ are $(-2, 495°)$ and $(2, 315°)$.

**(c)** Since
$x = r\cos\theta \Rightarrow x = (-2)\cos 135° = \sqrt{2}$ and
$y = r\sin\theta \Rightarrow y = (-2)\sin 135° = \sqrt{2}$, the
point is $\left(\sqrt{2}, -\sqrt{2}\right)$.

**7. (a)**

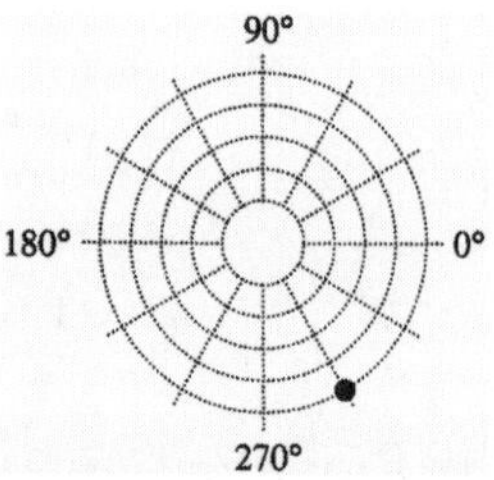

**(b)** Two other pairs of polar coordinates for $(5, -60°)$ are $(5, 300°)$ and $(-5, 120°)$.

**(c)** Since $x = r\cos\theta \Rightarrow x = 5\cos(-60°) = \dfrac{5}{2}$

and

$y = r\sin\theta \Rightarrow y = 5\sin(-60°) = -\dfrac{5\sqrt{3}}{2}$,

the point is $\left(\dfrac{5}{2}, -\dfrac{5\sqrt{3}}{2}\right)$.

**9. (a)**

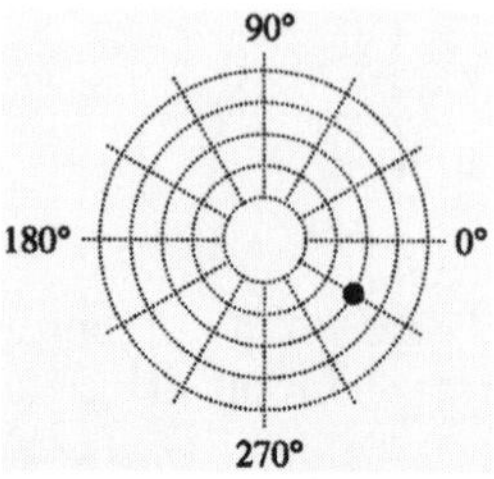

**(b)** Two other pairs of polar coordinates for $(-3, -210°)$ are $(-3, 150°)$ and $(3, -30°)$.

(c) Since

$x = r\cos\theta \Rightarrow x = (-3)\cos(-210°) = \dfrac{3\sqrt{3}}{2}$

and

$y = r\sin\theta \Rightarrow y = (-3)\sin(-210°) = -\dfrac{3}{2}$,

the point is $\left(\dfrac{3\sqrt{3}}{2}, -\dfrac{3}{2}\right)$.

**11. (a)**

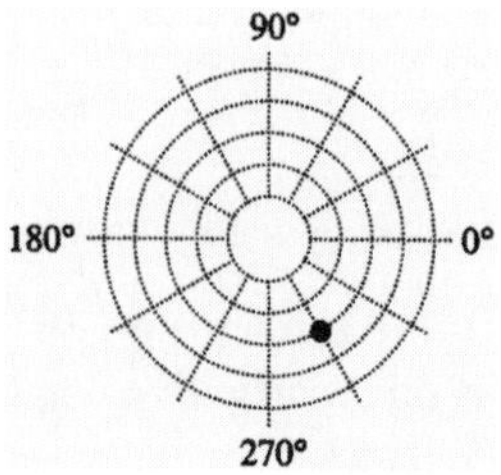

**(b)** Two other pairs of polar coordinates for
$\left(3, \dfrac{5\pi}{3}\right)$ are $\left(3, \dfrac{11\pi}{3}\right)$ and $\left(-3, \dfrac{2\pi}{3}\right)$.

**(c)** Since $x = r\cos\theta \Rightarrow x = 3\cos\dfrac{5\pi}{3} = \dfrac{3}{2}$ and

$y = r\sin\theta \Rightarrow y = 3\sin\dfrac{5\pi}{3} = -\dfrac{3\sqrt{3}}{2}$, the

point is $\left(\dfrac{3}{2}, -\dfrac{3\sqrt{3}}{2}\right)$.

For Exercises 13(b)–21(b), answers may vary.

**13. (a)**

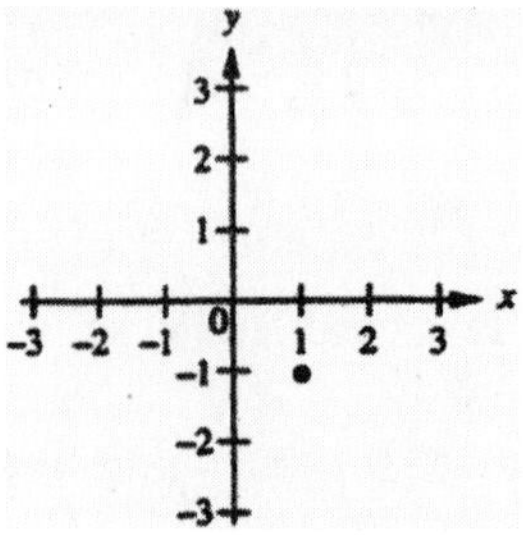

**(b)** $r = \sqrt{1^2 + (-1)^2} = \sqrt{1+1} = \sqrt{2}$ and

$\theta = \tan^{-1}\left(\dfrac{-1}{1}\right) = \tan^{-1}(-1) = -45°$, since

$\theta$ is in quadrant IV. Since
$360° - 45° = 315°$, one possibility is
$\left(\sqrt{2}, 315°\right)$. Alternatively, if $r = -\sqrt{2}$,
then $\theta = 315° - 180° = 135°$. Thus, a
second possibility is $\left(-\sqrt{2}, 135°\right)$.

**15. (a)**

**(b)** $r = \sqrt{0^2 + 3^2} = \sqrt{0+9} = \sqrt{9} = 3$ and $\theta = 90°$, since $(0,3)$ is on the positive $y$-axis. So, one possibility is $(3, 90°)$. Alternatively, if $r = -3$, then $\theta = 90° + 180° = 270°$. Thus, a second possibility is $(-3, 270°)$.

**17. (a)**

**(b)** $r = \sqrt{\left(\sqrt{2}\right)^2 + \left(\sqrt{2}\right)^2} = \sqrt{2+2} = \sqrt{4} = 2$

and $\theta = \tan^{-1}\left(\dfrac{\sqrt{2}}{\sqrt{2}}\right) = \tan^{-1} 1 = 45°$,

since $\theta$ is in quadrant I. So, one possibility is $(2, 45°)$. Alternatively, if $r = -2$, then $\theta = 45° + 180° = 225°$. Thus, a second possibility is $(-2, 225°)$.

**19. (a)**

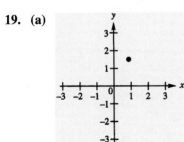

**(b)** $r = \sqrt{\left(\dfrac{\sqrt{3}}{2}\right)^2 + \left(\dfrac{3}{2}\right)^2} = \sqrt{\dfrac{3}{4} + \dfrac{9}{4}}$

$= \sqrt{\dfrac{12}{4}} = \sqrt{3}$ and $\theta = \arctan\left(\dfrac{3}{2} \cdot \dfrac{2}{\sqrt{3}}\right)$

$= \tan^{-1}\left(\sqrt{3}\right) = 60°$, since $\theta$ is in quadrant

I. So, one possibility is $\left(\sqrt{3}, 60°\right)$. Alternatively, if $r = -\sqrt{3}$, then $\theta = 60° + 180° = 240°$. Thus, a second possibility is $\left(-\sqrt{3}, 240°\right)$.

**21. (a)**

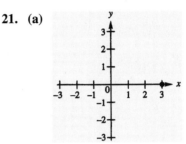

**(b)** $r = \sqrt{3^2 + 0^2} = \sqrt{9+0} = \sqrt{9} = 3$ and $\theta = 0°$, since $(3,0)$ is on the positive $x$-axis. So, one possibility is $(3, 0°)$. Alternatively, if $r = -3$, then $\theta = 0° + 180° = 180°$. Thus, a second possibility is $(-3, 180°)$.

**23.** $x - y = 4$

Using the general form for the polar equation of a line, $r = \dfrac{c}{a\cos\theta + b\sin\theta}$, with $a = 1, b = -1$, and $c = 4$, the polar equation is

$$r = \dfrac{4}{\cos\theta - \sin\theta}.$$

**25.** $x^2 + y^2 = 16 \Rightarrow r^2 = 16 \Rightarrow r = \pm 4$

The equation of the circle in polar form is $r = 4$ or $r = -4$.

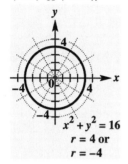

$x^2 + y^2 = 16$
$r = 4$ or
$r = -4$

**27.** $2x + y = 5$

Using the general form for the polar equation of a line, $r = \dfrac{c}{a\cos\theta + b\sin\theta}$, with $a = 2, b = 1,$ and $c = 5,$ the polar equation is

$$r = \frac{5}{2\cos\theta + \sin\theta}.$$

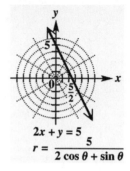

$2x + y = 5$
$r = \dfrac{5}{2\cos\theta + \sin\theta}$

**29.** $r\sin\theta = k$

**31.** $r = \dfrac{k}{\sin\theta} \Rightarrow r = k\csc\theta$

**33.** $r\cos\theta = k$

**35.** $r = \dfrac{k}{\cos\theta} \Rightarrow r = k\sec\theta$

**37.** $r = 3$ represents the set of all points 3 units from the pole. The correct choice is C.

**39.** $r = \cos 2\theta$ is a rose curve with $2 \cdot 2 = 4$ petals. The correct choice is A

**41.** $r = 2 + 2\cos\theta$ (cardioid)

| $\theta$ | 0° | 30° | 60° | 90° | 120° | 150° |
|---|---|---|---|---|---|---|
| $\cos\theta$ | 1 | .9 | .5 | 0 | −.5 | −.9 |
| $r = 2 + 2\cos\theta$ | 4 | 3.8 | 3 | 2 | 1 | .2 |

| $\theta$ | 180° | 210° | 240° | 270° | 300° | 330° |
|---|---|---|---|---|---|---|
| $\cos\theta$ | −1 | −.9 | −.5 | 0 | .5 | .9 |
| $r = 2 + 2\cos\theta$ | 0 | .3 | 1 | 2 | 3 | 3.7 |

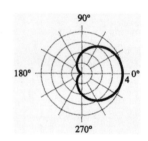

**43.** $r = 3 + \cos\theta$ (limaçon)

| $\theta$ | 0° | 30° | 60° | 90° | 120° | 150° |
|---|---|---|---|---|---|---|
| $r = 3 + \cos\theta$ | 4 | 3.9 | 3.5 | 3 | 2.5 | 2.1 |

| $\theta$ | 180° | 210° | 240° | 270° | 300° | 330° |
|---|---|---|---|---|---|---|
| $r = 3 + \cos\theta$ | 2 | 2.1 | 2.5 | 3 | 3.5 | 3.9 |

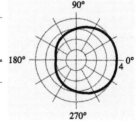

**45.**  $r = 4\cos 2\theta$  (four-leaved rose)

| $\theta$ | 0° | 30° | 45° | 60° | 90° | 120° | 135° | 150° |
|---|---|---|---|---|---|---|---|---|
| $r = 4\cos 2\theta$ | 4 | 2 | 0 | −2 | −4 | −2 | 0 | 2 |

| $\theta$ | 180° | 210° | 225° | 240° | 270° | 300° | 315° | 330° |
|---|---|---|---|---|---|---|---|---|
| $r = 4\cos 2\theta$ | 4 | 2 | 0 | −2 | −4 | −2 | 0 | 2 |

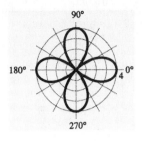

**47.**  $r^2 = 4\cos 2\theta \Rightarrow r = \pm 2\sqrt{\cos 2\theta}$  (lemniscate)

Graph only exists for [0°, 45°], [135°, 225°], and [315°, 360°] because $\cos 2\theta$ must be positive.

| $\theta$ | 0° | 30° | 45° | 135° | 150° |
|---|---|---|---|---|---|
| $r = \pm 2\sqrt{\cos 2\theta}$ | ±2 | ±1.4 | 0 | 0 | ±1.4 |

| $\theta$ | 180° | 210° | 225° | 315° | 330° |
|---|---|---|---|---|---|
| $r = \pm 2\sqrt{\cos 2\theta}$ | ±2 | ±1.4 | 0 | 0 | ±1.4 |

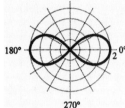

**49.**  $r = 4 - 4\cos\theta$  (cardioid)

| $\theta$ | 0° | 30° | 60° | 90° | 120° | 150° |
|---|---|---|---|---|---|---|
| $r = 4 - 4\cos\theta$ | 0 | .5 | 2 | 4 | 6 | 7.5 |

| $\theta$ | 180° | 210° | 240° | 270° | 300° | 330° |
|---|---|---|---|---|---|---|
| $r = 4 - 4\cos\theta$ | 8 | 7.5 | 6 | 4 | 2 | .5 |

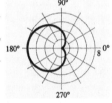

**51.**  $r = 2\sin\theta\tan\theta$  (cissoid)

$r$ is undefined at $\theta = 90°$ and $\theta = 270°$.

| $\theta$ | 0° | 30° | 45° | 60° | 90° | 120° | 135° | 150° | 180° |
|---|---|---|---|---|---|---|---|---|---|
| $r = 2\sin\theta\tan\theta$ | 0 | .6 | 1.4 | 3 | − | −3 | −1.4 | −.6 | 0 |

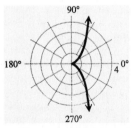

Notice that for [180°, 360°), the graph retraces the path traced for [0°, 180°).

**53.**  $r = 2 \sin \theta$

Multiply both sides by $r$ to obtain

$r^2 = 2r \sin \theta.$ Since

$r^2 = x^2 + y^2$ and $y = r \sin \theta, \ x^2 + y^2 = 2y.$

Complete the square on $y$ to obtain

$x^2 + y^2 - 2y + 1 = 1 \Rightarrow x^2 + (y-1)^2 = 1.$

The graph is a circle with center at $(0,1)$ and radius 1.

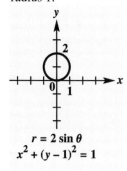

$r = 2 \sin \theta$
$x^2 + (y-1)^2 = 1$

**55.**  $r = \dfrac{2}{1 - \cos \theta}$

Multiply both sides by $1 - \cos \theta$ to obtain $r - r$

$\cos \theta = 2.$ Substitute $r = \sqrt{x^2 + y^2}$ to obtain

$\sqrt{x^2 + y^2} - x = 2 \Rightarrow \sqrt{x^2 + y^2} = 2 + x \Rightarrow$

$x^2 + y^2 = (2+x)^2 \Rightarrow x^2 + y^2 = 4 + 4x + x^2 \Rightarrow$

$y^2 = 4(1+x)$

The graph is a parabola with vertex at $(-1,0)$ and axis $y = 0.$

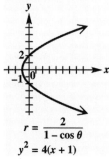

$r = \dfrac{2}{1 - \cos \theta}$
$y^2 = 4(x + 1)$

**57.**  $r + 2 \cos \theta = -2 \sin \theta$

$r + 2 \cos \theta = -2 \sin \theta \Rightarrow$

$r^2 = -2r \sin \theta - 2r \cos \theta \Rightarrow x^2 + y^2 = -2y - 2x$

$x^2 + 2x + y^2 + 2y = 0 \Rightarrow$

$x^2 + 2x + 1 + y^2 + 2y + 1 = 2 \Rightarrow$

$(x+1)^2 + (y+1)^2 = 2$

The graph is a circle with center $(-1,-1)$ and radius $\sqrt{2}.$

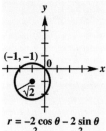

$r = -2 \cos \theta - 2 \sin \theta$
$(x + 1)^2 + (y + 1)^2 = 2$

**59.**  $r = 2 \sec \theta$

$r = 2 \sec \theta \Rightarrow r = \dfrac{2}{\cos \theta} \Rightarrow r \cos \theta = 2 \Rightarrow x = 2$

The graph is a vertical line, intercepting the $x$-axis at 2.

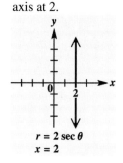

$r = 2 \sec \theta$
$x = 2$

**61.**  $r = \dfrac{2}{\cos \theta + \sin \theta}$

Using the general form for the polar equation

of a line, $r = \dfrac{c}{a \cos \theta + b \sin \theta}$, with

$a = 1, \ b = 1,$ and $c = 2,$ we have $x + y = 2.$

The graph is a line with intercepts $(0,2)$ and $(2,0).$

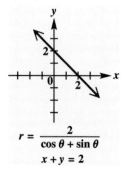

$r = \dfrac{2}{\cos \theta + \sin \theta}$
$x + y = 2$

**63.** Graph $r = \theta$, a spiral of Archimedes.

| $\theta$ | $-360°$ | $-270°$ | $-180°$ | $-90°$ | $0°$ |
|---|---|---|---|---|---|
| $\theta$ (radians) | $-6.3$ | $-4.7$ | $-3.1$ | $-1.6$ | $0$ |
| $r = \theta$ | $-6.3$ | $-4.7$ | $-3.1$ | $-1.6$ | $0$ |

| $\theta$ | $90°$ | $180°$ | $270°$ | $360°$ |
|---|---|---|---|---|
| $\theta$ (radians) | $1.6$ | $3.1$ | $4.7$ | $6.3$ |
| $r = \theta$ | $1.6$ | $3.1$ | $4.7$ | $6.3$ |

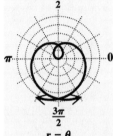

$r = \theta$

**65.** In rectangular coordinates, the line passes through $(1,0)$ and $(0,2)$. So

$$m = \frac{2-0}{0-1} = \frac{2}{-1} = -2 \text{ and}$$

$$(y - 0) = -2(x - 1) \Rightarrow y = -2x + 2 \Rightarrow$$

$2x + y = 2$. Converting to polar form

$$r = \frac{c}{a\cos\theta + b\sin\theta}, \text{ we have:}$$

$$r = \frac{2}{2\cos\theta + \sin\theta}.$$

**67.** **(a)** $(r, -\theta)$

**(b)** $(r, \pi - \theta)$ or $(-r, -\theta)$

**(c)** $(r, \pi + \theta)$ or $(-r, \theta)$

**69.**

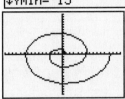

**71.**

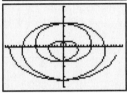

**73.** $r = 4\sin\theta$, $r = 1 + 2\sin\theta$, $0 \le \theta < 2\pi$
$4\sin\theta = 1 + 2\sin\theta \Rightarrow 2\sin\theta = 1 \Rightarrow$

$$\sin\theta = \frac{1}{2} \Rightarrow \theta = \frac{\pi}{6} \text{ or } \frac{5\pi}{6}$$

The points of intersection are

$$\left(4\sin\frac{\pi}{6}, \frac{\pi}{6}\right) = \left(2, \frac{\pi}{6}\right) \text{ and}$$

$$\left(4\sin\frac{5\pi}{6}, \frac{5\pi}{6}\right) = \left(2, \frac{5\pi}{6}\right).$$

**75.** $r = 2 + \sin\theta$, $r = 2 + \cos\theta$, $0 \le \theta < 2\pi$
$2 + \sin\theta = 2 + \cos\theta \Rightarrow \sin\theta = \cos\theta \Rightarrow$

$$\theta = \frac{\pi}{4} \text{ or } \frac{5\pi}{4}$$

$$r = 2 + \sin\frac{\pi}{4} = 2 + \frac{\sqrt{2}}{2} = \frac{4 + \sqrt{2}}{2} \text{ and}$$

$$r = 2 + \sin\frac{5\pi}{4} = 2 - \frac{\sqrt{2}}{2} = \frac{4 - \sqrt{2}}{2}$$

The points of intersection are

$$\left(\frac{4 + \sqrt{2}}{2}, \frac{\pi}{4}\right) \text{ and } \left(\frac{4 - \sqrt{2}}{2}, \frac{5\pi}{4}\right).$$

**77. (a)** Plot the following polar equations on the same polar axis in radian mode:

Mercury: $r = \dfrac{.39(1-.206^2)}{1+.206\cos\theta}$;

Venus: $r = \dfrac{.78(1-.007^2)}{1+.007\cos\theta}$;

Earth: $r = \dfrac{1(1-.017^2)}{1+.017\cos\theta}$;

Mars: $r = \dfrac{1.52(1-.093^2)}{1+.093\cos\theta}$.

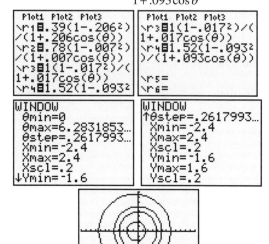

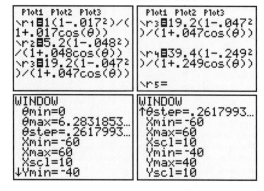

**(b)** Plot the following polar equations on the same polar axis:

Earth: $r = \dfrac{1(1-.017^2)}{1+.017\cos\theta}$;

Jupiter: $r = \dfrac{5.2(1-.048^2)}{1+.048\cos\theta}$;

Uranus: $r = \dfrac{19.2(1-.047^2)}{1+.047\cos\theta}$;

Pluto: $r = \dfrac{39.4(1-.249^2)}{1+.249\cos\theta}$.

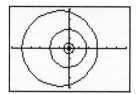

**(c)** We must determine if the orbit of Pluto is always outside the orbits of the other planets. Since Neptune is closest to Pluto, plot the orbits of Neptune and Pluto on the same polar axes.

Neptune: $r = \dfrac{30.1(1-.009^2)}{1+.009\cos\theta}$;

Pluto: $r = \dfrac{39.4(1-.249^2)}{1+.249\cos\theta}$

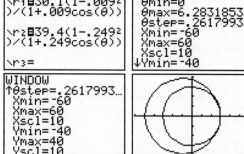

The graph shows that their orbits are very close near the polar axis. Use ZOOM or change your window to see that the orbit of Pluto does indeed pass inside the orbit of Neptune. Therefore, there are times when Neptune, not Pluto, is the farthest planet from the sun. (However, Pluto's average distance from the sun is considerably greater than Neptune's average distance.)

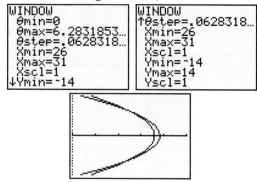

## Section 8.6: Parametric Equations, Graphs, and Applications

**1.** At $t = 2$, $x = 3(2) + 6 = 12$ and

$y = -2(2) + 4 = 0$. The correct choice is C.

**3.** At $t = 5$, $x = 5$ and $y = 5^2 = 25$. The correct choice is A.

**5.** **(a)** $x = t + 2$, $y = t^2$, for $t$ in $[-1, 1]$

| $t$ | $x = t + 2$ | $y = t^2$ |
|---|---|---|
| $-1$ | $-1 + 2 = 1$ | $(-1)^2 = 1$ |
| $0$ | $0 + 2 = 2$ | $0^2 = 0$ |
| $1$ | $1 + 2 = 3$ | $1^2 = 1$ |

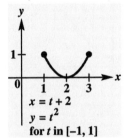

**(b)** $x - 2 = t$, therefore

$y = (x - 2)^2$ or $y = x^2 - 4x + 4$. Since $t$ is in $[-1, 1]$, $x$ is in $[-1 + 2, 1 + 2]$ or $[1, 3]$.

**7.** **(a)** $x = \sqrt{t}$, $y = 3t - 4$, for $t$ in $[0, 4]$.

| $t$ | $x = \sqrt{t}$ | $y = 3t - 4$ |
|---|---|---|
| $0$ | $\sqrt{0} = 0$ | $3(0) - 4 = -4$ |
| $1$ | $\sqrt{1} = 1$ | $3(1) - 4 = -1$ |
| $2$ | $\sqrt{2} = 1.4$ | $3(2) - 4 = 2$ |
| $3$ | $\sqrt{3} = 1.7$ | $3(3) - 4 = 5$ |
| $4$ | $\sqrt{4} = 2$ | $3(4) - 4 = 8$ |

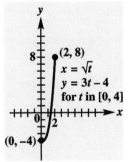

**(b)** $x = \sqrt{t}$, $y = 3t - 4$

Since $x = \sqrt{t} \Rightarrow x^2 = t$, we have

$y = 3x^2 - 4$. Since $t$ is in $[0, 4]$, $x$ is in $[\sqrt{0}, \sqrt{4}]$ or $[0, 2]$.

**9.** **(a)** $x = t^3 + 1$, $y = t^3 - 1$, for $t$ in $(-\infty, \infty)$

| $t$ | $x = t^3 + 1$ | $y = t^3 - 1$ |
|---|---|---|
| $-2$ | $(-2)^3 + 1 = -7$ | $(-2)^3 - 1 = -9$ |
| $-1$ | $(-1)^3 + 1 = 0$ | $(-1)^3 - 1 = -2$ |
| $0$ | $0^3 + 1 = 1$ | $0^3 - 1 = -1$ |
| $1$ | $1^3 + 1 = 2$ | $1^3 - 1 = 0$ |
| $2$ | $2^3 + 1 = 9$ | $2^3 - 1 = 7$ |
| $3$ | $3^3 + 1 = 28$ | $3^3 - 1 = 26$ |

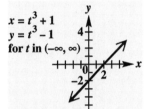

**(b)** Since $x = t^3 + 1$, we have $x - 1 = t^3$.

Since $y = t^3 - 1$, we have $y = (x - 1) - 1 = x - 2$.

Since $t$ is in $(-\infty, \infty)$, $x$ is in $(-\infty, \infty)$.

**11.** **(a)** $x = 2\sin t$, $y = 2\cos t$, for $t$ in $[0, 2\pi]$

| $t$ | $x = 2\sin t$ | $y = 2\cos t$ |
|---|---|---|
| $0$ | $2\sin 0 = 0$ | $2\cos\theta = 2$ |
| $\frac{\pi}{6}$ | $2\sin\frac{\pi}{6} = 1$ | $2\cos\frac{\pi}{6} = \sqrt{3}$ |
| $\frac{\pi}{4}$ | $2\sin\frac{\pi}{4} = \sqrt{2}$ | $2\cos\frac{\pi}{4} = \sqrt{2}$ |
| $\frac{\pi}{3}$ | $2\sin\frac{\pi}{3} = \sqrt{3}$ | $2\cos\frac{\pi}{3} = 1$ |
| $\frac{\pi}{2}$ | $2\sin\frac{\pi}{2} = 2$ | $2\cos\frac{\pi}{2} = 0$ |

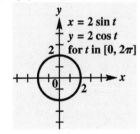

**(b)** Since $x = 2\sin t$ and $y = 2\cos t$, we have

$\dfrac{x}{2} = \sin t$ and $\dfrac{y}{2} = \cos t$. Since

$\sin^2 t + \cos^2 t = 1$, we have

$\left(\dfrac{x}{2}\right)^2 + \left(\dfrac{y}{2}\right)^2 = 1 \Rightarrow \dfrac{x^2}{4} + \dfrac{y^2}{4} = 1 \Rightarrow$

$x^2 + y^2 = 4$. Since $t$ is in $[0, 2\pi]$, $x$ is in $[-2, 2]$ because the graph is a circle, centered at the origin, with radius 2.

**13. (a)** $x = 3\tan t$, $y = 2\sec t$, for $t$ in $\left(-\dfrac{\pi}{2}, \dfrac{\pi}{2}\right)$

| $t$ | $x = 3\tan t$ | $y = 2\sec t$ |
|---|---|---|
| $-\dfrac{\pi}{3}$ | $3\tan\left(-\dfrac{\pi}{3}\right) = -3\sqrt{3}$ | $2\sec\left(-\dfrac{\pi}{3}\right) = 4$ |
| $-\dfrac{\pi}{6}$ | $3\tan\left(-\dfrac{\pi}{6}\right) = -\sqrt{3}$ | $2\sec\left(-\dfrac{\pi}{6}\right) = \dfrac{4\sqrt{3}}{3}$ |
| $0$ | $3\tan 0 = 0$ | $2\sec 0 = 2$ |
| $\dfrac{\pi}{6}$ | $3\tan\dfrac{\pi}{6} = \sqrt{3}$ | $2\sec\dfrac{\pi}{6} = \dfrac{4\sqrt{3}}{3}$ |
| $\dfrac{\pi}{3}$ | $3\tan\dfrac{\pi}{3} = 3\sqrt{3}$ | $2\sec\dfrac{\pi}{3} = 4$ |

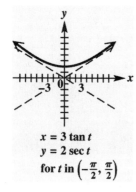

$x = 3\tan t$
$y = 2\sec t$
for $t$ in $\left(-\dfrac{\pi}{2}, \dfrac{\pi}{2}\right)$

**(b)** Since $\dfrac{x}{3} = \tan t$, $\dfrac{y}{2} = \sec t$,

and $1 + \tan^2 t = \left(\dfrac{y}{2}\right)^2 = \sec^2 t$, we have

$1 + \left(\dfrac{x}{3}\right)^2 = \left(\dfrac{y}{2}\right)^2 \Rightarrow 1 + \dfrac{x^2}{9} = \dfrac{y^2}{4} \Rightarrow$

$y^2 = 4\left(1 + \dfrac{x^2}{9}\right) \Rightarrow y = 2\sqrt{1 + \dfrac{x^2}{9}}$

Since this graph is the top half of a hyperbola, $x$ is in $(-\infty, \infty)$.

**15. (a)** $x = \sin t$, $y = \csc t$ for $t$ in $(0, \pi)$

Since $t$ is in $(0, \pi)$ and $x = \sin t$, $x$ is in $(0, 1]$.

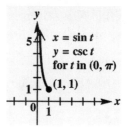

$x = \sin t$
$y = \csc t$
for $t$ in $(0, \pi)$
$(1, 1)$

**(b)** Since $x = \sin t$ and $y = \csc t = \dfrac{1}{\sin t}$, we

have $y = \dfrac{1}{x}$, where $x$ is in $(0, 1]$.

**17. (a)** $x = t$, $y = \sqrt{t^2 + 2}$, for $t$ in $(-\infty, \infty)$

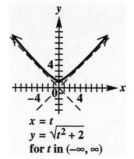

$x = t$
$y = \sqrt{t^2 + 2}$
for $t$ in $(-\infty, \infty)$

**(b)** Since $x = t$ and $y = \sqrt{t^2 + 2}$,

$y = \sqrt{x^2 + 2}$. Since $t$ is in $(-\infty\ \infty)$ and $x = t$, $x$ is in $(-\infty, \infty)$.

**19. (a)** $x = 2 + \sin t$, $y = 1 + \cos t$, for $t$ in $[0, 2\pi]$

Since this is a circle centered at $(2, 1)$ with radius 1, and $t$ is in $[0, 2\pi]$, $x$ is in $[1, 3]$.

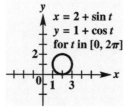

$x = 2 + \sin t$
$y = 1 + \cos t$
for $t$ in $[0, 2\pi]$

**(b)** Since $x = 2 + \sin t$ and $y = 1 + \cos t$,

$x - 2 = \sin t$ and $y - 1 = \cos t$. Since

$\sin^2 t + \cos^2 t = 1$, we have

$(x - 2)^2 + (y - 1)^2 = 1$.

**21. (a)**  $x = t + 2, \ y = \dfrac{1}{t+2}$, for $t \neq 2$

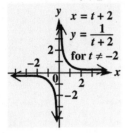

**(b)**  Since $x = t + 2$ and

$y = \dfrac{1}{t+2}$, we have $y = \dfrac{1}{x}$.  Since $t \neq -2$,

$x \neq -2 + 2, \ x \neq 0$. Therefore, $x$ is in

$(-\infty, 0) \cup (0, \infty)$.

**23. (a)**  $x = t + 2, \ y = t - 4$, for $t$ in $(-\infty, \infty)$

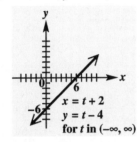

**(b)**  Since $x = t + 2$, we have $t = x - 2$. Since

$y = t - 4$, we have

$y = (x - 2) - 4 = x - 6$. Since $t$ is in $(-\infty,$

$\infty)$, $x$ is in $(-\infty, \infty)$.

**25.**  $x = 3\cos t, \ y = 3\sin t$

Since $x = 3\cos t \Rightarrow \cos t = \dfrac{x}{3}$,

$y = 3\sin t \Rightarrow \sin t = \dfrac{y}{3}$, and $\sin^2 t + \cos^2 t = 1$,

we have $\left(\dfrac{y}{3}\right)^2 + \left(\dfrac{x}{3}\right)^2 = 1 \Rightarrow \dfrac{y^2}{9} + \dfrac{x^2}{9} = 1 \Rightarrow$

$x^2 + y^2 = 9$

**27.**  $x = 3\sin t, \ y = 2\cos t$

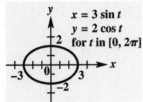

Since $x = 3\sin t \Rightarrow \sin t = \dfrac{x}{3}$,

$y = 2\cos t \Rightarrow \cos t = \dfrac{y}{2}$, and

$\sin^2 t + \cos^2 t = 1$, we have

$\left(\dfrac{x}{3}\right)^2 + \left(\dfrac{y}{2}\right)^2 = 1 \Rightarrow \dfrac{x^2}{9} + \dfrac{y^2}{4} = 1$

This is an ellipse centered at the origin with

axes endpoints $(-3,0), (3,0), (0,-2), (0,2)$.

In Exercises 29–31, answers may vary.

**29.**  $y = (x+3)^2 - 1$

$x = t, \ y = (t+3)^2 - 1$ for $t$ in $(-\infty, \infty)$;

$x = t - 3, \ y = t^2 - 1$ for $t$ in $(-\infty, \infty)$

**31.**  $y = x^2 - 2x + 3 = (x-1)^2 + 2$

$x = t, \ y = (t-1)^2 + 2$ for $t$ in $(-\infty, \infty)$;

$x = t + 1, \ y = t^2 + 2$ for $t$ in $(-\infty, \infty)$

**33.**  $x = 2t - 2\sin t, \ y = 2 - 2\cos t$, for $t$ in $[0, 4\pi]$

| $t$ | 0 | $\dfrac{\pi}{2}$ | $\pi$ | $\dfrac{3\pi}{2}$ |
|---|---|---|---|---|
| $x = 2t - 2\sin t$ | 0 | 1.4 | $2\pi$ | 11.4 |
| $y = 2 - 2\cos t$ | 0 | 2 | 4 | 2 |

| $t$ | $2\pi$ | $3\pi$ | $4\pi$ |
|---|---|---|---|
| $x = 2t - 2\sin t$ | $4\pi$ | $6\pi$ | $8\pi$ |
| $y = 2 - 2\cos t$ | 0 | 4 | 0 |

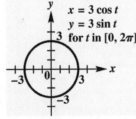

This is a circle centered at the origin with

radius 3.

Exercises 35–37 are graphed in parametric mode in the following window.

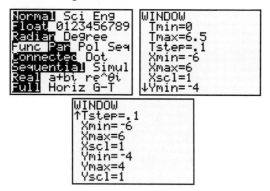

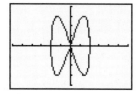

35.   $x = 2\cos t,\ y = 3\sin 2t$

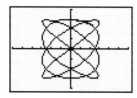

37.   $x = 3\sin 4t,\ y = 3\cos 3t$

For Exercises 39–41, recall that the motion of a projectile (neglecting air resistance) can be modeled by: $x = (v_0 \cos\theta)t,\ y = (v_0 \sin\theta)t - 16t^2$ for $t$ in $[0, k]$.

39.   (a)   $x = (v\cos\theta)t \Rightarrow$

$$x = (48\cos 60°)t = 48\left(\frac{1}{2}\right)t = 24t$$

$$y = (v\sin\theta)t - 16t^2 \Rightarrow$$

$$y = (48\sin 60°)t - 16t^2 = 48 \cdot \frac{\sqrt{3}}{2}t - 16t^2$$

$$= -16t^2 + 24\sqrt{3}t$$

(b)   $t = \dfrac{x}{24}$, so $y = -16\left(\dfrac{x}{24}\right)^2 + 24\sqrt{3}\left(\dfrac{x}{24}\right)$

$$= -\frac{x^2}{36} + \sqrt{3}x$$

(c)   $y = -16t^2 + 24\sqrt{3}t$

When the rocket is no longer in flight, $y = 0$. Solve $0 = -16t^2 + 24\sqrt{3}t \Rightarrow$

$$0 = t\left(-16t + 24\sqrt{3}\right).$$

$t = 0$ or $-16t + 24\sqrt{3} = 0 \Rightarrow$

$$-16t = -24\sqrt{3} \Rightarrow t = \frac{24\sqrt{3}}{16} \Rightarrow$$

$$t = \frac{3\sqrt{3}}{2} \approx 2.6$$

The flight time is about 2.6 seconds. The horizontal distance at $t = \dfrac{3\sqrt{3}}{2}$ is

$$x = 24t = 24\left(\frac{3\sqrt{3}}{2}\right) \approx 62\text{ ft}$$

41.   (a)   $x = (v\cos\theta)t \Rightarrow x = (88\cos 20°)t$

$$y = (v\sin\theta)t - 16t^2 + 2 \Rightarrow$$

$$y = (88\sin 20°)t - 16t^2 + 2$$

(b)   Since $t = \dfrac{x}{88\cos 20°}$, we have

$$y = 88\sin 20°\left(\frac{x}{88\cos 20°}\right)$$

$$-16\left(\frac{x}{88\cos 20°}\right)^2 + 2$$

$$= (\tan 20°)x - \frac{x^2}{484\cos^2 20°} + 2$$

(c)   Solving $0 = -16t^2 + (88\sin 20°)t + 2$ by the quadratic formula, we have

$$t = \frac{-88\sin 20° \pm \sqrt{(88\sin 20°)^2 - 4(-16)(2)}}{(2)(-16)}$$

$$= \frac{-30.098 \pm \sqrt{905.8759 + 128}}{-32} \Rightarrow$$

$t \approx -0.064$ or $1.9$

Discard $t = -0.064$ since it is an unacceptable answer.

At $t \approx 1.9\,\text{sec}$, $x = (88\cos 20°)t \approx 161\,\text{ft}$.

The softball traveled 1.9 sec and 161 feet.

**43. (a)** $x = (v\cos\theta)t \Rightarrow$

$$x = (128\cos 60°)t = 128\left(\frac{1}{2}\right)t = 64t$$

Since $t = \dfrac{x}{64}$,

$$y = 64\sqrt{3}\left(\frac{x}{64}\right) - 16\left(\frac{x}{64}\right)^2 + 8 \Rightarrow$$

$$y = -\frac{1}{256}x^2 + \sqrt{3}x + 8.$$

This is a parabolic path.

**(b)** Solving $0 = -16t^2 + 64\sqrt{3}t + 8$ by the quadratic formula, we have

$$t = \frac{-64\sqrt{3} \pm \sqrt{\left(64\sqrt{3}\right)^2 - 4(-16)(8)}}{2(-16)}$$

$$= \frac{-64\sqrt{3} \pm \sqrt{12,800}}{-32}$$

$$= \frac{-64\sqrt{3} \pm 80\sqrt{2}}{-32} \Rightarrow t \approx -.07,\ 7.0$$

Discard $t = -.07$ since it gives an unacceptable answer. At $t \approx 7.0$ sec, $x = 64t = 448$ ft. The rocket traveled approximately 7 sec and 448 feet.

**45. (a)** $x = (v\cos\theta)t \Rightarrow$

$$x = (64\cos 60°)t = 64\left(\frac{1}{2}\right)t = 32t$$

$$y = (v\sin\theta)t - 16t^2 + 3 \Rightarrow$$
$$y = (64\sin 60°)t - 16t^2 + 3$$

$$= 64\left(\frac{\sqrt{3}}{2}\right)t - 16t^2 + 3$$

$$= 32\sqrt{3}t - 16t^2 + 3$$

**(b)** Solving $0 = -16t^2 + 32\sqrt{3}t + 3$ by the quadratic formula, we have

$$t = \frac{-32\sqrt{3} \pm \sqrt{\left(32\sqrt{3}\right)^2 - 4(-16)(3)}}{2(-16)}$$

$$= \frac{-32\sqrt{3} \pm \sqrt{3264}}{-32} = \frac{-32\sqrt{3} \pm 8\sqrt{51}}{-32} \Rightarrow$$
$$t \approx -.05,\ 3.52$$

Discard $t = -.07$ since it gives an unacceptable answer. At $t \approx 3.52$ sec, $x = 32t \approx 112.6$ ft. The ball traveled approximately 112.6 feet.

**(c)** To find the maximum height, find the vertex of $y = -16t^2 + 32\sqrt{3}t + 3$.

$$y = -16t^2 + 32\sqrt{3}t + 3$$
$$= -16\left(t^2 - 2\sqrt{3}t\right) + 3$$
$$= -16\left(t^2 - 2\sqrt{3}t + 3 - 3\right) + 3$$
$$y = -16\left(t - \sqrt{3}\right)^2 + 48 + 3$$
$$= -16\left(t - \sqrt{3}\right)^2 + 51$$

The maximum height of 51 ft is reached at $\sqrt{3} \approx 1.73$ sec. Since $x = 32t$, the ball has traveled horizontally $32\sqrt{3} \approx 55.4$ ft.

**(d)** To determine if the ball would clear a 5-ft-high fence that is 100 ft from the batter, we need to first determine at what time is the ball 100 ft from the batter. Since $x = 32t$, the time the ball is 100 ft from the batter is $t = \dfrac{100}{32} = 3.125$ sec. We next need to determine how high off the ground the ball is at this time. Since $y = 32\sqrt{3}t - 16t^2 + 3$, the height of the ball is $y = 32\sqrt{3}(3.125) - 16(3.125)^2 + 3 \approx 20.0$ ft . Since this height exceeds 5 ft, the ball will clear the fence.

**47. (a)**

| WINDOW | | WINDOW |
|---|---|---|
| Tmin=0 | | ↑Tstep=.1 |
| Tmax=4.5 | | Xmin=0 |
| Tstep=.1 | | Xmax=400 |
| Xmin=0 | | Xscl=50 |
| Xmax=400 | | Ymin=0 |
| Xscl=50 | | Ymax=100 |
| ↓Ymin=0 | | Yscl=10 |

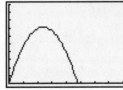

**(b)** $x = 56.56530965t = v(\cos\theta)t \Rightarrow$
$56.56530965 = v\cos\theta$

$$y = -16t^2 + 67.41191099t$$
$$= -16t^2 + v(\sin\theta)t \Rightarrow$$
$67.41191099 = v\sin\theta$

Thus, $\dfrac{67.41191099}{56.56530965} = \dfrac{v\sin\theta}{v\cos\theta} \Rightarrow$
$1.1918 = \tan\theta \Rightarrow \theta \approx 50.0°.$

**(c)** $67.41191099 = v\sin 50.0° \Rightarrow$
$v \approx 88.0$ ft/sec. Thus, the parametric
equations are $x = 88(\cos 50.0°)t$,

$$y = -16t^2 + 88(\sin 50.0°)t.$$

For Exercises 49–51, many answers are possible.

**49.** $y = a(x-h)^2 + k$

To find one parametric representation, let
$x = t$. We therefore have, $y = a(t-h)^2 + k$.
For another representation, let $x = t + h$. We
therefore have $y = a(t+h-h)^2 + k = at^2 + k$.

**51.** $\dfrac{x^2}{a^2} + \dfrac{y^2}{b^2} = 1$

To find a parametric representation, let
$x = a\sin t$. We therefore have

$$\frac{(a\sin t)^2}{a^2} + \frac{y^2}{b^2} = 1 \Rightarrow \frac{a^2\sin^2 t}{a^2} + \frac{y^2}{b^2} = 1 \Rightarrow$$

$$\sin^2 t + \frac{y^2}{b^2} = 1 \Rightarrow \frac{y^2}{b^2} = 1 - \sin^2 t$$

$$y^2 = b^2\left(1 - \sin^2 t\right) \Rightarrow y^2 = b^2\cos^2 t \Rightarrow$$

$$y = b\cos t$$

**53.** To show that $r\theta = a$ is given parametrically

by $x = \dfrac{a\cos\theta}{\theta}$, $y = \dfrac{a\sin\theta}{\theta}$,

for $\theta$ in $(-\infty,0) \cup (0,\infty)$, we must show that
the parametric equations yield $r\theta = a$, where
$r^2 = x^2 + y^2$.

$$r^2 = x^2 + y^2 \Rightarrow$$

$$r^2 = \left(\frac{a\cos\theta}{\theta}\right)^2 + \left(\frac{a\sin\theta}{\theta}\right)^2 \Rightarrow$$

$$r^2 = \frac{a^2\cos^2\theta}{\theta^2} + \frac{a^2\sin^2\theta}{\theta^2}$$

$$r^2 = \frac{a^2}{\theta^2}\cos^2\theta + \frac{a^2}{\theta^2}\sin^2\theta \Rightarrow$$

$$r^2 = \frac{a^2}{\theta^2}\left(\cos^2\theta + \sin^2\theta\right) \Rightarrow r^2 = \frac{a^2}{\theta^2} \Rightarrow$$

$$r = \pm\frac{a}{\theta} \text{ or just } r = \frac{a}{\theta}$$

This implies that the parametric equations
satisfy $r\theta = a$.

**55.** If $x = f(t)$ is replaced by $x = c + f(t)$, the
graph will be translated $c$ units horizontally.

## Chapter 8: Review Exercises

**1.** $\sqrt{-9} = i\sqrt{9} = 3i$

**3.** $x^2 = -81 \Rightarrow x = \pm\sqrt{-81} = \pm 9i$
Solution set: $\{\pm 9i\}$

**5.** $(1-i) - (3+4i) + 2i = -2 - 3i$

**7.** $(6-5i) + (2+7i) - (3-2i) = 5 + 4i$

**9.** $(3+5i)(8-i) = 24 - 3i + 40i - 5i^2$
$= 24 + 37i - 5(-1) = 29 + 37i$

**11.** $(2+6i)^2 = 4 + 2(2)(6i) + (6i)^2$
$= 4 + 24i - 36 = -32 + 24i$

**13.** $(1-i)^3 = 1^3 - 3\cdot 1^2 \cdot i + 3\cdot 1\cdot i^2 - i^3$
$= 1 - 3i - 3 - (-i) = -2 - 2i$

**15.** $\dfrac{25-19i}{5+3i} = \dfrac{25-19i}{5+3i} \cdot \dfrac{5-3i}{5-3i}$

$$= \frac{125 - 75i - 95i + 57i^2}{5^2 - (3i)^2}$$

$$= \frac{125 - 160i - 57}{25 - (-9)} = \frac{68 - 170i}{34} = 2 - 5i$$

**17.** $\dfrac{2+i}{1-5i} = \dfrac{2+i}{1-5i} \cdot \dfrac{1+5i}{1+5i} = \dfrac{2+10i+i+5i^2}{1^2 - 25i^2}$

$$= \frac{-3+11i}{26} = -\frac{3}{26} + \frac{11}{26}i$$

**19.** $i^{53} = i^{52} \cdot i = \left(i^4\right)^{13} \cdot i = i$

**21.** $\left[5(\cos 90° + i\sin 90°)\right]$

$$\cdot \left[6(\cos 180° + i\sin 180°)\right]$$

$$= 5\cdot 6\left[\cos(90° + 180°) + i\sin(90° + 180°)\right]$$

$$= 30(\cos 270° + i\sin 270°) = 30(0 - i)$$

$$= 0 - 30i \text{ or } -30i$$

**23.** $\dfrac{2(\cos 60° + i\sin 60°)}{8(\cos 300° + \sin 300°)}$

$$= \frac{2}{8}\left[\cos(60° - 300°) + i\sin(60° - 300°)\right]$$

$$= \frac{1}{4}\left[\cos(-240°) + i\sin(-240°)\right]$$

$$= \frac{1}{4}\left[\cos(240°) - i\sin(240°)\right]$$

$$= \frac{1}{4}\left[-\cos 60° + i\sin 60°\right]$$

$$= \frac{1}{4}\left(-\frac{1}{2} + \frac{\sqrt{3}}{2}\right) = -\frac{1}{8} + \frac{\sqrt{3}}{8}i$$

**25.** $\left(\sqrt{3}+i\right)^3$

$r = \sqrt{\left(\sqrt{3}\right)^2 + 1^2} = \sqrt{3+1} = \sqrt{4} = 2$ and since

$\theta$ is in quadrant I, $\tan\theta = \dfrac{1}{\sqrt{3}} = \dfrac{\sqrt{3}}{3} \Rightarrow$

$\theta = 30°$.

$\left(\sqrt{3}+i\right)^3 = \left[2\left(\cos 30° + i\sin 30°\right)\right]^3$

$= 2^3\left[\cos\left(3\cdot 30°\right) + i\sin\left(3\cdot 30°\right)\right]$

$= 8\left[\cos 90° + i\sin 90°\right]$

$= 8\left(0+i\right) = 0 + 8i = 8i$

**27.** $\left(\cos 100° + i\sin 100°\right)^6$

$= \cos\left(6\cdot 100°\right) + i\sin\left(6\cdot 100°\right)$

$= \cos 600° + i\sin 600° = \cos 240° + i\sin 240°$

$= -\cos 60° - i\sin 60° = -\dfrac{1}{2} - \dfrac{\sqrt{3}}{2}i$

**29.**

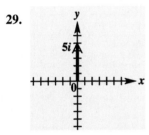

**31.**

**33.** $-2 + 2i$

$r = \sqrt{\left(-2\right)^2 + 2^2} = \sqrt{4+4} = \sqrt{8} = 2\sqrt{2}$

Since $\theta$ is in quadrant II,

$\tan\theta = \dfrac{2}{-2} = -1 \Rightarrow \theta = 135°$. Thus,

$-2 + 2i = 2\sqrt{2}\left(\cos 135° + i\sin 135°\right)$.

**35.** $2\left(\cos 225° + i\sin 225°\right)$

$= 2\left(-\cos 45° - i\sin 45°\right)$

$= 2\left(-\dfrac{\sqrt{2}}{2} - \dfrac{i\sqrt{2}}{2}\right) = -\sqrt{2} - i\sqrt{2}$

**37.** $1 - i$

$r = \sqrt{1^2 + \left(-1\right)^2} = \sqrt{1+1} = \sqrt{2}$ and

$\tan\theta = \dfrac{-1}{1} = -1 \Rightarrow \theta = 315°$, since $\theta$ is in

quadrant IV. Thus,

$1 - i = \sqrt{2}\left(\cos 315° + i\sin 315°\right)$.

**39.** $-4i$

Since $r = 4$ and the point $\left(0,-4\right)$ intersects

the negative $y$-axis, $\theta = 270°$ and

$-4i = 4\left(\cos 270° + i\sin 270°\right)$.

**41.** $z = x + yi$

Since the imaginary part of $z$ is the negative of

the real part of $z$, we are saying $y = -x$. This is

a line.

**43.** Convert $1 - i$ to polar form

$r = \sqrt{1^2 + \left(-1\right)^2} = \sqrt{1+1} = \sqrt{2}$ and

$\tan\theta = \dfrac{-1}{1} = -1 \Rightarrow \theta = 315°$, since $\theta$ is in

quadrant IV. Thus,

$1 - i = \sqrt{2}\left(\cos 315° + i\sin 315°\right)$. Since

$r^3\left(\cos 3\alpha + i\sin 3\alpha\right)$

$= \sqrt{2}\left(\cos 315° + i\sin 315°\right)$, then we have

$r^3 = \sqrt{2} \Rightarrow r = \sqrt[6]{2}$ and

$3\alpha = 315° + 360°\cdot k \Rightarrow$

$\alpha = \dfrac{315° + 360°\cdot k}{3} = 105° + 120°\cdot k$, $k$ any

integer. If $k = 0$, then $\alpha = 105° + 0° = 105°$.

If $k = 1$, then $\alpha = 105° + 120° = 225°$.

If $k = 2$, then $\alpha = 105° + 240° = 345°$.

So, the cube roots of $1 - i$ are

$\sqrt[6]{2}\left(\cos 105° + i\sin 105°\right)$,

$\sqrt[6]{2}\left(\cos 225° + i\sin 225°\right)$, and

$\sqrt[6]{2}\left(\cos 345° + i\sin 345°\right)$.

**45.** The number $-64$ has no real sixth roots

because a real number raised to the sixth

power will never be negative.

**47.** $x^4 + 16 = 0 \Rightarrow x^4 = -16$

We have, $r = 16$ and $\theta = 180°$.

$x^4 = -16 = -16 + 0i = 16(\cos 180° + i \sin 180°)$

Since $r^4(\cos 4\alpha + i \sin 4\alpha)$

$= 16(\cos 180° + i \sin 180°)$, then we have

$r^4 = 16 \Rightarrow r = 2$ and $4\alpha = 180° + 360° \cdot k \Rightarrow$

$\alpha = \dfrac{180° + 360° \cdot k}{4} = 45° + 90° \cdot k,\ k$ any

integer. If $k = 0$, then $\alpha = 45° + 0° = 45°$.

If $k = 1$, then $\alpha = 45° + 90° = 135°$.

If $k = 2$, then $\alpha = 45° + 180° = 225°$.

If $k = 3$, then $\alpha = 45° + 270° = 315°$.

Solution set:

$\{2(\cos 45° + i \sin 45°), 2(\cos 135° + i \sin 135°),$
$\quad 2(\cos 225° + i \sin 225°),$
$\quad 2(\cos 315° + i \sin 315°)\}$

**49.** $\left(-1, \sqrt{3}\right)$

$r = \sqrt{(-1)^2 + \left(\sqrt{3}\right)^2} = \sqrt{1+3} = \sqrt{4} = 2$ and

$\theta = \tan^{-1}\left(-\dfrac{\sqrt{3}}{1}\right) = \tan^{-1}\left(-\sqrt{3}\right) = 120°$, since

$\theta$ is in quadrant II. Thus, the polar
coordinates are $(2, 120°)$.

**51.** The angle must be quandrantal.

**53.** $r = 4 \cos \theta$ is a circle.

| $\theta$ | 0° | 30° | 45° | 60° | 90° | 120° | 135° | 150° | 180° |
|---|---|---|---|---|---|---|---|---|---|
| $r = 4\cos\theta$ | 4 | 3.5 | 2.8 | 2 | 0 | −2 | −2.8 | −3.5 | −4 |

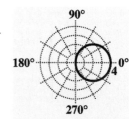

$r = 4 \cos \theta$
Graph is retraced in the
interval $(180°, 360°)$.

**55.** $r = 2 \sin 4\theta$ is an eight-leaved rose.

| $\theta$ | 0° | 7.5° | 15° | 22.5° | 0° | 37.5° | 45° |
|---|---|---|---|---|---|---|---|
| $r = 2\sin 4\theta$ | 0 | 1 | $\sqrt{3}$ | 2 | $\sqrt{3}$ | 1 | 0 |

| $\theta$ | 52.5° | 60° | 67.5° | 75° | 82.5° | 90° | 52.5° |
|---|---|---|---|---|---|---|---|
| $r = 2\sin 4\theta$ | −1 | $-\sqrt{3}$ | −2 | $-\sqrt{3}$ | −1 | 0 | −1 |

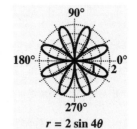

$r = 2 \sin 4\theta$
The graph continues to form
eight petals for the interval
$[0°, 360°)$.

**57.** $r = \dfrac{3}{1+\cos\theta}$

$r = \dfrac{3}{1+\cos\theta} \Rightarrow r(1+\cos\theta) = 3 \Rightarrow$

$r + r\cos\theta = 3 \Rightarrow \sqrt{x^2 + y^2} + x = 3 \Rightarrow$

$\sqrt{x^2 + y^2} = 3 - x$

$x^2 + y^2 = (3-x)^2 \Rightarrow x^2 + y^2 = 9 - 6x + x^2 \Rightarrow$

$y^2 = 9 - 6x \Rightarrow y^2 + 6x - 9 = 0 \Rightarrow y^2 = -6x + 9$

$y^2 = -6\left(x - \dfrac{3}{2}\right)$ or $y^2 + 6x - 9 = 0$

**59.** $r = 2 \Rightarrow \sqrt{x^2 + y^2} = 2 \Rightarrow x^2 + y^2 = 4$

**61.** $y = x^2 \Rightarrow r\sin\theta = r^2\cos^2\theta \Rightarrow$

$\sin\theta = r\cos^2\theta \Rightarrow r = \dfrac{\sin\theta}{\cos^2\theta} \Rightarrow$

$r = \dfrac{\sin\theta}{\cos\theta}\cdot\dfrac{1}{\cos\theta} = \tan\theta\sec\theta$

$r = \tan\theta\sec\theta$ or $r = \dfrac{\tan\theta}{\cos\theta}$

**63.** If $(r,\theta)$ lies on the graph, $(-r,\theta)$ would reflect that point into the quadrant diagonal to the original quadrant. The correct choice is A.

**65.** If $(r,\theta)$ lies on the graph, $(r,\pi-\theta)$ would reflect that point into the other quadrant on the same side of the $x$-axis as the original quadrant. Therefore, there is symmetry about the $y$-axis. The correct choice is B.

**67.** $x = 2$

$x = r\cos\theta \Rightarrow r\cos\theta = 2 \Rightarrow$

$r = \dfrac{2}{\cos\theta}$ or $r = 2\sec\theta$

**69.** $x + 2y = 4$

Since $x = r\cos\theta$ and $y = r\sin\theta$, we have

$(r\cos\theta)^2 + (r\sin\theta)^2 = 4 \Rightarrow$

$r^2\cos\theta^2 + r^2\sin^2\theta = 4 \Rightarrow$

$r^2\left(\cos\theta^2 + \sin^2\theta\right) = 4 \Rightarrow r^2 = 4 \Rightarrow$

$r = -2$ or $r = -2$

**71.** $x = t + \cos t,\ y = \sin t$ for $t$ in $[0, 2\pi]$

| $t$ | $0$ | $\dfrac{\pi}{6}$ | $\dfrac{\pi}{3}$ | $\dfrac{\pi}{2}$ | $\dfrac{3\pi}{4}$ | $\pi$ |
|---|---|---|---|---|---|---|
| $x = t + \cos t$ | $0$ | $\dfrac{\pi}{6} + \dfrac{\sqrt{3}}{2}$ $\approx 1.4$ | $\dfrac{\pi}{3} + \dfrac{1}{2}$ $\approx 1.5$ | $\dfrac{\pi}{2}$ $\approx 1.6$ | $\dfrac{3\pi}{4} - \dfrac{\sqrt{2}}{2}$ $\approx 1.6$ | $\pi - 1$ $\approx 2.1$ |
| $y = \sin t$ | $0$ | $\dfrac{1}{2} = .5$ | $\dfrac{\sqrt{3}}{2} \approx 1.7$ | $1$ | $\dfrac{\sqrt{2}}{2} \approx .7$ | $0$ |

| $t$ | $\dfrac{7\pi}{6}$ | $\dfrac{5\pi}{4}$ | $\dfrac{4\pi}{3}$ | $\dfrac{3\pi}{2}$ | $\dfrac{7\pi}{4}$ | $2\pi$ |
|---|---|---|---|---|---|---|
| $x = t + \cos t$ | $\dfrac{7\pi}{6} - \dfrac{\sqrt{3}}{2}$ $\approx 2.8$ | $\dfrac{5\pi}{4} - \dfrac{\sqrt{2}}{2}$ $\approx 3.2$ | $\dfrac{4\pi}{3} - \dfrac{1}{2}$ $\approx 3.7$ | $\dfrac{3\pi}{2}$ $\approx 4.7$ | $\dfrac{7\pi}{4} + \dfrac{\sqrt{2}}{2}$ $\approx 6.2$ | $2\pi + 1$ $\approx 7.3$ |
| $y = \sin t$ | $-\dfrac{1}{2} = -.5$ | $-\dfrac{\sqrt{2}}{2} \approx -.7$ | $-\dfrac{\sqrt{3}}{2} \approx -1.7$ | $-1$ | $-\dfrac{\sqrt{2}}{2} \approx -.7$ | $0$ |

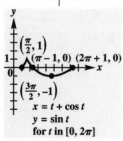

**73.** $x = \sqrt{t-1}, y = \sqrt{t}$, for $t$ in $[1, \infty)$

Since $x = \sqrt{t-1} \Rightarrow x^2 = t-1 \Rightarrow t = x^2 + 1$,

substitute $x^2 + 1$ for $t$ in the equation for $y$ to

obtain $y = \sqrt{x^2 + 1}$. Since $t$ is in $[1, \infty)$, $x$ is in

$[\sqrt{1-1}, \infty)$ or $[0, \infty)$.

**75.** $x = 5 \tan t, y = 3 \sec t$, for $t$ in $\left( -\dfrac{\pi}{2}, \dfrac{\pi}{2} \right)$

Since $\dfrac{x}{5} = \tan t, \dfrac{y}{3} = \sec t$, and

$1 + \tan^2 t = \sec^2 t$, we have

$$1 + \left( \frac{x}{5} \right)^2 = \left( \frac{y}{3} \right)^2 \Rightarrow 1 + \frac{x^2}{25} = \frac{y^2}{9} \Rightarrow$$

$$9 \left( 1 + \frac{x^2}{25} \right) = y^2 \Rightarrow y = \sqrt{9 \left( 1 + \frac{x^2}{25} \right)} \Rightarrow$$

$$y = 3 \sqrt{1 + \frac{x^2}{25}} \, .$$

$y$ is positive since $y = 3 \sec t > 0$ for $t$ in

$\left( -\dfrac{\pi}{2}, \dfrac{\pi}{2} \right)$. Since $t$ is in $\left( -\dfrac{\pi}{2}, \dfrac{\pi}{2} \right)$ and $x = 5$

$\tan t$ is undefined at $-\dfrac{\pi}{2}$ and $\dfrac{\pi}{2}$, $x$ is in

$(-\infty, \infty)$.

**77.** The radius of the circle that has center $(3, 4)$

and passes through the origin is

$$r = \sqrt{(3-0)^2 + (4-0)^2} = \sqrt{3^2 + 4^2}$$
$$= \sqrt{9 + 16} = \sqrt{25} = 5$$

Thus, the equation of this circle is

$(x-3)^2 + (y-4)^2 = 5^2$. Since

$\cos^2 t + \sin^2 y = 1 \Rightarrow$

$25 \cos^2 t + 25 \sin^2 t = 25 \Rightarrow$

$(5 \cos t)^2 + (5 \sin t)^2 = 5^2$, we can have

$5 \cos t = x - 3$ and $5 \sin t = y - 4$. Thus, a pair

of parametric equations can be $x = 3 + 5 \cos t$,

$y = 4 + 5 \sin t$, where $t$ in $[0, 2\pi]$

**79. (a)** $x = (v \cos \theta) t \Rightarrow x = (118 \cos 27°) t$ and

$y = (v \sin \theta) t - 16 t^2 + h \Rightarrow$
$y = (118 \sin 27°) t - 16 t^2 + 3.2$

**(b)** Since $t = \dfrac{x}{118 \cos 27°}$, we have

$$y = 118 \sin 27° \cdot \frac{x}{118 \cos 27°}$$
$$- 16 \left( \frac{x}{118 \cos 27°} \right)^2 + 3.2$$
$$= 3.2 - \frac{4}{3481 \cos^2 27°} x^2 + (\tan 27°) x$$

**(c)** Solving $0 = -16 t^2 + (118 \sin 27°) t + 3.2$ by

the quadratic formula, we have

$$t = \frac{-118 \sin 27° \pm \sqrt{(118 \sin 27°)^2 - 4(-16)(3.2)}}{2(-16)} \Rightarrow$$

$t \approx -.06, 3.406$

Discard $t = -0.06$ sec since it is an

unacceptable answer. At $t = 3.4$ sec, the

baseball traveled

$x = (118 \cos 27°)(3.406) \approx 358$ ft .

# Chapter 8 Test

**1. (a)** $\sqrt{-8} \cdot \sqrt{-6} = i\sqrt{8} \cdot i\sqrt{6} = i^2 \sqrt{48} = -4\sqrt{3}$

**(b)** $\dfrac{\sqrt{-2}}{\sqrt{8}} = \dfrac{i\sqrt{2}}{\sqrt{8}} = \dfrac{i}{\sqrt{4}} = \dfrac{1}{2} i$

**(c)** $\dfrac{\sqrt{-20}}{\sqrt{-180}} = \dfrac{i\sqrt{20}}{i\sqrt{180}} = \dfrac{1}{\sqrt{9}} = \dfrac{1}{3}$

**2.** $w = 2 - 4i, z = 5 + i$

**(a)** $w + z = (2 - 4i) + (5 + i) = 7 - 3i$

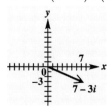

**(b)** $w - z = (2 - 4i) - (5 + i) = -3 - 5i$

**(c)** $wz = (2 - 4i)(5 + i) = 10 + 2i - 20i - 4i^2$
$= 10 - 18i - 4(-1) = 14 - 18i$

**(d)** $\dfrac{w}{z} = \dfrac{2 - 4i}{5 + i} = \dfrac{2 - 4i}{5 + i} \cdot \dfrac{5 - i}{5 - i}$

$$= \frac{10 - 2i - 20i + 4i^2}{5^2 - i^2} = \frac{10 - 22i + 4(-1)}{5^2 - (-1)}$$

$$= \frac{6 - 22i}{26} = \frac{3}{13} - \frac{11}{13} i$$

**3. (a)** $i^{15} = i^{12+3} = i^{12} \cdot i^3 = \left(i^4\right)^3 \cdot i^3 = 1(-i) = -i$

**(b)** $(1+i)^2 = (1+i)(1+i) = 1 + i + i + i^2$
$= 1 + 2i + (-1) = 2i$

**4.** $2x^2 - x + 4 = 0$
Use the quadratic formula with $a = 2$, $b = -1$, and $c = 4$:

$$x = \frac{-(-1) \pm \sqrt{(-1)^2 - 4(2)(4)}}{2(2)}$$

$$= \frac{1 \pm \sqrt{1 - 32}}{4} = \frac{1 \pm \sqrt{-31}}{4} = \frac{1 \pm i\sqrt{31}}{4}$$

$$= \frac{1}{4} \pm \frac{\sqrt{31}}{4}i$$

Solution set: $\left\{ \dfrac{1}{4} \pm \dfrac{\sqrt{31}}{4}i \right\}$

**5. (a)** $3i$

$$r = \sqrt{0^2 + 3^2} = \sqrt{0 + 9} = \sqrt{9} = 3$$

The point $(0,3)$ is on the positive $y$-axis, so, $\theta = 90°$. Thus,
$$3i = 3(\cos 90° + i\sin 90°).$$

**(b)** $1 + 2i$

$$r = \sqrt{1^2 + 2^2} = \sqrt{1 + 4} = \sqrt{5}$$

Since $\theta$ is in quadrant I,

$$\theta = \tan^{-1}\left(\frac{2}{1}\right) = \tan^{-1} 2 \approx 63.43°. \text{ Thus,}$$

$$1 + 2i = \sqrt{5}\left(\cos 63.43° + i\sin 63.43°\right).$$

**(c)** $-1 - \sqrt{3}i$

$$r = \sqrt{(-1)^2 + (-\sqrt{3})^2} = \sqrt{1 + 3} = \sqrt{4} = 2$$

Since $\theta$ is in quadrant III,

$$\theta = \tan^{-1}\left(\frac{-\sqrt{3}}{-1}\right) = \tan^{-1}\sqrt{3} = 240°.$$

Thus, $-1 - \sqrt{3}i = 2(\cos 240° + i\sin 240°)$

**6. (a)** $3(\cos 30° + i\sin 30°) = 3\left(\dfrac{\sqrt{3}}{2} + \dfrac{1}{2}i\right)$
$$= \frac{3\sqrt{3}}{2} + \frac{3}{2}i$$

**(b)** $4\operatorname{cis} 40° = 3.06 + 2.57i$

**(c)** $3(\cos 90° + i\sin 90°) = 3(0 + 1 \cdot i)$
$$= 0 + 3i = 3i$$

**7.** $w = 8(\cos 40° + i\sin 40°)$,
$z = 2(\cos 10° + i\sin 10°)$

**(a)** $wz$
$$= 8 \cdot 2\left[\cos(40° + 10°) + i\sin(40° + 10°)\right]$$
$$= 16(\cos 50° + i\sin 50°)$$

**(b)** $\dfrac{w}{z} = \dfrac{8}{2}\left[\cos(40° - 10°) + i\sin(40° - 10°)\right]$
$$= 4(\cos 30° + i\sin 30°) = 4\left(\frac{\sqrt{3}}{2} + \frac{1}{2}i\right)$$
$$= 2\sqrt{3} + 2i$$

**(c)** $z^3 = \left[2(\cos 10° + i\sin 10°)\right]^3$
$$= 2^3(\cos 3 \cdot 10° + i\sin 3 \cdot 10°)$$
$$= 8(\cos 30° + i\sin 30°)$$
$$= 8\left(\frac{\sqrt{3}}{2} + \frac{1}{2}i\right) = 4\sqrt{3} + 4i$$

**8.** Find all the fourth roots of
$-16i = 16(\cos 270° + i\sin 270°)$.

Since $r^4(\cos 4\alpha + i\sin 4\alpha)$
$= 16(\cos 270° + i\sin 270°)$, then we have

$r^4 = 16 \Rightarrow r = 2$ and $4\alpha = 270° + 360° \cdot k \Rightarrow$

$$\alpha = \frac{270° + 360° \cdot k}{4} = 67.5° + 90° \cdot k, \ k \text{ any}$$

integer. If $k = 0$, then $\alpha = 67.5°$.

If $k = 1$, then $\alpha = 157.5°$.
If $k = 2$, then $\alpha = 247.5°$.
If $k = 3$, then $\alpha = 337.5°$.

The fourth roots of $-16i$ are
$2(\cos 67.5° + \sin 67.5)$,

$2(\cos 157.5° + i\sin 157.5°)$,
$2(\cos 247.5° + i\sin 247.5°)$, and
$2(\cos 337.5° + i\sin 337.5°)$.

**9.** Answers may vary.

**(a)** $(0,5)$

$$r = \sqrt{0^2 + 5^2} = \sqrt{0 + 25} = \sqrt{25} = 5$$

The point $(0,5)$ is on the positive $y$-axis.
Thus, $\theta = 90°$. One possibility is $(5, 90°)$.
Alternatively, if $\theta = 90° - 360° = -270°$, a second possibility is $(5, -270°)$.

**(b)** $(-2, -2)$

$r = \sqrt{(-2)^2 + (-2)^2} = \sqrt{4+4} = \sqrt{8} = 2\sqrt{2}$

Since $\theta$ is in quadrant III,

$\theta = \tan^{-1}\left(\dfrac{-2}{-2}\right) = \tan^{-1} 1 = 225°$. One

possibility is $(2\sqrt{2}, 225°)$. Alternatively, if $\theta = 225° - 360° = -135°$, a second possibility is $(2\sqrt{2}, -135°)$.

**10. (a)** $(3, 315°)$

$x = r\cos\theta \Rightarrow$

$x = 3\cos 315° = 3 \cdot \dfrac{\sqrt{2}}{2} = \dfrac{3\sqrt{2}}{2}$ and

$y = r\sin\theta \Rightarrow$

$y = 3\sin 315° = 3\left(-\dfrac{\sqrt{2}}{2}\right) = \dfrac{-3\sqrt{2}}{2}$

The rectangular coordinates are

$\left(\dfrac{3\sqrt{2}}{2}, \dfrac{-3\sqrt{2}}{2}\right)$.

**(b)** $(-4, 90°)$

$x = r\cos\theta \Rightarrow x = -4\cos 90° = 0$ and
$y = r\sin\theta \Rightarrow y = -4\sin 90° = -4$

The rectangular coordinates are $(0, -4)$.

**11.** $r = 1 - \cos\theta$ is a cardioid.

| $\theta$ | 0° | 30° | 45° | 60° | 90° | 135° |
|---|---|---|---|---|---|---|
| $r = 1 - \cos\theta$ | 0 | .1 | .3 | .5 | 1 | 1.7 |

| $\theta$ | 180° | 225° | 270° | 315° | 360° |
|---|---|---|---|---|---|
| $r = 1 - \cos\theta$ | 2 | 1.7 | 1 | .3 | 0 |

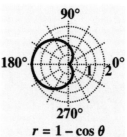

$r = 1 - \cos\theta$

**12.** $r = 3\cos 3\theta$ is a three-leaved rose.

| $\theta$ | 0° | 30° | 45° | 60° | 90° | 120° | 135° | 150° | 180° |
|---|---|---|---|---|---|---|---|---|---|
| $r = 3\cos 3\theta$ | 3 | 0 | −2.1 | −3 | 0 | 3 | 2.1 | 0 | −3 |

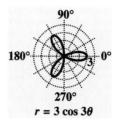

$r = 3\cos 3\theta$

Graph is retraced in the interval $(180°, 360°)$.

**13. (a)** Since

$r = \dfrac{4}{2\sin\theta - \cos\theta} = \dfrac{4}{-1 \cdot \cos\theta + 2\sin\theta}$,

we can use the general form for the polar equation of a line, $r = \dfrac{c}{a\cos\theta + b\sin\theta}$,

with $a = -1$, $b = 2$, and $c = 4$, we have $-x + 2y = 4$ or $x - 2y = -4$. The graph is a line with intercepts $(-4, 0)$ and $(0, 2)$.

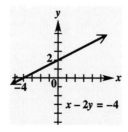

**(b)** $r = 6$ represents the equation of a circle centered at the origin with radius 6, namely $x^2 + y^2 = 36$.

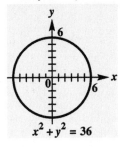

$$x^2 + y^2 = 36$$

**14.** $x = 4t - 3,\ y = t^2$ for $t$ in $[-3, 4]$

| $t$ | $x$ | $y$ |
|-----|-----|-----|
| $-3$ | $-15$ | $9$ |
| $-1$ | $-7$ | $1$ |
| $0$ | $-3$ | $0$ |
| $1$ | $1$ | $1$ |
| $2$ | $5$ | $4$ |
| $4$ | $13$ | $16$ |

Since $x = 4t - 3 \Rightarrow t = \dfrac{x+3}{4}$ and $y = t^2$, we have

$$y = \left(\frac{x+3}{4}\right)^2 = \frac{1}{4}(x+3)^2,\ \text{where } x \text{ is in } [-15, 13]$$

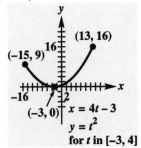

$x = 4t - 3$
$y = t^2$
for $t$ in $[-3, 4]$

**15.** $x = 2\cos 2t,\ y = 2\sin 2t$ for $t$ in $\left[0, 2\pi\right]$

| $t$ | $0$ | $\frac{\pi}{8}$ | $\frac{\pi}{4}$ | $\frac{3\pi}{8}$ | $\frac{\pi}{2}$ | $\frac{5\pi}{8}$ |
|-----|-----|-----|-----|-----|-----|-----|
| $x$ | $2$ | $\sqrt{2}$ | $0$ | $-\sqrt{2}$ | $-2$ | $-\sqrt{2}$ |
| $y$ | $0$ | $\sqrt{2}$ | $2$ | $\sqrt{2}$ | $0$ | $-\sqrt{2}$ |

| $t$ | $\frac{3\pi}{4}$ | $\pi$ | $\frac{5\pi}{4}$ | $\frac{3\pi}{2}$ | $\frac{7\pi}{4}$ | $2\pi$ |
|-----|-----|-----|-----|-----|-----|-----|
| $x$ | $0$ | $2$ | $0$ | $-2$ | $0$ | $2$ |
| $y$ | $-2$ | $0$ | $2$ | $0$ | $-2$ | $0$ |

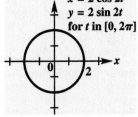

$x = 2\cos 2t$
$y = 2\sin 2t$
for $t$ in $[0, 2\pi]$

Since $x = 2\cos 2t \Rightarrow \cos 2t = \dfrac{x}{2}$,

$y = 2\sin 2t \Rightarrow \sin 2t = \dfrac{y}{2}$, and

$\cos^2(2t) + \sin^2(2t) = 1$, we have

$$\left(\frac{x}{2}\right)^2 + \left(\frac{y}{2}\right)^2 = 1 \Rightarrow \frac{x^2}{4} + \frac{y^2}{4} = 1 \Rightarrow x^2 + y^2 = 4,$$

where $x$ is in $[-1, 1]$.

**16.** $z = -1 + i$

$$z^2 - 1 = (-1+i)^2 - 1 = 1 - i - i + i^2 - 1$$
$$= -2i - 1 = -1 - 2i$$

Since $r = \sqrt{(-1)^2 + (-2)^2} = \sqrt{1+4} = \sqrt{5} > 2$, $z$ is not in the Julia set.

# Appendices

## Appendix A: Equations and Inequalities

**1.** Solve the equation $2x + 7 = x - 1$

$$2x + 7 = x - 1$$
$$2x + 7 - x = x - 1 - x$$
$$x + 7 = -1$$
$$x + 7 - 7 = -1 - 7$$
$$x = -8$$

Moreover, replacing $x$ with $-8$ in $2x + 7 = x - 1$ yields a true statement. Therefore, the given statement is true.

**3.** The equations $x^2 = 4$ and $x + 2 = 4$ are not equivalent. The first has a solution set $\{-2, 2\}$ while the solution set of the second is $\{2\}$.

Since the equations do not have the same solution set, they are not equivalent. The given statement is false.

**5.** B cannot be written in the form $ax + b = 0$. A can be written as $15x - 7 = 0$ or $15x + (-7) = 0$, C can be written as $2x = 0$ or $2x + 0 = 0$, and D can be written as $-.04x - .4 = 0$ or $-.04x + (-.4) = 0$.

**7.**
$$5x + 4 = 3x - 4$$
$$2x + 4 = -4$$
$$2x = -8 \Rightarrow x = -4$$
Solution set: $\{-4\}$

**9.**
$$6(3x - 1) = 8 - (10x - 14)$$
$$18x - 6 = 8 - 10x + 14$$
$$18x - 6 = 22 - 10x$$
$$28x - 6 = 22$$
$$28x = 28 \Rightarrow x = 1$$
Solution set: $\{1\}$

**11.**
$$\frac{5}{6}x - 2x + \frac{4}{3} = \frac{5}{3}$$
$$6 \cdot \left[\frac{5}{6}x - 2x + \frac{4}{3}\right] = 6 \cdot \frac{5}{3}$$
$$5x - 12x + 8 = 10$$
$$-7x + 8 = 10$$
$$-7x = 2 \Rightarrow x = -\frac{2}{7}$$
Solution set: $\left\{-\frac{2}{7}\right\}$

**13.**
$$3x + 5 - 5(x + 1) = 6x + 7$$
$$3x + 5 - 5x - 5 = 6x + 7$$
$$-2x = 6x + 7$$
$$-8x = 7 \Rightarrow x = \frac{7}{-8} = -\frac{7}{8}$$
Solution set: $\left\{-\frac{7}{8}\right\}$

**15.**
$$2[x - (4 + 2x) + 3] = 2x + 2$$
$$2(x - 4 - 2x + 3) = 2x + 2$$
$$2(-x - 1) = 2x + 2$$
$$-2x - 2 = 2x + 2$$
$$-2 = 4x + 2$$
$$-4 = 4x \Rightarrow -1 = x$$
Solution set: $\{-1\}$

**17.**
$$\frac{1}{14}(3x - 2) = \frac{x + 10}{10}$$
$$70 \cdot \left[\frac{1}{14}(3x - 2)\right] = 70 \cdot \left[\frac{x + 10}{10}\right]$$
$$5(3x - 2) = 7(x + 10)$$
$$15x - 10 = 7x + 70$$
$$8x - 10 = 70$$
$$8x = 80 \Rightarrow x = 10$$
Solution set: $\{10\}$

**19.**
$$.2x - .5 = .1x + 7$$
$$10(.2x - .5) = 10(.1x + 7)$$
$$2x - 5 = x + 70$$
$$x - 5 = 70 \Rightarrow x = 75$$
Solution set: $\{75\}$

**21.**
$$-4(2x - 6) + 8x = 5x + 24 + x$$
$$-8x + 24 + 8x = 6x + 24$$
$$24 = 6x + 24$$
$$0 = 6x \Rightarrow 0 = x$$
Solution set: $\{0\}$

**23.**
$$4(2x + 7) = 2x + 22 + 3(2x + 2)$$
$$8x + 28 = 2x + 22 + 6x + 6$$
$$8x + 28 = 8x + 28$$
$$28 = 28 \Rightarrow 0 = 0$$
identity; $\{\text{all real numbers}\}$

**25.**
$$2(x - 8) = 3x - 16$$
$$2x - 16 = 3x - 16$$
$$-16 = x - 16 \Rightarrow 0 = x$$
conditional equation; $\{0\}$

**27.** $4(x+7) = 2(x+12) + 2(x+1)$
$4x + 28 = 2x + 24 + 2x + 2$
$4x + 28 = 4x + 26$
$\quad\quad 28 = 26$
contradiction; $\varnothing$

**29.** **(a)** $x^2 = 25$
$\quad\quad x = \pm\sqrt{25} = \pm 5;\ E$

**(b)** $x^2 - 5 = 0$
$\quad\quad x^2 = 5$
$\quad\quad\quad x = \pm\sqrt{5};\ C$

**(c)** $x^2 = 20$
$\quad\quad x = \pm\sqrt{20} = \pm 2\sqrt{5};\ A$

**(d)** $x - 5 = 0$
$\quad\quad x = 5;\ B$

**(e)** $x + 5 = 0$
$\quad\quad x = -5;\ D$

**31.** B is the only one set up for direct use of the square root property.
$(2x+5)^2 = 7$
$2x + 5 = \pm\sqrt{7}$
$2x = -5 \pm\sqrt{7} \Rightarrow x = \dfrac{-5\pm\sqrt{7}}{2}$
Solution set: $\left\{\dfrac{-5\pm\sqrt{7}}{2}\right\}$

**33.** $x^2 - 5x + 6 = 0$
$(x-2)(x-3) = 0$
$x - 2 = 0 \Rightarrow x = 2\ $ or $\ x - 3 = 0 \Rightarrow x = 3$
Solution set: $\{2, 3\}$

**35.** $5x^2 - 3x - 2 = 0$
$(5x+2)(x-1) = 0$
$5x + 2 = 0 \Rightarrow x = -\frac{2}{5}\ $ or $\ x - 1 = 0 \Rightarrow x = 1$
Solution set: $\left\{-\frac{2}{5}, 1\right\}$

**37.** $-4x^2 + x = -3$
$\quad\quad 0 = 4x^2 - x - 3$
$\quad\quad 0 = (4x+3)(x-1)$
$4x + 3 = 0 \Rightarrow x = -\frac{3}{4}\ $ or $\ x - 1 = 0 \Rightarrow x = 1$
Solution set: $\left\{-\frac{3}{4}, 1\right\}$

**39.** $x^2 = 16 \Rightarrow x = \pm\sqrt{16} = \pm 4$
Solution set: $\{\pm 4\}$

**41.** $27 - x^2 = 0 \Rightarrow 27 = x^2 \Rightarrow x = \pm\sqrt{27} = \pm 3\sqrt{3}$
Solution set: $\left\{\pm 3\sqrt{3}\right\}$

**43.** $(x+5)^2 = 40$
$x + 5 = \pm\sqrt{40} = \pm 2\sqrt{10}$
$\quad\quad x = -5 \pm 2\sqrt{10}$
Solution set: $-5 \pm 2\sqrt{10}$

**45.** $(3x-1)^2 = 12$
$3x - 1 = \pm\sqrt{12}$
$3x = 1 \pm 2\sqrt{3} \Rightarrow x = \dfrac{1\pm 2\sqrt{3}}{3}$
Solution set: $\left\{\dfrac{1\pm 2\sqrt{3}}{3}\right\}$

**47.** $x^2 - 4x + 3 = 0$
Let $a = 1,\ b = -4,\ c = 3.$
$x = \dfrac{-b \pm \sqrt{b^2 - 4ac}}{2a}$
$\ = \dfrac{-(-4) \pm \sqrt{(-4)^2 - 4(1)(3)}}{2(1)}$
$\ = \dfrac{4 \pm \sqrt{16 - 12}}{2} = \dfrac{4 \pm \sqrt{4}}{2} = \dfrac{4 \pm 2}{2} = 3 \text{ or } 1$
Solution set: $\{1, 3\}$

**49.** $x^2 - x - 1 = 0$
Let $a = 1, b = -1,$ and $c = -1.$
$x = \dfrac{-b \pm \sqrt{b^2 - 4ac}}{2a}$
$\ = \dfrac{-(-1) \pm \sqrt{(-1)^2 - 4(1)(-1)}}{2(1)}$
$\ = \dfrac{1 \pm \sqrt{1 + 4}}{2} = \dfrac{1 \pm \sqrt{5}}{2}$
Solution set: $\left\{\dfrac{1\pm\sqrt{5}}{2}\right\}$

**51.** $x^2 - 6x = -7$
$x^2 - 6x + 7 = 0$
Let $a = 1, b = -6,$ and $c = 7.$
$x = \dfrac{-b \pm \sqrt{b^2 - 4ac}}{2a}$
$\ = \dfrac{-(-6) \pm \sqrt{(-6)^2 - 4(1)(7)}}{2(1)} = \dfrac{6 \pm \sqrt{36 - 28}}{2}$
$\ = \dfrac{6 \pm \sqrt{8}}{2} = \dfrac{6 \pm 2\sqrt{2}}{2} = 3 \pm \sqrt{2}$
Solution set: $\left\{3 \pm \sqrt{2}\right\}$

**53.** $\frac{1}{2}x^2 + \frac{1}{4}x - 3 = 0$

$4\left(\frac{1}{2}x^2 + \frac{1}{4}x - 3\right) = 4 \cdot 0$

$2x^2 + x - 12 = 0$

Let $a = 2, b = 1,$ and $c = -12.$

$x = \dfrac{-b \pm \sqrt{b^2 - 4ac}}{2a} = \dfrac{-1 \pm \sqrt{1^2 - 4(2)(-12)}}{2(2)}$

$= \dfrac{-1 \pm \sqrt{1 + 96}}{4} = \dfrac{-1 \pm \sqrt{97}}{4}$

Solution set: $\left\{\dfrac{-1 \pm \sqrt{97}}{4}\right\}$

**55.** $.2x^2 + .4x - .3 = 0$

$10\left(.2x^2 + .4x - .3\right) = 10 \cdot 0$

$2x^2 + 4x - 3 = 0$

Let $a = 2, b = 4,$ and $c = -3.$

$x = \dfrac{-b \pm \sqrt{b^2 - 4ac}}{2a}$

$x = \dfrac{-4 \pm \sqrt{4^2 - 4(2)(-3)}}{2(2)} = \dfrac{-4 \pm \sqrt{16 + 24}}{4}$

$= \dfrac{-4 \pm \sqrt{40}}{4} = \dfrac{-4 \pm 2\sqrt{10}}{4} = \dfrac{-2 \pm \sqrt{10}}{2}$

Solution set: $\left\{\dfrac{-2 \pm \sqrt{10}}{2}\right\}$

**57.** $(4x - 1)(x + 2) = 4x$

$4x^2 + 7x - 2 = 4x \Rightarrow 4x^2 + 3x - 2 = 0$

Let $a = 4, b = 3,$ and $c = -2.$

$x = \dfrac{-b \pm \sqrt{b^2 - 4ac}}{2a} = \dfrac{-3 \pm \sqrt{3^2 - 4(4)(-2)}}{2(4)}$

$= \dfrac{-3 \pm \sqrt{9 + 32}}{8} = \dfrac{-3 \pm \sqrt{41}}{8}$

Solution set: $\left\{\dfrac{-3 \pm \sqrt{41}}{8}\right\}$

**59. (a)** $x < -6$

The interval includes all real numbers less than $-6$ not including $-6$. The correct interval notation is $(-\infty, -6)$, so the correct choice is F.

**(b)** $x \leq 6$

The interval includes all real numbers less than or equal to 6, so it includes 6. The correct interval notation is $(-\infty, 6]$, so the correct choice is J.

**(c)** $-2 < x \leq 6$

The interval includes all real numbers from $-2$ to 6, not including $-2$, but including 6. The correct interval notation is $(-2, 6]$, so the correct choice is A.

**(d)** $-3 \leq x \leq 3$

The interval includes all real numbers between $-3$ and 3, including $-3$ and 3. The correct interval notation is $[-3, 3]$, so the correct choice is H.

**(e)** $x \geq -6$

The interval includes all real numbers greater than or equal to $-6$, so it includes $-6$. The correct interval notation is $[-6, \infty)$, so the correct choice is I.

**(f)** $6 \leq x$

The interval includes all real numbers greater than or equal to 6, so it includes 6. The correct interval notation is $[6, \infty)$, so the correct choice is D.

**(g)** The interval shown on the number line includes all real numbers between $-2$ and 6, including $-2$, but not including 6. The correct interval notation is $[-2, 6)$, so the correct choice is B.

**(h)** The interval shown on the number line includes all real numbers between 0 and 8, not including 0 or 8. The correct interval notation is $(0, 8)$, so the correct choice is G.

**(i)** The interval shown on the number line includes all real numbers less than $-3$, not including $-3$, and greater than 3, not including 3. The correct interval notation is $(-\infty, -3) \cup (3, \infty)$, so the correct choice is E.

**(j)** The interval includes all real numbers less than or equal to $-6$, so it includes $-6$. The correct interval notation is $(-\infty, -6]$, so the correct choice is C.

**61.** $2x + 8 \leq 16 \Rightarrow 2x + 8 - 8 \leq 16 - 8 \Rightarrow$

$2x \leq 8 \Rightarrow \dfrac{2x}{2} \leq \dfrac{8}{2} \Rightarrow x \leq 4$

Solution set: $(-\infty, 4]$

Graph:

**63.**
$$-2x - 2 \le 1 + x$$
$$-2x - 2 + 2 \le 1 + x + 2$$
$$-2x \le x + 3 \Rightarrow -2x - x \le 3 \Rightarrow$$
$$-3x \le 3 \Rightarrow \frac{-3x}{-3} \ge \frac{3}{-3} \Rightarrow x \ge -1$$

Solution set: $[-1, \infty)$

Graph:

**65.** $2(x + 5) + 1 \ge 5 + 3x$
$$2x + 10 + 1 \ge 5 + 3x \Rightarrow 2x + 11 \ge 5 + 3x \Rightarrow$$
$$2x + 11 - 3x \ge 5 + 3x - 3x \Rightarrow -x + 11 \ge 5 \Rightarrow$$
$$-x + 11 - 11 \ge 5 - 11 \Rightarrow \frac{-x}{-1} \le \frac{-6}{-1} \Rightarrow x \le 6$$

Solution set: $(-\infty, 6]$

Graph:

**67.** $8x - 3x + 2 < 2(x + 7)$
$$5x + 2 < 2x + 14$$
$$5x + 2 - 2x < 2x + 14 - 2x$$
$$3x + 2 < 14 \Rightarrow 3x + 2 - 2 < 14 - 2 \Rightarrow$$
$$3x < 12 \Rightarrow \frac{3x}{3} < \frac{12}{3} \Rightarrow x < 4$$

Solution set: $(-\infty, 4)$

Graph:

**69.**
$$\frac{4x + 7}{-3} \le 2x + 5$$
$$(-3)\left(\frac{4x + 7}{-3}\right) \ge (-3)(2x + 5)$$
$$4x + 7 \ge -6x - 15$$
$$4x + 7 + 6x \ge -6x - 15 + 6x$$
$$10x + 7 \ge -15$$
$$10x + 7 - 7 \ge -15 - 7 \Rightarrow 10x \ge -22 \Rightarrow$$
$$\frac{10x}{10} \ge \frac{-22}{10} \Rightarrow x \ge -\tfrac{11}{5}$$

Solution set: $\left[-\tfrac{11}{5}, \infty\right)$

Graph:

**71.**
$$\frac{1}{3}x + \frac{2}{5}x - \frac{1}{2}(x + 3) \le \frac{1}{10}$$
$$30\left[\frac{1}{3}x + \frac{2}{5}x - \frac{1}{2}(x + 3)\right] \le 30\left[\frac{1}{10}\right]$$
$$10x + 12x - 15(x + 3) \le 3$$
$$10x + 12x - 15x - 45 \le 3$$

**—**

$$7x - 45 \le 3$$
$$7x - 45 + 45 \le 3 + 45$$
$$7x \le 48$$
$$\frac{7x}{7} \le \frac{48}{7} \Rightarrow x \le \tfrac{48}{7}$$

Solution set: $\left(-\infty, \tfrac{48}{7}\right]$

Graph:

**73.**
$$-5 < 5 + 2x < 11$$
$$-5 - 5 < 5 + 2x - 5 < 11 - 5$$
$$-10 < 2x < 6$$
$$\frac{-10}{2} < \frac{2x}{2} < \frac{6}{2}$$
$$-5 < x < 3$$

Solution set: $(-5, 3)$

Graph:

**75.**
$$10 \le 2x + 4 \le 16$$
$$10 - 4 \le 2x + 4 - 4 \le 16 - 4$$
$$6 \le 2x \le 12$$
$$\frac{6}{2} \le \frac{2x}{2} \le \frac{12}{2}$$
$$3 \le x \le 6$$

Solution set: $[3, 6]$

Graph:

**77.**
$$-11 > -3x + 1 > -17$$
$$-11 - 1 > -3x + 1 - 1 > -17 - 1$$
$$-12 > -3x > -18$$
$$\frac{-12}{-3} < \frac{-3x}{-3} < \frac{-18}{-3}$$
$$4 < x < 6$$

Solution set: $(4, 6)$

Graph:

**79.**
$$-4 \le \frac{x + 1}{2} \le 5$$
$$2(-4) \le 2\left(\frac{x + 1}{2}\right) \le 2(5)$$
$$-8 \le x + 1 \le 10$$
$$-8 - 1 \le x + 1 - 1 \le 10 - 1 \Rightarrow -9 \le x \le 9$$

Solution set: $[-9, 9]$

Graph:

## Appendix B: Graphs of Equations

**1.–7.**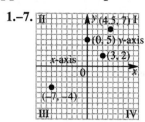

**9.** $(-5, 1.25)$ has a negative $x$-coordinate and a positive $y$-coordinate. The point lies in quadrant II.

**11.** $(-1.4, -2.8)$ has a negative $x$-coordinate and a negative $y$-coordinate. The point lies in quadrant III.

**13.** **(a)** Since $(a, b)$ lies in quadrant II, $a < 0$ and $b > 0$. Therefore, $-a > 0$ and $(-a, b)$ lies in quadrant I.

   **(b)** Since $(a, b)$ lies in quadrant II, $a < 0$ and $b > 0$. Therefore, $-a > 0$ and $-b < 0$, so $(-a, -b)$ lies in quadrant IV.

   **(c)** Since $(a, b)$ lies in quadrant II, $a < 0$ and $b > 0$. Therefore, $-b < 0$ and $(a, -b)$ lies in quadrant III.

   **(d)** Since $(a, b)$ lies in quadrant II, $a < 0$ and $b > 0$. The point $(b, a)$ has a positive $x$-coordinate and a negative $y$-coordinate. Therefore, $(b, a)$ lies in quadrant IV.

**15.** False. The expression should be
$$\sqrt{\left(x_2 - x_1\right)^2 + \left(y_2 - y_1\right)^2}\ .$$

**17.** True. The midpoint has coordinates
$$\left(\frac{a + 3a}{2}, \frac{b + \left(-3b\right)}{2}\right) = \left(\frac{4a}{2}, \frac{-2b}{2}\right)$$
$$= (2a, -b).$$

**19.** $(a, b, c) = (9, 12, 15)$
$$a^2 + b^2 = 9^2 + 12^2$$
$$= 81 + 144$$
$$= 225 = 15^2 = c^2$$
The triple is a Pythagorean triple since $a^2 + b^2 = c^2$.

**21.** $(a, b, c) = (5, 10, 15)$
$$a^2 + b^2 = 5^2 + 10^2$$
$$= 25 + 100$$
$$= 125 \neq 225 = 15^2 = c^2$$
The triple is not a Pythagorean triple since $a^2 + b^2 \neq c^2$.

**23.** Let $(x_1, y_1) = (5, -6)$. Let $(x_2, y_2) = (5, 0)$ since we want the point on the $x$-axis with $x$-value 5.
$$d = \sqrt{(5 - 5)^2 + \left[0 - (-6)\right]^2} = \sqrt{0^2 + 6^2}$$
$$= \sqrt{0 + 36} = \sqrt{36} = 6$$
Thus, the point $(5, -6)$ is 6 units from the $x$-axis.

**25.** $P(-5, -7)$, $Q(-13, 1)$

   **(a)** $d(P, Q) = \sqrt{\left[-13 - (-5)\right]^2 + \left[1 - (-7)\right]^2}$
   $$= \sqrt{\left(-8\right)^2 + 8^2} = \sqrt{128} = 8\sqrt{2}$$

   **(b)** The midpoint $M$ of the segment joining points $P$ and $Q$ has coordinates
   $$\left(\frac{-5 + (-13)}{2}, \frac{-7 + 1}{2}\right) = \left(\frac{-18}{2}, \frac{-6}{2}\right)$$
   $$= (-9, -3).$$

**27.** $P(8, 2)$, $Q(3, 5)$

   **(a)** $d(P, Q) = \sqrt{(3 - 8)^2 + (5 - 2)^2}$
   $$= \sqrt{\left(-5\right)^2 + 3^2}$$
   $$= \sqrt{25 + 9} = \sqrt{34}$$

   **(b)** The midpoint $M$ of the segment joining points $P$ and $Q$ has coordinates
   $$\left(\frac{8 + 3}{2}, \frac{2 + 5}{2}\right) = \left(\frac{11}{2}, \frac{7}{2}\right).$$

**29.** $P(-8, 4)$, $Q(3, -5)$

   **(a)** $d(P, Q) = \sqrt{\left[3 - (-8)\right]^2 + \left(-5 - 4\right)^2}$
   $$= \sqrt{11^2 + (-9)^2} = \sqrt{121 + 81}$$
   $$= \sqrt{202}$$

   **(b)** The midpoint $M$ of the segment joining points $P$ and $Q$ has coordinates
   $$\left(\frac{-8 + 3}{2}, \frac{4 + (-5)}{2}\right) = \left(-\frac{5}{2}, -\frac{1}{2}\right).$$

**31.** $P\left(3\sqrt{2}, 4\sqrt{5}\right)$, $Q\left(\sqrt{2}, -\sqrt{5}\right)$

   **(a)** $d(P, Q)$

$$= \sqrt{\left(\sqrt{2} - 3\sqrt{2}\right)^2 + \left(-\sqrt{5} - 4\sqrt{5}\right)^2}$$

$$= \sqrt{\left(-2\sqrt{2}\right)^2 + \left(-5\sqrt{5}\right)^2}$$

$$= \sqrt{8 + 125} = \sqrt{133}$$

   **(b)** The midpoint $M$ of the segment joining points $P$ and $Q$ has coordinates

$$\left(\frac{3\sqrt{2} + \sqrt{2}}{2}, \frac{4\sqrt{5} + (-\sqrt{5})}{2}\right)$$

$$= \left(\frac{4\sqrt{2}}{2}, \frac{3\sqrt{5}}{2}\right) = \left(2\sqrt{2}, \frac{3\sqrt{5}}{2}\right).$$

**33.** Find $x$ such that the distance between $(x, 7)$ and $(2, 3)$ is 5. Use the distance formula and solve for $x$:

$$5 = \sqrt{(2-x)^2 + (3-7)^2} = \sqrt{(2-x)^2 + (-4)^2}$$

$$= \sqrt{(2-x)^2 + 16}$$

$$25 = (2-x)^2 + 16$$

$$9 = (2-x)^2 \Rightarrow \pm 3 = 2 - x \Rightarrow x = 2 \pm 3 \Rightarrow$$

$$x = -1 \text{ or } x = 5$$

**35.** Find $y$ such that the distance between $(3, y)$ and $(-2, 9)$ is 12. Use the distance formula and solve for $y$:

$$12 = \sqrt{(-2-3)^2 + (9-y)^2}$$

$$= \sqrt{(-5)^2 + (9-y)^2}$$

$$= \sqrt{25 + (9-y)^2}$$

$$144 = 25 + (9-y)^2$$

$$119 = (9-y)^2 \Rightarrow \pm\sqrt{119} = 9 - y \Rightarrow$$

$$y = 9 \pm \sqrt{119} \Rightarrow$$

$$y = 9 + \sqrt{119} \text{ or } x = 9 - \sqrt{119}$$

**37.** Let $P = (x, y)$ be a point 5 units from $(0, 0)$. Then, by the distance formula,

$$5 = \sqrt{(x-0)^2 + (y-0)^2} \Rightarrow 5 = \sqrt{x^2 + y^2} \Rightarrow$$

$$25 = x^2 + y^2$$

This is the required equation. The graph is a circle with center $(0, 0)$ and radius 5.

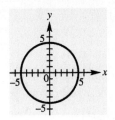

**39.** The endpoints of the segment are $(1990, 20.3)$ and $(2006, 28.0)$.

$$M = \left(\frac{1990 + 2006}{2}, \frac{20.3 + 28.0}{2}\right) = (1998, 24.15)$$

The estimate is 24.15%. This is close to the actual figure of 24.4%.

In exercises 41–45, other ordered pairs are possible.

**41.**  **(a)**

| $x$ | $y$ | |
|---|---|---|
| 0 | −2 | $y$-intercept: $x = 0 \Rightarrow$ $6y = 3(0) - 12 \Rightarrow$ $6y = -12 \Rightarrow y = -2$ |
| 4 | 0 | $x$-intercept: $y = 0 \Rightarrow$ $6(0) = 3x - 12 \Rightarrow$ $0 = 3x - 12 \Rightarrow$ $12 = 3x \Rightarrow 4 = x$ |
| 2 | −1 | additional point |

   **(b)**

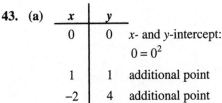

**43.**  **(a)**

| $x$ | $y$ | |
|---|---|---|
| 0 | 0 | $x$- and $y$-intercept: $0 = 0^2$ |
| 1 | 1 | additional point |
| −2 | 4 | additional point |

   **(b)**

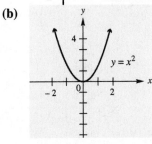

**45.** **(a)**

| x | y | |
|---|---|---|
| 0 | 0 | x- and y-intercept: |
| | | $0 = 0^3$ |
| −1 | −1 | additional point |
| 2 | 8 | additional point |

**(b)**

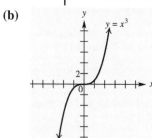

**47.** **(a)** Center (0, 0), radius 6

$$\sqrt{(x-0)^2 + (y-0)^2} = 6$$
$$(x-0)^2 + (y-0)^2 = 6^2$$
$$x^2 + y^2 = 36$$

**(b)**

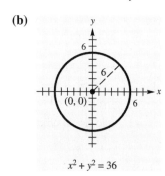

$x^2 + y^2 = 36$

**49.** **(a)** Center (2, 0), radius 6

$$\sqrt{(x-2)^2 + (y-0)^2} = 6$$
$$(x-2)^2 + (y-0)^2 = 6^2$$
$$(x-2)^2 + y^2 = 36$$

**(b)**

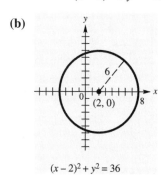

$(x-2)^2 + y^2 = 36$

**51.** **(a)** Center (−2, 5), radius 4

$$\sqrt{\left[x-(-2)\right]^2 + (y-5)^2} = 4$$
$$[x-(-2)]^2 + (y-5)^2 = 4^2$$
$$(x+2)^2 + (y-5)^2 = 16$$

**(b)**

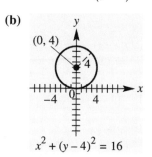

$(x+2)^2 + (y-5)^2 = 16$

**53.** **(a)** Center (0, 4), radius 4

$$\sqrt{(x-0)^2 + (y-4)^2} = 4$$
$$x^2 + (y-4)^2 = 16$$

**(b)**

$x^2 + (y-4)^2 = 16$

**55.** The center of the circle is located at the midpoint of the diameter determined by the points (1, 1) and (5, 1). Using the midpoint formula, we have $C = \left(\dfrac{1+5}{2}, \dfrac{1+1}{2}\right) = (3,1)$.

The radius is one-half the length of the diameter: $r = \dfrac{1}{2}\sqrt{(5-1)^2 + (1-1)^2} = 2$

The equation of the circle is

$$(x-3)^2 + (y-1)^2 = 4$$

**57.** The center of the circle is located at the midpoint of the diameter determined by the points (−2, 4) and (−2, 0). Using the midpoint formula, we have

$$C = \left(\dfrac{-2+(-2)}{2}, \dfrac{4+0}{2}\right) = (-2,2).$$

The radius is one-half the length of the diameter: $r = \dfrac{1}{2}\sqrt{\left[-2-(-2)\right]^2 + (4-0)^2} = 2$

The equation of the circle is

$$(x+2)^2 + (y-2)^2 = 4$$

# Appendix C: Functions

**1.** The relation is a function because for each different $x$-value there is exactly one $y$-value. This correspondence can be shown as follows.

$\{5, 3, 4, 7\}$ $x$-values

↓ ↓ ↓ ↓

$\{1, 2, 9, 8\}$ $y$-values

**3.** Two ordered pairs, namely $(2, 4)$ and $(2, 6)$, have the same $x$-value paired with different $y$-values, so the relation is not a function.

**5.** The relation is a function because for each different $x$-value there is exactly one $y$-value. This correspondence can be shown as follows.

$\{-3, 4, -2\}$ $x$-values

$\{1, 7\}$ $y$-values

**7.** Two sets of ordered pairs, namely $(1, 1)$ and $(1, -1)$ as well as $(2, 4)$ and $(2, -4)$, have the same $x$-value paired with different $y$-values, so the relation is not a function.
domain: $\{0, 1, 2\}$ ; range: $\{-4, -1, 0, 1, 4\}$

**9.** The relation is a function because for each different $x$-value there is exactly one $y$-value.
domain: $\{2, 3, 5, 11, 17\}$ ; range: $\{1, 7, 20\}$

**11.** The relation is a function because for each different $x$-value there is exactly one $y$-value. This correspondence can be shown as follows.

$\{0, -1, -2\}$ $x$-values

↓ ↓ ↓

$\{0, 1, 2\}$ $y$-values
Domain: $\{0, -1, -2\}$; range: $\{0, 1, 2\}$

**13.** This graph represents a function. If you pass a vertical line through the graph, one $x$-value corresponds to only one $y$-value.
domain: $(-\infty, \infty)$; range: $(-\infty, \infty)$

**15.** This graph does not represent a function. If you pass a vertical line through the graph, there are places where one value of $x$ corresponds to two values of $y$.
domain: $[3, \infty)$; range: $(-\infty, \infty)$

**17.** This graph does not represent a function. If you pass a vertical line through the graph, there are places where one value of $x$ corresponds to two values of $y$.
domain: $[-4, 4]$; range: $[-3, 3]$

**19.** $y = x^2$ represents a function since $y$ is always found by squaring $x$. Thus, each value of $x$ corresponds to just one value of $y$. $x$ can be any real number. Since the square of any real number is not negative, the range would be zero or greater.

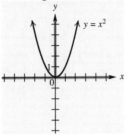

domain: $(-\infty, \infty)$; range: $[0, \infty)$

**21.** The ordered pairs $(1, 1)$ and $(1, -1)$ both satisfy $x = y^6$. This equation does not represent a function. Because $x$ is equal to the sixth power of $y$, the values of $x$ are nonnegative. Any real number can be raised to the sixth power, so the range of the relation is all real numbers.

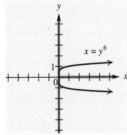

domain: $[0, \infty)$ range: $(-\infty, \infty)$

**23.** $y = 2x - 5$ represents a function since $y$ is found by multiplying $x$ by 2 and subtracting 5. Each value of $x$ corresponds to just one value of $y$. $x$ can be any real number, so the domain is all real numbers. Since $y$ is twice $x$, less 5, $y$ also may be any real number, and so the range is also all real numbers.

*(continued on next page)*

(*continued from page 245*)

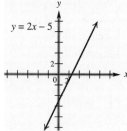

domain: $(-\infty, \infty)$; range: $(-\infty, \infty)$

**25.** For any choice of $x$ in the domain of $y = \sqrt{x}$, there is exactly one corresponding value of $y$, so this equation defines a function. Since the quantity under the square root cannot be negative, we have $x \geq 0$. Because the radical is nonnegative, the range is also zero or greater.

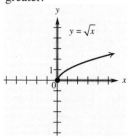

domain: $[0, \infty)$; range: $[0, \infty)$

**27.** For any choice of $x$ in the domain of $y = \sqrt{4x+1}$ there is exactly one corresponding value of $y$, so this equation defines a function. Since the quantity under the square root cannot be negative, we have $4x+1 \geq 0 \Rightarrow 4x \geq -1 \Rightarrow x \geq -\frac{1}{4}$. Because the radical is nonnegative, the range is also zero or greater.

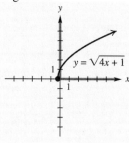

domain: $\left[-\frac{1}{4}, \infty\right)$; range: $[0, \infty)$

**29.** Given any value in the domain of $y = \frac{2}{x-3}$, we find $y$ by subtracting 3, then dividing into 2. This process produces one value of $y$ for each value of $x$ in the domain, so this equation is a function. The domain includes all real numbers except those that make the denominator equal to zero, namely $x = 3$. Values of $y$ can be negative or positive, but never zero. Therefore, the range will be all real numbers except zero.

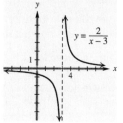

domain: $(-\infty, 3) \cup (3, \infty)$;

range: $(-\infty, 0) \cup (0, \infty)$

**31.** B

**33.** $f(x) = -3x + 4$
$f(0) = -3 \cdot 0 + 4 = 0 + 4 = 4$

**35.** $g(x) = -x^2 + 4x + 1$
$g(-2) = -(-2)^2 + 4(-2) + 1$
$\qquad = -4 + (-8) + 1 = -11$

**37.** $f(x) = -3x + 4$
$f\left(\frac{1}{3}\right) = -3\left(\frac{1}{3}\right) + 4 = -1 + 4 = 3$

**39.** $g(x) = -x^2 + 4x + 1$
$g\left(\frac{1}{2}\right) = -\left(\frac{1}{2}\right)^2 + 4\left(\frac{1}{2}\right) + 1$
$\qquad = -\frac{1}{4} + 2 + 1 = \frac{11}{4}$

**41.** $f(x) = -3x + 4$
$f(p) = -3p + 4$

**43.** $f(x) = -3x + 4$
$f(x+2) = -3(x+2) + 4$
$\qquad = -3x - 6 + 4 = -3x - 2$

**45.** **(a)** $f(2) = 2$     **(b)** $f(-1) = 3$

**47.** **(a)** $f(2) = 15$     **(b)** $f(-1) = 10$

**49.** **(a)** $f(2) = 3$     **(b)** $f(-1) = -3$

**51. (a)** $f(-2)=0$  **(b)** $f(0)=4$

**(c)** $f(1)=2$  **(d)** $f(4)=4$

**53. (a)** $f(0)=11$  **(b)** $f(6)=9$

**(c)** Since $f(-2)=0$, $a=-2$

**(d)** Since $f(2)=f(7)=f(8)=10$,
$x=2,7,8$

**(e)** We have $(8,f(8))=(8,10)$ and
$(10,f(10))=(10,0)$. Using the distance formula, we have
$$d=\sqrt{(10-8)^2+(0-10)^2}=\sqrt{2^2+10^2}$$
$$=\sqrt{4+100}=\sqrt{104}$$

**55. (a)** $[4,\infty)$  **(b)** $(-\infty,-1]$

**(c)** $[-1,4]$

**57. (a)** $(-\infty,4]$  **(b)** $[4,\infty)$

**(c)** none

**59. (a)** none  **(b)** $(-\infty,-2]; [3,\infty)$

**(c)** $(-2,3)$

**61. (a)** Yes, it is the graph of a function.

**(b)** $[0,24]$

**(c)** When $t=8$, $y=1200$ from the graph. At 8 A.M., approximately 1200 megawatts is being used.

**(d)** The most electricity was used at 17 hr or 5 P.M. The least electricity was used at 4 A.M.

**(e)** $f(12)\approx1900$; At 12 noon, electricity use is about 1900 megawatts.

**(f)** increasing from 4 A.M. to 5 P.M.; decreasing from midnight to 4 A.M. and from 5 P.M. to midnight

## Appendix D: Graphing Techniques

**1. (a)** B; $y=(x-7)^2$ is a shift of $y=x^2$ 7 units to the right.

**(b)** E; $y=x^2-7$ is a shift of $y=x^2$ 7 units downward.

**(c)** F; $y=7x^2$ is a vertical stretch of $y=x^2$ by a factor of 7.

**(d)** A; $y=(x+7)^2$ is a shift of $y=x^2$ 7 units to the left.

**(e)** D; $y=x^2+7$ is a shift of $y=x^2$ 7 units upward.

**(f)** C; $y=\frac{1}{7}x^2$ is a vertical shrink of $y=x^2$.

**3. (a)** B; $y=x^2+2$ is a shift of $y=x^2$ 2 units upward.

**(b)** A; $y=x^2-2$ is a shift of $y=x^2$ 2 units downward.

**(c)** G; $y=(x+2)^2$ is a shift of $y=x^2$ 2 units to the left.

**(d)** C; $y=(x-2)^2$ is a shift of $y=x^2$ 2 units to the right.

**(e)** F; $y=2x^2$ is a vertical stretch of $y=x^2$ by a factor of 2.

**(f)** D; $y=-x^2$ is a reflection of $y=x^2$ across the $x$-axis.

**(g)** H; $y=(x-2)^2+1$ is a shift of $y=x^2$ 2 units to the right and 1 unit upward.

**(h)** E; $y=(x+2)^2+1$ is a shift of $y=x^2$ 2 units to the left and 1 unit upward.

**(i)** I; $y=(x+2)^2-1$ is a shift of $y=x^2$ 2 units to the left and 1 unit down.

**5.** $y=4x^2$

| $x$ | $y=x^2$ | $y=4x^2$ |
|---|---|---|
| $-2$ | 4 | 16 |
| $-1$ | 1 | 4 |
| 0 | 0 | 0 |
| 1 | 1 | 4 |
| 2 | 4 | 16 |

*(continued on next page)*

*(continued from page 247)*

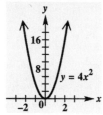

**7.** $y = \frac{2}{3}x^2$

| $x$ | $y = x^2$ | $y = \frac{2}{3}x^2$ |
|---|---|---|
| $-3$ | 9 | 6 |
| $-2$ | 4 | $\frac{8}{3}$ |
| $-1$ | 1 | $\frac{2}{3}$ |
| 0 | 0 | 0 |
| 1 | 1 | $\frac{2}{3}$ |
| 2 | 4 | $\frac{8}{3}$ |
| 3 | 9 | 6 |

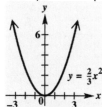

**9.** $y = -3x^2$

| $x$ | $y = x^2$ | $y = -3x^2$ |
|---|---|---|
| $-3$ | 9 | $-27$ |
| $-2$ | 4 | $-12$ |
| $-1$ | 1 | $-3$ |
| 0 | 0 | 0 |
| 1 | 1 | $-3$ |
| 2 | 4 | $-12$ |
| 3 | 9 | $-27$ |

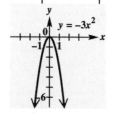

**11.** $y = \sqrt{-2x}$

| $x$ | $-2x$ | $y = \sqrt{-2x}$ |
|---|---|---|
| 0 | 0 | 0 |
| $-\frac{1}{2}$ | 1 | 1 |
| $-2$ | 4 | 2 |
| $-\frac{9}{2}$ | 9 | 3 |
| $-8$ | 16 | 4 |

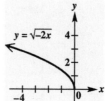

**13. (a)** $y = \frac{1}{4}f(x)$ is a vertical shrinking of $f$, by a factor of $\frac{1}{4}$. The point that corresponds to $(8, 12)$ on this translated function is $\left(8, \frac{1}{4} \cdot 12\right) = (8, 3)$.

**(b)** $y = 4f(x)$ is a vertical stretching of $f$, by a factor of 4. The point that corresponds to $(8, 12)$ on this translated function is $(8, 4 \cdot 12) = (8, 48)$.

**15. (a)** The point that is symmetric to $(5, -3)$ with respect to the $x$-axis is $(5, 3)$.

**(b)** The point that is symmetric to $(5, -3)$ with respect to the $y$-axis is $(-5, -3)$.

**(c)** The point that is symmetric to $(5, -3)$ with respect to the origin is $(-5, 3)$.

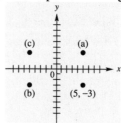

**17. (a)** The point that is symmetric to $(-4, -2)$ with respect to the $x$-axis is $(-4, 2)$.

**(b)** The point that is symmetric to $(-4, -2)$ with respect to the $y$-axis is $(4, -2)$.

**(c)** The point that is symmetric to $(-4, -2)$ with respect to the origin is $(4, 2)$.

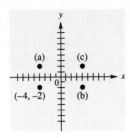

**19.** $y = x^2 + 2$

Replace $x$ with $-x$ to obtain
$y = (-x)^2 + 2 = x^2 + 2$. The result is the same as the original equation, so the graph is symmetric with respect to the $y$-axis. Since $y$ is a function of $x$, the graph cannot be symmetric with respect to the $x$-axis. Replace $x$ with $-x$ and $y$ with $-y$ to obtain
$-y = (-x)^2 + 2 \Rightarrow -y = x^2 + 2 \Rightarrow y = -x^2 - 2$.
The result is not the same as the original equation, so the graph is not symmetric with respect to the origin. Therefore, the graph is symmetric with respect to the $y$-axis only.

**21.** $x^2 + y^2 = 10$

Replace $x$ with $-x$ to obtain
$(-x)^2 + y^2 = 10 \Rightarrow x^2 + y^2 = 10$.
The result is the same as the original equation, so the graph is symmetric with respect to the $y$-axis. Replace $y$ with $-y$ to obtain
$x^2 + (-y)^2 = 10 \Rightarrow x^2 + y^2 = 10$
The result is the same as the original equation, so the graph is symmetric with respect to the $x$-axis. Since the graph is symmetric with respect to the $x$-axis and $y$-axis, it is also symmetric with respect to the origin.

**23.** $y = -3x^3$

Replace $x$ with $-x$ to obtain
$y = -3(-x)^3 \Rightarrow y = -3(-x^3) \Rightarrow y = 3x^3$.
The result is not the same as the original equation, so the graph is not symmetric with respect to the $y$-axis. Replace $y$ with $-y$ to obtain $-y = -3x^3 \Rightarrow y = 3x^3$.
The result is not the same as the original equation, so the graph is not symmetric with respect to the $x$-axis. Replace $x$ with $-x$ and $y$ with $-y$ to obtain
$-y = -3(-x)^3 \Rightarrow -y = -3(-x^3) \Rightarrow$
$-y = 3x^3 \Rightarrow y = -3x^3$.

The result is the same as the original equation, so the graph is symmetric with respect to the origin. Therefore, the graph is symmetric with respect to the origin only.

**25.** $y = x^2 - x + 7$

Replace $x$ with $-x$ to obtain
$y = (-x)^2 - (-x) + 7 \Rightarrow y = x^2 + x + 7$.
The result is not the same as the original equation, so the graph is not symmetric with respect to the $y$-axis. Since $y$ is a function of $x$, the graph cannot be symmetric with respect to the $x$-axis. Replace $x$ with $-x$ and $y$ with $-y$ to obtain $-y = (-x)^2 - (-x) + 7 \Rightarrow$
$-y = x^2 + x + 7 \Rightarrow y = -x^2 - x - 7$.
The result is not the same as the original equation, so the graph is not symmetric with respect to the origin. Therefore, the graph has none of the listed symmetries.

**27.** $y = x^2 - 1$

This graph may be obtained by translating the graph of $y = x^2$ 1 unit downward.

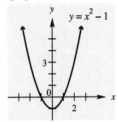

**29.** $y = x^2 + 2$

This graph may be obtained by translating the graph of $y = x^2$ 2 units upward.

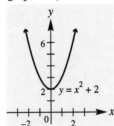

**31.** $y = (x-1)^2$

This graph may be obtained by translating the graph of $y = x^2$ 1 unit to the right.

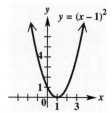

**33.** $y = (x+2)^2$

This graph may be obtained by translating the graph of $y = x^2$ 2 units to the left.

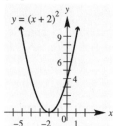

**35.** $y = (x+3)^2 - 4$

This graph may be obtained by translating the graph of $y = x^2$ 3 units to the left, and then 4 units down.

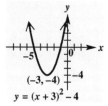

**37.** $y = 2x^2 - 1$

This graph may be obtained by translating the graph of $y = x^2$, 1 unit down. It is then stretched vertically by a factor of 2.

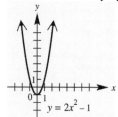

**39.** $f(x) = 2(x-2)^2 - 4$

This graph may be obtained by translating the graph of $y = x^2$, 2 units to the right and 4 units down. It is then stretched vertically by a factor of 2.

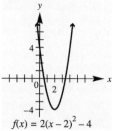

**41.** It is the graph of $f(x) = |x|$ translated 1 unit to the left, reflected across the *x*-axis, and translated 3 units up. The equation is $y = -|x+1| + 3$.

**43.** It is the graph of $f(x) = \sqrt{x}$ translated 4 units left, stretched vertically, and then translated 4 units down. The equation is $y = 2\sqrt{x+4} - 4$.

**45.** $f(x) = 2x + 5$: Translate the graph of $f(x)$ up 2 units to obtain the graph of $t(x) = (2x+5) + 2 = 2x + 7$.

Now translate the graph of $t(x) = 2x + 7$ left 3 units to obtain the graph of $g(x) = 2(x+3) + 7 = 2x + 6 + 7 = 2x + 13$.

(Note that if the original graph is first translated to the left 3 units and then up 2 units, the final result will be the same.)

**47.** Answers will vary.

There are four possibilities for the constant, *c*.

**i)** $c > 0$  $|c| > 1$  The graph of $F(x)$ is stretched vertically by a factor of *c*.

**ii)** $c > 0$  $|c| < 1$  The graph of $F(x)$ is shrunk vertically by a factor of *c*.

**iii)** $c < 0$  $|c| > 1$  The graph of $F(x)$ is stretched vertically by a factor of $-c$ and reflected over the *x*-axis.

**iv)** $c < 0$  $|c| < 1$  The graph of $F(x)$ is shrunk vertically by a factor of $-c$ and reflected over the *x*-axis.